Donated by
Ian Thomson

Not for loan

# The Economics of Antitrust and Regulation in Telecommunications

# The Economics of Antitrust and Regulation in Telecommunications

## Perspectives for the New European Regulatory Framework

*Edited by*

Pierre A. Buigues

*Economic Adviser (DG Competition), European Commission, Belgium*

*and*

Patrick Rey

*Professor of Economics, Institut d'Economie Industrielle, Université des Sciences Sociales, France*

**Edward Elgar**

Cheltenham, UK • Northampton, MA, USA

Published by
Edward Elgar Publishing Limited
Glensanda House
Montpellier Parade
Cheltenham
Glos GL50 1UA
UK

Edward Elgar Publishing, Inc.
136 West Street
Suite 202
Northampton
Massachusetts 01060
USA

A catalogue record for this book is available from the British Library

**Library of Congress Cataloguing in Publication Data**
The economics of antitrust and regulation in telecommunications : perspectives for the new European framework / edited by Pierre Buigues and Patrick Rey.
p. cm.
"This work documents a conference, 'The economics of antitrust and regulation in the telecommunication sector,' … held in Brussels on 16 September 2002 … in which officials from the telecoms unit of the European Commission's DG Competition and a group of high-level European economists worked together"—P. .
1. Telecommunication—Government policy—Europe. 2. Trade regulation—Europe. 3. Competition—Europe. I. Buigues, Pierre-André. II. Rey, Patrick, 1957–

HE8085.E25 2004
384'.041—dc22

2003061588

ISBN 1 84376 510 1 (cased)

Typeset by Manton Typesetters, Louth, Lincolnshire, UK.
Printed and bound in Great Britain by MPG Books Ltd, Bodmin, Cornwall.

# Contents

# List of figures

# List of tables

# List of contributors

| | |
|---|---|
| MONTI, Mario | *European Commissioner* |
| BUIGUES, Pierre A. | *Head of Unit, DG Competition, European Commission* |
| CANOY, Marcel | *CPB, The Netherlands* |
| CAVE, Martin | *Warwick Business School, Warwick, UK* |
| CHOUMELOVA, Dessislava | *EFTA Surveillance Authority* |
| DELGADO, Juan | *DG Competition, European Commission* |
| DE BIJL, Paul | *TILEC, Tilburg University, The Netherlands* |
| FEHRENBACH, Jérôme | *DG Competition, European Commission* |
| GABATHULER, David | *DG Competition, European Commission* |
| GUAL, Jordi | *IESE Business School, Barcelona* |
| HOCEPIED, Christian | *Head of Sector, DG Competition, European Commission* |
| KATZ, Michael L. | *Deputy Assistant Attorney-General, US Department of Justice, Antitrust Division* |
| KEMP, Ron | *EIM, Zoetermeer* |
| KLOTZ, Robert | *DG Competition, European Commission* |
| KRUSE, Jörn | *University of Hamburg* |
| LOWE, Philip | *Director General, DG Competition, European Commission* |
| PAPADIAS, Lambros | *DG Competition, European Commission* |
| REY, Patrick | *IDEI, University of Toulouse* |
| SAUTER, Wolf | *DG Competition, European Commission* |
| TIROLE, Jean | *MIT, IDEI, CERAS* |

# List of abbreviations

| | |
|---|---|
| 2G | second generation |
| 3G | third generation |
| ADSL | asynchronous digital subscriber line |
| ATC | average total cost |
| ATM | asynchronous transfer mode |
| AUC | authentication centre |
| AVC | average variable cost |
| BSCs | base station controllers |
| BTS | base station transceivers |
| CEPT | European Conference of Post and Telecommunications Administration |
| CFI | Court of First Instance |
| CN | core network |
| CPP | calling party pays |
| CPS | carrier pre-selection |
| CS | carrier selection |
| DCS | digital communication system |
| DECT | digital European cardless telecommunications |
| DSL | digital subscriber line |
| DSLAM | digital subscriber line access module |
| EDC | embedded direct cost |
| EEA | European Economic Area |
| EIR | equipment identity register |
| EPMU | equal proportionate mark-up |
| ERG | European Regulators Group |
| ERMES | European radio messaging system |
| ESA | EFTA Surveillance Authority |
| ETSI | European Telecommunications Standards Institute |
| FCC | Federal Communications Commission |
| FCFS | first come first served |
| FPSB | first-price sealed-bid (auction) |
| GSM | global system for mobile communications |
| HLR | home location register |
| IBP | Internet backbone provider |
| ICU | International Telecommunications Union |
| IMTS | international mobile telecommunications standard |
| IOT | inter-operator tariff |
| IP | (IT) (legal) intellectual property |
| IP | Internet protocol |

| | |
|---|---|
| IRG | Independent Regulators Group |
| ISDN | integrated services digital network |
| ISP | Internet service provider |
| ISO | independent service operator |
| LAN | local access network |
| LLP | local loop provider |
| LRAIC | long-run average incremental cost |
| LRIC | long-run incremental cost |
| MDF | main distribution frame |
| MMS | multimedia message service |
| MNO | mobile network operator |
| MNP | mobile number portability |
| MSC | mobile switching centre |
| MVNO | mobile virtual network operator |
| NARA | National Administrative and Regulatory Authority |
| NRA | national regulatory authority |
| ONP | open network provision |
| PSTN | public switched telephone network |
| RAN | radio access network |
| RNC | radio network controller |
| RPP | receiving party pays |
| RUO | reference unbundling offer |
| S-PCS | satellite personal communication services |
| SLA | service-level agreement |
| SME | small and medium-sized enterprise |
| SMP | significant market power |
| SMS | short messaging service |
| SP | service provider |
| SSC | site support cabinet |
| SSNIP | small but significant non-transitory increase in price |
| STIRA | standard international roaming agreement |
| ULL | unbundled local loop |
| UMTS | universal mobile telecoms system |
| USO | universal service obligation |
| WCDMA | wide code division multiple access |
| WLAN | wide local area network |
| xDSL | any DSL application |

# Foreword and acknowledgements

This book is the result of close co-operation between EU antitrust officials and academic economists. As a result, it is neither a comprehensive overview of the economic literature on the telecommunications industry, nor a detailed presentation of antitrust principles and case-law for that sector. Instead, it aims to address the integration of economics with the factual specificities of the telecommunications industry in order to further enhance the treatment of antitrust cases in the sector.

This aim is reflected in the fact that half of the contributions are from EU officials (mainly from the antitrust unit in charge of the telecommunications industry), while the other half are from academic economists. As a result, the reader will find, in each chapter, both an economist's perspective and the view of a Commission case-handler. The opinions expressed by Commission officials are the personal view of the speakers and do not necessarily reflect the position of the Commission.

We hope that this work will help foster the dialogue between outside academics and insider experts in Europe and contribute to a convergence of legal and economic approaches to EU antitrust policy.

This work documents a conference, 'The economics of antitrust and regulation in the telecommunications sector', held in Brussels on 16 September 2002. That conference was the culmination of a one-year project in which officials from the telecoms unit of the European Commission's DG Competition and a group of high-level European economists worked together on a number of key issues in the telecommunications sector. The topics, namely market definition, collective dominance, access to networks, and allocation of scarce resources, were chosen on the basis of the key problems arising out of the antitrust cases pending at the time.

The aim of the project was to incorporate modern economic theory and tools into day-to-day antitrust casework. To that aim, a number of meetings took place, which were devoted to discussion of the economic 'toolbox' available on the various issues and of the main characteristics of the relevant cases. These meetings then progressively led to a closer interaction between advisers and case-handlers and for a dialogue more tailored to specific cases.

The last meeting took the form of an open conference in which the experts highlighted the main theoretical and empirical findings relating to the above-mentioned topics, while DG Competition officials presented specific situations where this economic analysis was particularly relevant.

The conference was introduced by Commissioner Monti and structured around five thematic sessions and a closing roundtable, chaired by Philip Lowe, Michael Katz and Jean Tirole. Each session consisted of two presentations, given by an academic expert and a DG Competition official in turn, followed by a discussion with comments and questions from the floor (consisting of some fifty invited competition lawyers, consultants, economists and academics as well as Commission staff).

Part I begins with the opening remarks by the European Commissioner for Competition, Mario Monti, who presents the main challenges of competition and regulatory policies in the electronic communications sector. The contributions of the experts and the Commission officials are structured in five sections:

**1. The New Regulatory Telecommunications Framework**

Pierre Buigues (Head of Unit, DG Competition, European Commission) presents the approach of the new regulatory framework. Based on competition law, the new framework provides for, *inter alia*: the convergence of the concept of 'significant market power' (used in the sector-specific legislation) with the (competition-law-based) notion of 'dominance'; the use of competition-law analysis for the definition of markets; and the determination of the list of markets to be regulated (namely those markets where competition-law remedies are considered insufficient).

Martin Cave (Warwick Business School) comments on the economic aspects of the new regulatory framework. He views the new framework as a road towards deregulation and reliance on competition law, and considers its remedies as less intrusive and more effective than those under the previous framework. He points out, however, the complexity of its implementation, due mainly to the multiplication of markets to be analysed, the poor guidance on the choice of remedies, and the complex and potentially divisive institutional interaction.

**2. Market Definition in Telecommunications Markets**

Jordi Gual (IESE Business School) criticizes the Commission for using overly narrow market definitions in its antitrust and regulatory practices, and for failing to take sufficient account of supply-side substitution, complementary products and bundles of goods when determining the scope of a relevant market for antitrust purposes.

Christian Hocepied (European Commission) outlines the market delineation process contained in the new regulatory framework, first by reference to the markets contained in the Recommendation on relevant markets to be regulated, and, second, by reference to the Guidelines on market analysis and the assessment of significant market power in the electronic communications sector.

**3. Economic Analysis of Collective Dominance in Telecoms Markets**

Patrick Rey (University of Toulouse) presents the necessary ingredients (including entry barriers, frequent interaction and low innovation) and important factors (such as the number of firms, symmetries, structural links) for the existence of a collective dominant position in a given market. He compares the concept of collective dominance with the economic concept of tacit collusion, and considers the difficulties that antitrust authorities might encounter in applying this concept.

Lambros Papadias (European Commission) discusses the legal aspects of collective dominance by reference to the existing case-law and the recently adopted Commission Guidelines on market analysis and the assessment of significant market power in the electronic communications sector.

**4. Economic Aspects of Access to Networks**

Marcel Canoy (CPB), Paul de Bijl (TILEC) and Ron Kemp (EIM) describe the problems relating to access to telecommunications bottlenecks, and analyse the various solutions available to regulators and competition authorities. They explain the drawbacks and benefits of using static and dynamic pricing strategies to achieve short-term service-based competition and long-term infrastructure-based competition. They also stress the difficulties created by asymmetric information for the implementation of regulatory and competition policy in the case of access problems.

Juan Delgado, Jérôme Fehrenbach and Robert Klotz (European Commission) describe the current situation of local loop unbundling in Europe. They note the poor results achieved after the entry into force of the unbundling Regulation, in part caused by the incumbents' behaviour and the current price structure. Local loop unbundling prices are deemed to be too high, as evidenced by the existence of margin squeezes in certain Member States. Different potential solutions, such as adapting the pricing methodology and the cost methodology, are discussed.

**5. Competition and Allocation of Scarce Resources: the Case of UMTS**

Jörn Kruse's (University of Hamburg) chapter presents the different methodologies that can be used to allocate (intramodal and intermodal) spectrum frequencies and their appropriateness for achieving certain goals (efficiency, social and fiscal objectives, and so on). He supports infrastructure-sharing as a way of reducing costs and speeding up UMTS (universal mobile telecoms system) development, and maintains that detrimental effects on competition will be limited (provided that only the radio access network is shared).

David Gabathuler and Wolf Sauter (European Commission) examine the present and future regulatory framework for frequency allocation and assignment in the EU, in particular in relation to mobile communications, and discuss recent developments towards the introduction of spectrum trading. They also consider the regulatory and competition policy implications of the introduction of 3G (third generation)/UMTS, with a focus on network-sharing and MVNO (mobile virtual network operator) access.

**Roundtable on The Economics of Antitrust and Regulation in the Telecoms Sector**

Philip Lowe (Director General, DG Competition, European Commission) introduces the roundtable by reviewing the role of sector-specific regulation and competition law in the context of the new regulatory framework for the telecoms sector.

Michael Katz (Deputy Assistant Attorney-General, US Department of Justice, Antitrust Division) considers the respective roles of antitrust enforcement and regulation. He describes the US situation regarding regulation and the application of antitrust rules, and notes that regulation can go further than competition policy on some issues (for example with regard to essential facilities) and may pursue different goals (such as universal service subsidies). He points out that in some markets competition cannot take place in the absence of regulation, so that antitrust rules will never be a substitute of regulation.

Jean Tirole (IDEI – University of Toulouse) addresses two issues: the role of antitrust in the telecommunications sector and the conceptual challenges faced by competition policy in this sector. He pleads for a balanced approach to regulatory and antitrust enforcement, and argues that there is no need to adapt the antitrust rules to the specific characteristics of the telecommunications sector, but that, instead, a more sophisticated analysis is required. He illustrates his position with three examples: the essential facilities doctrine, the economics of networks (and the problem of access), and the necessity for competitors to co-operate in telecommunications markets. In each case, he compares the problems arising in the telecommunications sector with those of similar nature in other sectors (e.g. patents in research-intensive sectors, compatibility of standards, and so on).

In Part II, Dessislava Choumelova (ESA) and Juan Delgado (European Commission) present an evaluation of the sector inquiry instrument – a sector-wide investigation into commercial practices relating to an entire sector of the EU economy – which has been used by the Commission in the telecommunications sector for three markets: leased lines, mobile roaming and local loop access. The main results, and a summary of the main findings of the three sector inquiries, are presented.

Finally, Part III reproduces the main regulatory texts of the new regulatory framework and in particular the Directives, the Decision on the European Regulators Group, the guidelines on market analysis and the calculation of significant market power, and the Recommendation on relevant products and service markets susceptible to *ex ante* regulation.

The academic papers presented in this book and the conference were funded by the European Commission Directorate General for Competition. We are grateful to Juan Delgado, Yves Moulin, Natacha Muller and Valérie Leclercq, whose work was of crucial importance in the co-ordination and production of this book.

The publishers wish to thank the following, who have kindly given permission for the use of copyright material:

Office des Publications Officielles des Communautes Européennes for Main Regulatory Texts. Only European Communities legislation printed in the *Official Journal of the European Communities* is deemed to be authentic.

Every effort has been made to trace all the copyright holders but if any have been inadvertently overlooked the publishers will be pleased to make the necessary arrangements at the first opportunity.

P.A. Buigues
P. Rey

# PART I

# Telecommunications and Antitrust

# 1. Introduction

**Mario Monti**

## 1. INTRODUCTION

This book is the final step of an initiative taken by the Directorate General for Competition of the European Commission to invite several distinguished economists to examine a number of economic issues related to antitrust in the telecommunications sector. They were asked to examine how standard economic theory could shed more light on what is primarily a legal assessment of certain forms of anticompetitive behaviour and practices. The idea was to gather around a table what often appear to be two 'independent' schools of thought, economics and antitrust law, and together to try to foster a mutual understanding of how modern economic thinking assesses market behaviour in technology-driven markets, and the fundamental legal premises upon which such an assessment should take place.

If one looks back to almost ten years of antitrust enforcement in the telecoms sector, when liberalization of the sector began, it becomes apparent that the Commission has not only applied competition law with sensitivity to the special characteristics of the telecommunications industry, but has also recognized that, as in all other sectors of the economy, competition plays an important role in spurring innovation and conveying the benefits of that innovation to consumers.

The telecoms sector is one of the most important sectors in today's economy. It is a 'network' industry which directly affects the daily lives of each of us and plays a critical role in the economic development of Europe as a whole. It is, however, an industry that started as a monopoly and has gradually moved towards full liberalization. The importance of competition law and policy in making this transition a reality has recently been underscored by the adoption of a new regulatory framework for electronic communications which aligns regulation with the basic concepts of competition law.

## 2. COMPETITION RULES AND REGULATION: THE NEW EUROPEAN REGULATORY FRAMEWORK

Competition rules and sector-specific regulation are complementary tools to achieve a single objective: the promotion of competition. This is one of the main principles underlying the new regulatory framework for electronic communications. The application of fundamental competition-law notions, such as market definition and dominance, in an *ex ante* environment represents the best means to ensure a smooth transition towards a fully liberalized electronic communications market.

Sector-specific regulation is, however, indispensable today for certain areas where competition is not likely to arise from the mere application of competition rules. This is, for example, the case of the 'local loop', where incumbents enjoy an incontestably dominant position, built during former monopoly times. Competition law cannot always guarantee effective access to facilities which are crucial for the development of competition.

Under the new framework, regulation remains nevertheless the exception, while the application of antitrust rules is the rule. The imposition of regulation has to be justified on a case-by-case basis. This is why the new regulatory framework is flexible and limits regulatory intervention to those activities where it is (for the time being) necessary. However, as soon as a market becomes effectively competitive and regulation is no longer necessary to sustain competition, then regulation should be phased out.

In any case, it is important to note that the new regulatory framework does not affect the powers of the Commission under the competition rules of the Treaty and the Merger Regulation. The fact that undertakings are subject to regulation does not prevent the application of competition law to the electronic communications industry.

## 3. SHOULD ANTITRUST LAW APPLY TO FAST-MOVING, TECHNOLOGY-DRIVEN INDUSTRIES?

Given that the telecoms industry is largely innovation-driven, antitrust enforcement has not been exempt from criticism or controversy. Some have questioned whether the antitrust laws that were originally designed to apply to traditional manufacturing and distribution industries should be applied *mutatis mutandis* to fast-moving industries where products rapidly become obsolete and market shares may be ephemeral. Operators often claim that their market is an 'emerging market' and therefore forbearance from antitrust enforcement is warranted. Others complain that insufficient account is taken of the level of investment needed to roll out new advanced or alternative networks. What these voices are saying is 'leave us alone for a while and competition will flourish'.

Yet antitrust principles of general application must be applied to technology-driven industries as to any other industry. As was once said, 'the competitive struggle without effective antitrust enforcement is like a fight without a referee'.

It is also important to remember that competition principles remain valid irrespective of financial trends. Competition policy is primarily designed to benefit consumers who need protection regardless of 'business cycles' or 'stock market valuations'.

## 4. THE NEED TO APPLY CONVENTIONAL ANTITRUST THEORIES

Despite earlier promises that the miracles and wonders of technology would open the way for new competitors to overturn, overnight, the dominance of the incumbents, the situation we see today is somewhat different. In most 'electronic communications' markets, large sunk costs, high risks and other entry barriers may mean that, while product characteristics may change rapidly, the identity of the dominant players can remain unchanged for long periods of time.

Experience shows that anticompetitive practices such as 'price squeezes', 'leveraging of market power', 'aggressive bundling' have not yet been abandoned by incumbents. It is imperative therefore that antitrust enforcement remains vigilant and that entry by new competitors is not rendered illusionary. The importance of the communications industry to the economy requires an increased antitrust scrutiny in order to ensure that growth and innovation are not obstructed by market power abuses or other anticompetitive practices.

I do not mean to suggest, however, that the telecoms industry does not have distinctive features. What I mean is that sound antitrust enforcement in network industries requires careful attention to the particular characteristics of the industry. It is in this context that the use of economics can greatly enhance our understanding of how such industries work. To my mind, the real challenge of economic analysis in antitrust investigations is to shed light on evolving, often complex, issues of market definition, barriers to entry and competitive effects, to name but a few current important issues.

## 5. THE APPROPRIATE PLACE FOR ECONOMICS

Economics helps us to understand the fundamental and not always easy question of how markets work, a fundamental question any competition policy-maker should consider before planning any action. Economics is crucial for the analysis of the market's structure and dynamics, and for the analysis of the market players' behaviour and their incentives to behave in certain ways. These are fundamental issues in complex and dynamic markets, such as the electronic communications market, which involve a variety of elements needing sophisticated analysis (for example, the speed of technological innovation and the existence of network effects).

In the last few decades, we have seen a continuous increase in the role of economics in competition and regulatory policies. The Commission's recent 'Guidelines on market definition and the assessment of Significant Market Power' for the purposes of *ex ante* regulation in the electronic communications sector should be seen as a concrete example of incorporation of tested and proven economic theory into standard competition analysis, be it for the purposes of delineating the boundaries of the relevant market or for assessing the existence of single or collective dominance.

## 6. CONCLUDING REMARKS

The Commission will continue its vigilance in enforcing the antitrust rules in the electronic communications markets; and, in doing so, it will always take into consideration the special characteristic of this sector and base its approach on sound, well-grounded economic analysis.

This book highlights the importance of balanced economic and legal analysis in the debate on the role of antitrust and regulation in the telecommunications industry.

# Section 1

# The New Regulatory Telecommunications Framework

# 2. The competition policy approach

**Pierre A. Buigues**[1]

## 1. INTRODUCTION

One of the most important recent developments in the telecoms sector in the EU has been the adoption of the new regulatory framework for electronic communications networks and services, which should be transposed by Member States into national law by mid-2003. The aim of this chapter is to contribute to the open debate in Europe about this new regulatory framework, which is a major step in the process of liberalization of the telecommunications industry, itself a major sector of the EU economy, and to discuss in some more detail the competition-law aspects of this new regulatory framework. The chapter is structured as follows. Following this short introduction, the lessons of the 'old' regulatory framework are presented, with particular focus on the European Commission's conclusions set out in the 1999 Review, and the results of the Implementation Reports, where the Commission has each year examined the transposition and the implementation of the directives when incorporated into national law. The chapter then focuses on the analysis of the different components of the new regulatory environment and tries to give preliminary answers to the main critical questions surrounding the new regulatory regime: first, the markets to be regulated (fewer or more markets to be regulated?), second, the designation of significant market power operators (fewer or more operators to be regulated?), third, the remedies to be imposed on operators by regulators (more strict or more lenient?), and, fourth, the degree of co-ordination between the NRAs (national regulatory authorities) and the Commission (more centralized or more fragmented?). Finally, the conclusion summarizes the main message of the chapter and considers the respective roles of competition policy and regulatory policy in the new regulatory package.

## 2. THE LESSONS OF THE 1999 REVIEW

The 1998 legislative package of liberalization and harmonization directives was introduced in order to achieve the goal of liberalizing the telecommunications industry. At the early stages of liberalization, certain key instruments were introduced, as set out by Cave and Prosperetti (2001):

- *Control of retail prices*: in a traditionally controlled tariff industry, the structure of tariffs (monthly fees, call charges) was not related to the structure of costs. Moreover, without retail price controls, the historical operator, with nearly 100 per cent market share, could exercise its market power by increasing retail prices after liberalization.

Regulatory intervention was therefore needed, in particular for cost orientation and price capping.

- *Control of access prices*: interconnection prices were a key issue of the development of price competition. New entrants required access to the historical operator's network. Regulatory intervention on access prices was therefore needed to ensure fair conditions for new entrants.
- *Universal service obligations*: the historical operator was obliged to provide the same services at uniform prices throughout the whole country, but new entrants without such an obligation could focus just on the most profitable parts of the territory. Regulatory intervention was needed to define and to impose universal service obligations on all parties.

The first phase of liberalization having been achieved, a new approach was needed and the 1999 Review and the Implementation Reports formed the basis for the preparation of the new regulatory package.

## 2.1 The 1999 Review

The aim of the 1999 Review was to propose the main elements of a new legislative framework to be introduced in 2003. The new framework is designed to address the shortcomings of the old one, in particular by clarifying and simplifying the existing rules and by introducing a 'light' regulatory approach based on competition law. The major conclusions of the 1999 Review were as follows:

### Clarification and simplification of existing regulation

As stated in the 1999 Review, 'the existing legislative framework was primarily designed to manage the transition to competition and was therefore focused on the creation of a competitive market and the rights of new entrants'. Removing obligations of the existing framework which are no longer necessary is therefore one of the key principles underpinning the new regulatory framework.

The reduction of the number of legal instruments from twenty to six represents a substantial simplification. The future Community sector-specific legislation was designed to consist of one framework directive and four specific directives on licensing, access and interconnection, universal services, privacy and data protection. Non-binding measures such as recommendations and guidelines were considered to be important complementary measures to ensure flexible adaptation to market changes.

### Greater reliance on the general competition rules

As competition becomes more effective, much of the sector-specific regulation will be replaced by the application of the general competition rules of the EC Treaty. Regulation should only be proposed where there is a market failure, i.e. when it is absolutely indispensable. Regulation should therefore be kept to the strict minimum and should be designed according to competition-law concepts.

The 1999 Review also proposed improved co-operation between sector-specific regulators and competition authorities at national and European level to ensure that the decisions of the NRAs are compatible with the competition rules.

**No need for European Regulatory Authority**

In the 1999 Review, the Commission considered that the creation of a 'European Regulatory Authority' would not provide sufficient added value to justify the likely costs. In addition, it could lead to duplication of responsibilities, resulting in more, rather than less, regulation. It was felt that the issues that might be better dealt with at EU level could be addressed through an adaptation and improvement of the NRAs' existing regulatory structures. The potential disparities between national application of telecommunications directives could be solved by stronger co-ordination between the NRAs and the Commission.

## 2.2 Implementation Reports

In a series of Implementation Reports, the Commission has examined the transposition of the directives of the regulatory package and their implementation when incorporated into national law. These reports focus on progress in implementing regulatory requirements, licensing, interconnection, local access, universal services, mobile services, tariffs, cost accounting, leased lines, numbering, data protection and access rights. These reports were prepared on the basis of meetings between the Commission and the market players, users and NRAs.

Certain key conclusions can be drawn from an examination of the implementation of the old framework.

**Continuing regulatory fragmentation**

The current regulatory framework is not applied consistently across the EU, with the result that there is still not a level playing field for all market players throughout Europe. That regulatory fragmentation deters large enterprises from putting in place a coherent pan-European business strategy.

In some Member States, political considerations intervened to delay or even impede the rebalancing of tariffs by keeping monthly fees for local voice telephony at artificially low levels. In practice, the process of tariff rebalancing is politically sensitive as it primarily involves an increase of the line rental tariff. However, many Member States' regulators introduced price cap regimes (covering a basket of prices including line rental but also call charges) with the effect that any increase in line rental charges should normally be at least offset by decreases in call charges. Moreover, the impact of increases in line rental charges can be softened by well-targeted tariff measures such as measures in favour of low-volume users and users with special social needs. The economic importance of rebalancing can be seen, for example, with the situation whereby, with tariff imbalances, wholesale prices for local loop unbundling were set in certain Member States at a higher level than the monthly fee for local voice telephony, creating a margin squeeze for new entrants.

On 19 December 2001, the Commission went to the European Court of Justice with regard to the issue of tariff rebalancing in Spain after a three-year infringement proceeding. Spain was the Member State where the discrepancy between the retail tariff and wholesale tariff was the largest. Spain was not, however, the only country involved, and the Commission has also launched infringement proceedings against three other Member States at the same time. Moreover, on 6 December 2001, the Court condemned France for infringing the directives regarding universal service provision but also with regard to tariff rebalancing.

The case of leased lines is another example of the fragmentation of the single market. Leased lines are a vital element for the new economy, since they provide the underlying transmission capacity for the Internet, data services and voice telephony. If prices are excessive, access to the Internet will remain too expensive. The Commission therefore took action and launched a sector inquiry, which found that incumbents' prices may be excessive. The main conclusion was that national regulators must ensure that the principles set out in the directives are, in fact, respected.

In particular, the results of the leased lines sector inquiry demonstrated that the existing sector-specific regulation had not been implemented in full as regards the requirement of cost orientation. Delayed and/or incomplete application of the cost-orientation principle was reported by a number of national regulatory authorities. The Commission opened five *ex officio* procedures to investigate leased line pricing policies in Belgium, Greece, Italy, Portugal and Spain.

Both the NCAs and NRAs in those Member States discussed the findings of the leased lines sector inquiry, and, in several regulators' meetings, stipulated their readiness to investigate, where necessary, the high prices in question and possible discrimination, and to take appropriate measures to improve the situation.

In the *ex officio* procedures, following months of negotiations, all five national telecoms regulators opened a review of the national markets with a view to creating better conditions of competition and remedying existing regulatory failures. The Italian regulator launched a

## BOX 2.1 LEASED LINE SECTOR INQUIRY

**Price evolution**

The main problem in the early years of liberalization, and the reason for launching the sector inquiry, was high international leased line prices. In 1999 prices in Europe were as least three times higher than prices in the USA. Such high prices were detrimental to the competitiveness of European companies and have hampered development of vital links between companies and Internet networks.

Despite the fact that the pattern of decrease differs significantly per country, average international leased line prices showed a clear downward trend since 1999.

*Table 2.1 Price evolution of leased line prices (%)**

| | 1999 | 2000 | 2001 |
|---|---|---|---|
| 2 Mbs to near EU | –11 | –23 | –30 |
| 2 Mbs to far EU | –12 | –29 | –40 |
| 2 Mbs to USA | –16 | –28 | –33 |

*Note*: *Average price variations since 1998

*Source*: European Commission (2001).

**Pricing of interconnection and retail leased lines**
However, the sector inquiry results demonstrated that short-distance leased line prices were of more serious concern. Short-distance leased lines form part of the international leased infrastructure (tail ends or terminating circuits) and represent a means of local access which is typically still not subject to competitive pressure. The same applies to interconnection leased lines that are vital for providing access to a competitor's network. Prices charged for interconnection and retail leased lines of different distances still diverge considerably between Member States.

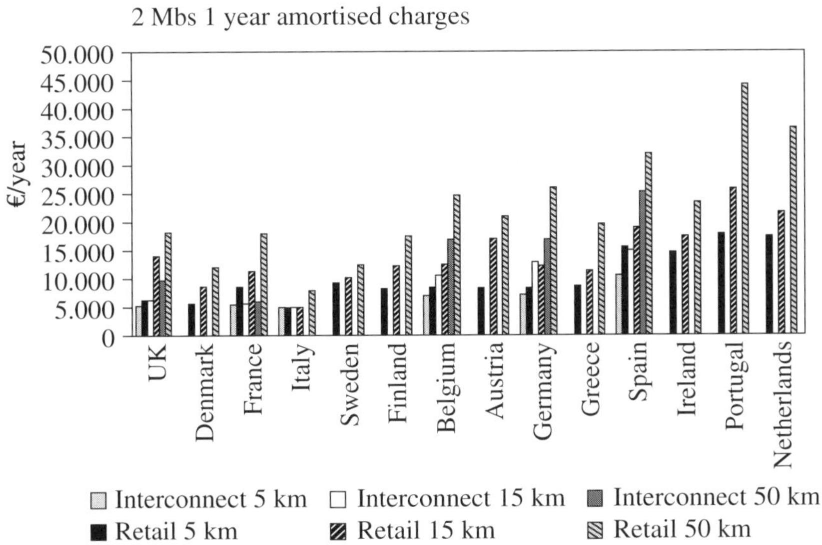


*Source*: European Competitive Telecommunications Associations (2001), Annual Scoreboard.

*Figure 2.1 Annual amortised rates (install + monthly + charges), interconnect vs retail*

**Non-price-related issues**
Incumbent operators maintain a *de facto* monopoly in the tail end or terminating leased line segments. This is exactly where the new entrants depend on the correct implementation of the sector-specific regulation by the national regulators, which is meant to guarantee a level playing field for all operators.

Non-price-related issues like timely deliveries, quality of service and performance indicators are now covered by the yearly open network provision performance reports produced by the Commission on the basis of data provided by the national regulators. There is considerable room for improvement of these reports so that they can meet the requirements of the new entrants and provide an objective picture of the non-price-related aspects of leased line provision in the EU.

public consultation on the introduction of separate wholesale and retail offers of leased lines and a market study related to the market for international leased lines. A number of steps were taken by the national regulators in Belgium, Germany, Spain, the Netherlands, France and the UK relating to both prices and non-price issues surrounding leased line provision. However, the disparities in price between these Member States clearly indicate a lack of adequate supervision by certain national regulators (see Box 2.1). However, following these investigations by the Commission, the national regulators have taken steps to address competition problems and in December 2002, the Commission decided to close its leased lines inquiry.

The transposition of the concept of non-discrimination regarding the Interconnection Directive is also very different between the Member States, and in some, incumbents give differential treatment to new entrants and their own downstream operating arms. In particular, provision of wholesale offerings to the incumbents' downstream arm is made in advance of giving notice to competitors.

**Continuing domination of incumbents in their national markets**

In 2001, it was estimated that in more than ten Member States, incumbent operators had market shares exceeding 90 per cent in the local call segment, and, in the majority of Member States, the incumbent operators had market shares exceeding 65 per cent, notwithstanding the fact that new entrants charge much lower rates than incumbent operators. For example, for long-distance calls in Belgium, France, Spain and the UK, new entrants offer tariffs 40 per cent lower than the respective incumbent operators. For local calls, new entrants offer smaller reductions (around 20 per cent in many Member States). Thus, while consumer sensitivity to price decreases exists, it is still limited, the incumbent operators benefiting from other key assets such as their presence in the entire geographical territory and their image based on their historical presence in the market.

Moreover, if incumbent operators have been losing limited market shares in their core business, the fixed lines, they have been very active in other telecommunication businesses such as the mobile telephony market and the Internet to recover the lost turnover. In these new markets, characterized by much higher growth than the fixed lines business, the incumbent operators are the market leaders in many Member States. The incumbents' business strategy of diversification into high-growth segments was the prime reason for the increase in their levels of debt.

A good example of the strategy used by incumbents to dominate new markets is that of broadband services. In some Member States, the political goal of increasing the penetration of broadband Internet has allowed incumbents to provide their own DSL (digital subscriber line) services at low, or even predatory, retail prices, which has kept new entrants out of the broadband Internet market. A good example is the case of Wanadoo in France. At the end of November 2001, Wanadoo had 90 per cent of the French market for ADSL (asynchronous digital subscriber line) Internet access. The European Commission sent Wanadoo, a subsidiary of France Telecom, a Statement of Objections on 21 December 2001 on the ground that the preliminary ADSL Internet access services were currently being charged below cost. Finally, the Commission has adopted a decision against Wanadoo Interactive (16 July 2003) for abuse of a dominant position in form of predatory pricing in ADSL-based Internet access services for the general public. Maintaining below-cost prices for these services makes the economic situation extremely difficult for competitors. These competitors, who do not, unlike Wanadoo, have the backing of France Telecom, are unable to withstand the price pressure imposed by Wanadoo on ADSL and are in practice forced out of the market. The aim of predatory pricing is to pre-empt competitors, the dominant operator being able to use its market position in the medium term to charge high prices (for instance for access to content) to recover losses incurred during its predatory pricing policy.

In conclusion, the fragmentation of the European market before full liberalization, when each incumbent enjoyed a monopoly in its Member State, has not evolved as had been expected into a scenario of pan-European operators active in a large number of Member States. On the contrary, the same fragmentation is reproduced in new mobile and subsidiary markets, since the incumbent is very often the market leader.

## 3. THE NEW REGULATORY FRAMEWORK

The main elements of the new framework are in place: the Framework Directive, the Authorisation Directive, the Access Directive, the Universal Service Directive and the Spectrum Directive. These directives all came into force on the day of their publication in the *Official Journal*, and, the Member States having 15 months for their transposition into national law, the effective date of their national transposition is 25 July 2003. Two other directives, the Commission Directive on competition in the markets for electronic communications networks and services (based on Article 86 of the Treaty) and the Data Protection Directive, are subject to a different timing in the legislative process. The new regulatory package also contains 'soft law': first, a Recommendation on Relevant Product and Service Markets (see Article 15(1) of the Framework Directive) and, second, the Guidelines on market analysis and the assessment of significant market power (SMP). The Recommendation and Guidelines are reproduced in Part III of this book.

The first thing that needs to be stressed is that the new regulatory framework represents the Community's response to the so-called convergence challenge. In today's communications world, voice telephony is delivered not only over wireline telecommunications networks but also over wireless, cable and Internet networks. Likewise, radio and television programming are delivered over radio, cable and, to a limited but growing degree, the Internet. In other words, in today's marketplace applications and services are to a large extent decoupled from the underlying mechanisms of transmission. A living proof of the convergence phenomenon is the fact that most telecoms operators are today present in more than one segment of the overall electronic communications market, offering their services on an array of alternative and competing infrastructures.

The real challenge for the Community legislator was restricted to finding new ways to address such convergence challenges, but needed to avoid, as had happened under the old framework, creating conditions for an asymmetric regulation, where different rules applied for essentially the same type of service due to the use of different underlying infrastructures. Such a situation contrasted sharply with a competition-law approach which defines the relevant product or services market by reference to demand and supply substitutability.

Another important question facing the Community legislator was whether, in devising a new framework, sector-specific regulation should continue in parallel to but separately from general competition law, or whether a bridge would need to be built between *ex ante* regulation, on the one hand, and *ex post* enforcement of competition law, on the other. The fact that under the old framework, undertakings were subject to two different sets of legal rules (*ex ante* regulation, on the one hand, and *ex post* competition-law enforcement, on the other hand) had created more market distortions than those that both sets of rule were supposed to remedy.

In the words of Cave and Larouche, 'The new regulatory framework is a major step down the transition path between monopoly and normal competition, governed exclusively by generic competition law' (2001). The objective is to propose an approach consistent with competition law but in a 'pre-emptive ex-ante form'. The new regime covers, in particular, two fundamental issues: first, the markets to be regulated, and, second, the designation of operators enjoying significant market power on which *ex ante* obligations may be imposed.

### 3.1 Markets to be Regulated

In the 'old framework', NRAs could only intervene in 'markets' referred to in the directives by designating undertakings as having significant market power. These 'markets' were not defined according to competition-law criteria. The old framework therefore introduced a limited number of constraints in the list and a definition of activities that could be regulated. In the new framework, markets to be regulated are defined in accordance with competition-law criteria and principles. Article 14 of the new Framework Directive requires the Commission to adopt a 'Recommendation on Relevant Product and Service Markets within the electronic communication sector susceptible to *ex ante* regulation in accordance with Directive 2002/21/EC on a common regulatory framework for electronic communications networks and services' (hereafter 'the Recommendation') which will identify those product and service markets within the electronic communications sector, the characteristics of which may be such as to justify the *ex ante* imposition of regulatory obligations. Article 14 specifies that 'the Commission shall define markets in accordance with the principles of competition law'.

Therefore, the proposed Recommendation constitutes an important document which should provide NRAs with the appropriate tools for carrying out their regulatory obligations. In particular, NRAs will have to determine whether the markets listed in the Recommendation are effectively competitive, that is, whether there are any dominant undertakings, and, accordingly, impose, amend, or withdraw regulatory obligations.

Under the new regulatory framework, *ex ante* regulation is warranted only in those markets for which competition-law remedies are insufficient to effectively redress possible market failures. The guiding principle is to avoid over-regulation. The governing criteria are presented in the Recommendation in order to provide sufficient reasoning as to the market segmentation used to justify the imposition of *ex ante* regulatory obligations.

#### Static criteria

In principle, electronic communications markets require regulation because of their particular characteristics, in essence the existence of high and non-transitory entry barriers and the existence of natural monopolies (static criteria). With respect to these static criteria, two types of barriers to entry are relevant: structural barriers and legal or regulatory barriers. Structural barriers to entry exist when the technology and the cost structures are such that they create asymmetric conditions between incumbents and new entrants, preventing or even impeding market entry by the latter. Structural barriers include, for example, substantial economies of scale and high sunk costs. Legal or regulatory barriers are not based on economic conditions but have a direct effect on the conditions of entry, for example the number of undertakings having access to spectrum.

**Dynamic criteria**

The existence or likely existence of entry barriers, although a necessary condition by itself, ought not, however, be sufficient to warrant inclusion of a given market in the Recommendation. Given the dynamic character and functioning of most of the electronic communications markets, the possibility of overcoming barriers within a relevant time must also be taken into consideration when carrying out a prospective analysis to identify the relevant markets for possible *ex ante* regulation (dynamic criteria). In particular, entry barriers may also become less relevant in the case of innovation, where competitive constraints may result from innovations from potential competitors not currently in the market.

**Criteria for adequate competition law remedies**

The decision to impose *ex ante* regulatory obligations in certain electronic communications markets should depend on whether competition-law remedies are or are not sufficient to redress market failures and to ensure effective and sustainable competition over a foreseeable time in markets characterized by high and non-transitory entry barriers. In particular, where the intervention needed to redress a market failure is extensive, for example because an assessment of costs based on detailed accounting is required, or because frequent intervention is needed, *ex ante* regulation could constitute an appropriate complement to competition law.

In summary, the decision to subject a market to *ex ante* regulation should be based on an overall assessment of the state of the effectiveness of competition within such a market, taking into consideration not only static but also dynamic criteria. Markets which fulfil these two criteria should also be made subject to *ex ante* regulation if competition law remedies are deemed insufficient for ensuring effective and sustainable competition.

**List of markets in the Recommendation**

Annex I of the Framework Directive provides a list of markets to be included in the first version of the Recommendation. The Commission has, therefore, simply reformulated and segmented, where appropriate, the markets mentioned in the Directive. The Commission will, however, regularly update the list of markets contained in the Recommendation.

This list does not prevent an NRA from regulating other markets. The methodology presented in the Recommendation is therefore crucial to show the reasoning behind the market segmentation choices. The Recommendation is not binding, and NRAs may define a relevant market which is not listed in it. However, the NRA would need the Commission's prior approval before imposing *ex ante* regulatory obligations on the undertaking(s) concerned (Article 7 of the Framework Directive, the so-called 'veto' power of the Commission). NRAs need therefore to know which are the Community-law criteria for *ex ante* regulation before they extend regulation to markets other than those listed in the Recommendation, since the criteria for regulating national markets should be the same as those relied upon for the markets already listed in the Recommendation.

**More markets to be regulated?**

While the application of a competition-law approach may increase the number of markets to be regulated, these markets will be defined much more narrowly than the markets currently subject to *ex ante* regulation under the 'old' framework.

A good example is that of mobile call termination, which is under the current framework part of the large interconnection market in relation to the Interconnection Directive. At present, there is no regulation of mobile call termination in Germany in the absence of a mobile operator with SMP in the interconnection market. In France, Belgium, Sweden, Italy, Finland, Spain and Ireland, however, this market is regulated and two mobile operators have been designated as SMP operators in the interconnection market, while in Sweden three SMP mobile operators have been designated.

The market for call termination on individual mobile networks is presently identified in the list of markets of the Commission Recommendation as a candidate market for *ex ante* regulation by national regulatory authorities. This list will be reconsidered by the end of 2004, based among other things on an assessment of whether problems in the markets concerned can be solved *ex post* according to the general competition rules. In this context, it is important to stress that competition law and regulatory instruments are not mutually exclusive. As the antitrust rules always apply, the only question is whether they are likely to be sufficient in themselves or need to be complemented with regulatory intervention. (See Box 2.2.)

The Commission has come to the preliminary conclusion that the provision of terminating access services on public mobile telecommunications network constitutes a separate market (KPN case). This conclusion is supported by an analysis of demand and supply, both at wholesale and at retail levels. At the wholesale level, all public network operators are obliged, and have a strong commercial incentive, to offer calls to other networks. To be able to do so they must purchase wholesale terminating access services on each other's network as a necessary input, services for which there are no substitutes because, for a call to reach a subscriber of network A, the network operator of A must provide the terminating access. At the retail level, users cannot choose which network they call when they want to reach a particular mobile user. Given the 'calling party pays' (CPP) principle, mobile users are not sensitive to the price that others pay to call them. On the supply side, there is no substitution between network operators since only the network operator can offer terminating access services on its own network. Moreover, the very low price-elasticity and the lack of competitive response to significant price changes clearly show that mobile operators do not compete on call termination services. Accordingly, the provision of call terminating services on individual mobile networks should be regarded as separate markets.

The UK Competition Commission has published a Statement of Hypothetical Remedies dealing with the mobile phone inquiry on mobile call termination. The major conclusions are: first the full support for the market definition for wholesale call termination on individual mobile networks; second an assessment of lack of competitive constraints in the foreseeable future on call termination; and finally a preference indicated for a price cap remedy on all operators but taking into account a number of factors influencing cost (Competition Commission, 2002).

With respect to wholesale international roaming, the Recommendation adopts the market definition contained in the Framework Directive. Wholesale international roaming is provided by a domestic network operator in country A to a mobile network operator in country B when a subscriber of this latter B network is travelling in country A. In other words, wholesale international roaming services give access and capacity (airtime minutes) to a foreign mobile network operator for the purposes of enabling its subscribers to make and

## BOX 2.2 TERMINATION CHARGES ON MOBILE NETWORKS

Studies from the OECD (2000) emphasize that termination charges on mobile networks are high (the EU average being around cent 18.16), ten times higher in Europe than the average charge for fixed-to-fixed interconnections (the EU average being around cent 1.86). Indeed, in its Seventh Implementation Report (European Commission, 2001), the Commission highlighted that the 'range of peak time call termination charges in mobile of cent 12.44 to cent 22.69 as between Member States cannot be justified in terms of the actual costs of terminating calls, nor in terms of its level in relation to the average fixed to fixed tariff'. For 2001, as regards the range of prices between the Member States, the peak time charges in three Member States (Greece, Italy and Portugal) were around twice as high as in the Member State with the lowest tariffs (Sweden); in three others (Spain, France and Finland) they were close to the highest, while in three others (Luxembourg, Austria and the Netherlands) they were close to the lowest. Since mobile operators are monopolists in relation to the supply of termination services on their networks, they have insufficient incentives to reduce such charges and, absent any regulation, are even likely to raise them (for example the Netherlands). As pointed out by the OECD, it is clear that 'prices in this market appear to be well above cost and should be reviewed by regulators who should require cost-based pricing'.

A number of national regulatory and competition authorities have already taken action on domestic prices. These include Belgium, France, Italy, Sweden and the UK. These authorities have decided to cut the average price of charges for call termination substantially over a number of years. Despite these initiatives, it is clear that differences remain between several Member States. For example, the level of termination charges in Austria and Belgium in 2002 (approximately cent 12.00) will only be attained in France in 2004 and in the UK in 2005. Moreover, the termination charges in those Member States where no effective regulatory action has been taken remain high. For example, in January 2002, the termination charges in the Netherlands, Italy and Spain were still at a level of approximately cent 17.00 to 18.00, despite some voluntary reductions decided by mobile operators in Italy and Spain. According to some fixed operators, an indication of the level of mobile termination charges in a competitive market can be measured by halving the level of the cost of an on-net call in the mobile network concerned, and it is claimed that this test reveals discriminatory levels of charging. Mobile operators, however, contest these findings by arguing, for example, that there are inherent cost-savings on on-net calls.

receive calls while on another operator's network abroad. International wholesale roaming services are thus provided by a domestic mobile network operator (visited network) to a mobile network operator in another country (home network).

The Recommendation notes 'the absence of substitution for wholesale international roaming services', which implies that there is a separate national wholesale market for international roaming services.

This concern about high prices for using a mobile phone when abroad is relatively serious in Europe. The OECD's 2000 report (OECD, 2000) underlines that prices for international roaming appear to bear little relation to cost and that this reflects insufficient competition in the market, due partly to its relatively small size. Oftel stated in November 2001 that 'roaming prices are relatively high and static' and that 'the main reasons are the lack of consumer awareness of different prices and tariffs and barriers to consumers switching networks' (Oftel, 2001).

Those findings have largely supported the conclusion reached in the course of the Commission's roaming sector inquiry that national markets for wholesale international roaming suffered from a number of market failures. More specifically, it found that those markets were concentrated, and were characterized by high barriers to entry (in the form of licensing), high and rigid prices, similar pricing structures and suspicions of price-fixing. Even though national markets for wholesale international roaming services are included in the initial list of the markets to be considered for regulation by the national regulators, sector-specific regulation would be superfluous should competition-law remedies prove to be sufficient to address the market failures at hand.

That is why the outcome of the Commission's pending competition-law investigations (in the form of the sector inquiry and individual cases) into the prices of roaming across the EU will be of great importance in deciding whether sectoral regulation or competition law should take precedence in addressing the problems in the wholesale international roaming markets.

## 3.2 The Designation of SMP Operators

Under the old regulatory regime applied in the telecommunications field, NRAs can impose *ex ante* obligations on operators with SMP whose market share in the relevant market exceeds 25 per cent. However, a certain flexibility was introduced in the Interconnection Directive, as it provides the possibility to deviate from the 25 per cent rule by taking into account other criteria such as the turnover of the undertaking relative to the size of the market, the financial resources of the undertaking and so on.

Under the new framework, NRAs should impose *ex ante* obligations on operators only if the latter are in a dominant position within the meaning of Article 82 of the EC Treaty. That new definition of SMP is clarified in the 'Guidelines on Market Analysis and the Assessment of Significant Market Power' in order to assist NRAs in applying the new definition of SMP. This new definition will have the effect of raising the existing threshold for *ex ante* regulation, while at the same time creating the necessary conditions for legal certainty throughout the Community. What is even more important, it will put an end to the existing asymmetric treatment of companies which in the past, under the old framework, were deemed to have significant market power and were thus being subject to *ex ante* regulation if their market share was just above 25 per cent, but were still short of dominance within the meaning of Article 82 if their market share was below 40 per cent. The application of fundamental competition law notions, such as market definition and dominance, in an *ex ante* environment represents the best means to ensure a smooth transition towards a fully liberalized electronic communications market (Hocepied, 2002).

The Guidelines deal essentially with two main issues, market definition and assessment of dominance. This document reflects not only the relevant case-law of the Court of Justice and of the Court of First Instance, but also the Commission's own experience in defining the relevant market and in applying the competition rules, in particular, to the telecommunications sector.

As regards market definition, Section 2 of the Guidelines sets out the main principles for defining the relevant market and also lists a number of product and geographical markets which have already been identified by the Commission in a number of previous competition decisions relating to the telecoms sector. The Guidelines present the main criteria for defining the relevant market, and, in particular, the 'hypothetical-monopolist test' which is one way of assessing the existence of demand-side and supply-side substitutability. The likely responses by consumers to a price increase of between 5 and 10 per cent will aid in determining whether substitutable products exist and, if so, where the boundaries of the relevant product market should be delineated.

As far as dominance is concerned, Section 3 of the Guidelines sets out the relevant criteria for designating an undertaking as dominant. These criteria derive from the jurisprudence of the Court of Justice and the Court of First Instance. The assessment of a dominant position is not a mechanical, automatic process but one based on a case-by-case analysis which depends on the particular characteristics of each market. Therefore, the Guidelines present a set of fundamental principles which are indicative of market power. While high market shares can, in principle, be a prime indication of a dominant position, other criteria can also be used to measure market power. These criteria are listed and discussed in detail in the Guidelines. The Guidelines also acknowledge that an *ex ante* assessment of a market situation differs from an *ex post* application of the competition rules in so far as NRAs must conduct a forward-looking analysis, similar to that carried out by the Commission when applying the Merger Control Regulation.

As far as collective or 'oligopolistic' markets are concerned, where competition may not yet be effective (the market for mobile telephony often being cited as a prime example in this regard), the Guidelines consider that the concept of collective or oligopolistic dominance is sufficiently developed in the case-law or in the Commission's own practice. In particular, over the last five years, both the Court of First Instance and the Court of Justice have rendered a number of important judgments which have applied, defined and clarified the concept of collective and oligopolistic dominance. Furthermore, the Commission has already adopted a number of decisions where the concept of collective and oligopolistic dominance has been applied.

### 3.3 Appropriate Remedies

Under the new regulatory framework, once an operator is designated as having SMP, the NRAs shall impose appropriate remedies on the basis of the relevant obligations listed in the directives.

The ascending hierarchy of obligations listed in the directives relates to transparency, non-discrimination, accounting separation, access, price control and cost accounting (Articles 9 to 13). Moreover, Article 8 provides that 'in exceptional circumstances, when a national regulatory authority intends to impose on operators with significant market power other obligations for access on interconnection than those set out in Articles 9 to 13 in this

Directive it shall submit this request to the Commission. The Commission shall take a decision authorising or preventing the national regulatory authority from taking such measures'. In principle, NRAs shall not impose the obligations set out in Articles 9 to 13 of the Access Directive on operators that have not been designated as SMP operators.

However, Article 8(3) of the Access Directive also refers to provisions containing obligations on undertakings other than those designated as having SMP in relation to the Framework Directive, the Authorisation Directive, the Universal Service Directive and the Directive on processing of personal data and the protection of privacy in the telecommunications sector. For example, in Article 12 of the Framework Directive, NRAs may impose the sharing of facilities on an operator because of 'the need to protect the environment, public health, public security, or to meet town and country planning objectives'.

When NRAs have decided on the appropriate remedy to be imposed on operators, the only recourse available to operators against the choice of this remedy is through the national courts. On the other hand, the Commission may intervene when NRAs impose on SMP operators obligations other than those listed in the Access Directive, and may intervene under Article 226 for failure to correctly apply the provision on the choice of remedies (see Table 2.2).

*Table 2.2 Regulatory remedies – situation from 25 July 2003 onwards*

| Definition markets by recommendation or NRAs | Market assessment | Regulatory remedies |
|---|---|---|
| Triggers market analysis by NRA | If dominant undertakings, NRA must impose remedies | 1. Transparency<br>2. Non-discrimination<br>3. Accounting separation<br>4. Unbundled access and resale of facilities<br>5. Price control<br>*Any other obligation with Commission approval* |
| Other markets left only to DG Comp., NCAs and national courts | DG Comp., NCAs and national courts must prove abuse | |

Two further issues are also discussed within the context of the new regulatory framework:

- *Designation as SMP without imposition of regulatory obligations*: Is the NRA obliged to impose obligations on operators with SMP or does it have the discretion not to do so where appropriate? Recital 27 of the Framework Directive underlines that '[i]t is essential that ex-ante regulatory obligations should only be imposed … in markets where there is one or more undertakings with significant market power and where national and Community competition law remedies are not sufficient to address the problem'. If the appropriate remedy is non-discrimination (Article 10 of the Access Directive), NRAs may decide not to duplicate existing non-discrimination of competition law.
- *Differential obligations on SMP operators within an oligopoly*: Where several SMP operators are in a situation of collective dominance but are significantly different in

size, the question of whether the Commission and NRAs are obliged to impose the same remedies on each of them arises. NRAs are required to present the Commission with the reasoning on which their proposed remedy is based (Article 7 of the Framework Directive), with particular regard to the principle of proportionality.

## 3.4 Consultation with other NRAs and the Commission

The NRAs must analyse the product and services markets presented in the Recommendation, determine whether or not there is an SMP operator in these markets, and, finally, withdraw, amend or impose regulatory obligations as appropriate. An NRA may nevertheless challenge the Commission's assessment and show that certain relevant markets within its territory have characteristics justifying *ex ante* obligations even if the markets are not included in the Recommendation. The assessment by NRAs in different Member States of SMP operators could diverge since SMP is defined on the basis of dominance in the sense of Article 82 (a 'position of economic strength affording an undertaking the power to behave to an appreciable extent independently of competitors, customers and ultimately consumers'). Even if the guidelines provide substantial background, it is not obvious that all NRAs, some with limited experience of applying competition-law concepts, will use the same approach to dominance, in particular in cases of collective dominance.

The procedure contained in Article 7(4) of the Framework Directive should none the less avoid unjustified divergences between Member States and therefore fragmentation of the internal market. Under this procedure, the Commission is able to veto draft NRA decisions on SMP designation where interpretations of the concept of dominance diverge and in new markets not included in the Recommendation. This procedure will avoid NRAs defining an operator as dominant under sector-specific regulations when it is not dominant under competition law, and will oblige them to justify the regulation of markets other than those listed in the Recommendation.

Under Article 7 of the Framework Directive, NRAs will have to notify to the Commission and to the other NRAs all draft measures designating either telecoms operators as being dominant, or new service product market definitions they intend to apply.

In a 'first phase', other NRAs and the Commission have one month to react to the measures notified. As regards measures which define a product market which differs from those listed in the Commission Recommendation or designate operators as having SMP when these measures could affect trade between Member States, the Commission may, within this time period, raise 'serious doubts'. If it does so, it has another month to take a ('second phase') decision requiring the NRA in question to withdraw the measure when it is not compatible with Community law.

The Commission will therefore carefully oversee the market analysis carried out by the NRAs, that is, market definition, SMP assessment, and the type of remedies prescribed by the NRAs.

### The European Regulators Group (ERG)

This group has been established by the Commission as an advisory group for the new regulatory framework and succeeds the current informal NARA (National Administration and Regulatory Authorities) and IRG (Independent Regulators Group). The informal character of the IRG and the absence of effective institutional arrangements have meant that, although it has grown in importance over recent years, the IRG has not succeeded in presenting itself as a true regulators' group capable of speaking with one voice in all matters related to the current telecommunications framework.

The objective of the ERG is to provide an interface for assisting and advising the Commission so as to contribute to the development of a true internal market. It will also constitute a suitable mechanism for encouraging co-operation and co-ordination of NRAs in order to achieve consistent application in all Member States of the directives making up the new regulatory framework. The ERG shall be composed of the head of each NRA in each Member State, with the Commission as secretariat. The ERG may organize sub-groups or working groups as appropriate, and shall extensively consult market participants, consumers and end-users in an open and transparent way.

Once the new framework is put in place in the various Member States, the ERG will thus assume an important role, seeking in particular to harmonize, on a voluntary basis, the approach of the NRAs as regards the *ex ante* obligations ('remedies') imposed on dominant undertakings in regulated markets. In particular, the ERG will promote 'best practices' and will monitor closely any possible failings by NRAs in carrying out their regulatory tasks. Such pressure may prove more rapid and effective than the lengthy Article 226 infringement procedure against Member States which fail to comply with their Community-law obligations.

## 4. CONCLUSION

The new framework aims to adapt the regulatory environment to the changes taking place as a result of both completion of the first stage of the liberalization process (the '1999 package') and technological progress. This new framework is intended therefore to reduce the degree of regulation in the process of transition towards a free competitive environment, and to increase the degree of harmonization, in order to ensure a level playing field in the single market for all EU undertakings. It is too early to have a clear view on how the new regulatory framework will be implemented in practice, but a first broad assessment is possible, in particular in relation to the major concerns and criticisms expressed to date (Larouche, 2002).

### Degree of Regulation

Certain experts have stressed that, in the new regulatory framework, market definition should not be too narrow, and should take into account foreseeable changes in the market. On this point, the Recommendation identifies the characteristics of markets which may be such as to justify the imposition of regulatory obligations without prejudice to markets that may be defined in specific cases under competition law. In any event, the message conveyed by the Guidelines to the NRAs is that market definition should be assessed on the basis of a prospective forward-looking assessment.

Even if, with the competition-law approach presented in the new regulatory framework, market definition may be narrower than the markets defined in the directives and currently subject to *ex ante* regulation, there is no risk of further regulation. The Recommendation is clear that the decision to impose *ex ante* obligations should depend on whether competition-law remedies are insufficient to ensure effective and sustainable competition.

Moreover, the volume of telecommunications activities (in terms of turnover, for example) subject to regulation under the new regime will be much smaller than at present, even if the number of regulated markets may increase and their definition may be narrower. With a narrower market definition, for example, SMP operators may be regulated only in the specific segment fulfilling the characteristics set out in the Recommendation, instead of being regulated in a much broader market. Moreover, a narrower definition of the relevant market will facilitate the application of proportionate remedies adapted to the specificity of that market.

## Degree of Legal Certainty

Certain commentators have argued that the new regulatory framework may create a high degree of uncertainty for firms since NRAs may apply concepts of competition law where European jurisprudence is not yet clear, particularly as regards collective dominance. Certainly, competition-law concepts are more complex to apply than the mechanical threshold of 25 per cent of the old framework, but the Guidelines contain a long list of criteria to be considered in assessing cases of collective dominance and a clear methodology to be followed. Any additional uncertainty brought about by the approach to market definition in the new regulatory framework is, in any event, no different from that which merging telecommunications enterprises face when confronted with the Merger Control Regulation. Ultimately, the application of competition-law notions such as dominance in an *ex ante* environment represents the best way to ensure a smooth transition to a fully liberalized telecommunications market.

## Competition and Regulation

As one commentator has noted, '[at] the highest level of abstraction which necessarily involves some simplification, the difference between Competition policy and regulation is that while competition policy tells firms what not to do, regulation involves telling them what they are to do' (Ergas, 2002). Thus, competition policy proscribes and regulation prescribes. Even if this distinction is somewhat simplistic, it shows the very basic functions of both approaches.

The regulatory approach is preventive and based on detailed control of business behaviour, such as, for example, the imposition of price controls which supposes extensive information on the present costs of the incumbents. The regulatory approach also requires taking forward-looking views on the evolution of the market to prevent possible adverse effects resulting from regulatory interference with a firm's behaviour. The *ex post* approach does not impose specific obligations, but companies can be fined *ex post* if their conduct has had anticompetitive effects. Competition law does not prohibit dominance in itself, but prevents abuse of dominance, and can be viewed as an *ex post* regulation of the markets to contain abuses of market power.

The regulatory solution is costly for society since regulators need substantial information to have the specialist expertise to monitor the market. This implies that the regulator must be in almost continuous contact with the regulated firm, since the latter has the best knowledge of the market. However, given that regulated firms have a very high incentive to seek to influence the regulator, and governments are not always in a position to fully monitor the regulator, some economists argue that 'regulatory capture' (regulators being controlled by the regulated firm) is a clear threat. Thus 'it is essential that ex ante regulatory obligations should only be imposed where national and Community law remedies are not sufficient to address the problem' (Recital 27, Framework Directive).

Therefore, the general objective in the telecommunications sector with the new regulatory framework was to find the appropriate balance between *ex ante* and *ex post* regulatory approaches in the specific context of the network industry.

## NOTE

1. I would like to acknowledge all those who contributed with useful internal papers and comments in producing this chapter, in particular C. Hocepied and L. Papadias, D. Choumelova and B. Claes for the boxes on specific issues, and S. Bone for his assistance with the drafting.

## REFERENCES

Cave, M. and P. Larouche (2001), 'European Communications at the Crossroads', report of a CEPS Working Party, Brussels, October.

Cave, M. and L. Prosperetti (2001), 'European Telecommunications infrastructure', *Oxford Review of Economic Policy*, no. 4.

Competition Commission (2002), 'Mobile Phones Inquiry Statement of Hypothetical Remedies', Competition Commission, IP/39/02.

Ergas, H. (2002), 'Regulation and essential facilities: some key issues', The network economics consulting group, Rome.

European Commission (1999), 'Towards a new framework for electronic communications infrastructure and associated services: the 1999 Communication Review', *COM(539)*, November.

European Commission (2001), 'Seventh report on the implementation of the Telecommunications regulatory package', *COM(706)*, November.

Hocepied, C. (2002), 'The new EU regulatory Framework for electronic communications: from sector specific regulation to Competition law – Provisional working', IBA/ABA Communications and Competition, Developments at the Crossroad, 20–21 May, Washington, USA.

Larouche P. (2002), 'A closer look at some assumptions underlying EC regulation of electronic communications', *Journal of Network Industries*, **3**: 129–49.

OECD (2000), 'Cellular Mobile Pricing Structures and Trends', OECD Working Party on Telecommunications and Information Service Policies', DSTI/ICCP/TISP(99) 11/final, Paris, France.

Oftel (2001), 'International Benchmarking Study of Mobile Services', Oftel, 7 November.

# 3. Economic aspects of the new regulatory regime for electronic communications services

**Martin Cave**

In this chapter I outline and comment on the economic basis of the new regime for regulating electronic communications services, coming into force in July 2003. I first make some comments on the current arrangements (the 1998 package) and set out the new regime in summary form. Then I address two specific issues: the basis for choosing markets subject to *ex ante* regulation, and the choice of remedies. Many other major issues, including the specifics of market definition and analysis, are treated briefly as they are considered elsewhere in this volume.

## 1. LESSONS OF THE 1998 PACKAGE

It is useful to describe briefly the 1998 legislative package, which aimed at liberalizing telecommunications in the European Union.[1] For ten years or more before that, a series of green papers, directives, recommendations and other interventions has imposed obligations on Member States with respect to equipment markets, regulatory structures, value-added services and the regulation of infrastructure and service competition where it existed. But in 1998, the obligation was imposed on governments to liberalize entry into their telecommunications markets (except for those few countries for which extensions were granted).

The 1998 framework was inevitably fragmentary, as it developed over time. It also embodied rough-hewn remedies, designed to 'take on' the incumbents. Overall it represented a crude but potentially effective toolkit. The problem lay as much with implementation as with its own conceptual inadequacies. Liberalization and harmonization directives first have to be transposed into national legislation to take effect in the Member States. The transposition process took nearly two years, but the Commission was able to report in October 1999 that it was largely complete. But, as noted above, the directives give Member States considerable latitude in implementation. For example, the Licensing Directive (Directive 97/13) permits ample variation on the requirements imposed on new entrants, and despite a requirement in the Interconnection Directive (Directive 97/33) that interconnection charges be cost-based, such charges within the EU varied significantly.

Achieving the goals of liberalizing the industry against the (initial) wishes of most incumbent operators and of many Member States required substantial regulatory intervention from the Commission. Moreover, the regulatory framework had to be flexible enough to cover Member States proceeding at quite different rates. In the early stages of liberalization – the transition to competition – it was necessary to constrain the former monopolists considerably. Gradually this restraint can give way to a state of normalization, as competition is established and regulatory intervention diminishes.

In markets where there are no technical restrictions on entry, arising, for example, from spectrum scarcity, the key structural requirement is that licences should be granted to operators subject to a minimum of conditions.

For behavioural regulation, three instruments are required in the early stages of liberalization:

- *Control of retail prices.* This is necessary only when the historical operator exercises market power at the retail level and where in the absence of retail price controls, customers will be significantly disadvantaged. Member States have historically fulfilled this consumer protection function, though under monopoly conditions, in a way that had the result that the controlled tariffs were seriously unbalanced with respect to cost.
- *Universal service obligation (USO).* Governments have typically imposed a universal service obligation requiring the historical telecommunications operator to provide service to all parts of the country at a uniform price, despite the presence of significant cost differences. Firms entering the market without such an obligation have a strong incentive to focus on low-cost, 'profitable' customers, putting the USO operator at a disadvantage; or the incumbent uses this as an argument against entry. There are, however, other ways to fulfil a USO, such as tendering for its provision.
- *Control of access prices.* In order to keep all subscribers connected with each other in the presence of competing networks, operators require access to one another's networks to complete their customers' calls. This requires a system of inter-operator wholesale or network access prices. Especially in the early stages of competition, entrants will require significant access to the dominant incumbent's network, and this relationship will almost inevitably necessitate regulatory intervention. As infrastructure is duplicated (at different rates in different parts of the network), the need for direct price regulation of certain network facilities diminishes. Interconnection has been central to the development of competition within the EU, and the Commission has been heavily involved.

  The Interconnection Directive (Directive 97/33) requires that charges for interconnection follow the principles of transparency and cost orientation. The first principle implies the publication of a reference interconnection offer. As a corollary, operators with significant market power (SMP) – defined as a 25 per cent market share of a pre-specified national market – are required to keep separate accounts for their wholesale or network activity and for other activities, including retailing. Cost orientation turned out to be an excessively vague phrase, permitting excessive interconnection charges. Until adequate cost data were available, the Commission published recommended 'best current practice' interconnection charges, based upon the average of the Member States with the lowest charges. Charges dropped quickly in some countries, but not in all.

The level of access charges is likely to have had a major impact on entry strategy. A recent study commissioned by OPTA, the Dutch regulator, investigated the relationship between access pricing in the Netherlands and the development of infrastructure competition.[2] It concluded that entrants in Holland typically adopted the policy of first replicating those assets which involved a relatively small amount of sunk costs. Thus a *de novo* entrant may typically begin purely as a re-seller, investing primarily in marketing and advertising. It may then switch to making investments in switching and conveyance at the national level, before contemplating investment in the loop.

A cable operator entering into telephony or the provision of Internet services has a different inheritance and is likely to adopt a different strategy. Further investment is required in the loop, and the cable operator has to buy call termination from the incumbent – a service that is wholly non-replicable. The cable operator also needs access to long-distance conveyance, which it can either replicate itself or, for example, utilize facilities built by a former re-seller now investing in the national infrastructure. Finally, hypothetical non-cable entrants into the high-bandwidth market will encounter the loop as a non-replicable asset, making unbundling of the loop indispensable.

This analysis identifies two ways in which regulatory influence can be brought to bear to affect investment by infrastructure competitors. The first relates to the relative access prices of different assets. An entrant's investment decisions are more concerned with the price of replicable assets than of non-replicable assets. A regulatory policy that imposes a low price (relative to cost) for the latter thus encourages infrastructure investment. Second, because entrants take time to develop their competing asset base and begin with those assets that are the easiest to replicate, a policy of prices that rise over time will facilitate the gradual development by entrants of their own comprehensive network. The most investment-friendly policy is one of initially lower access prices for all network services, followed by a rising price trend applied successively to assets in descending order of replicability. Although the discussion above has been couched in favour of the price at which mandatory access is available, it can be translated into the alternative dimension of where access is mandated.

On this basis, the falling trajectory, relative to cost, of access prices in Europe would seem to sustain a 'service bias' in European telecommunications regulation. The regime has sought to open up all avenues of competition but in practice has been more preoccupied with opening existing infrastructures than with encouraging the construction of new ones. Entry of re-sellers has been vigorous and prices have been dropping fast, but this does not seem to have provided medium-term incentives for an extensive deployment of alternative infrastructures. This situation has, of course, been aggravated by the collapse in telecoms share prices since early 2001, which has eliminated or weakened many alternative carriers. As a result the new regime is coming into operation against the background of weakening competition in important parts of the fixed telecommunications marketplace.

Competition in mobile markets is more equal and more vibrant. As second and successive operators came into the market, prices fell and demand grew, assisted by innovative tariffs such as handset subsidies and pre-pay. The sector was lightly regulated apart from two possible pinch points, mobile call termination and international roaming charges. The absence of regulation, on balance, was highly beneficial, although the mobile industry's future development is affected by high licence fees paid by operators for 3G licences. The new regulatory regime is thus being developed against a background of significant successes in mobile but much more limited competition in fixed services.

## 2. THE NEW REGIME IN OUTLINE

At one level, the new regime is a major step down the transition path between monopoly and normal competition, governed exclusively by generic competition law. Its provisions are applied across the range of 'electronic communications services', ignoring pre-convergence distinctions. It represents an ingenious attempt to corral the NRAs down the path of normalization – allowing them, however, to proceed at their own speed (but within the uniform framework necessary for the internal market). Since the end state is one governed by competition law, the Commission proposes to move away from the rather arbitrary and piecemeal approach of the current regulatory package towards something consistent with that law. However, competition law is to be applied (in certain markets) not in a responsive *ex post* fashion, but in a pre-emptive *ex ante* form. The new regime therefore relies on a special implementation of the standard competition triple of: market definition, identifying dominance, and formulating remedies. We examine these in turn.[3]

According to the underlying logic, a list of markets where *ex ante* regulation is permissible is first established, the markets being defined according to normal competition-law principles. These markets are analysed with the aim of identifying dominance (on a forward-looking basis). Where no dominance is found, no remedy can be applied. Where dominance is found, the choice of an appropriate remedy can be made from a specified list, except that no remedy need be adopted, and any remedy may be applied only to part of the market. The effect of this is to create a series of market-by-market 'sunset clauses' as the scope of effective competition expands.

### 2.1 Market Definition

The Commission will issue a Recommendation on relevant markets – defined broadly in the manner of competition policy. NRAs may depart from the Recommendation subject to objection by the Commission. The draft Recommendation currently under consultation incorporates flexibility by allowing related 'technical services' to be aggregated within a market definition and by permitting segmentation (and differential remedies) in certain markets. Member States can also add or subtract markets, using specified (and quite complex) procedures.

A key issue, to which I return in Section 3 below, is how the relevant markets are chosen. In principle, all markets could be included. Yet the Commission, in the draft Recommendations, has adopted a different approach. This choice goes to the core of the new regime.

Both NRAs and NCAs and the Commission and European Court of Justice have undertaken many market definition exercises already, often using the now conventional competition policy approach. This often involves applying the so-called hypothetical-monopolist test, under which the analysts seek to identify the smallest set of goods or services with the characteristic that, if a monopolist gained control over them, it would be able to raise prices by 5 to 10 per cent over a sustained period, normally taken to be about a year. The monopolist's ability to force through a price increase obviously depends on the extent to which consumers can switch away from the good or service in question (demand substitution) and the extent to which firms can quickly adapt their existing productive capacity to enhance supply (supply substitution).

A consequence of the reliance of the proposed new regime on pre-emptive regulation is that it is necessary to adopt a forward perspective. This approach is endorsed in the draft Recommendation on market definition and the Guidelines on market analysis.

## 2.2 Dominance

After first proposing an even more interventionist proposal, the Commission adopted, in its initial set of proposals of July 2000, the classical 'dominance' as a threshold for intervention, which it calls 'significant market power'.[4] The dominance can be exercised by a single firm, or collectively, or leveraged into a vertically related market.

In addition, Article 14 of the Framework Directive contained a prohibition on intervention in markets that are effectively competitive – implicitly defining markets where dominance is absent as effectively competitive. This is a change of fundamental deregulatory significance, and a major step along the route towards convergence with competition law.

Although single-firm dominance has come to be well understood, joint dominance has been one of the more elusive components of European competition law. The Airtours judgment may now have clarified this. However, what is more noteworthy is the relative lack of candidates for joint dominance in telecommunications markets. This arises because (i) fixed markets are typically effectively competitive or dominated (singly) by the historic monopolist, and (ii) mobile markets (subject to the possible exceptions noted above of call termination and wholesale international roaming are generally competitive. Concerns about the likelihood of tacitly collusive behaviour by operators in setting bilateral termination have now abated.[5]

The communications industry, like many industries, consists of a series of activities that can be performed either individually or by vertically integrated firms. There are well-established benign motives for firms to become vertically integrated. In particular, doing so may reduce production costs, by eliminating the costs of transactions between two separate firms. It has also been argued that vertical integration by itself does not add market power. Thus, if a firm held a monopoly of an activity or process at any stage in an industry it would be able to extract maximum profit from that monopoly, and would have no desire to engage in other activities, unless it was extremely efficient in performing them.

This simplified approach has now given way to more complex modelling of situations in which a vertically integrated firm may find it advantageous to distort competition downstream as a means of bolstering its upstream market power.[6] This is achieved by a variety of means involving the interaction of particular features of each market. For example, in one market (say, for delivery platforms) there may be consumer switching costs, because consumers need to make significant investments in equipment. The second market may exhibit service differentiation (for example, a live broadcast from one soccer league may be a poor substitute for a broadcast from another). In such circumstances, making the service exclusive to the delivery platform may strengthen consumer lock-in and give the firm an ability to distort competition. In the broadcasting industry, particular activities (including the operation of networks and of technical services) are likely to be undertaken by small numbers of operators that are themselves vertically integrated into services. Significant disadvantages may arise for consumers in circumstances where a finding of single dominance would be difficult. In these cases, the issues of joint dominance and leveraging interact, creating considerable uncertainty about the regulatory outcome.

In addition, a firm can enhance its position by bringing together degrees of market power at different levels in the production process, in such a way that the whole becomes greater than the sum of the parts.

## 2.3 Remedies

Under the directives, NRAs will have the power to impose obligations on firms found to enjoy significant market power in a relevant market. The NRAs will act within a framework of duties set out in Article 8 of the Framework Directive. The measures they take shall be proportionate to the policy objectives identified. This can be construed as meaning that the intervention is appropriate, no more than is necessary, and, by implication, satisfies a cost–benefit test, in the sense that the expected benefits from the intervention exceed the expected costs. Article 8 additionally specifies policy objectives, which determine the weights appropriate for use in the cost–benefit analysis. For example, Article 8(2) requires NRAs to promote competition for electronic communications networks and services by maximizing users' choice and value for money, eliminating distortions or restrictions to competition and encouraging efficient investment and infrastructure. Article 7(4) requires NRAs to promote the interest of EU citizens by, *inter alia,* providing consumers with protection in their dealings with suppliers, and requiring transparency of tariffs and conditions for use of publicly available electronic communications services. NRAs must also contribute to the development of the internal market by avoiding different approaches to regulation within the EU. These provisions provide an important context in which NRAs must hone their interventions.

In the light of the application of the concept of dominance in the proposed directive (in particular in Article 14 of the Framework Directive), the NRAs must also ensure that they act in accordance with European case-law. One possible, if restrictive, way of interpreting this requirement is that the NRAs should intervene in relation to an operator with SMP to produce behaviour consistent with that which would exonerate a dominant firm from the charge of abuse before the European Court of Justice. Regulatory interventions should also take account of other aspects of the framework to be established by the proposed directives. Notably, the existence of a procedure for resolving disputes between undertakings (set out in Article 20 of the Framework Directive) provides an additional mechanism for resolving disputes. The existence of this alternative approach should be taken into account in determining whatever interventions are required.

Little attention has been paid to the question of the action to be taken by an NRA in relation to a firm or firms judged to exercise SMP. While the circumstances in which intervention is required are set out in the proposed Framework Directive, discussion of the nature of the regulatory response is principally confined to the Access Directive.[7] Putting on one side proposals relating to conditional access systems and other associated facilities, Articles 8 to 13 of the proposed Access Directive outline the NRA's options. Thus Article 8 (Imposition, Amendment or Withdrawal of Obligations) reads as follows:

> 1: Where an operator is designated as having significant market power on a specific market … national regulatory authorities shall impose one or more of the obligations set out in Articles 9–13 of this Directive as appropriate.
> 4: Obligations imposed in accordance with this Article shall be based on the nature of the problem identified, and shall be proportionate and justified in the light of the objectives laid down in Article 8 of the [Framework Directive] …

However, while Articles 9–13 of the Access Directive and the associated recitals (Nos 16–21) contain some indications of the circumstances in which, in the Commission's view, a particular obligation should be imposed, the mapping from the existence of SMP to the appropriate regulatory solution is, at the least, incomplete. This opens the door to the possibility of over-regulation, under-regulation or regulation which, because the intervention does not deal with the problem in question, is simultaneously unnecessary and ineffective. Moreover, as the Framework Directive gives the Commission no explicit power to participate in the selection of remedies, it is unlikely to offer guidance on this aspect, at least until some experience has been accumulated. My own views on remedies are given in Section 4 below.

## 3. CRITERIA FOR 'RELEVANT MARKETS'

Inclusion in the Recommendation's list of relevant markets is normally a necessary condition for *ex ante* regulatory intervention[8] but not a sufficient condition, as a finding of SMP in the market in question is also required. The criteria for inclusion thus provide the fulcrum or starting-point for the system as a whole.

The draft Recommendation identifies three (cumulative) criteria: barriers to entry and to the development 'of competition'; 'dynamic aspects' – related to the development of competition in the period in question; and 'the relative efficiency of competition law and complementary *ex ante* regulation'. The second of these is straightforward in principle but is difficult to apply in practice. The logic of the regime thus rests heavily on the combined operation of the first and third criteria. These are now reviewed.

It is worth noting that, in standard competition-law practice, barriers to entry figure in the stage of market analysis rather than market definition. They are relevant to the question: what kind of competitive constraint is imposed on players in the (defined) market by those outside it? In this respect, then, the process of market definition looks forward to market analysis.[9] In the process intended under the directive, however, the height of barriers to entry is estimated on general criteria, not on the facts of any particular Member State's market situation.

Among entry barriers, the least problematic are those created by legal or regulatory rules. A legal barrier might take the form of a requirement that firms have a licence in circumstances where additional licences are not available. Alternatively, there may be a legal limitation on the availability of a particular input which is necessary to produce a relevant electronic communications service. These two considerations come together in the case of wireless markets, for which operators require both a licence and access to spectrum. As a consequence, entry by network providers into wireless (fixed and mobile) markets may be effectively blocked.

Consideration has to be given to whether legal barriers to entry can be removed. In recent years, competition has been facilitated by changes to operators' rights to undertake the civil works necessary to construct networks and gain access to numbers. The legal barrier to entry in relation to wireless operators could be alleviated by changes in spectrum allocation methodologies (and the ability to trade spectrum), thereby allowing greater economic efficiency and lowering entry barriers. In summary, legal barriers to entry justify inclusion as a criterion by which to identify relevant markets, when they are expected to continue to operate, unless a significantly large number of operators are already operating behind the barrier to entry to ensure effective competition.

A regulatory barrier arises when, as a result of regulatory policy or previous practice, entry into a particular market is made financially unprofitable and this situation is expected to persist. This state of affairs arises, for example, when the NRA, in pursuit of other objectives – typically relating to ensuring the affordability of retail services – imposes a retail pricing structure which means that some services are provided at below cost. This does not imply that the operator providing such services is doing so on a non-commercial basis, since its customers may buy other services from it as well which in aggregate make the package profitable. It does, however, make entry into the 'loss-making' segment of the market risky and unattractive. The most obvious example of this phenomenon is the retail pricing of access. Although it is widely recognized that tariff rebalancing can address this problem, experience suggests that it is likely to occur slowly, so that this regulatory barrier may persist longer than anticipated.

Unfortunately, even the prospect of such a regulatory constraint may be enough to create a barrier to entry. Regulatory uncertainty may make investors reluctant to enter new markets, especially where investment entails large sunk costs and assets are long-lived.

Economies of scale do not of themselves create a barrier to entry that cannot be overcome by entrants. For scale-related barriers to exist, it is necessary that the economy of scale operate over a range which is large in relation to the market as a whole. It is also necessary that exit from the market be costly because the firm must incur substantial sunk expenditures which are not recoverable on exit.

Establishing whether an entrant would incur substantial sunk costs requires an examination of the kind of investments which have to be made. Investments in saleable capital equipment, such as switches, are not sunk. Investments in civil works and underground plant can only be sold *in situ* and, as they have few if any other uses, they tend to have limited resale value. The wireline local loop, where the proportion of expenditure on trenches, ducts and underground plant is particularly high, contains an especially high proportion of sunk costs.

The three barriers listed above are all non-strategic barriers to entry – that is, barriers to entry that are not artificially 'manufactured' by the firm which enjoys them (except to the extent that firms can shape legal and regulatory policy).

Strategic barriers to entry include many types of possible conduct. They may take the form of excessive investment in product capacity, R&D or advertising, which creates in the minds of entrants the rational expectation that entry will be subject to a strongly competitive response by an incumbent. It may also raise rivals' costs, for example by seeking to pre-empt inputs, raising the price of inputs, or artificially inflating consumers' switching costs by contractual or other means.

A final barrier to entry may be associated with demand-side network effects. Network effects in certain circumstances, give a firm an enduring competitive advantage over its smaller rivals, causing the market to 'tip' in its favour. However, such markets may still be subject to competitive pressures over the longer term as competitors seek to replace the dominant operator through technological advances. Other regulatory measures, such as requirements for inter-operability and end-to-end connectivity, may alleviate the risk of tipping.

The above is an (incomplete) list of barriers to entry and the development of competition. Recall that, for markets to be subject to *ex ante* regulation, persistence of the barrier must be expected. How does the third criterion – relative efficiency of competition law and complementary *ex ante* regulation – apply in relation to these types of entry barrier? The draft Recommendation does not address this question directly, confining itself to the observations that extensive compliance requirements or an indispensable need for frequent and timely intervention may tip the balance in favour of *ex ante* regulation.

In my view, however, *ex ante* regulation should be confined to the first three categories of barrier to entry identified above – legal, regulatory and scale-related. These are (by the second criterion) persistent. They are relatively easy to identify (compared with strategic barriers), and the remedies, discussed below, may well be intrusive and difficult. The same does not apply to strategic entry barriers, where intervention is likely to be idiosyncratic and episodic.

## 4. REMEDIES

This section examines in more detail the remedies envisaged in the directives.

The ascending hierarchy of obligations in Articles 9–13 of the Access Directive can be summarized and evaluated as follows:

### Article 9 of the Access Directive

*Obligation of transparency. This involves publishing data on technical specifications, network characteristics, terms and conditions for supplying use, and prices. In particular, NRAs may require the publication of a reference offer.*

This contains a range of different disclosures. Some of them can be construed as necessary for other firms to achieve interconnection or network access, for example the location and current technical specifications of points of interconnection. In the case of a firm with SMP, these may be fairly unexceptional, although the standardization process may have already placed much of the information in the public domain. Others relate to transparent commercial conditions for supply, including price. Without the obligation of non-discrimination (see Article 10), logic requiring the publication of a reference offer is diminished.

Disclosure of price data is a possible instrument of collusion or price leadership. For this reason, price disclosure may be particularly unsuitable as a remedy where joint dominance is in question. The same problem applies in a moderated form to single dominance.

The implication of this analysis is that NRAs should address the question of transparency in a nuanced way. While the case for disclosure of technical information for a firm's access to technical facilities may be strong, the same reasoning does not extend to the mandatory disclosure of pricing information, which is rarely desirable of and by itself even though it may be an element of more vigorous obligations.

### Article 10 of the Access Directive

*Obligation of non-discrimination. This requires the operator to provide equivalent conditions in equivalent circumstances to other undertakings providing similar services, and to provide services and information to others under the same conditions and of the same quality as it provides for its own services, or those of its subsidiaries or partners.*

Since obligations relating to mandatory and cost-oriented access appear later in the Access Directive (see Articles 12 and 13 discussed below), it seems reasonable to conclude that Article 10 does not of itself trigger these obligations.

This is primarily relevant to cases of an SMP operator which is vertically integrated into a competitive market, and the obligation is said to be needed to prevent exclusionary behaviour by the firm with SMP, through the foreclosure of competition in the upstream and downstream market. The key issue, therefore, is whether this is a matter that can adequately be dealt with under competition law, on the basis that such behaviour would be a clear and foreseeable breach of Article 82 EC, for which significant case-law exists.

This case-law is capable of dealing with blatant discriminations. However, NRAs may need to guard themselves against more subtle departures from 'similar conditions'.[10] These are made possible by a number of features of the communications industry. First, networks are often characterized by fixed costs and economies of scale. If access prices are set on an 'average cost' basis, such as long-run incremental cost plus equiproportionate mark-up, they will exceed marginal cost. An entrant will thus price downstream on the basis of a higher input price than the implicit marginal cost used by the network owner's downstream affiliate. It may be possible to overcome this difficulty by mandating access on the basis of the supply of capacity, rather than on a per-minute basis, but this has well-known difficulties. Second, the cost drivers in the provision of complex network services are often varied, and this creates opportunities for pricing by the network operator which may disadvantage competitors.

### Article 11 of the Access Directive

*Obligation of accounting separation. An NRA may require a vertically integrated company to make its wholesale prices and its internal transfer prices transparent, especially where the companies which it supplies and itself compete in the same downstream market.*

This represents a considerable ratcheting up of the regulatory burden on the firm with SMP. NRAs should ask themselves whether the additional burden is justified. The principal justification would be a situation in which there was a persistent network monopoly enjoying an entrenched competitive advantage (which may itself be the result of regulatory policies on retail tariffs – see below), which merits ring-fencing.

It has been argued that separate accounts may also play a role in facilitating the subsequent detection of particular abuses using competition law instruments. An NRA would, however, have to give serious thought to the question of whether this provides a legitimate justification for imposing such a wide-ranging obligation rather than the obligation to be able to demonstrate non-discrimination, or to respond to charges of abuse of dominance.

### Article 12 of the Access Directive

*Obligations meet reasonable requests for access to, and use of specific network facilities. An NRA may impose obligations on operators to grant access to specific facilities or services, including in situations when the denial of access would hinder the emergence of a competitive retail market, or would not be in the end-user's interests.*

This represents an obligation to be implemented in circumstances similar to, but significantly broader than, those in which the essential facilities doctrine is applied under competition law. The extension of the test lies in the replacement of the precondition under competition law for mandatory access, that the asset is essential and cannot be replicated, by a much broader condition that NRAs can mandate access in circumstances where its denial 'would hinder the emergence of a sustainable competitive market at the retail level, or would not be in the end-user's interest'.

There is a risk that the last phrase in particular ('would not be in the end-user's interest') might open the door for extensive regulatory intervention. The obvious problem is the conflict between users' short-run and long-run interests. Short-run interests might best be furthered by the adoption of mandatory access on a wide scale. However, such a policy clearly reduces incentives to invest in competing facilities, and is likely to stifle innovation in the long term.

Article 12(2) gives a list of factors to be taken into account, including 'the initial investment of the facility owner, bearing in mind the risks involved in making the investment'.

The obligation is silent about the pricing of such access, except to the extent that it prohibits 'unreasonable terms and conditions' having a similar effect on denials of access. The range of pricing principles may therefore encompass 'retail minus'.

## Article 13 of the Access Directive

*Price control and cost accounting obligations. This deals with situations where a potential lack of competition means that the operator concerned might be capable of sustaining prices at an excessive level, or applying a price squeeze, to the detriment of end users. National regulators should take into account the investment made by the operator and the risks involved.*

It is generally accepted that the finding that a facility is essential implies the application of some appropriate pricing rule. The nature of that pricing rule is, however, by no means clear. In this context, Article 13 can be conceived as imposing the obligation of cost-oriented prices, the operator assuming the burden of proving that 'the charges are derived from costs including a reasonable rate of return on investment …'

The circumstances identified as appropriate for the application of this rule are 'situations where a market analysis indicates that a potential lack of effective competition means that the operator concerned might be capable of sustaining prices at an excessively high level, or applying a price squeeze to the detriment of end users'. There is, however, a major distinction between these two cases. In the case of excessive pricing, the question is whether SMP is being exploited directly through an excessive price. In the case of a margin squeeze, the hypothesis is that the operator with SMP is foreclosing competition in a vertically related area by pricing that fails to cover costs. The appropriate way of dealing with the latter issue is discussed above.

Cost-oriented pricing for interconnection or access to customers should only be considered when dealing with an operator with SMP which is both persistent and incapable of being dealt with by other remedies, including particularly structural remedies. A classic case for its application might therefore be access to the local loop, either for call termination or for the purposes of leasing unbundled loops – provided that one operator enjoys a monopoly or position of super-dominance in the relevant geographical area. Even here the impact on incentives for infrastructure investment by competitors must be taken into account. Setting prices which are low in relation to forward-looking costs, or even creating an expectation that this will happen, can deal a fatal blow to infrastructure competition.[11]

It is far more questionable whether the principle should be applied where there is network duplication. This question has arisen particularly in connection with the mobile market, in relation both to indirect access and access to the network by virtual network operators.

In the former connection, it has correctly been argued by Oftel (and other NRAs) that offering cost-plus access to mobile call origination to indirect access providers would have the consequence of imposing a particular configuration of mobile call charges, prohibiting tariff innovation, such as the use of handset subsidies to expand the market, and reducing incentives to invest in competing networks. On the other hand, Oftel has argued that the obligation to provide indirect access on a retail-minus basis may be appropriate in the interests of expanding the number of competitive suppliers in a sector where spectrum limitations place a constraint upon the number of network operators.

In other words, NRAs should be extremely parsimonious in imposing the obligation of cost-oriented prices. It should be confined to cases where the monopoly is persistent and incapable of being addressed by structural remedies. The option of retail-minus pricing should be retained.

These provisions relate to the terms on which one operator obtains access to another's network or customers. It is arguable that, if appropriate regulatory arrangements are in place in this respect, then the need for additional retail price regulation is limited, on the basis that firms would be able to buy interconnection and access services and perform their own retailing functions, where entry into retailing is freer – absent retail margin squeezes which the interventions above are designed to prohibit. However, the Directive on universal service and users' rights offers an additional regulatory intervention relating to retail tariffs (see below).

### Article 17 of the Universal Service and Users' Rights Directive

*Retail Price Regulation. Where an NRA determines that a retail market is not effectively competitive and that carrier pre-selection will not suffice to solve the problem, it shall ensure that undertakings with significant market power in that market orient their tariffs towards costs, avoiding excessive pricing, predatory pricing, undue preference to specific users or unreasonable bundling of services. This may be done by appropriate retail price cap measures. Where there is retail tariff regulation, appropriate cost accounting systems must be implemented. Retail tariff control may not be applied in geographical or user markets where the NRA* is *satisfied that there is effective competition.*

The implicit assumption of many regulators of and commentators on the communications industry is that if wholesale markets can be regulated to avoid the harmful effects of SMP, then regulation of retail markets can be confined to solving residual consumer protection problems, rather than problems relating to the abuse of market power. The underlying hypothesis is that entry into retailing activities will be sufficiently free from barriers to permit deregulation.

This proposition has not been properly tested, as many NRAs have been reluctant to accept the fundamental implication of the policy, which is that cost-based wholesale prices and competition in the retail market will bring retail prices into line with costs, thereby eliminating distributionally important pricing distortions associated with regulator-driven cross-subsidies (or departure from cost orientation), involving either different services, such as fixed line access and usage, or different groups of customers, such as high-cost rural and low-cost urban consumers.

The above discussion has suggested indications and contra-indications for the application of the various interventions considered, which are presented in Table 3.1.

*Table 3.1 Indications and contra-indications for the application of interventions*

| Obligation | Indication | Contra-indications or adverse effects |
|---|---|---|
| Transparency | Technical information indispensable to successful interconnection | Price disclosure may ensure excessive/rigid prices |
| Non-discrimination | Partial remedy against margin squeeze | Too broad a prohibition may reduce consumer welfare; conditions for discrimination may not exist |
| Separate accounting | Potentially useful for persistent monopoly | Costly and not essential for many price squeeze investigations |
| Mandatory access | Useful for dealing with persistent network monopoly | Reduces incentives to invest and innovate |
| Cost-oriented pricing | Useful for dealing with persistent network monopoly | Reduces incentives to invest and innovate |
| Retail price control | Can maintain distorted retail price structure; possible approach to consumer protection issues (e.g. ignorance) | Widespread mandatory access by re-sellers an alternative |

Perhaps the most striking feature of this analysis is that three of the five 'access and interconnection' remedies are principally appropriate to cases of persistent monopoly, of the kind found primarily in the fixed local loop. On this basis, the scale of desirable regulatory intervention under the new arrangement is quite limited. This highlights the fact that sector-specific regulation should be confined to cases where some durable feature of the market creates a persistent and high barrier to entry. In some cases, the barrier may be legal or regulatory – for example, limited allocation of spectrum may represent an absolute barrier to entry in mobile markets, although the latter may still be effectively competitive provided enough operators are licensed, even if entry is then prohibited. In other cases, notably the local loop, the barrier may arise from the combination of significant economies of scale and sunk investment, which together ensure that the market is not contestable.

## 5. CONCLUSIONS

This chapter has given a comprehensive but necessarily brief account of the economic issues raised by the new regime. That regime comes into effect on 25 July 2003, and the period up to that date will be crucial for the implementation of the new system by NRAs. Some have made careful plans for introducing the new regime, and have begun to put them into effect. Others have not, and run the risk of missing the deadlines.

At present, an assessment has to be based on *a priori* reasoning, rather than experience. Given the scope for competitive entry into almost all communications markets, it is right that the new regime looks forward to the progressive withdrawal of sector-specific regulation. This is achieved by the market-by-market 'sunset clauses'. Although the market analysis which triggers the sunsets is less reliable than movements of the heavenly bodies, this represents a flexible solution to the problem of different rates of progress in Member States. It is also clear from the last two years that the general market environment – boom or slump – has a major impact on the scope for growth of competition via entrants' ease of access to capital markets. The regime implicitly takes this into account.

What are the main risks of the regime? Experience in Europe and elsewhere suggests that regulators find it hard to 'let go'. Existing beneficiaries of regulation fight harder to retain it than potential beneficiaries of deregulation – who may not know what lies in store – fight to eliminate it. Regulators will have to take modest risks, especially in the relaxation of controls on retail prices. Damage can be limited by the maintenance of 'safety-net' price controls in certain cases, subject to a finding of dominance.

Second, there is a risk that pressure by entrants to keep access prices low may chill incentives to invest in alternative infrastructure, including new wireless technology. Broadband roll-out is a pervasive goal of policy in the EU, and one way of achieving it is to set low access prices to the historic operator's facilities without consideration of the effect on alternative platforms based on cable, terrestrial wireless and satellite. Guidance from the Commission in relation to remedies would be helpful here.

Finally, the multi-level nature of the regime, involving interactions between the Commission and NRAs, as well as among NRAs, runs the risk of creating delays and raising tensions, which incumbents or others will seek to exploit.

## NOTES

1. This section draws on M. Cave and L. Prosperetti, 'European telecommunications infrastructure', *Oxford Review of Economic Policy*, No. 4, 2001, pp. 416–31.
2. M. Cave et al., *Access Prices, Investment and Entry in Telecommunications*, available at www.http://users.wbs.warwick.ac.uk/cmur/publications/research_papers.htm
3. The first two of these processes are elaborated on respectively, a draft Recommendation and a Guideline. The absence of guidance from the Commission on remedies is discussed below.
4. The same term was used in the 1997 Interconnection Directive, but with a different meaning (see Section 1 above).
5. See M. Armstrong, 'The theory of access pricing and interconnection', in M. Cave et al. (eds), *Handbook of Telecommunications Economics*, Vol. 1, Elsevier, 2002.
6. For an account of models of this kind, see P. Rey and J. Tirole, *A Primer on Foreclosure*, mimeo, 1997.
7. Article 16 of the Universal Service Directive also considers retail price control.
8. There is also a procedure for NRAs to 'add' markets.
9. I owe this point to Patrick Rey.

10. For a discussion of these problems, see M. Cave, *Can Regulators Get Smart about Discrimination?*, available at http://users.wbs.warwick.ac.uk/cmur/publications/research_papers.htm
11. The opinion of Justice Stephen Breyer, delivered in a US Supreme Court judgment, is relevant here: 'Increased sharing by itself does not automatically mean increased competition. It is in the unshared, not in the shared portions of the enterprise that meaningful competition would likely emerge. Rules that force firms to share every resource or element of the business would create, not competition, but pervasive regulation, for the regulators, not the market place, would set the relevant terms'. (Quoted in Alfred E. Kahn, *Whom the Gods Would Destroy or How not to Deregulate*, AEI-Brookings Joint Center for Regulatory Studies, Washington, 2001, p. 7.)

# Section 2

# Market Definition in Telecommunications Markets

# 4. Market definition in the telecoms industry

**Jordi Gual**

## 1. INTRODUCTION

Market definition for antitrust purposes is by now firmly rooted in economic analysis both in the USA and the EU, even if the approaches are slightly different. This chapter examines the theoretical basis for the legal definitions and assesses whether the general principles need to be adapted when dealing with the telecommunications services industry.

The chapter finds that the conventional antitrust methodology for market definition can be, to a large extent, readily applied to the telecoms industry but points out some key adjustments that have to be made to this methodology to ensure that the antitrust and regulatory authorities end up defining markets which capture adequately the nature of the competitive interaction in this industry.

Three adjustments are needed. First, careful consideration should be given not only to demand substitutabilities, but also to the competitive interaction with all potential suppliers. Second, demand complementarities and the economies of joint production should be properly recognized since they imply that in this industry bundles or systems of services may become a significant unit of antitrust and regulatory analysis. Third, market definition should take into account that the telecoms service industry is characterized by fixed and continuous sunk costs of service provision, and that this will often imply the need to revise the conventional concepts of market power and substitutability based on price elasticities.

The chapter also argues that if some competitive problems are detected with particular services (for example mobile call termination), that should not condition the market definition exercise. The definition of markets should be based on a detailed analysis of demand (both complementarities and substitutabilities) and the consideration of all companies which have the assets and capabilities to satisfy these consumer needs. Such an exercise should be done first, and distinguished from the subsequent analysis of the competitive conditions in the markets defined as relevant.

The chapter is organized as follows. Section 2 develops the general conventional framework. Section 3 looks at the issues specific to telecoms and analyses how the question of market definition in this industry is handled in the main antitrust jurisdictions. The chapter considers first (Section 3.1) how the conventional market definition framework is used to define the relevant product and geographic markets in the USA and the EU, and how it is applied in particular to several common problems of market definition. For example, the distinction between fixed and mobile telephony, mass and business customers or wholesale and retail services. Some specific attention is also given to other telecoms industry features such as the importance of networks and bottlenecks. Section 3.2 analyses in detail the implications of complementarities in demand for the conventional market definition model, an issue which is central to the specific discussion on mobile call termination in Section 3.3. Finally, Section 3.4 considers the adjustments that are needed to the traditional model due to the fact that some segments of the telecoms industry are characterized by a rapid pace of service and technological innovation. Section 4 summarizes the main conclusions of the chapter.

## 2. THE ECONOMICS OF MARKET DEFINITION

The definition of the relevant market for competition policy purposes is based on the principle of the hypothetical monopoly. This concept is already well established in antitrust legislation, both in the EU and in the USA,[1] and provides the standard framework for market definition analysis in competition policy cases.

The principle states that a product (or geographical) market should be defined as the minimum set of products (or areas) that could be successfully monopolized, that is, the range of products or services whose provision, if it were in the hands of a single firm, could profitably be restricted. The principle is also known as the SSNIP (small but significant non-transitory increase in price) test, since it refers to a small but significant non-transitory increase in price by the hypothetical monopolist.

There are several features of this definition which are worth pointing out. The first is that if one of the key goals of antitrust analysis is to assess whether the relative position of one or more firms in a market gives them the power to raise prices, it makes sense to measure that position in the marketplace relative to an aggregate which – if fully controlled by a single entity – would provide a monopoly position and therefore full control over prices.

The second key feature is that the process of determining the relevant market is carried out through a gradual increase of the number of products or areas under consideration, starting from the smallest possible set. This again makes sense, since increasing the collection of products controlled by a firm or a group of firms acting jointly increases the potential for monopoly as alternative substitute products are eliminated by their gradual inclusion in the set under examination.

A third important feature is that the framework is explicitly designed for quantitative analysis. The profitability of higher prices refers to a 5 or 10 per cent mark-up, sustained for a period of about one year. Even if the quantitative information needed to compute this is not available, the precise definition offers in practice a useful framework which guides the analysis of the data at hand.

A fourth remark is that the hypothetical-monopolist principle requires a thought experiment which refers to the competitive benchmark. That is, the analyst should undertake a counterfactual experiment, trying to assess what would happen if all these goods were provided by a single firm. The goal is to determine whether the hypothetical monopolist would be able to profitably increase prices by 5 per cent or 10 per cent relative to those that would prevail in a situation of perfect competition. Since current prices may not correspond to those of a perfectly competitive market, the analysis will often imply a twofold counterfactual: assessing the pricing of both the hypothetical monopoly and the competitive regime.

Given the practical difficulties of undertaking a statistical analysis which follows rigorously the hypothetical-monopolist framework, the concept is very often applied by considering informally, for a given group of products or areas, the forces which could restrain a firm from increasing prices if it were to control the sale of all the products under consideration. These forces are typically classified under the headings of demand and supply substitution.

Demand substitution refers to the products, services or geographical areas to which consumers could turn for a substitute of the monopolized good or collection of goods. If close substitutes are available, the ability of a potential monopoly to raise prices will be diminished. Note that this is a pure demand-side perspective. The goods to be considered could be delivered by a completely different production method. What matters is that they are reasonably good alternatives in consumption.

Supply substitution is meant to include all the producers which, if the hypothetical monopolist were to exercise its power by raising prices, would have the capability – in the short run – to enter the market, providing new output and thus limiting the ability of the monopoly to restrain production. Note that supply substitution means that the definition of the market should comprise all firms that can, without incurring sunk costs, be relevant competitors at short notice (that is, firms that can easily rearrange their production facilities and serve some or all of the relevant goods). These potential entrants must be distinguished from those that could enter in the medium run (around two years) by deploying new capacity (and incurring sunk costs). In fact, in the US Guidelines these producers are considered when assessing the relevance of entry barriers.

Informal assessment of the forces that would restrain the market power of a potential monopolist is a far superior method for determining the relevant market to relying on market boundaries based on production process similarities or marketing and industry practices. Nevertheless, it is an exercise fraught with difficulties, particularly because it will usually lead to an informal gathering of evidence about the substitutability of products in consumption. Data on choice patterns and reactions to changes in market conditions (prices, quality and availability) can be useful, but the lack of a proper statistical experiment will clutter any attempt to draw reliable inferences when assessing the behaviour of consumers. For example, when we analyse substitution in consumer surveys, are we sure that all influences on consumer decisions other than price changes have been taken care of? The shortcomings of an informal approach have triggered the development of a more systematic framework, to which we now turn.

## 2.1 A Formal Approach to Market Power and the Relevant Market

Economic analysis provides a clear methodology that, under certain conditions, allows a rigorous determination of the degree of market power and can be used for market delineation purposes. This section summarizes the main principles for the simple case where the technology is characterized by constant returns to scale and the number of firms is given.

### The measure of market power

Consider, first, a market where all output is supplied by a single firm $i$. Let the constant marginal cost be $c_i$ and the market demand be expressed by $Q_i = D(p_i)$ where $p_i$ is the price charged by the monopolist and $Q_i$ the quantity sold. Let $\varepsilon_M$ be the price elasticity of this demand function, which we will refer to as the market demand elasticity ($\varepsilon_M = -(\delta D(p_i)/\delta p_i)/(Q_i/p_i)$). As is well known, the choice of the optimal price by the monopolist implies the following equilibrium relationship:

$$\frac{p_i - c_i}{p_i} = \frac{1}{\varepsilon_M}. \tag{4.1}$$

The left-hand side of this equation, the relative mark-up, also known as the Lerner index, provides a measure of market power to the extent that it assesses the ability of the firm to charge a price above marginal cost. The ability to increase price will be inversely related to the elasticity of demand, with a larger mark-up when the market demand is less elastic (close to 1). For example, if the elasticity is 4, the equilibrium mark-up will be 0.25 (or 25 per cent). If the market is perfectly competitive, the elasticity faced by the firm is extremely large, and the mark-up collapses to zero. A very low elasticity (say 1.5) leads to very large mark-ups (66 per cent). The monopolist will never sell at a point where $\varepsilon_M < 1$, since this is not consistent with profit maximization (revenues – and profits – can be increased by reducing output since the price increase more than compensates for the reduction in volume).

A benchmark of a 5 per cent (or 10 per cent) mark-up implies, in particular, that markets where the elasticity of demand is above 20 (or 10) are considered to be fairly competitive even if controlled by a single firm. Indeed, the high elasticity of demand in fact captures the existence of many alternative or substitute products. Note, also, that in a monopolized market, observed prices will already be above marginal costs, and by definition it will not be profitable for a firm to increase them further. The observed mark-up in equilibrium already reflects the market power that is being exercised.

Equation (4.1) provides a valid starting-point which can be used for the formal definition of market power in situations where there is more than one provider.

Consider first the case of a market with a dominant firm which faces a fringe of competitive suppliers: a large number of firms which take prices as given and have an aggregate supply function defined by $Q = S(p)$. Let $\varepsilon_S$ be the elasticity of this supply function ($\varepsilon_S(p) = (\delta S(p)/\delta p)/(Q/p)$).

The dominant firm will choose its output taking into account the supply of the competitive fringe. The corresponding Lerner index can be shown to be:[2]

$$\frac{p_i - c_i}{p_i} = \frac{s_i}{\varepsilon_M + \varepsilon_S(1 - s_i)}, \tag{4.2}$$

where $s_i$ corresponds to the share of the market served by the dominant firm. It is interesting to note that equation (4.2) can also be derived by defining the problem of the dominant firm as one of profit maximization when facing a 'residual demand function' computed as the difference between the market demand and the supply of the competitive fringe. From this point of view, the market power of the dominant firm will be measured (in the equilibrium) by the elasticity ($\varepsilon_i^R$) of its residual demand function. That is:

$$\frac{p_i - c_i}{p_i} = \frac{1}{\varepsilon_i^R}, \tag{4.3}$$

where

$$\varepsilon_i^R = \frac{\varepsilon_M + \varepsilon_S(1 - s_i)}{s_i}.$$

That is, from an empirical point of view, the extent of market power can be directly assessed by estimating the elasticity of the residual demand faced by the dominant firm. This elasticity captures the joint effect of the existence of substitute products (the elasticity of market demand), and the presence and importance of alternative providers ($\varepsilon_S$ and $s_i$ in equation (4.3)).

Consider next the case of a market where $n$ oligopolists compete through the provision of slightly differentiated products. Let the (inverse) market demand system be:[3] $p_i = p_i(q_1, q_2, \ldots, q_i, \ldots, q_n)$ $(i = 1, 2, \ldots, n)$, and define

$$\mu_i = -\frac{\delta p_i}{\delta q_i}\frac{q_i}{p_i} \qquad \mu_{ij} = -\frac{\delta p_i}{\delta q_j}\frac{q_j}{p_i},$$

where $\mu_i > 0$ and $\mu_{ij} \geq 0$, indicating that the products are gross substitutes $((\delta p_i/\delta q_j) \leq 0$, for all $j \quad i)$.

With constant marginal costs ($c_i = 1, \ldots, n$), the equilibrium conditions of the oligopolistic competition model are defined by a set of $n$ first-order conditions as follows[4] (where we assume constant elasticities, following Baker and Bresnahan, 1985):

$$\frac{p_i - c_i}{p_i} = \mu_i^R = \mu_i + \sum\nolimits_j \mu_{ij}\rho_{ij}.$$

The parameter $\rho_{ij}$ captures the extent to which firms compete aggressively:

$$\rho_{ji} = \frac{\delta q_j}{\delta q_i}\frac{q_i}{q_j}.$$

That is, $(\delta q_j/\delta q_i)$ is the marginal change in the quantity of $j$ when the quantity of $i$ changes. The parameter $\rho$ expresses this as an elasticity. A negative $\rho$ corresponds to aggressive behaviour (reacting to a quantity reduction – which would tend to raise price – with an expansion of output – which lowers price) and a positive parameter reflects co-operation. The rivals react with output contractions when one firm reduces production.

Typically, we would expect that the elasticity of the (inverse) residual demand would be smaller[5] than the own price elasticity: $\mu_i^R < \mu_i$. That is, the presence of strong substitutes (large $\mu_{ij}$ in absolute value) and aggressive competitors (large and negative $\rho$) limits market power and reduces the elasticity of the (inverse) residual demand ($\mu_i^R$), which measures directly the mark-up.

**From market power to market definition**

The elasticity of the residual demand curve provides a convenient measure of market power which can be readily applied to the definition of markets for antitrust purposes. The hypothetical monopoly test with a 5 per cent (or a 10 per cent) mark-up can be applied with the use of an econometric estimate of the appropriate residual demand elasticity. The procedure involves an iterative estimation process, beginning with the estimation of the residual demand at the level of one individual firm, and gradually enlarging the set of products under consideration if the null hypothesis, $\mu_i^R < 0.05$ for a 5 per cent benchmark, cannot be rejected. The relevant market will correspond to the set of firms/products which find it profitable to sustain a mark-up larger than 5 per cent (the group of firms/products which face a direct residual demand with elasticity below 20). If the 10 per cent mark-up is used as a benchmark, the relevant elasticity will be 10.[6]

The framework implies, of course, that the firms/products which should be added first are the closest substitutes (large cross-price elasticity $\mu_{ij}$) provided by very aggressive competitors (large and negative $\rho$). Removing this kind of firm as a competitor is most likely to enhance the market power of the merged (or co-ordinating) entity and is the appropriate way to establish the minimum set of firms which can (if acting jointly) monopolize the market.

## 2.2 Fixed Costs and the Relevant Market

This formal approach to market definition runs into difficulties when we consider markets characterized by the presence of fixed costs. The previous model is based upon the definition of market power as pricing over marginal cost, because marginal cost pricing is taken to be the competitive benchmark. Mark-ups above 5 per cent or 10 per cent would correspond to a non-competitive market and result in abnormally high profitability. Yet, under conditions of increasing returns to scale, a positive mark-up (possibly above 10 per cent) may in fact be necessary if a firm is to be viable, and the same may hold at the industry level. What this implies is that under increasing returns the SSNIP benchmark could lead to excessively narrow markets. As the set of products is enlarged, the potential mark-up grows, but it does not make sense to stop at 5 per cent or 10 per cent, since under increasing returns to scale these percentages could correspond to mark-ups which barely allow the industry to recover its fixed costs.

Markets with fixed costs are pervasive and, unfortunately, economic analysis provides much less guidance as to what the proper benchmark is in these situations. Ideally, the benchmark should correspond to the mark-up which generates sufficient revenues to cover the fixed costs without leading to excess profitability. Since this endogenous determination of the benchmark is likely to be difficult, one way to deal with the problem is to consider changes in the mark-up (say, 5 per cent increases) above the pre-existing level. This boils down to assuming that current prices correspond to 'competitive' levels, where 'competitive' is defined now as the mark-up over marginal cost which allows the recovery of industry fixed costs. Indeed, both the European Commission (EC) Notice on the definition of the relevant market (paragraph 19) and the US Merger Guidelines (section 1.1) admit that current prices can be used as benchmarks, in the absence of indications of insufficient competition (EC) or co-ordinated interaction (US).

If the SSNIP test is applied to a situation where current prices are above marginal costs, two potential sources of bias should be borne in mind when assessing the increase in the mark-up. Assume, for the moment, that the own and cross-price elasticities are constant. Since prices are kept above marginal cost, possibly thanks to a certain degree of co-ordination among firms, the assessment of the impact of the joint price-setting for a group of products results in a small increase in the mark-up. This is so because co-ordination had already led to a higher price level to begin with (in terms of the formal framework, $\rho_{ij}$ is positive). In this case, in general, adding the product will not increase the mark-up much, and the analyst will be led to consider incorrectly a wider range of products.

If the elasticities increase with prices (decrease with quantities), which is the common case, there will be an additional effect. Due to the high prices, the starting-point will be a higher direct elasticity and this will imply that even small increases in the mark-up will generate significant substitution. This is the idea that at a higher price the degree of substitutability increases (the cellophane fallacy) and so does the overall own-price elasticity.[7] This effect leads also to the incorrect establishment of excessively large markets.

The problems of applying the hypothetical monopoly test in industries where increasing returns to scale are significant do not end with the adjustments mentioned so far. Not only do we have to ensure that we choose the appropriate competitive benchmark, check the absence of excess profitability and take into account the potential upward biases in the definition of the market. It is also important to consider what is the exact source of the increasing returns and its expected impact on market structure.

The previous qualifications may be appropriate for industries where the increasing returns are the result of large, non-sunk fixed costs. These are industries which can be well approximated by models of monopolistic competition or contestable markets. These qualifications could also be appropriate in industries that correspond to what John Sutton (1991) classifies as industries with exogenous sunk costs. That is, industries with technical scale economies where firms invest in sunk (non-recoverable) assets but where the magnitude of these investments is determined by the nature of the technology and not by the strategic interaction between competitors. In these industries, one can think of the positive mark-up as the margin that is needed to pay (over the life of the assets) for the up-front investment costs.

However, these qualifications and the overall framework prove less useful in industries with endogenous sunk costs, where competition does not take place through prices or production capacities, but rather through continuous expenditures on non-recoverable advertising and R&D. These are dynamic industries where companies compete through product choice, product development or the introduction of new network products. William Baumol[8] has recently used the term 'continuing sunk costs' for these industries and argued that 'you need new guidelines for even old-line industries with heavy but continuing sunk costs', since these industries 'cannot be expected to live up to marginal cost pricing'.

In these industries we not only have problems of identifying the appropriate 'competitive' mark-up. More importantly, companies compete along new dimensions which are often much more important than price. In markets where competition is driven by innovation, where firms do not compete *in* the market but rather *for* the market, the use of methodological instruments such as the SSNIP test can be misleading and the concepts of market power and substitutability may have to be reinterpreted.

Indeed, there is a long tradition of analysis relating the extent of product market competition (the Lerner index which we have used so far as the measure of market power) to the rate of innovation. Recent research in this area points to an inverted U relationship (Aghion et al., 2002), showing that there is an optimal level of product market competition in terms of its impact on industry innovation and, therefore, dynamic efficiency. In other words, if there is too much product market competition, too little innovation takes place; but the same happens if product market competition is too soft. For our purposes what this implies is that the mark-up (the static measure which assesses the degree of product market competition and is also used by Aghion et al., 2002) is not correlated with 'dynamic efficiency'. It may be an adequate measure of static efficiency but it is a very poor guide in emerging industries where dynamic efficiency gains should be the key welfare criterion.

Since in these markets competition takes place through numerous dimensions other than price and the Lerner index is a poor proxy of efficiency, some authors[9] have argued that the conventional antitrust tools of market definition and market power should be adapted to the specific features of these rapidly changing industries.

Much of the discussion has centred on the case of high-tech industries where firms compete through product innovation. Teece and Coleman (1998) argue that in these markets the SSNIP will tend to define markets which are too narrow. Since in these industries consumer demand is driven by a set of performance variables rather than price, without changes in performance there might not be substitutability across products even for price increases as large as 20 per cent. What this literature suggests is that the analysis should focus on these new dimensions of competition (what Pleatsikas and Teece, 2001 call product performance) and not on price. Some of the indicators that are suggested include: the depth of technology competition, the effect of product innovation on customer response, the extent of market share shifts over time, the extent of spending on R&D and of actual innovation, even if the same company out-innovates its rivals and does not lose the leadership position, and so on. We will review later (in Section 3.4) the implications of these criticisms of the standard antitrust approach for market definition in the telecoms industry.

## 3. DEFINING MARKETS IN THE TELECOMS INDUSTRY

The formal model outlined in Section 2.1 provides a rigorous set-up for the empirical determination of the relevant market in antitrust cases. It is certainly quite demanding in terms of the information available to the analysts, but nevertheless it has been applied to several specific industries by academic and professional economists.[10] The framework provides a robust technique, rooted in standard economic analysis, which can be deployed if sufficient data are available. If information is poor, the formal model constitutes a good conceptual apparatus to evaluate scattered data such as information on cross-price elasticities and consumer behaviour (survey data).

When considering the market definition issue in an industry such as telecommunications, however, the formal framework can only shed partial light on the analysis. This is due to three facts. First, the telecommunications industry is characterized by the pervasive presence of fixed (sunk) costs. Second, the telecommunications industry comprises a wide array of very different services, ranging from conventional local or long-distance voice telephony to high-speed Internet access. For some of these services, in particular those which are provided with mature technologies, the framework will prove fairly adequate, once the fixed costs problem is properly taken into account. For others, however, the rapid pace of technological change means that quite often competition takes place through dimensions other than price (for example the introduction of new services with improved performance). In those instances, the framework of static oligopoly, which is the basis of the model sketched in Section 2.1, may not be appropriate and we will have to adopt a broader perspective. This question is considered in Section 3.4.

The third relevant fact is that telecommunications services are typically, although not necessarily, consumed in bundles. Moreover, they are often provided by multi-service firms in a joint production process where costs of stand-alone services may be hard to ascertain. We will consider the implications of these facts later (see Sections 3.2 and 3.3). Before that, we will review in some detail how the standard framework can be applied in practice to the more conventional aspects of the telecoms industry.

## 3.1 Conventional Market Definition Problems in Telecoms

Many of the market definition questions raised in the telecoms sector can, in principle, be handled with a straightforward application of the conventional framework developed in Section 2.1. This approach could be used to assess the antitrust relevance of the telecommunications submarkets which are usually distinguished in the industry. For example, local and long-distance calls, mass and business markets, fixed versus mobile calls, pre-paid versus contract mobile markets, and so on.

Indeed, both the USA and the EU recognize the concept of the SSNIP test as an appropriate instrument for the delineation of the relevant product and geographical markets. Nevertheless, the lack of sufficiently good data implies that in practice the establishment of antitrust markets is undertaken focusing on the concept of substitutability, with an informal use of the hypothetical monopoly idea. As we will see, the methodology used by the regulators in the two jurisdictions is substantially different, even if in practice the actual market delimitations do not vary very much.

Let us consider first the question of defining the relevant product/service market. In the USA, the FCC (Federal Communications Commission) introduced in 1996 (in the LEC In-Region Interexchange Order,[11] henceforth the LEC Order) a new methodology which purported to apply the 1992 Merger Guidelines. The goal was to define markets explicitly on the basis of demand substitutability, discarding previous procedures which took into consideration the substitutability in supply. The new principle was later on used in a set of landmark cases involving several mergers between major carriers and local operating companies (NYNEX/Bell Atlantic, MCI/WorldCom, Ameritech/SBC and Bell Atlantic/GTE).

As stated in the LEC Order, the methodology bases market definition explicitly on the hypothetical monopoly principle, focusing on the determination of substitute products from the demand side, and enlarging the product group until a large enough set of products is included so that market power could hypothetically be exercised. The LEC Order does not refer to a specific percentage, but it certainly discusses the absence of viable substitutes which could prevent a potential monopolist from increasing price.[12] Following the Merger Guidelines, the LEC Order uses supply substitutability as an instrument to assess which companies have the incentive and the ability to enter the market if the price increase takes place.[13]

The FCC recognizes, however, that in practice the use of substitutability in demand could lead to the impractical definition of a very large number of markets. As a consequence, the agency argues that it is not necessary to consider all of these markets unless there is an indication of lack of competition in a particular individual service.

This means that, in practice, the agency resorts to the 'similarity of competitive conditions' faced by different consumers as an aggregating device to define markets. For example, in the case of the definition of service markets, all long-distance services (including, for example, individual services such as directory assistance, foreign exchange service, discount plans, and so on[14]) are aggregated into one single service. The FCC asserts that 'we need not delineate particular product markets … unless there is credible evidence suggesting that there is or there could be a lack of competitive performance with respect to a particular service or group of services'.[15]

However, a more robust aggregation procedure would be achieved with the explicit use of the extent of supply substitutability. Indeed, in its comments to the LEC Order, AT&T

pointed out[16] that the FCC was using a rather narrow interpretation of the Merger Guidelines. According to AT&T the Guidelines support an aggregate product market where 'production substitution among a group of products is nearly universal among the firms selling one or more of the products' (footnote 14 of the Merger Guidelines). The same company remarks that there 'is no difference between the facilities used to provide different services'.

In the same Order, in fact, the criterion of similarity of competitive conditions is used as the basis for the aggregation of services which had been previously defined as separate markets (IMTS [International message telephone service] and non-IMTS service[17]), but as Sprint points out,[18] the true reason one can aggregate those services is the fact that players in one market can very easily be suppliers in the other, that is, supply substitution.

The approach introduced in the LEC Order is used in subsequent decisions. For example, in the *NYNEX/Bell Atlantic* decision, markets for local exchange and exchange access and long distance are distinguished on the basis of the aggregation of product markets for which customers face the same competitive conditions.[19] It is interesting to note that this methodology allows the FCC to recognize as a relevant market the joint provision, as a bundle, of local and long-distance calls, on the grounds of consumers facing comparable competitive alternatives. We will return to bundling in Section 3.2.

In the EU, the approach put forward in the market definition notice (European Commission, 1997) and used in many cases is based from a methodological standpoint on the SSNIP test. In practical applications it implies the simultaneous assessment of demand and supply substitutability as recently reasserted by the Guidelines on market analysis of the new regulatory framework (henceforth the Guidelines).[20] For most of the cases, however, what drives the delineation of markets is demand substitutability, and no detailed market breakdown is attempted when the investigations of the Commission determine that the decision would not change by a more refined market definition, that is, when the competitive conditions would not be altered by a narrower definition of markets.[21] This is not very different from the practice of the FCC.

Consider next the case of fixed versus mobile calls and whether or not they belong to the same relevant market. The analysis of this question has increasingly focused on the degree of substitutability between the two services. This is exemplified clearly by the debate in the UK market, where the regulator has repeatedly argued, on the basis of survey data, that the two services are not close substitutes.[22] What is remarkable is that the quantitative information is reported without a clear sense of what degree of substitution is considered to be sufficient. Econometric evidence on substitutability has been presented by the operators to Oftel, but has been rejected as technically unreliable by the agency.[23] In the end, Oftel bases its conclusions on its periodic surveys of consumer behaviour.

Even if it is true that the econometric studies may be the subject of substantial technical controversy, the use of survey data is even more open to debate since that information can hardly constitute evidence about demand substitutability. Oftel has based its separation of the two markets on this type of evidence. It remarks, for example, that survey data show that only 11 per cent of customers are willing to make less use of their mobiles in response to a substantial reduction of fixed line prices.[24]

This type of analysis raises at least two issues. First, has the change in conditions been undertaken maintaining everything else constant? Second, as pointed out by several scholars, under certain conditions the number of marginal consumers needed to discipline pricing can be rather small. As argued by Hausman and others,[25] in industries where prices have to be above marginal cost due to the existence of fixed costs, the elasticity of the direct residual demand will tend to be comparatively lower (the mark-up is higher) and a price increase will become non-profitable even if only a few clients change provider. This caveat shows very clearly that this type of survey evidence should not be a substitute for the analysis of substitutability based on a fully specified econometric model.

The standard approach therefore has a very important role to play in the determination of telecoms markets. In fact, it has been used systematically in several prominent cases, even if not exactly in the way it was presented in Section 2.1. For example, Jerry Hausman has estimated, for the case of the Sprint/MCI WorldCom[26] merger and the effects on the long-distance market, the own-price and cross-price elasticities corresponding to the main providers (AT&T, MCI WorldCom, Sprint and an aggregate of non-branded suppliers) as well as the expected increase in the price of services provided by the new firm (which is equivalent to the computation of the residual demand of the merged firm).[27] The analysis of the cross-price elasticities allows a determination of how close substitutes firms are, and the computation of the elasticities of the residual demand function takes into account the pre-existing degree of rivalry between the competitors involved in the merger (the parameter denoted as $\rho$ in Section 2.1.).

Another recurrent market definition problem is the distinction between mass and business markets in conventional voice telephony. The broad definition of the market, including both segments, would make sense if the key characteristics of the services (quality, coverage, capacity and so on) were similar and/or if the providers present in one market segment, but not in the other, could easily deploy capacity and commercial presence in the neighbouring market segment. In practice, these two markets have very often been considered as separate (for example the MCI WorldCom cases), but not because the framework of the hypothetical monopolist has been systematically applied.

In the USA, the two markets can in principle be distinguished on the basis of aggregating users which face similar competitive conditions,[28] but the imperfect substitutability between the different services offered to households and companies was already highlighted as important in the NYNEX/Bell Atlantic case.[29] This is, in fact, fully consistent with the FCC approach, based on disregarding supply substitutability. However, in a later decision, the MCI/WorldCom case, the FCC Order highlights as relevant the fact that the assets owned by producers have to be different if they want to tackle the two different sorts of users. In the Ameritech/SBC case, the FCC goes on to add not only different assets but also different firm capabilities. It is clear, therefore, that this implies a non-explicit consideration of supply substitutability. This is all the more contradictory if one notes the contents of more recent decisions such as Bell Atlantic/GTE.[30] In that Order the FCC asserts that ‘defining relevant markets involves identifying and aggregating consumers with similar demand patterns’, and there is no mention of how important it is whether those different consumer groups can be served by the same group of companies or not.

In the EU, the distinction between the two markets is based on the existence of significant differences in demand ‘and supply’.[31] The EC decisions stress, however, that the two markets are characterized by substantial differences in the demand of both types of customers, although some reference is also made to differences in supply.

Finally, the recent Guidelines consider also the ability to discriminate as reason enough to have two markets.[32] That is, even if the two markets are served by the same producers, they could be different markets on account of price discrimination. Of course, the ability to price-discriminate requires that the hypothetical monopolist be able to distinguish the two market segments and avoid arbitrage, and this has to be shown for the two markets to be separated.[33] As the European Commission rightly stresses,[34] price differences are not enough, since they can reflect quality differences and need not lead to the consideration of separate markets.

With regard to the geographical determination of the market, both the USA and the EU apply in principle a methodological approach based on the SSNIP test. However, as with the definition of the relevant product market, the FCC focuses on demand substitutability. This leads to markets that should be narrowly defined (on a point-to-point basis) on the grounds of the limited substitutability between calls with different originations and destinations. Since this approach could yield a very large number of markets, aggregation proceeds by considering jointly all consumers that face a comparatively similar competitive situation. Even if long-distance calls out of Miami and Los Angeles may be considered as different services, the fact that the supply conditions are similar allows the agency to consider an aggregate US market for long-distance calls:[35] 'We conclude that when a group of point-to-point markets exhibit sufficiently similar competitive characteristics (i.e. market structure), we will examine the group of markets using aggregate data.'[36] As argued before, this is in fact an indirect way to consider supply substitutability. This is exemplified when considering the markets for long-distance calls, where a distinction is made between in-region and out-of-region calls. These two are considered to be different markets[37] on the grounds that, for any particular region, the control of the local loop by the local Baby Bell leads to potential competition problems for in-region calls. It is clear that the same conclusion would be achieved if supply substitutability were explicitly considered. Given the *de facto* monopoly of the local loop, there is no supply alternative to the incumbent for in-region calls.

The EC Guidelines approach the issue of geographical determination by grouping all areas where 'competitive conditions are similar'.[38] Consider, however, the decision on the joint venture between BT and AT&T.[39] In this decision the Commission recognizes that country pairs could be the relevant markets for international carrier services, but sees no need to define any route in detail since it does not find any evidence of the 'creation or strengthening of any dominant position in the area of international carrier services resulting directly from the proposed operation … the question of whether the provision of carrier services should be looked at in terms of country pairs routes or on a more global basis may be left open'.[40] This is not what happens with international voice telephony services, where the US–UK route is defined as a relevant market on the basis of the non-existence of good substitutes (rerouting, calling cards, call-back) and the share of capacity owned by the players in the proposed operation.[41] This market is not defined using demand and supply substitutability, but rather using an evaluation of market power even before the market has been defined. In practice, though, both approaches yield the same result: if prices were to go up, given the capacities installed, no other players would have the incentive and ability to provide service in that particular route.

The general approach discussed so far, which is in theory based on the SSNIP test but less so in practice, does not change significantly when we look at some of the more specific features of the telecoms industry, such as (1) the existence of vertical relations with a small number of upstream firms, (2) the presence of networks with varying degrees of capilarity and (3) the existence of bottlenecks.

A recurrent problem in telecommunications markets is whether a distinction should be drawn between wholesale and retail services. This applies to market segments such as the origination of mobile calls, access to fixed voice telephony or Internet access.

Wholesale markets comprise network services provided to service providers, and retail markets include the services aimed at end-users. As usual, the extent to which these markets should be distinguished for competition policy purposes depends on the degree of substitutability. Demand-side substitutability is in principle not possible, since the two types of services are aimed at different groups of users. However, the extent to which the wholesale and retail service are technically different is sometimes not clear (think, for example, of access to the public switched telephone network, PSTN). Even if we assume that from the demand side the two services are not substitutes, they could still be in the same market if there were ease of entry from one market into the other. This is usually the case for the retail market, where wholesalers can easily enter if there is a price rise. But not the other way around, because typically the provision of wholesale services requires building costly infrastructure and/or owning expensive operator's licences (entry, if possible at all, involves substantial sunk costs).

This methodological approach is the one adopted by Oftel in the case of the mobile market.[42] The UK agency distinguishes the two markets on the basis of the absence of demand substitutability and the asymmetry in supply substitutability. The key issue is, of course, the existence of very significant barriers to entry in the provision of wholesale services due to the limited number of mobile operators. Yet, if mobile virtual network operators are allowed a 'regulated' access to the market, the distinction between wholesale and retail may cease to be relevant since the two types of services would have (potentially) the same competitors.

The EC has distinguished some wholesale services such as the lease of transmission capacity and the provision of related services,[43] or the pan-European market for access (SMS – short messaging service) to mobile infrastructure.[44] The EC argues that in fixed services there is a difference between subscriber (retail) access to infrastructure and operator (wholesale) access,[45] but in the case of mobile services this is to be decided on a case-by-case basis.[46] As argued above, regulation plays a key role in determining the conditions of entry in the provision of infrastructure (in fixed networks with measures such as the unbundling of the local loop and carrier preselection) and therefore it will be the key determinant if a distinction is to be drawn between wholesale and retail markets. Note finally that the wholesale/retail distinction in telecoms is different from its equivalent in other industries to the extent that in telecoms the question overlaps with the distinction between network and final services, and the number of networks is limited by technology and/or regulation.

As for network effects, which are so pervasive in the telecoms industry, they appear to be easily incorporated into the standard framework of analysis. Indeed, a network which becomes very large will tend to have fewer substitutes and may end up constituting a market in itself. In fact, this is the type of reasoning underlying the concerns expressed by the antitrust authorities with regard to the effect on the Internet backbone market of the failed merger between MCI WorldCom and Sprint. The US decision[47] refers to tier 1 Internet backbone providers (IBPs), while the EU[48] talks about top-level or universal connectivity, but in both cases the discussion focuses on the insufficient pricing constraint that second- and third-tier providers impose on the leading IBPs. Insufficient demand and supply substitutability due to network effects leads to the definition of two separate markets.

Another distinctive feature of the telecommunications industry is the existence of bottlenecks that arise due to technology or regulation. For example, access to residential customers has been for many years a bottleneck in many countries, since the home was reached only by the PSTN and, given the existence of substantial (literally) sunk costs, it was not clear that it was economical to have more than one access to the home. Therefore, the local loop constituted a bottleneck.

By definition, a bottleneck is akin to an essential facility, a key resource for which there is no alternative. Access to this resource is usually regulated in order to prevent the potential abuse of market power by its owner.[49] From the market definition viewpoint, of course, a bottleneck is a market on its own, since it has no good substitutes in use, and cannot be reproduced easily.

Changes in regulation and technology may, however, modify the boundaries of this market. For example, in the local loop the widespread availability of upgraded cable networks, fixed wireless access and satellite connectivity can lead to the disappearance of the bottleneck and to a wider definition of the market, embracing several technologies. This possibility has been explicitly recognized by the EC in its decisions in this area.[50]

An interesting and controversial bottleneck arises in the mobile segment. In mobile telephony, as opposed to what happens in fixed telephony, there are usually several alternative networks. Access to the network does not, therefore, constitute a bottleneck, since the user can choose between several alternatives and the fixed (and sunk) costs of access for the user are low.

Yet, in many mobile markets the way service payments are structured leads to the appearance of 'a bottleneck problem' in call termination. In countries where calls are paid only by the calling party (the 'calling party pays', or CPP, principle) the receiving party will have little incentive to change providers when the price of finalizing calls goes up. The only way[51] to reach a particular end-user is to call and finalize the call on the network to which that user is subscribed. If the end-user were paying for receiving the call, it would react to a potential abuse of monopoly power by changing providers, but the fact that it does not pay for that service generates an inelastic demand and a result akin to a bottleneck.

From the point of view of market definition, it is clear that under these conditions call termination on each network can easily become a market on its own and, of course, each network operator will have a 100 per cent market share of its own network. It is obvious, however, that these bottlenecks – which lead to very narrow definitions of markets – are the result of the CPP principle and the subsequent price insensitivity. As bottlenecks they may require regulation, but it is less clear that they can be properly defined as relevant markets. As we will see later, the termination of calls is a service which is usually acquired together with a set of other mobile services (call origination, SMS, and so on). The relevant market for antitrust purposes could well be a bundle of mobile services (see Section 3.2 below), and this need not be incompatible with a regulated termination rate.

The EU Guidelines allow the potential definition of two separate mobile markets, one for call origination and one for call termination,[52] and recognize the problems of low incentives to compete on prices for terminating traffic.[53]

Oftel has gone further than that, and in fact concluded that call termination on each mobile network is a relevant market where price regulation is needed due to the existence of insufficient competitive pressures.[54] We will take up this argument in detail later (Section 3.3).

## 3.2 Market Definition and Bundles of Services

Telecommunications services, like many other goods and services, can be defined in a very narrow way. This involves decomposing the overall, broader service, which is more directly perceived by the user, into a collection of components, considering individually each service which could conceivably be consumed or supplied independently. Some complex goods such as automobiles can also be assessed in this fashion. One can certainly think of a car as a bundle of components, including diverse items such as tyres, seats, rear-view mirrors and the like. Similarly, one can think of fixed voice telephony as an overall service, which includes a collection of narrowly defined component services such as access, call termination, local call origination, long-distance call origination, directory services, and so on. These component services may sometimes not be consumed independently (for example, no consumer wishes to have access to the network without using it, much as there is no independent demand for steering wheels), and they will have varying degrees of complementarity (very strong between access and use, more moderate between long-distance and local call origination). How fine a distinction should be made of the different component services becomes a difficult decision for antitrust analysis.

Moreover, if one takes a broad approach to product/service definition, the problems do not disappear. As with automobiles, it is not clear what the client perceives as 'the service or product'. Does it include 'extras' such as air conditioning or fog lights? Are these components part of a car, or should they be considered as different products?

Let us consider first an approach which starts with a narrow definition of telecommunications services, assuming that a distinguishable service exists if it can be supplied by a firm independently (even if the firm has to acquire complementary or input services from other firms to be able to provide that particular service). This approach leads to the determination of very narrowly defined services. For example, in mobile we could have markets such as access, origination of calls, termination, SMS, roaming and so on. Conceivably, a consumer can have different providers for all these services and the operators could, in principle, provide a few or only one of these services. There are, of course, significant complementarities in demand and scope economies in supply which will influence the actual choices of consumers and service providers, but it is instructive to discuss briefly the antitrust implications of this narrow approach to service definition.

Consider a narrowly defined service such as retail call termination in the mobile industry. The first thing to note is that the existence of strong complementarities with other mobile services reduces the own-price elasticity for this individual service,[55] that is, a price increase for this service will not diminish demand significantly if the prices of complementary services do not change. In other words, the existence of complementarities increases the costs of switching for an individual service. Moreover, the extent to which the consumer will change to an alternative provider for call termination will be limited by the reduced number of providers of call termination (we assume here that there is no mandated call-by-call or carrier preselection for this service) and the fact that changing providers may also lead – due to the same complementarities – to a change in the provision of the other services.

In this context, the conventional method of market definition does not work. Starting from one individual product and enlarging the product set will yield unreasonable results. First, the own-price elasticity for a given service such as call termination will typically be very low due to the existence of complements and the limited number of substitutes. Call termination by a single provider may then be in itself a relevant market! Second, if the collection of services is enlarged to include complementary services such as access or call origination, we will usually find that the residual demand faced by the firm becomes more elastic, reflecting the loss of market power.[56] That is, starting from an individual service, market power increases if comparable services provided by other operators are included, but declines if what is added are complementary services.

It is, therefore, apparent that if complementarity effects are important (as well as scope economies from the supply side), the standard market definition approach will make sense only when applied to a system of services. Indeed, we can say that a bundle of goods will constitute a system if the goods are complements.[57]

If firms indeed produce these services jointly and consumers buy them as a system, the analysis of observed mark-ups can be a very misleading indicator of market power and, as a consequence, lead to very poor decisions in terms of defining markets. Firms producing a collection of goods in the presence of fixed costs will optimally charge higher mark-ups in those services which face more inelastic demand. A high mark-up in an individual service need not reflect an overall high level of market power.[58]

As shown in Table 4.1, the available evidence with regard to the complementarity of telecoms services is very significant (see figures in bold), in particular with regard to pairs of services such as access (subscription) and use, whether in mobile or in fixed telephony.

*Table 4.1 Complementarities in telecommunications*

| | |
|---|---|
| *Mobile telephony* | |
| Terminating mobile calls [Access Economics, 1999]* | |
| own-price | –0.1 |
| Mobile subscription [Dotecon, 2001] | |
| own-price | –0.37 |
| cross-price mobile calls | **–0.25** |
| cross-price fixed to mobile | **–0.21** |
| Mobile calls [Dotecon, 2001] | |
| own-price | –0.62 |
| cross-price mobile subscriptions | **–0.48** |
| cross-price fixed to mobile | **–0.27** |
| Originating mobile calls [Access Economics, 1999]* | |
| own-price | –0.8 |
| Originating mobile calls [Hausman, 1999] | |
| own-price | –0.5/–0.6 |
| | |
| *Fixed telephony* | |
| Basic Access Service [Hausman et al., 1993] | |
| own-price | –0.005 |
| cross-price intraLATA | **–0.0086** |
| cross-price interstate | **–0.0055** |

*Note*: * Quoted by Hausman (1999).

In practice, this close complementarity and the fact that services such as access are not typically consumed independently means that the service definition will require the determination of a system. Of course, not all the services that can be supplied over a network need be components of that system. Services may be considered as stand-alone services when the complementarity with the rest of the components of the system is low[59] and/or when there are alternative (independently supplied) substitute services (this in turn will be affected by the importance of scope economies on the supply side and other market characteristics). A good example of a separate service market is international roaming, where the service is supplied as an option by the operators and users can choose between their own operator and alternative roaming-only service providers.[60]

It is therefore clear that the strong complementarities in demand and the characteristics of supply could justify the analysis of market definition at the level of bundles of services. This is, in fact, something which is well established in antitrust practice. For example, in *US* v. *Philadelphia National Bank* the Court defines the product market as 'the cluster of products (various kinds of credit) and services (such as checking accounts and trust administration) denoted by the term "commercial banking"'. In the NYNEX/Bell Atlantic[61] case the FCC indicates that 'to the extent that consumer demand for bundled service packages forces carriers to offer such bundles, the bundling of local exchange and exchange access services with long distance services may well become a relevant product market'.

The bundling concept is also acknowledged by Oftel for the case of mobiles, even if half-heartedly. In the mobile effective competition review (Oftel, 2001b) the operators argued that access is not a separate retail market since users would not wish to purchase the ability to make calls without actually making them. Oftel accepts that these two services are consumed jointly[62] but dismisses the importance of this fact in terms of market definition since 'theoretically for access there are no substitutes on the demand and supply side, so a hypothetical monopolist could feasibly raise the price of access'. Oftel bases its argument (and its market definition) not on actual market practice but rather on a hypothetical situation: 'innovations such as indirect access would suggest that calls *could* be supplied by a number of suppliers, not just the access supplier' (own italics).

## 3.3 The Debate on Call Termination to Mobiles

The difficulties posed by market definition in the telecommunications industry are exemplified by the ongoing debate about mobile call termination, particularly in the UK. Oftel reviewed in 2001 the UK regulation of call termination. It issued a consultative document in February and a final statement (September) after assessing the views of the different parties. The UK regulator ended up proposing a CPI–12 per cent price cap on termination charges for the next four years. Since this determination was not accepted by the mobile operators, the issue was addressed to the Competition Commission, which has to decide before early 2003.

The Oftel approach can be summarized as follows. Oftel starts by highlighting that the CPP principle creates few incentives for price competition between service providers, since customers do not directly pay for the service that they use. Oftel next analyses the importance of several potential offsetting factors such as the significance of closed user groups, the possibility of switching networks, substitution for other types of calls and the countervailing power of the calling network. The agency finds[63] that these factors are ineffective in terms of leading to downward pressures on the pricing of termination.

Oftel also compares the rates for call termination with its estimates of long-run marginal costs for this service and finds them to be substantially above, with a large mark-up of up to 70 per cent. The industry claims that the high prices for call termination are compensated by the low prices for call origination and other services, but Oftel dismisses this argument, countering that these cross-subsidies are neither efficient nor fair and that the broadly defined mobile market is still not effectively competitive and therefore will not ensure that high termination charges will be compensated by low prices elsewhere.[64]

After assessing the non-existence of sufficient price competition in call termination, Oftel goes on to delineate the relevant markets and concludes that – due to insufficient demand and supply substitutability – each mobile network constitutes a relevant antitrust market.

The analytical approach of Oftel, however, does not follow the usual antitrust methodology. The UK agency assesses first the degree of competition in call termination, and then moves on to determine the relevant markets. Yet the procedure should be precisely the reverse. Start with an analysis which determines the relevant markets, and then proceed, for each of those markets, to an examination of the competitive conditions and the potential for abuse of market power. If the analysis of the mobile market is done in this fashion the results may be strikingly different.

First, an analysis of the nature of demand relationships, without mixing it up with competitive conditions, brings forcefully to the fore the importance of demand complementarities between individual services. This has, in fact, important implications for the study of the overall effect of an increase in call termination prices.

Indeed, if termination charges go up, there is an indirect but important effect on the called party, even if it does not suffer the price increase directly. When the termination charge increases, the costs of the calling network go up and this will prompt an increase in its own termination charges,[65] thus leading to an increase in the price of the outgoing calls of the called party.

Oftel's analysis dismisses this effect because substitutability is assessed under the assumption of 'all other prices constant'. However, as argued in Section 2.1, a complete analysis of supply substitutability should take into account the (price) reaction of the rivals. This is further reinforced in this case because the complementarities between the services imply that it will not make sense for any of the providers to consider what happens to the demand for one particular service when other prices are kept constant. If the provider controls those prices, the proper benchmark is an estimation of how this firm is going to adjust them accordingly (like the Cournot monopolist of complementary products) and we have to consider the reaction of demand to the whole new set of prices determined by the monopolist – given the strategies of its competitors. That Oftel's analysis is far from this framework is clear in subsequent documents issued by the regulator. For example, in a recent paper entitled 'Oftel's further documents on market definition' the agency stresses that 'the prices of other services are held constant' and it argues that the companies have 'not identified termination and retail services as either demand or supply substitutes'. This is not surprising, since they are complements.[66]

To conclude, it is worth pointing out that the whole question of market definition in mobile call termination should not be mixed up with the issue of whether a regulation of termination charges is required. Even if the market is broadly defined as the bundle of mobile services offered by operators and jointly acquired and consumed by end-users, there may still be scope for regulating one or more prices within that bundle. This could be so in several circumstances.

First, one could argue that the overall level of competition in the mobile industry is too low, so that overall mark-ups are excessive, although in an industry with high fixed and sunk costs some degree of mark-up is needed and the determination of excess profitability is bound to be complex. Since the CPP principle creates an artificial bottleneck for the calling party, one can argue that, as with other essential facilities, access to this bottleneck has to be guaranteed to all market players under fair conditions. And furthermore, one could argue that the small number of players and the reciprocal nature of termination charges make it natural for operators to use these charges as 'facilitating mechanisms' for non-cooperative price co-ordination (see Armstrong, 1998 and Laffont et al., 1998).

All these are serious reasons for possibly regulating termination charges, without necessarily requiring the establishment of termination in each network as a relevant antitrust market.[67] In fact, the Competition Commission of Australia has regulated termination charges, while recognizing both the relevance of the pricing interdependencies from the point of view of mobile operators and a broader definition of the market.[68]

## 3.4 Market Definition in Rapidly Changing Markets

As discussed in Section 2.2, the SSNIP test is less useful in industries where fixed costs are significant, specially if these costs are sunk and related to strategic competition in service or product innovation.

We also stressed above that in dynamic industries the mark-up becomes a misleading tool for measuring the efficiency of the market and that competition should be assessed across several performance dimensions.

As we have seen, in actual practice regulatory agencies and competition authorities use the SSNIP test as a reference concept and base their analysis on the consideration of demand substitutability and, to a lesser degree, substitution in supply.

In dynamic markets, the concept of substitutability has to be interpreted broadly, since substitution will take place not only on the basis of price differences but also depending on other relative performance indicators. From the demand side this often means taking a forward-looking view of the substitutability between different technologies, considering how different technical solutions may end up satisfying comparable consumer needs. From the supply side, a broad perspective implies that the analysis should take into account a larger range of potential suppliers of the services.

The focus on the substitutability between technologies is an approach similar to the one suggested by authors such as Teece and Coleman (1998) or Evans and Schmalensee (2001) when they claim that in dynamic high-tech markets market delineation should basically assess 'technology' competition.

To a certain degree, this is taken into account by the EC approach to market definition in telecoms. For example, the recent Guidelines on market analysis recognize the importance of technological convergence[69] and consider the extent to which the availability of alternative technological solutions may diminish the control of the local loop by the incumbent PSTN operators.[70] In the Sprint/MCI WorldCom case, for example, the EC considered in detail new technologies (caching, mirroring and so on) that could reduce the importance of the first-tier Internet backbone. Regulators like Oftel have also recently examined how technological advances may reduce the importance of the call termination bottleneck.[71] In all these cases, the potential technologies are at least carefully considered, even if the final decision has almost always involved the definition of a fairly narrow market.

As for a broad approach to potential suppliers, the experience is less encouraging, since the perspective of the regulators has usually been quite restrictive. In rapidly changing markets, which are being created by firms exploiting first-mover advantages with new technologies or network effects, it is not surprising that firms enjoy very large market shares, since markets are in their infancy. In these contexts, one can argue that the agencies should adopt a broad view of potential suppliers, taking into account, first, all firms that own assets which could lead to the development of substitute technologies and second, a longer time horizon to assess the potential supply response.[72]

This forward-looking view of markets is usually missing in the definition of markets in the telecoms industry. Consider, for example, the Sprint/MCI WorldCom case. In the rapidly changing backbone market, the merging firms enjoyed a large market share, but one that was predicted to deteriorate rapidly in the forthcoming years.[73] This was not taken into account by the antitrust authorities in the establishment of the relevant market.

The lack of a forward-looking approach may also lead to a different sort of problem. As new markets are being created, some firms lead the way with the introduction of innovative services. Early recognition by the regulator or the antitrust authority of this type of market is bound to detect competition problems since, by definition, the firms that master-mind the creation of a new market start by enjoying very large market shares. In the limit, an innovative firm will have a 100 per cent market share when it introduces a successful new service category.

In the EU, the Commission has recognized very quickly the existence of new relevant markets, such as the market for global broadband data communications services,[74] the pan-European market for wholesale access to mobile infrastructure[75] and the emerging EU-wide seamless retail mobile market.[76] The problem also arises in the USA. If we consider the approach set forth in the LEC Order, it implies the absence of regulation of individual services unless the agency suspects lack of competition. Again, with a 100 per cent market share, the suspicion of competition problems is likely to arise very soon.

In sum, the regulatory authorities have tended to be fairly conservative when establishing relevant markets in the emerging segments of the telecoms industry. In case of uncertainty, they have preferred to err on the side of the early (and as a consequence, narrow) definition of new markets for antitrust and regulatory purposes. This is a pity, since in this type of industries there are sound reasons to believe that such an approach may easily lead to the imposition of undue restrictions on the activity of innovative companies and may not promote economic efficiency in the longer term.

## 4. CONCLUSIONS

The definition of markets in the telecommunications industry for antitrust purposes is based, in the main jurisdictions, on the principle of the 'hypothetical monopolist'. Both in the USA and in the EU (as well as in other areas such as the UK and Australia), the framework is also used when delineating markets with regulatory objectives.

The 'hypothetical-monopolist' framework, or the small but significant non-transitory increase in price (SSNIP) test, can be closely linked to a well-grounded model of oligopolistic competition. It provides thus a rigorous statistical methodology for the empirical determination of markets. This methodology, however, requires a large amount of data and, in practice, it is most useful as a conceptual framework which guides the analysis of the available quantitative and qualitative information.

When applied to the telecommunications industry, the framework sheds useful light on some of the issues that clutter market definition in this and related industries. Yet this chapter argues that the standard approach needs to be carefully adapted along three basic dimensions.

The first concern is that supply substitutability should be fully incorporated in the market definition exercise. Giving full consideration to supply substitutability has two important implications. It means enlarging the relevant market to include all firms which have the assets and capabilities to provide, within a reasonable amount of time, a good substitute service to the one being considered. When defining the market in telecommunications, supply substitutability provides a powerful aggregation device, so that authorities do not have to analyse the competitive conditions for a very large number of individual services.

Full consideration of supply substitutability also means that when we undertake the thought experiment of considering substitute services that can impose competitive constraints on the pricing of a particular service, not only do we have to consider first the best substitutes in use. We also have to take into account the types of firms that are supplying those services and whether they are currently more or less aggressive competitors. The more aggressive they are, the more likely it is that they belong to the same relevant market.

The second area where the conventional framework has to be adapted when applied to telecoms is the analysis of demand. It must be recognized that systems of services constitute a potentially relevant unit of market analysis. The importance of systems in this industry arises both from the nature of user demand and the characteristics of supply. The analysis should fully recognize the existence of very significant complementarities in demand, which imply that consumers demand many of the services in bundles (and in some cases there is, in fact, no independent demand for some particular services, such as access). Similarly, from the supply side, the importance of scope economies implies that the services can be more efficiently provided as joint products and this is the way they are offered to consumers by the operators.

The importance of complementarities and the technological advantages of joint production are bound to change over time and be influenced by regulatory changes and technological progress, but for each geographical market at any point in time a careful distinction must be made between those services which are linked by strong complementarities and provided and consumed as systems, and those which are consumed and produced as stand-alone services. In the mobile segment, for example, depending on the regulatory and technological situation, the core group of services (access, call termination and origination, SMS and so on) can form a single system provided by all competing operators, while other services such as roaming can be separate services that can be bought independently from a non-coincident set of suppliers. The fact that some competitive problems may be detected which affect a few individual services within a system (for example access or mobile call termination) should not obscure the analysis at the market definition stage, where the focus should be on the nature of consumer behaviour and the existence of alternative sources of supply given what consumers demand.

The use of the traditional market definition framework in telecoms also has to be adapted because this is an industry characterized by the existence of significant fixed and sunk costs. This is relevant at least for two reasons. First, it implies that the analysis of market power – and as a consequence that of a hypothetical monopolist – cannot be based on the use of marginal cost pricing as a competitive benchmark. In industries with fixed costs, a positive mark-up is fully compatible with a competitive market and if this is not taken into account, markets may be defined too narrowly. Moreover, to the extent that the sunk costs incurred by telecommunications service providers are of a continuous nature, the result of competition in service innovation and technology, it becomes harder to apply a methodology based upon a static measure of competition such as the Lerner index. If telecoms firms compete through new services and better performance (and this may be the case in some segments of the industry), the conventional analysis of demand and supply substitutability based upon cross-price elasticities may become irrelevant and the analysis of the boundaries of the market should take into account several performance variables from the point of view of demand, as well as a broader set of potential firms engaged in the introduction of new services and technologies that could end up satisfying the same consumer needs.

Finally, the chapter discusses the market definition problems associated with the existence of bottlenecks, with a particular focus on the mobile call termination bottleneck caused by the 'calling party pays' principle. This discussion is specially relevant in view of the new EU regulatory framework. The new EU approach is based upon the idea that regulation will be based on competition policy principles. That is, regulatory action will be justified only when there is insufficient competition. The absence of significant competition constraints will be assessed using well-established competition policy tools, the first of them being the definition of the relevant market.

This regulatory procedure poses, however, significant methodological problems, illustrated by the mobile call termination controversy. The widespread perception of insufficient competitive pressures in the provision of this service has led in some jurisdictions to the definition of a very narrow relevant market: call termination in each network. This market definition does not take into account some of the key characteristics of the telecoms industry (the complementarity between services) and runs counter to some of the main principles that should guide market definition (what services, or bundles of services, are in fact perceived by users as constituting viable alternatives in consumption and are being or could be offered by service providers). Nevertheless, the overriding objective of regulating call termination and the need to follow the conventional antitrust methods have led to the definition of call termination in each mobile network as a relevant – and by definition monopolized – market. There may be sound reasons to regulate a bottleneck such as mobile call termination. But these should not lead to a biased determination of the relevant market.

## NOTES

1. See the Notice on the definition of the relevant market (European Commission, 1997) and the Merger Guidelines in the USA (FTC, 1997).
2. See Landes and Posner (1980) for a detailed exposition.

3. It is more convenient now to use this demand system rather than the direct system (see Baker and Bresnahan, 1985, for details).
4. Following Baker and Bresnahan (1985) we assume constant elasticities. These authors also show that the residual demand curve approach can encompass both quantity and price competition, as well as different degrees of non-cooperative and co-operative behaviour (p. 431).
5. The relationship between the corresponding direct demand elasticities will be reversed. That is, $\varepsilon_i^R > \varepsilon_i$. The elasticity of the direct residual demand will usually be larger than the direct own-price elasticity (and, again, the mark-up smaller).
6. Strictly speaking, with more than one firm the analyst will have to compute a set of partial residual demands and add their elasticities (see Baker and Bresnahan, 1988). A residual demand curve faced by one firm has a slope which depends (in part) on how the customers of the firm defect to all other firms as prices are increased. The slope of a partial residual demand curve takes into account that some of the competing firms also raise prices.
7. See Landes and Posner (1981), p. 961 and White (1999). As Posner (2001), p. 156 points out, 'the fact that a further price increase though modest would cause substantial substitution either in consumption or production is not a reason to define the market more broadly'.
8. See Krueger (2001), p. 220, 'An Interview with William J. Baumol'.
9. Teece and Coleman (1998), Pleatsikas and Teece (2001) and Evans and Schmalensee (2001).
10. See Baker and Bresnahan (1985, 1988), Scheffman and Spiller (1987) and Tollison et al. (1991).
11. *Regulatory Treatment of LEC Provision of Interexchange Services Originating in the LEC's Local Exchange Area*, Second Report & Order, CC Dkt No. 96-61, FCC 97-142, 1997 WL 193831 (rel. 18 April 1997).
12. See LEC Order, #28. (# = paragraph number)
13. Ibid., # 28.
14. Ibid., # 36.
15. Ibid., # 42.
16. Ibid., # 34.
17. Ibid., # 55.
18. Ibid., # 53.
19. Case *NYNEX/Bell Atlantic*, Docket 97-286, # 50.
20. See European Commission (2002).
21. See, for example, the Case No. IV/JV.15 BT/AT&T, # 73 and 79.
22. See the Oftel positions as presented in Oftel (2001b and 2002a). Several other documents on the Oftel web site reinforce this view.
23. See the paper by Dotecon (2001) and the comments by Hunter and Ioannidis (2002), both available on the Oftel web site.
24. See Oftel (2001b), # A.1.9 to A.1.11.
25. See Hausman et al. (1996), pp. 368–9.
26. Declaration of Professor J. Hausman on behalf of SBC in the case *Sprint/MCI WorldCom*. Docket 99–333.
27. See Hausman's declaration in the *Sprint/MCI WorldCom* case.
28. LEC Order # 42.
29. Case *NYNEX/Bell Atlantic*, FCC 97-286, # 53.
30. Case *BA/GTE*, FCC 00-221, # 102.
31. See footnote 58 in the Guidelines, and the cases cited therein.
32. See the Guidelines, # 46.
33. See Hausman et al. (1996).
34. See the Guidelines, # 46.
35. LEC Order # 64–6.
36. Ibid., # 66.
37. Ibid., # 76.
38. See the Guidelines, # 55–60.
39. Case No. IV/JV.15 *BT/AT&T*.
40. Ibid., # 78.
41. Case No. IV/JV.15, *BT/AT&T*, # 83.
42. See Oftel (2001b), p. 46.
43. See the Guidelines, footnote 65.
44. Ibid., footnote 64.
45. Ibid., # 65 and footnote 56.
46. Ibid., footnote 61.
47. *DoJ* v. *WorldCom and Sprint*, # 28 to 30, 26 June 2000. Docket 99-333.
48. See case COMP M. 1741 *MCI WorldCom/Sprint* # 53. The EU distinguishes between tier 1 and the other

backbone providers not only on the basis of network effects but also taking into account the nature of the relationships between firms (peering arrangements vs customer relationships).

49. See Laffont and Tirole (2000), pp. 97–8.
50. See the Guidelines, # 68 and footnote 62.
51. Of course, one possibility would be to introduce for call termination a regulation similar to carrier selection (call by call) or carrier preselection, whereby the caller is able to choose on which network to terminate the call, which is then rerouted to the network of the called party.
52. See the Guidelines, # 69.
53. Ibid., footnote 69.
54. Oftel (2001c and d).
55. This is the case even in the absence of the CPP principle. Of course, with the CPP principle the price increase is borne by the calling party and this only indirectly may lead to a price increase of the services paid for by the called party, say.
56. This is, of course, a variant of the result highlighted long ago by Cournot. A monopolist producing a range of complementary goods will tend to set prices lower than those that would be established by a collection of independent monopolists selling the same set of complementary goods.
57. See, for example, Katz and Shapiro (1994).
58. It could be, of course, that firms use a service such as termination which constitutes a cost that each firm imposes on each other and is not directly charged to customers, as a vehicle for collusion (see the discussion in Section 3.4).
59. The literature on aftermarkets (see Shapiro, 1994) has tackled a related issue: should durables and their associated complementary goods (photocopiers and parts in the famous Kodak case) be considered a single product? Similarly, Katz and Shapiro (1999) discuss complementarity in software and hardware markets. They conclude that strong complementarity warrants the definition of the system as the relevant market.
60. See European Commission (2000).
61. Case *NYNEX/Bell Atlantic*, # 52.
62. Oftel (2001b), A.1. # 29.
63. Oftel refers to quantitative evidence when in fact all the analysis is qualitative in nature. The agency uses data (survey evidence) but it does not attempt to measure the key variable: the extent of substitutability between different types of calls. (See Oftel 2001b, pp. 7–9.) Using survey evidence leads to an informal assessment of the importance of consumers switching demand under certain circumstances (for example Oftel, 2001d, Annex 2, p. 58), but as we have seen (see above, Section 3.1) this is a particularly troublesome approximation to elasticities of residual demand in industries with fixed costs.
64. Oftel recognizes (see Oftel 2002b, # 16 and 17) that to pay for the fixed costs, a positive mark-up will be required, but it is less clear what the mark-up level compatible with effective competition is. In any case, its criticisms of Ramsey prices miss an important point. Part of the inefficiencies created by the cross-subsidies of the operators may be the result of other regulations: that is, the CPP principle, or the fact that the incumbent fixed operator is subject to regulation.
65. This is the case, of course, for mobile networks, and not for fixed networks due to regulation.
66. See Oftel (2002b). The same document (# 32) argues also that access and calls are not substitutes but does not recognize the complementarity between the two. Indeed, these services would be complements even if indirect mobile access were available. The analysis of Oftel mixes up the competitive conditions with the nature of the demand relationships between services.
67. The overriding objective of regulating mobile call termination may undermine the analysis. For example, in Oftel (2001d), # 3.6 the agency argues that demand for call termination is so inelastic, even below one, that revenues would fall if prices were lower. However, no single product monopolist would ever set such a price. This highlights the fact that the competitors should be analysed as oligopolists simultaneously setting the prices of a group of services which are complements in demand.
68. See Australian Competition Commission (2001), pp. 24–31. Other regulators, such as OPTA in the Netherlands, follow an approach similar to Oftel.
69. See the Guidelines, # 47.
70. Ibid., footnote 62.
71. 'Oftel's response to the Competition Commission's letter of 7 May 2002 (Oftel 20) entitled "Technological Change"', Oftel, 24 May 2002.
72. See Pleatsikas and Teece (2001).
73. Supplemental Internet Submission to the *MCI WorldCom/Sprint* case, CC Docket No. 99-333.
74. See the Guidelines, footnote 56.
75. Ibid., footnote 64.
76. See the case COMP M.1975, *Vodafone Airtouch Mannesman*, # 12 to 21.

# BIBLIOGRAPHY


Aghion, P. et al. (2002), 'Competition and Innovation: An Inverted U Relationship', mimeo, Harvard University, February.

Armstrong, M. (1998), 'Network Interconnection', *Economic Journal*, **108**, pp. 545–64.

Australian Competition Commission (2001), 'Pricing Methodology for the GSM Termination Service', Final Report, July.

Baker, J. and T.F. Bresnahan (1985), 'The Gains from Merger or Collusion in Product Differentiated Industries', *Journal of Industrial Economics*, **33**, pp. 427–44.

Baker, J. and T.F. Bresnahan (1988), 'Estimating the Residual Demand Curve Facing a Single Firm', *International Journal of Industrial Organization*, **6**, pp. 283–300.

Dotecon (2001), 'Ramsey Prices for Mobile Services', a report prepared for BTCellnet, 4 May.

Economides, N. and S. Salop (1992), 'Competition and integration among complements, and network market structure', *Journal of Industrial Economics*, **40** (1), March, pp. 105–23.

European Commission (1997), 'Commission Notice on the definition of the relevant market for the purpose of Community competition policy', OJ C 372, 9/12/1997.

European Commission (2000), 'On the Initial Findings of the Sector Inquiry into Mobile Roaming Charges', Brussels, Working Document, 13 December.

European Commission (2002), 'Commission guidelines on market analysis and the assessment of significant market power under the Community regulatory framework for electronic communications networks and services', OJ C 165/03, 11/07/2002.

Evans, D. and R. Schmalensee (2001), 'Some economic aspects of antitrust analysis in dynamically competitive industries', NBER Working Paper 8268, May.

Federal Trade Commission (1997), *Merger Guidelines*.

Hausman, J. (1999), Declaration in response to the paper by Prof. Gans (University of Melbourne), 'An Evaluation of Regulatory Pricing Options for Mobile Termination Service', of 9 December.

Hausman, J. et al. (1993), 'The Effects of the Break-up of AT&T on Telephone Penetration in the United States', *American Economic Review Papers and Proceedings*, **83** (2), May, pp. 178–84.

Hausman, J. et al. (1996), 'Market definition under price discrimination', *Antitrust Law Journal*, **64**, pp. 367–86.

Hunter, J. and C. Ioannidis (2002), 'A review of Ramsey Prices for Mobile Services, produced for BTCellnet by Dotecon', 15 June.

Katz, M. and C. Shapiro (1994), 'Systems Competition and Network Effects', *Journal of Economic Perspectives*, **18**, pp. 93–115.

Katz, M. and C. Shapiro (1999), 'Antitrust in Software Markets', in J.A. Eisenach and T.M. Lenard, *Competition, Innovation and the Microsoft monopoly: antitrust in the digital marketplace*, Boston: The Progress and Freedom Foundation.

Krueger, Alan B. (2001), 'An Interview with William J. Baumol', *Journal of Economic Perspectives*, **15** (3), Summer, pp. 211–31.

Laffont, J.-J. and J. Tirole (2000), *Competition in Telecommunications*, Munich Lectures in Economics, Cambridge, MA: MIT Press.

Laffont, J.-J., P. Rey and J. Tirole (1998), 'Network Competition I: Overview and Non-discriminatory Pricing', *Rand Journal of Economics*, **29**, pp. 1-37.

Landes, W.M. and R.A. Posner (1981), 'Market power in antitrust cases', *Harvard Law Review*, **94** (5).

LECG (2000), *The economics of mobile call termination*, report prepared on behalf of BTCellnet, One2One, Orange and Vodafone.

Neven, D., R. Nuttall and P. Seabright (1993), *Merger in Daylight*, London: CEPR.

Oftel (2001a), 'Effective competition review: Mobile', February.

Oftel (2001b), 'Effective competition review: Mobile', September.

Oftel (2001c), 'Review of the Price Control on Calls to Mobiles', February.

Oftel (2001d), 'Review of the Price Control on Calls to Mobiles', September.

Oftel (2002a), 'Protecting Consumers by Promoting Competition', January.

Oftel (2002b), 'Oftel's Further Comments on Market Definition', 27 March.

Pleatsikas, C. and D. Teece (2001), 'The analysis of market definition and market power in the context of rapid innovation', *International Journal of Industrial Organization*, **19**, pp. 665–93.
Posner, R.A. (2000), 'Antitrust in the new economy', in *Tech Law Journal*, available at www.techlawjournal.com/atr/20000914posner.asp, 14 September. Address at the Conference of the American Law Institute.
Posner, R.A. (2001), *Antitrust Law*, 2nd edn, Chicago: University of Chicago Press.
Scheffman, D.T. and P.T. Spiller (1987), 'Geographic Market Definition under the U.S. Department of Justice Merger Guidelines', *Journal of Law and Economics*, **XXX**, pp. 123–46.
Shapiro, C. and D.J. Teece (1994), 'Systems Competition and Aftermarkets: An Economic Analysis of Kodak', *Antitrust Bulletin*, **39** (1), pp. 135–62.
Slade, M.E. (1986), 'Exogeneity tests of market boundaries applied to petroleum products', *The Journal of Industrial Economics*, **XXXIV** (3), pp. 291–303.
Stigler, G.J. and R.A. Sherwin (1985), 'The Extent of the Market', *Journal of Law and Economics*, **XXVIII**, pp. 555–85.
Sutton, J. (1991), *Sunk costs and market structure: price competition, advertising and the evolution of concentration*, Cambridge, MA: MIT Press.
Teece, D.J. and M. Coleman (1998), 'The meaning of monopoly: antitrust analysis in high-technology industries', *The Antitrust Bulletin*, **43**, pp. 801–57.
Tollison, R., D. Kaplan and R. Higgins (1991), *Competition in the Carbonated Soft Drink Industry*, Lexington, MA: Lexington Books.
US Department of Justice (1997), '1992 Horizontal Merger Guidelines (with April 8, 1997, Revisions to section 4 on efficiencies)', Source: FTC web site.
White, L.J. (1999), 'Wanted: A Market Definition Paradigm for Monopolization Cases', *The Computer Industry Newsletter*, Spring. Available at www.abanet.org/antitrust/committees/computer/spr99white.html

# 5. The approach to market definition in the Commission's Guidelines and Recommendation

**Christian Hocepied***

Directive 2002/21/EC on a common regulatory framework for electronic communication networks and services (the Framework Directive – FWD) is based on the premise that there is a continued need for *ex ante* obligations in certain circumstances in order to ensure the development of a competitive market (see, for example, recital 25). The experience of the liberalization of the telecommunications markets in the European Union since 1998 is that *ex ante* regulation is indispensable as a complement to competition law to allow market entry following liberalization. Sector-specific *ex ante* regulation was used to ensure that access conditions on the networks belonging to the incumbents were published and fair so that entrants could serve customers connected to the latter's networks (as well as to the networks of the other entrants). Specific access obligations were initially imposed on operators having significant market power (SMP) in four market areas: voice telephony, interconnection, mobile telephony, leased lines and, later, extended to local loop unbundling.

However, the *ex ante* definition of the market areas in which obligations are imposed could not be maintained in an industry where new markets emerge due to technical and regulatory innovation.[1] One of the aims of the 1999 Review was to address this shortcoming of the previous framework. The Commission Communication on the 1999 Review therefore proposed that national regulators should have responsibility for defining markets for the purposes of *ex ante* regulation, on the basis of Commission Recommendations.[2]

Most commentators to the 1999 Review consultation agreed with this approach in principle, although some argued that it also created the potential for divergence of regulatory decision-making across the EU. In this context, the importance of Commission Recommendations was emphasized, as well as the need for effective co-operation between NRAs and national competition authorities. Moreover, some entrants expressed concern over the methodologies used for market definition.[3]

The Framework Directive (FWD)[4] subsequently replaced the *ex ante* definition of markets to be regulated by the flexible – albeit somewhat complex[5] – approach set out by Pierre Buigues in this book.[6] The aim of this contribution is to examine in more detail the market definition approach followed by the Commission under this Directive.

It will first set out how the markets for sector-specific regulation were delineated. It will in this context summarize the main points made during the public consultation as well as the conclusions drawn by the Commission. It will finally set out the differences between the delineation of markets for the purpose of sectoral regulation and for the application of the competition rules.

# 1. THE GENESIS OF THE LIST OF MARKETS ESTABLISHED BY THE COMMISSION

## 1.1 The Initial Recommendation of 11 February 2003

The Framework Directive is based on the premise that there is a need for *ex ante* obligations in certain circumstances in order to ensure the development of a competitive market (see, for example, recital 25). Article 15(1) FWD requires that the Commission adopt a Recommendation identifying, in accordance with Annex I of that Directive, those product and service markets within the electronic communications sector, the characteristics of which may be such as to justify the imposition of regulatory obligations set out in the specific directives. This Recommendation was adopted on 11 February 2003.[7] In this Recommendation[8], the Commission identified 18 markets in the electronic communications sector that national regulators have to investigate in order to decide whether these markets continue to justify sector-specific regulation (Table 5.1).

The markets listed in the Recommendation of 11 February do not necessarily justify *ex ante* regulation. Annex I FWD is entitled 'List of Markets *to be* included in the initial Commission Recommendation on Relevant Product and Service Markets referred to in Article 15'. There-

*Table 5.1 List of markets to be regulated* ex ante *in the Recommendation of 11 February 2003*

| Market areas listed in Annex I of the Framework Directive | Corresponding markets listed by the initial Recommendation |
|---|---|
| **Retail** | |
| Provision of connection to and use of the public telephone network at fixed locations | 1. Access to the public telephone network at a fixed location – residential |
| | 2. Access to the public telephone network at a fixed location – business |
| | 3. Publicly available local/and or national telephone services provided at a fixed location – residential |
| | 4. Publicly available local/and or national telephone services provided at a fixed location – business |
| | 5. Publicly available international telephone services provided at a fixed location – business |
| | 6. Publicly available international telephone services provided at a fixed location – residential |
| Provision of leased lines to end-users | 7. The minimum set of leased lines |
| **Wholesale** | |
| Call origination in the fixed public telephone network | 8. Call origination on the public telephone network provided at a fixed location |

*Table 5.1 continued*

| Market areas listed in Annex I of the Framework Directive | Corresponding markets listed by the initial Recommendation |
|---|---|
| Call termination in the fixed public telephone network | 9. Call termination on individual public telephone networks provided at a fixed location |
| Transit services in the fixed public telephone network | 10. Transit services in the public telephone network provided at fixed locations |
| Call origination on public mobile telephone networks | 15. Access and call origination on public mobile telephone networks |
| Call termination on public mobile telephone networks | 16. Voice call termination on individual mobile networks |
| Leased line interconnection (interconnection of part circuits) | 13. Wholesale terminating segments of leased lines<br>14. Wholesale trunk segments of leased lines |
| Access to the fixed public telephone network, including unbundled access to the local loop | 11. Wholesale unbundled access (including shared access) to metallic loops and sub-loops, for the purpose of providing broadband and voice services<br>12. Wholesale broadband access |
| Access to public mobile telephone networks, including carrier selection | 15. Access and call origination on public mobile telephone networks |
| Wholesale provision of leased line capacity to other suppliers of electronic communications networks or services | 13. Wholesale or terminating segments of leased lines<br>14. Wholesale trunk segments of leased lines |
| Services provided over unbundled (twisted metallic pair) loops | 11. Wholesale unbundled access (including shared access) to metallic loops and sub-loops, for the purpose of providing broadband Internet services |
| National market for international roaming services on public mobile telephone networks | 17. Wholesale national market for international roaming on public mobile networks |
| | 18. Broadcasting services to deliver broadcast content to end users |

fore, even if the Commission considered that certain markets of Annex I did not expressly exhibit characteristics generally justifying *ex ante* regulation,[9] it had to include them in the list.

This means that the list of markets might be longer than what would have been justified according to Recital 27 of the FWD, which indicates that *ex ante* regulatory obligations should only be imposed where there is no effective competition, that is, in markets where there are one or more undertakings with SMP, within the meaning of Article 14 of that Directive, and where national and Community competition-law remedies are not sufficient to address the problem. One of the tasks of the Commission in the subsequent revisions of the list will be to ascertain those markets which became competitive or where competition-law instruments are sufficient to monitor anticompetitive behaviour, in order to remove them from the markets listed in the initial Recommendation of 11 February 2003.

### 1.2 The Expert Study

The starting-point to delineate markets was Annex I of the FWD instead of inquiries into the competitive constraints faced by a hypothetical undertaking which serves the smallest possible market, as standard competition-law practice would suggest. The Commission therefore did not focus on the behaviour of merely hypothetical undertakings which could for example provide collocation space or access to roaming services. The Commission limited itself to the market situation of the existing markets listed, thereby using for identifying retail markets the perception of average users[10] and, for wholesale markets, assessing what inputs are necessary to provide the relevant retail services[11] and which are the actual and potential competitive pressures on the providers of such inputs.

The Commission analysis was based on a study prepared by independent experts, which was commissioned in 2001[12] by DG Information Society. A key element of the methodology of the study team was the use of a starting-point that

> narrow product market definition is more appropriate for ex post intervention, especially because competition regulators have the benefit of identifying the parameters of the relevant market in the context of particular identified abuses or a particular anti-competitive arrangement, and because competition rules can be relied on to address instances of leveraging of market power into related or neighbouring markets. In particular, competition rules are best used to address 'virtual' or nascent markets (usually at the wholesale level) which might arise from an existing local access monopoly. By contrast, an aggregated market approach provides a more realistic appraisal of market power in a forward-looking context; it also provides a broader basis upon which one can take into consideration all potentially substitutable services and platforms, including those that may act as chain substitutes.

The study provided detailed relevant market analyses of wholesale and retail networks and services in the fixed sector (chapter 4) and in the mobile sector (chapter 5), and also wholesale services to broadcasters which were likely to fall within the proposed new regulatory package for 'electronic communications networks and services' (chapter 6). It was followed by preliminary conclusions and recommendations (chapter 7). The Study Team proposed the following list of markets for regulation:

1. Wholesale fixed network access to end-users
2. Fixed interconnection services including origination, termination and single transit interconnection services

3. Wholesale and retail access to dedicated capacity for local tails
4. Wholesale access to fixed networks and low-speed switched voice (and related) services
5. Wholesale mobile voice termination (without deciding whether on an individual network basis or on the basis of mobile networks taken in the aggregate)
6. Wholesale broadcasting transmission services.

### 1.3 The Draft Recommendation

On 18 June 2002, the Commission launched a public consultation on a draft 'Recommendation on relevant product and service markets within the electronic communications sector susceptible to ex ante regulation in accordance with Article 15 of Directive 2002/21/EC (2002/C 168/09)', in accordance with Article 15(1) FWD.[13] The draft was made available on the web site of the Commission.[14]

The Commission draft identified 12 markets corresponding to the aggregate market approach proposed by the expert study. The Commission mentioned separately retail and wholesale markets. It added four markets: two retail fixed telephony markets, access to the local fixed networks for broadband services as well as access and call origination on mobile networks (see Table 5.2).

*Table 5.2 Markets for* ex ante *regulation identified in the draft Recommendation of 17 June 2002*

**Retail**
- The provision of access to the public telephone network at a fixed location
- The provision of services over the public telephone network at fixed locations
- The provision of leased lines to end-users

**Wholesale**
- Call origination on the public telephone network provided at a fixed location
- Call termination on individual public telephone networks provided at a fixed location
- Transit services in the fixed public telephone network
- Access and call origination on public mobile telephone networks
- Call termination on individual mobile telephone networks
- Local dedicated capacity (or terminating segments of leased lines)
- Access to the local fixed public telephone network for broadband services, including unbundled access to the local loop
- Broadcasting transmission services and distribution networks
- The national market for international roaming services on public mobile telephone networks

The draft Recommendation acknowledged that additional market segmentation could be justified 'on the basis of demand or supply characteristics and the possibilities for substitution'.[15] It stated moreover that 'In the light of the consultation, alternative or additional segmentation may be considered' and that 'The finally adopted Recommendation will specify which market segmentations are identified'.[16]

Market definitions in accordance with the principle of competition law do not encompass 'market segments'. Two products or services are either part of the same market or they are not, depending on whether they are substitutes either from a demand- or a supply-side point of view. The concept of 'market segments' was only introduced provisionally as a consequence of the lack of sufficient information to decide on the market definitions before the public consultation and seek further information from the relevant interested parties.

The possible narrower market definitions mentioned were the following:

1. As regards retail access to the public telephone network at a fixed location, residential and business markets, and possibly further narrower markets.
2. As regards retail publicly available telephone services provided at a fixed location, possible narrower market definitions were first retail markets for local, regional or national and international calls. In addition, separate markets could be envisaged for specially tariffed calls, such as those provided using number translation services, so that a market is defined with respect to the terminating end of the call.
3. As regards wholesale local access for the purpose of providing broadband Internet services, the market for wholesale bit-stream access was considered as a possible distinct market from the market for unbundled access (including shared access) to metallic loops and sub-loops.
4. As regards local dedicated capacity (wholesale local or terminating segments of leased lines), the draft Recommendation mentions the possibility of having narrower market definitions according to capacity, such as lower and higher bandwidth, 'particularly for example where there are constraints with respect to supply substitution in providing bandwidth in excess of 2 Mb/sec'.
5. As regards broadcasting transmission services and distribution networks, to deliver broadcast content to end-users, the possible existence of a specific market for the carriage of broadcasting (audio and visual) services between transmission masts was considered 'where such a market is distinct from a relevant leased lines market'.
6. As regards the market for wholesale transit services in the fixed public telephone network, the following more narrowly defined markets were considered: the market for local tandem conveyance (conveyance between local exchanges and switches to which the interconnected operator is connected); the market for interconnection circuits; third-party transit services (for example the conveyance service provided by an operator to convey calls from one network to another when the networks are not directly interconnected with each other).

## 1.4 The Public Consultation

Following the publication of the draft Recommendation, the Commission received 81 comments from NRAs (12), broadcasters (5), from fixed telecoms operators or associations representing them (28), mobile operators or associations of such operators (17), from cable operators and their association (5) as well as from ISPs, consultants, a user association and other associations with broader membership (14).

The comments regarding the definition of markets were to a large extent 'polluted' by considerations on whether to include or not these markets in the Recommendation. The market players seemed sensitive to this aspect rather than to the precise definition of product or service markets.

Among others, the following comments were made.

First, as regards the markets for access to fixed telephony and services over the fixed telephony network, most of the respondents (9) spoke in favour of more narrowly defined markets at the retail level, based on the type of customer (for example, business, SME – small and medium-sized enterprise – or residential users) and/or the type of service (local, national or international call, directory queries, operator-assisted, emergency call, specific cities and /or regions).

Second, in respect of the provision of retail leased lines, most of the respondents (11 in total) were supportive of further market definitions according to the type of route (local, metropolitan, rural, national, specific cities and/or regions), the bandwidth (above or below a certain threshold, for many set at 2 Mb/s), or type of line (analogue, digital). Many respondents asked for clarification on the technical details for the markets defined.

Third, in relation to the fixed telephony call origination and the fixed telephony call termination markets, many respondents (8) asked for clarification on the technical/geographical boundaries of the market, for example, whether origination would include tandem switches or not. Also, it was pointed out that local-tandem conveyance might be defined as a separate market, or as part of origination or termination, depending on the circumstances.

Fourth, as regards wholesale local access, almost all of the respondents (27) pointed out that referring only to Internet broadband services is misleading, as broadband services encompass a number of different applications of which access to the Internet is merely one. Furthermore, it was pointed out by some respondents that bit-stream access cannot be defined as a segment of the broader wholesale local access for broadband services, as it represents an alternative to xDSL (digital subscriber line applications) and uses different parts of the network.

Fifth, relating to the market for the provision of wholesale leased lines, most of the respondents (8) asked for clarification on the term 'local', in particular in respect of whether trunk segments would be included in the market. One respondent asked that a market for trunk dedicated capacity be defined.

Sixth, as regards the proposed market for access and call origination to public mobile telephone networks, including carrier selection, it was stated that mobile access and call origination are separate markets.

Seventh, in relation to the proposed market for call termination on individual public mobile telephone networks, a number of respondents asked for a definition of a separate market for mobile data telephony (because of the growing importance of SMS).

Eighth, as regards the market for transit services in the fixed public telephone network, a number of respondents (13) asked for fixed voice telephony call conveyance to be added to the wholesale origination and termination markets. Others pointed out that it should be defined on a route-by-route basis.

Ninth, for what concerns the market for broadcasting transmission services, some of the respondents (8 in total) asked for it to be split into terrestrial transmission and other types of transmission.

Finally, one comment highlighted that the Commission services applied an implicit criterion – admitted at the public hearing organized by the Commission on 3 July 2002 – that is, that they had not included markets they deemed competitive in a majority of Member States. This criterion was challenged by another comment stating that in circumstances where markets become competitive in some Member States and not in others, the relevant markets should be kept on the list.

## 1.5 Main Lessons from the Public Consultation

It appears from a comparison of the list in the draft Recommendation published on 18 June 2002 and the wording adopted on 11 February 2003 that the Commission followed a number of requests to define markets more narrowly.

This was first the case with the markets for call origination on the public telephone network provided at a fixed location and for the provision of services over the public telephone network at fixed locations. These markets were split between residential and business markets.[17] The services markets were further split between local/and or national and international.

The market for access to the local fixed public telephone network for broadband services, including unbundled access to the local loop, was split in two markets: the market for wholesale unbundled access (including shared access) to metallic loops and sub-loops, for the purpose of providing broadband and voice services and the market for wholesale broadband access. In addition, the market for terminating segments of leased lines was complemented with a market for wholesale trunk segments of leased lines.

In narrowing its market definitions the Commission moved towards the classical competition-law analysis which starts from the narrowest possible market definition.[18] Starting from an incorrect point (such as, for example, a market defined too broadly and encompassing all publicly available telephone services provided at fixed location) could have led to the wrong conclusion at the subsequent stage of the analysis of market power. At that stage, an incumbent could be considered as no more dominant on that market because it lost substantial market shares as regards international telephone services and on the business segment.[19] Or vice versa, an operator which is no longer dominant in a narrow market section of the aggregated market may still be subject to obligations which relate to a greater market, because the narrower market has not been defined. This was in particular the fear of the incumbents regarding the regulatory obligations imposed on them relating to the provision of bit-stream access to their xDSL services. This is why a separate 'wholesale broadband access' market was defined. This market covers, according to the Recommendation, not only 'bit-stream' access provided by the incumbents but also wholesale access provided over any other infrastructures, if and when they offer facilities equivalent to bit-stream access. Consequently, in those Member States where the latter provide a genuine alternative to access to the incumbent's networks, the NRA will have to assess whether the incumbent remains dominant. This could be the case when, due to limited geographical coverage of the competing platforms, service providers have no real alternative way of reaching the final customers in major parts of the country concerned.

However, the Commission did not narrow its market definitions as far as requested by certain comments. The result is that the list contains on the one hand very narrow market definitions as that of call termination on individual fixed and mobile telephone networks[20] and on the other hand broad definitions as that of 'the broadcasting transmission services to deliver broadcast content to end user'.

As regards the latter market, the Commission decided not to divide it into distinct markets according to the platform used, notwithstanding clear requests made by interested parties.

The explanatory memorandum, however, acknowledges that demand and supply substitution conditions between the different delivery platforms may be such that the feasibility of switching between platforms is limited. It states that in that situation a hypothetical monopolist on one platform may not necessarily be constrained by the activities of operators of other platforms. In such cases the NRA concerned will have to subdivide the market into distinct markets. This could occur in several Member States given that satellite broadcasting is not a substitute for terrestrial broadcasting in all of them.

## 2. SPECIFICITY OF THE MARKET DELINEATION APPROACH FOLLOWED

### 2.1 The Market Definitions in the Recommendation are Genuine Market Definitions

Recommendations under Article 249(5) of the Treaty are not legally binding. In this case, Article 15(3) of the Framework Directive requires NRAs to, *inter alia*, take utmost account of the Recommendation when defining the relevant markets. Although the Commission Guidelines for market analysis and the assessment of significant market power (hereafter 'the Guidelines') acknowledge that NRAs enjoy considerable discretion in the exercise of their powers due to complexity of interrelated factors that must be assessed concerning the economic, factual and legal elements of identified markets,[21] they will have the burden of proof in case they envisage diverging market definitions.

But may NRAs limit themselves to using the product and services market definitions of the Recommendation to start assessing whether there are dominant operators in these markets? Or have they to start a new product market analysis? Article 15(3) FWD is not very helpful in this regard since it only states that NRAs 'shall, taking the utmost account of the recommendation and the guidelines, define relevant markets appropriate to national circumstances, in particular relevant geographic markets within their territory'. On the one hand, NRAs must take utmost account of the list of the Recommendation. On the other hand, the words 'define .. in particular relevant geographic markets' implies *a contrario* that they must also define the product and service markets.

As a matter of fact, Article 15(3) FWD must be read in conjunction with Article 6 of the Directive. This Article requires NRAs to 'give interested parties the opportunity to comment on the draft measures [which have a significant impact on the relevant market] within a reasonable period'. The definition of a relevant product or service market for the sake of imposing obligations on the dominant player in that market will have a significant impact on that market. One can assume that an NRA may start its market analysis on the basis of the market definitions in the list. However, given that using the market definitions of the Recommendation is a measure that falls within the scope of Article 6, it will have to submit this decision to comments of interested parties. If some of those invoke arguments showing that market definitions of the Recommendation do not correspond to the national circumstances, the NRA will have to assess these claims. It will therefore have to deepen the assessment of the Recommendation. Consequently, one can conclude that Article 15(3) FWD introduces a legal presumption *juris tantum*. NRAs are not bound to redefine services and product markets, except if operators bring in evidence showing that the definitions of the Recommendation do not correspond to the situation of the relevant national market.

Obviously, if NRAs have indications that this is the case, they may redefine the relevant markets *ex officio*.

This interpretation of Article 15(3) FWD is the only one possible. Why would Article 15(1) FWD indeed require the Commission to 'define markets in accordance with the principles of competition law' if NRAs would in any case have to redefine these product and services markets? The requirement would in this case lead to duplication and waste of resources, which can obviously not be the aim of Parliament and Council. In addition, NRAs have limited resources. Asking them to assess all possible alternative market definitions of the markets listed in the Recommendation would lengthen and render more complex a process aimed at increasing market certainty. Finally, if NRAs had to redefine all markets, this would necessarily lead to divergences and go against the objective of further harmonizing the national approaches.

Consequently, where an NRA uses the product definition of the Recommendation, it is not bound to provide a justification for the market definition used. It will be up to the undertaking challenging the market definition to show that another is required.

## 2.2 Use of the SSNIP Test

The delineation of the markets of the draft Recommendation is not based on SSNIP tests, nor based on real market data. The Commission defined markets in its Recommendation based on what it considered to be the characteristics of supply of the relevant services most common to the Member States. Indeed, the Commission could not, for this exercise, as it can under its antitrust powers, request undertakings to provide it with all necessary information to proceed to its analysis. No such power was given to the Commission for the sake of the defining product and services markets in its Recommendation. In addition, the Commission had less than a year to list all markets for *ex ante* regulation and could not afford – as in antitrust cases – to continue refining its analysis on the basis of a contradictory procedure.

Dr Sjoerd Van Geffen,[22] in a comment made during the consultation period, considered that such an approach is not in line with the principles of competition law.

> The European Commission obviously did not define product markets by applying the SSNIP-test for the initial Recommendation, in sharp contrast to the explicitly stated goal of market delineation according to normal competition law principles. A product market delineation exercise typically results in a package of products which are close substitutes, whereas Annex I of the Framework Directive seems to contain only single products.

On the other hand, he recalls that a

> danger in calling an aggregated set of similar relevant markets a (single) 'relevant market' itself, is that it may not be obvious enough that this is where the analysis should stop, e.g. even if there is a very high degree of substitution between such an aggregated set of relevant markets and other products/services when applying a SSNIP-test, the latter products/services should not be added to the 'relevant market', since these other products/services are not substitutes for the products/services in the real relevant markets.

However, the Guidelines state that

> According to settled case-law, the relevant product/service market comprises all those products or services that are sufficiently interchangeable or substitutable, not only in terms of their objective characteristics, by virtue of which they are particularly suitable for satisfying the constant needs of consumers, their prices or their intended use, but also in terms of the conditions of competition and/or the structure of supply and demand on the market in question. Products or services which are only to a small, or relative degree interchangeable with each other do not form part of the same market. NRAs should thus commence the exercise of defining the relevant product or service market by grouping together products or services that are used by consumers for the same purposes (end use).[23]

The Guidelines thus do not exclude a theoretical analysis based on common sense. Point 48 confirms this:

> an NRA, in addition to considering products or services whose objective characteristics, prices and intended use make them sufficiently interchangeable, should also examine, *where necessary* [author's emphasis], the prevailing conditions of demand and supply substitution by applying the hypothetical monopolist test.

Point 49 makes explicit that

> *Where available* [author's emphasis], an NRA should examine historical price fluctuations in potentially competing products, any records of price movements, and relevant tariff information. In such circumstances evidence showing that consumers have in the past promptly shifted to other products or services, in response to past price changes, should be given appropriate consideration. In the absence of such records, and where necessary, NRAs will have to seek and assess the likely response of consumers and suppliers to a relative price increase of the service in question.[24]

However, the comments of Dr Van Geffen were relevant at least in one regard: the market definitions listed in the draft were reviewed and narrower market definitions – as advocated by the comments quoted – introduced (see above).

## 2.3 The Markets are Delineated with a Specific Purpose

The list of the markets in the Recommendation has a specific role in the new regulatory framework for electronic communications in Europe. The markets delineation is the starting-point of the process provided for in the new framework to phase out sector-specific regulation. Indeed, 'Sector specific regulation should', as Commissioner Monti said, 'be the exception, antitrust rules should be the norm. For an operator, being subject to ex ante regulation is a heavy burden because it limits commercial freedom without any evidence of an abuse having occurred.'[25] Article 16(3) of FWD lays down that where a market is deemed to be effectively competitive, NRAs shall not impose or maintain any sector-specific regulatory obligations and shall, where applicable, remove existing obligations. Article 16(3) has been seen as a 'deregulation mechanism'.[26]

Given this difference in purpose, there was no need for the Commission to identify all possible, narrow markets of the electronic communications sector. It could limit itself to analysing, with a competition-law approach, the areas currently regulated. It is therefore easy for Larouche to claim that the 'market definition exercise provided for (in) the Framework Directive bears little more than terminological resemblance to market definition under competition law'.[27]

Another consequence of the specific purpose of the listing of markets in the Recommendation is that it will be reviewed regularly. Conversely, market analysis under the antitrust rules are not intended to be reviewed, besides the case of legal challenges. The market delineation under the new framework must be operational. It need not be as exhaustive as in, for example, merger cases. The Commission will indeed be able to refine its market delineation on the basis of the experience in Member States and in particular of notifications under Article 7(4) FWD by NRAs who, because of national specificities, deviate from the market delineation in the list of the Recommendation.

## 3. THE APPLICATION OF THE INITIAL RECOMMENDATION

### 3.1 The NRAs will Complement the Market Delineation of the Recommendation

The market delineation process is not completed with the publication of the Recommendation. The power to identify the relevant geographical markets is entrusted to the NRAs under Article 15(3) of the FWD.[28] Under the 'classic' competition-law market analysis, product and geographical market analysis are closely intertwined. For example, an NRA could identify, in case demand and supply substitution exist in certain areas and not in others, distinct geographical retail business markets for access to the public telephone network at a fixed location and publicly available local/and or national telephone services provided at a fixed location.

Furthermore, NRAs can be induced to define narrower markets than those listed in the Recommendation on the basis of the situation in their country. For example, the wholesale transit services in the fixed public telephony network market cover among other things the conveyance of calls between networks in the same Member States which are not directly interconnected with each other (indirect interconnection) where incumbent telecom operators will remain dominant, and the conveyance of international calls, which is probably already competitive in most Member States. An NRA could conclude from its assessment that these services are part of distinct markets and phase out regulation on the latter.

The list of the Recommendation is thus only a first step. Besides, in addition to the power of the NRAs to refine the list of markets, they have also the power to impose remedies on activities differing from the market as defined.[29] For example, no market for collocation is listed and defined in the Recommendation, but NRAs are entitled to maintain collocation requirements on operators dominant in call origination or termination markets.

### 3.2 The Second 'Deregulation Mechanism'

The assessment by the NRAs of the competitive situation of the markets listed in the Recommendation will probably not lead to less regulation. The reason is that the FWD applies a narrow interpretation of the concept of 'effective competition'. According to its recital 27, a finding that an operator enjoys a single or joint dominant position on a market means that no effective competition exists on that market. Even if an undertaking were not likely to abuse its dominant position, NRAs would be required to impose on such undertakings appropriate specific obligations, or maintain such obligations where they already exist. This interpretation was criticized, since it would find 'no support in the Court's case law on the concept of dominance as entailed in Article 82 EC ... According to the Court's case law on the concept of dominance it is the abuse and not the dominant position as such which hinders effective competition.'[30] Requiring, on the one hand, NRAs to impose obligations on any dominant undertaking in markets that are not effectively competitive and defining, on the other, the latter as markets where there is a dominant undertaking could lead to unjustified regulation. To avoid this, the new framework contains another 'deregulation mechanism', the periodic review by the Commission of the list of markets to be regulated *ex ante*.

Under this second 'deregulation mechanism' the Commission will, on the occasion of future revisions, exclude those markets from the list that do not meet the following three criteria: (a) existence of high barriers to entry, (b) absence of dynamic competition in spite of these barriers, and (c) the fact that competition rules are not sufficient to address the perceived market failures, and this even if incumbents are still dominant on those markets.[31]

By providing the criteria for future revision in the initial Recommendation, the Commission provided legal certainty to the market players concerned. Their application will probably not affect the listing in the short and medium terms of the markets in the initial Recommendation, except when they do not already meet these criteria:

1. The existence of high and non-transitory entry barriers will only change in the case of major innovations or technological progress. The initial Recommendation gives the example of the provision of local access networks to fixed locations. One can assume that these networks will not be duplicated in the foreseeable future in the large majority of Member States. Similar observations can be made for nearly all markets in the list of the initial Recommendation.
2. The second criterion admits only those markets whose structure does not tend towards effective competition within the relevant time horizon. The initial Recommendation mentions that this might be the case in markets with a limited but sufficient number of undertakings having diverging cost structures and facing price-elastic market demands. Indicators are whether market shares change over time and/or whether falling prices may be observed. As regards the markets listed in the initial Recommendation, and taking into account the current financial position of the entrants in the various Member States, one can doubt that such an effect may be observed in the short term.
3. The third criterion is that application of competition law alone would not adequately address the market failure concerned. The application of this criterion will depend on the assessment of the sufficiency of competition law in reducing or removing the above-mentioned barriers or in restoring effective competition. As highlighted during the public consultation, the latter assessment is not easy. Several requests for clarification were made. Concerns were expressed that this test of insufficiency was not well defined.

As a matter of fact, it is not possible to define *a priori* the circumstances in which competition-law remedies might be sufficient. The explanatory memorandum only gives[32] a number of examples:

> such circumstances would for example include situations where the compliance requirements of an intervention to redress a market failure are extensive (e.g. the need for detailed accounting for regulatory purposes, assessment of costs, monitoring of terms and conditions including technical parameters etc) or where frequent and/or timely intervention is indispensable, or where creating legal certainty is of paramount concern.

This is in line with the perception that 'issues such as unfairly high or low prices are better dealt with when tariffs are in review before their introduction (ex-ante regulation)'.[33]

The explanatory memorandum provides a practical solution to decide on the matter: 'NRAs should consult with their NCA and take into account that body's opinion when deciding whether use of both complementary regulatory tools is appropriate to deal with a specific issue, or whether competition law instruments are sufficient.' Indeed, while the wording of the initial Recommendation casts doubts on the efficiency of competition-law instruments, the efficiency of sectoral intervention may also not be overstated. NRAs are not in a position to assess systematically all tariffs from the regulated operators. This was illustrated by the outcome of the Commission competition inquiries relating to interconnection tariffs, accounting rates and leased lines. This intervention was necessary to induce 'companies adjusting their tariffs or national regulatory authorities taking over the case'.[34] As Braun and Capito stated, 'there remain areas where Member States have not implemented all provisions of the relevant directives satisfactorily. This clearly demonstrates the need for applying the competition rules alongside sector-specific regulation. This need might even be increased with the new regulatory package.'[35]

## 3.3 The 2004 Review

The initial Recommendation announces a first review no later than 30 June 2004,[36] less than a year after the new framework becomes operational. It is therefore likely that, besides in the case of the wholesale market for access and call origination on public mobile telephone networks, neither of the first two criteria will justify the removal of any market off the list. The issue will be the application of the third criterion, or as Larouche rightly states, 'considering that competition law is there and applies in any event, does the use of sector specific regulation bring added value?',[37] that is, is sector-specific intervention indispensable?

This test is quite different from reviewing whether there have been competition-law cases regarding the relevant market. Under the current framework, it occurred that parallel procedures under *ex ante* regulation and competition law were opened regarding the same problems. Such overlapping procedures addressing the same concern are not efficient and complaints were often referred to the NRAs. In most cases, complainants only went to the NRAs, which they considered as the easiest way to obtain a remedy, since it did not require the proof of an abuse.

As a result, a kind of specialization emerged between both sets of procedures. For example, mobile termination and roaming tariffs were challenged under the competition rules because they did not fall within the scope of sector-specific rules. The latter resulted from the fact that under the former directives, sector-specific rules only applied to well-defined market areas. Under the new framework, the list of markets to be regulated has been extended. There is therefore a risk that the latter would be used to address even those situations that can be dealt with by Article 82 prohibiting abuses of dominant position, only because the burden of proof is lighter for the application of sector-specific regulation. Indeed, to be subject to *ex ante* regulation it is enough to be dominant.

This could lead to a 'vicious circle'. If the number of markets on which sector-specific regulation is applied is extended, and competition authorities accept a *modus vivendi* whereby the NRA is given priority to deal with the cases, the third criterion could possibly never become applicable. Indeed, given the absence of any case-law relating to these markets as well as the existence of a corpus of sectoral obligations, it will become difficult to show that the latter can be phased out.

The application of the third criterion will mainly fall upon the Commission, either when reviewing the Recommendation or when assessing notifications by NRAs under Article 7 FWD. Such reviews will be accompanied with a detailed justification and will imply a public consultation. This procedure will therefore be crucial to further specify the third criterion, taking into account the specifics of the relevant market.

## 4. CONCLUSION

This chapter has provided a brief outline of the rationale behind the approach taken by the Commission on delineating electronic communications markets for *ex ante* regulation.

While it is difficult to evaluate whether the current list of markets can represent a way to address the concerns of all interested parties (an objective which would seem very hard to achieve in any case), the approach to market definition established by the new framework provides the opportunity for *ex ante* regulation to evolve with the help of competition analysis tools. This will provide legal certainty and consistency across the EU. The latter will contrast with the situation in the USA, which is characterized by a 'patchwork of interconnection regimes, which is based on existing distinctions between local and long distance service, interstate and intrastate service, and basic and enhanced services'.[38] The periodic review of the list of markets will allow further refining the delineation of markets for *ex ante* regulation, taking into account the evolution of each specific market.

The assessment by the Commission of the possible sufficiency of competition-law remedies to remove markets for *ex ante* regulation or veto Member States adding markets will also be determinant for the future regulation of the sector. Only a forward-looking application of this test will ultimately ensure that sector-specific intervention in the economy can be further fine-tuned and limited to the situations where it is indispensable to ensure a level playing field as well as the other objectives of the EU electronic communications policy.

## NOTES

* The views expressed by the author are purely personal and do not necessarily reflect the official position of the European Commission.

1. An example of a new service resulting from technical innovation is pre-paid international roaming, on the other hand, regulatory innovation ('collecting model') leads in several Member States to the emergence of a specific market of number translation services provided by telcos to ISPs.
2. European Commission, 'The results of the public consultation on the 1999 Communications Review and Orientations for the new Regulatory Framework', COM (2000)239 (26 April 2000), p. 18.
3. Ibid.
4. European Commission, *Directive 2002/21/EC of the European Parliament and of the Council of 7 March 2002 on a common regulatory framework for electronic communications networks and services*, OJ L 108, 24/04/2002), p. 33.
5. Larouche (2002), p. 137.
6. See Chapter 2.
7. IP/03/221 of 12 February 2003, 'Commission puts in place final piece of new framework for electronic communications'.
8. OJ L 114 (08.05.2003), Commission Recommendation of 11 February 2003 on relevant product and service markets within the electronic communication sector susceptible to *ex ante* regulation in accordance with

Directive 2002/21/EC of the European Parliament and the Council on a common regulatory framework for electronic communication networks and services, pp. 45–9.
9. Cf. on p. 30 of the explanatory memorandum, as regards the market for mobile call origination: 'However, the level of competition generally observed in this market at the retail level indicates that ex-ante regulatory intervention at a wholesale level may not be warranted. The Commission does not at this stage anticipate that this market will be included in future revisions of this Recommendation.'
10. For example on p. 16 of the explanatory memorandum: 'In respect of outgoing telephone calls, end users do not perceive local, national and international calls to be substitutes for each other. Therefore, it can be argued that these constitute separate retail markets.'
11. For example on p. 23 of the explanatory memorandum: 'Given the preceding description of retail markets, it is possible to define and identify wholesale markets for *ex ante* regulatory analysis. In order for access to data and related services to be supplied to an end-user at a fixed location, an appropriate transmission channel is required, capable of passing data in both directions and at rates that are appropriate for the service demanded. Therefore, an undertaking providing services to end-users needs to build, establish or obtain access to a transmission channel to an end-user location'.
12. Contract No. 48548, 'Market Definitions For Regulatory Obligations in Communications Markets', prepared by Squire, Sanders & Dempsey L.L.P., WIK and the Centre for Management under Regulation of the Warwick Business School, University of Warwick, under the direction of Professor Martin Cave.
13. European Commission, OJ C 168, 13/07/2002), p. 19.
14. http://europa.eu.int/information_society/topics/telecoms/news/index_en.htm
15. European Commission, 'Draft Recommendation on relevant product and service markets' (17 June 2002), p. 8.
16. Ibid.
17. Braun and Capito (2002b, p. 353) express their reservations against such distinction between residential and business, which was also rejected by RegTP as regards the market for calls between Germany and Denmark.
18. See Gual (2003). See also, for example, the current guidelines issued by antitrust authorities, such as OFT, *Market definition, Guidelines on the Competition Act 1998* (1998), or US Department of Justice, *1992 Horizontal Merger Guidelines – Revisions to Section 4*, FTC 1997.
19. See also Pleatsikas and Teece (2001). A typical example of a misleading market definition is the one based on the inclusion of products or services using artificially inflated prices. This problem is known in economic theory as the 'cellophane fallacy', because it arose for the first time when the US Department of Justice was looking at a case which involved producers of wrapping material. In this landmark case, the Supreme Court erroneously concluded that Du Pont did not have significant market power because the Court evaluated the elasticity of demand for cellophane at the monopoly equilibrium, at which the elasticity was far higher than at the competitive equilibrium. The Court's error is commonly referred to as the 'cellophane fallacy'. In other words, if in a given market the prevailing price is not the competitive one, starting from a broad market definition may lead to the *de facto* inclusion of products which are non-competitively priced, without questioning the degree to which those products would be considered substitutes looking at narrower markets.
20. Although this is questioned by a number of economists. See, for example, P. Crocioni (2001): 'This view is more an assertion than a result of economic analysis' (p. 47).
21. OJ C 165/6, 11/07/2002, Commission Guidelines, para. 21.
22. Researcher, Institute for Information Law (IViR, http://www.ivir.nl/index-english.html); SEO Amsterdam Economics (http://www.fee.uva.nl/seo/english/index.htm).
23. OJ C 165/6, 11/07/2002, Commission Guidelines, point 44, p. 11.
24. Ibid., p. 12.
25. IP/O3/221 p. 1.
26. Braun and Capito (2002b), p. 322; see also European Commission Press Release IP/00/749 (12 July 2000).
27. Larouche (2002), p. 137.
28. Except as regards transnational markets (Art. 15(4)).
29. Obligations can be imposed on any ancillary service of those of the listed markets. According to the draft Recommendation 'metered and un-metered calls are generally considered to be in the same retail market, and the wholesale call origination market is considered to include both telephone calls and dial-up Internet calls … an obligation concerning the provision of un-metered wholesale call origination may be an important remedy to deal with SMP on the wholesale call origination market. Also relevant for any obligations imposed are ancillary services such as billing' (p. 9).
30. Braun and Capito (2002b), p. 322.
31. See recitals 9–15 of the Recommendation of 11 February 2003.
32. pp. 11–12.
33. Braun and Capito (2002), p. 67.
34. Ibid., p. 68.

35. Braun and Capito (2002a), p. 65.
36. Recital 21 of the Recommendation.
37. Larouche (2002), p. 135.
38. Uri (2001), p. 613.

## REFERENCES

Braun, J.D. and R. Capito (2002a), 'The Emergence of EC Telecommunications Law as a New Self-Standing Field Within Community Law', in Christian Koenig et al. (eds), *EC Competition and Telecommunications Law*, Boston: Kluwer Law International, pp. 51–69.

Braun, J.C. and R. Capito (2002b), 'The Framework Directive', in Christian Koenig et al. (eds), *EC Competition and Telecommunications Law*, Boston: Kluwer Law International, pp. 309–58.

Crocioni, P. (2001), 'Should telecoms liberalisation stop at call termination?', *Telecommunications Policy*, **25** (1–2), pp. 39–58.

Gual, J. (2003), 'Market definition in the telecoms industry', in P. Buigues and P. Rey (eds), *The Economics of Antitrust and Regulation in Telecommunications*, Cheltenham, UK and Northampton, MA: Edward Elgar.

Larouche, P. (2002), 'A closer look at some assumptions underlying EC regulation of electronic communications', *Journal of Network Industries*, pp. 129–50.

Pleatsikas, C. and D. Teece (2001), 'The analysis of market definition and market power in the context of rapid innovation', *International Journal of Industrial Organization*, **19**, pp. 665–93.

Uri, N.D. (2001), 'Monopoly power and the problem of CLEC access charges', *Telecommunications Policy*, **25** (8–9), pp. 611–23.

# Section 3

# Economic Analysis of Collective Dominance in Telecoms Markets

# 6. Collective dominance and the telecommunications industry*

**Patrick Rey**

## 1. INTRODUCTION

The development of the concept of collective dominance fills an important gap in European competition policy, since threats to competition can arise even in the absence of any single dominant firm. This is particularly the case when firms engage in what economists refer to as tacit collusion, since their behaviour may then approximate that of a single dominant firm.

While some practices can involve positive as well as negative aspects, thereby generating arguably difficult tradeoffs, there is instead a consensus – at least in practice – that collusion should be forbidden.[1] Unfortunately, it does not suffice to 'forbid' collusion, which can be 'implicit' rather than 'explicit'. Explicit collusion, where firms engage in written or oral agreements, organize meetings to design and implement collusive mechanisms, and so forth, can be caught under Article 81 if the appropriate evidence is recovered. Implicit collusion instead does not involve any explicit agreement, and arises from the mere repetition of competitive interactions. The concept of collective dominance provides a basis for antitrust intervention in such cases.

This chapter discusses the scope for antitrust intervention against such tacit collusion. Section 2 first studies the circumstances in which such collusion can arise, and the forms it can take. Section 3 discusses the alternative courses of action available to competition authorities, and Section 4 focuses on specific issues for the telecommunications industry.

## 2. RELEVANT FACTORS FOR TACIT COLLUSION

### 2.1 The Logic of Tacit Collusion

Tacit collusion may arise when the same firms repeatedly compete in the same markets. A firm may then have an incentive to maintain high prices because it expects that if it does not do so, the rivals will lower their own prices in the future. Whether firms can in fact maintain high prices depends on four main factors:

1. *How much each firm gains from undercutting its rivals*. Tacit collusion is clearly easier to sustain when the gains from undercutting are low. This puts limits on the level of collusive prices that can be sustained, since the gains from undercutting depend among other things on the price cost margin and the elasticity of the firm demand.[2] The degree

of product differentiation may thus matter, as well as the nature of competition – for example, the benefits from undercutting are typically smaller when firms compete in quantities than when they compete in prices.

2. *How much such a firm would lose in the future if its rivals retaliate.* The long-term profit loss from a deviation is the difference between the long-run collusive profit that the firm would obtain by sticking to collusion, and the long-run profit it obtains under the market conditions that may prevail if undercutting occurs. The reaction of firms to perceived undercutting of the collusive price is often referred to as 'retaliation', although it needs not always take the form of aggressive actions against the firm.

   Indeed, a simple form of retaliation consists in the breakdown of collusion and the restoration of 'normal' competition and profits. Firms then anticipate that collusive prices will be maintained as long as none of them deviates, but if one attempts to reap short-term profits by undercutting prices, there will be no more collusion in the future, at least for some time. Firms may then adhere to the current collusive prices in order to keep the collusion going, in which case collusion is self-sustaining. This form of collusion has a simple interpretation: firms trust each other to maintain collusive prices; but if one of them deviates, trust vanishes and all firms start acting in their short-term interest.

   However, more sophisticated forms of retaliation may inflict tougher punishments and thereby allow sustaining higher collusive prices. For example, retaliation may include temporary price wars, leading to profits below 'normal' levels for some period of time.[3] It may also include actions that are specifically targeted at reducing the profits of the deviant firm.[4] Alternatively, firms may refuse to co-operate on other joint policies (such as joint ventures or joint distribution arrangements) or in standard-setting processes. The retaliatory power of rivals thus depends on market specificity, and determines to a large extent the ability of all parties to maintain tacit collusion.

3. *How likely is undercutting by the firm to lead to such retaliation.* Clearly, if there is little chance that undercutting triggers retaliation, the fear of losing the collusive profit will be an ineffective deterrent. The probability that undercutting by one firm triggers retaliation depends mostly on firms' ability to monitor each other's behaviour, and thus on market transparency.[5] The extent of publicly available information on prices and quantities is thus highly relevant, but other dimensions such as market stability (demand and cost volatility, frequency of innovation and so forth) or the degree of similarity in cost and demand conditions matter as well.

4. *How much the firm discounts future profit losses relative to today's gains.*[6] If firms care mostly about current profits, they tend to focus on the short term and thus 'ignore' the consequences of retaliation. They thus have a strong incentive to undercut and collusion is uneasy to sustain. The relative weight of current and future profits in the firm's objectives depends among other things on the market real interest rate. Future profits matter more with low interest rates, which facilitates collusion. Another key determinant is the delay before competitors react, which depends on monitoring and structural factors such as adjustment costs, long-term contractual arrangements, and so forth.

## 2.2 Relevant Market Characteristics

The multiplicity of retaliation and collusive mechanisms creates a potential for collusion in many industries. The main issue is how large this potential is, that is, how credible and effective are the collusive mechanisms and to what extent is collusion likely to emerge.

Economic analysis allows a better understanding of the basic nature of retaliation mechanisms and their common features. It so provides key insights into the structural characteristics that determine the effectiveness of collusive and retaliatory conducts. We shall concentrate on these aspects and discuss the various factors to be considered when evaluating the potential for collusion.

The four factors described above determine the sustainability of tacit collusion. These factors in turn depend on market-specific characteristics, which can be grouped into three categories. The first category includes necessary ingredients for collusion. The second category covers important characteristics that determine whether collusion can be sustained. The last category corresponds to characteristics that are still relevant, although to a lesser extent.

**Necessary ingredients**

Some characteristics have a decisive impact on firms' ability to sustain tacit collusion. These include *entry barriers*, the *frequency of interaction* and the role of *innovation*.

*Entry barriers*

Collusion cannot be sustained in the absence of entry barriers and it is more difficult to sustain, the lower the entry barriers. In the absence of entry barriers any attempt to maintain supra-competitive prices would trigger entry (and so-called short-term or 'hit and run' entry strategies), which would erode the profitability of collusion. In addition, the prospect of future entry tends to reduce the scope for retaliation, since firms have less to lose from future retaliation if entry occurs anyway. More precisely, the prospect of future entry does not affect the short-run benefit that a firm can obtain from a deviation, but it reduces the potential cost of deviation: retaliation is less costly for the deviating firm if entry occurs and dissipates profits, irrespective of past behaviour. Firms are then more tempted to undercut collusive prices and the ability to collude thus declines when the likelihood of entry increases.

*Frequency of interaction*

Frequent interaction and frequent price adjustments facilitate collusion. As already noted, firms could not tacitly collude if they did not anticipate interacting again in the future. Similarly, collusion is unlikely when firms interact only infrequently, since the short-term gains from undercutting a collusive price could then be 'punished' only in a far future.[7] Collusion is instead easier when firms interact more frequently, since they can then react more quickly to deviations and retaliation can thus come sooner.

A similar idea applies to the frequency of price adjustments. When prices adjust more frequently, retaliation can again come sooner; in addition, a cheating firm is unable to take advantage for as long a time as before of its cheating behaviour. Both factors contribute to hinder collusion. Thus, what matters most is not whether the firms are 'selling' in each period or only every now and then, but how frequently they can adjust their prices. The more frequent price adjustments are, the easier it is to sustain collusion.

*Innovation*

Collusion is easier to sustain in mature markets where innovation plays little role than in innovation-driven markets. The reason is that innovations, particularly drastic ones, may allow one firm to gain a significant advantage over its rivals. This prospect reduces both the value of future collusion and the amount of harm that rivals will be able to inflict if the need arises.

Consider for example an industry where, in the absence of any innovative activity, the incumbents would benefit from a secure, stable situation. They would then hesitate before cheating on a collusive conduct, which would trigger a price war and dissipate their future rents. Suppose now that, with some probability, one incumbent makes a drastic innovation, which drives its rival out of the market. If the probability of successful innovation is substantial, the incumbents then anticipate that their market position is short-lived (at least in expected terms); they thus put less emphasis on the cost of future retaliation and are more tempted to cheat on collusion. A similar reasoning applies to the case when the innovation comes from an outsider: retaliation is again less effective when an innovator arrives, whether the innovator is an incumbent or an outsider.

Clearly, there is little scope for collusion in the absence of entry barriers, or if firms interact very infrequently, or else in innovation-driven markets. Therefore, whenever an industry presents one of these features, collusion is unlikely to constitute a significant concern.

**Important factors**

The second group of market characteristics includes key determinants of the scope for collusion. These factors include of course the *number of market participants*, but also the degree of *symmetry* among those participants, the existence of *maverick firms*, of *structural links and co-operative agreements*.

*Number of participants*

Collusion is more difficult when there are more competitors. For one thing, co-ordination is more difficult, the larger the number of parties involved, in particular when co-ordination is only based on a tacit common understanding of the collusive market conducts. For example, identifying a 'focal point' in terms of prices and market shares becomes less and less obvious, particularly when firms are not symmetric.[8]

There is another reason that makes it difficult to collude with many competitors. Since firms must share the collusive profit, as the number of firms increases each firm gets a smaller share of the pie. This has two implications. First, the gain from deviating increases for each firm since, by undercutting the collusive price, a firm can steal market shares from all its competitors; that is, having a smaller share, each firm would gain more from capturing the entire market. Second, for each firm the long-term benefit of maintaining collusion is reduced, precisely because it gets a smaller share of the collusive profit. Thus the short-run gain from deviation increases, while at the same time the long-run benefit of maintaining collusion is reduced. It is thus more difficult to prevent firms from deviating.[9] This impact of the number of competitors is likely to be particularly important when there are few competitors.[10]

*Symmetry*

It is easier to collude among equals, that is, among firms that have similar cost structures, similar production capacities, or offer similar ranges of products. Suppose for example that firms have different marginal costs. The presence of such cost asymmetry has several implications.[11] First, firms may find it difficult to agree to a common pricing policy, since firms with a lower marginal cost will insist on lower prices than those the other firms would wish to sustain.[12] More generally, the diversity of cost structures may rule out 'focal points' for pricing policies and so exacerbate co-ordination problems. In addition, technical efficiency would require allocating market share to low-cost firms, but this would clearly be difficult to sustain in the absence of explicit agreements and side-transfers.[13]

Second, even if firms agree on a given collusive price, low-cost firms will be more difficult to discipline, both because they might gain more from undercutting their rivals and because they have less to fear from a possible retaliation from high-cost firms.[14] Retaliation is indeed less effective when exerted by an inefficient firm against an efficient one, since the ability of the former to compete against the latter is limited. In particular, the inefficient firm will not be able to induce a substantial profit loss on the efficient one without imposing on itself an even larger burden. This means that the retaliation that the inefficient firm will be rationally willing to undertake will impose little discipline on the efficient firm. Thus the incentive to deviate from the collusive conduct of the low-cost firm will be larger than if it faced another low-cost firm.

A low-cost firm will be more willing to collude if it gets a larger share of the collusive profits. However, while this helps to provide incentives for the low-cost firms, at the same time it affects the incentives of the high-cost firms. Thus there is a limit to the possible reallocation of market shares and, while it may help to sustain collusion, it does not restore the same collusive possibilities as if the cost structure were symmetric: there is less scope for collusion with an asymmetric cost structure.

A similar reasoning applies when one firm has a superior product quality, since such a firm has less to fear from retaliation. The argument also extends to other types of cost differences, such as asymmetric production capacities. Capacity constraints potentially affect the sustainability of collusion in two ways. First, a capacity-constrained firm has less to gain from undercutting its rivals. Second, capacity constraints limit firms' retaliatory power. At first glance, capacity constraints may appear to have an ambiguous effect on collusion, since they reduce both the incentives to deviate and the ability to punish such deviations. And indeed, studies that have focused on symmetric capacities[15] have confirmed this apparent ambiguity.[16] The impact of asymmetric capacities is, however, less ambiguous. Compared with a situation where all firms face the same capacity constraints, increasing the capacity of one firm at the expense of the others both increases the first firm's incentive to undercut the others and limits these other firms' retaliatory power. Overall, therefore, introducing such asymmetry hinders collusion.

This insight has been hinted at by several studies[17] and has recently been explored in more detail by Compte et al. (2002), who show that asymmetric capacities indeed make collusion more difficult to sustain when the aggregate capacity is itself quite limited. As for asymmetric unit costs, giving larger shares of the collusive profits to the larger firms alleviates somewhat the tension generated by asymmetric capacities, but it does not eliminate it entirely, and the scope for collusion is none the less reduced.

The fact that the most effective collusive conducts involve asymmetric market shares, reflecting firms' costs or capacities, has an interesting implication: it shows that, while market shares are highly endogenous variables, *market share asymmetry* may provide indirect evidence of more profound asymmetry that tends to hinder collusion.

### *Maverick firms*

The intuition that 'it is easier to collude among equals' also explains the role of so-called 'mavericks.' A maverick firm can be interpreted as a firm with a drastically different cost structure, which is thus unwilling to participate to a collusive action.[18] Consider for example a firm that has a drastically different cost structure, production capacity, or that is affected by different factors than the other market participants.[19] Very often such a firm will exhibit a market conduct that differs from the others, reflecting its different supply conditions. This firm may then be unwilling to be party to collusive conduct – put another way, it would do so only under terms that would not be acceptable or sustainable for the other firms. Of course, this 'asymmetry' can be along other dimensions. For example, a firm may have a stronger preference for the short term and be therefore more tempted to undercut the rivals.[20]

The existence of such a 'maverick' clearly tends to make collusion difficult if not impossible to sustain. It is, however, necessary to identify carefully the origin of the 'maverick' character, in order to determine whether it is an inherent, long-lasting characteristic, or only reflects a transitory situation.

### *Structural links and co-operative agreements*

Structural links can facilitate collusion among firms. For example, cross ownership reduces the gains derived from undercutting the other firm. Joint venture agreements can also enlarge the scope for retaliation – a firm can then for example punish a deviating partner by investing less in the venture.[21] For these reasons, collusion is more likely to appear in markets where competitors are tied through structural links.

In the absence of structural links, simple co-operation agreements can also contribute to fostering collusion. As in the case of joint ventures, these agreements can for example enlarge the scope for retaliation, thereby enhancing the ability to punish deviating partners. This may be particularly relevant for industries such as the telecommunications industry, where competitors need to reach interconnection agreements in order to offer good services. These agreements not only enlarge the scope for retaliation, they also have a direct impact on the operators' pricing strategies. Competitors may then design these interconnection agreements so as to facilitate collusion.

More generally, firms may alter their contractual agreements, either between themselves or with third parties, so as to facilitate collusion. Marketing agreements can constitute good tools to that effect. Jullien and Rey (2002) show for example that producers of consumer goods can resort to Resale Price Maintenance to impose more uniform prices across local retail markets, thereby making it easier to detect deviations from a collusive price. Record companies have been accused of marketing their discs according to simple pricing grids (with only a few categories, instead of personalized prices for each author or composition) for a similar purpose.

**Other relevant factors**

Other factors can have an influence on the sustainability of collusion, although possibly to a lesser extent or in a more ambiguous way. Among these, the degree of market transparency appears to be a key factor. Other factors include product differentiation, the characteristics of demand (demand trend and fluctuations, as well as demand elasticity and buying power), multi-market contact, club effects or the organization of particular markets such as bidding markets. These dimensions are relevant to assessing the plausibility of collusion, particularly when the factors of the first two groups do not suffice to send a clear signal.

*Market transparency*

Collusion is easier when firms observe each other's prices and quantities. Frequent price adjustments give firms the physical possibility to retaliate quickly when one market participant undercuts the others, but such deviation must first be identified by the other participants. As a result, collusion can be difficult to sustain when individual prices are not readily observable and cannot be easily inferred from readily available market data. This, in turn, supposes that some uncertainty affects the market: otherwise any deviation would be detected by the rivals, who would perceive a reduction in their market share.

This observability problem was first stressed by Stigler's (1964) classic paper, and formally analysed by Green and Porter (1984) and Abreu et al. (1985): the lack of transparency on prices and sales does not necessarily prevent collusion completely, but makes it both more difficult to sustain and more limited in scope.

We can illustrate this with Tirole's (1988) version of Green and Porter's model. Starting with the base duopoly model, suppose now that: (i) each firm only observes its own price and sales, but not the others'; and (ii) with some probability, demand vanishes (is equal to zero). Therefore, when a firm is unable to sell in a given period, it can either be because of 'bad luck' (adverse shock on demand), or because another market participant has 'cheated' (undercut the collusive price). As a consequence, perfect collusion is no longer possible. Perfect collusion would require firms to go on with the monopoly price, even after a shock on demand. But then, each firm would have an incentive to undercut the others – and blame the fact that the others did not sell on bad luck.

The best collusive scheme consists in: (i) starting with the monopoly price, and maintaining this price as long as each firm maintains its market share; (ii) whenever a firm is unable to sell, launching a price war for a limited number of time, namely, $T$ periods, before reverting to the monopoly price. The price war is needed and must be sufficiently lengthy (and thus costly) to deter potential cheaters. But this price war can be triggered by pure bad luck, that is, simply because of an adverse shock on demand; firms have thus an incentive to limit the duration of the price war to what is just sufficient to discipline the tacit conduct.

What matters here, however, is not what is directly observed by the firms, but what information firms can *infer* from the available market data. For example, inferring deviations from collusive conduct is easier and requires less market data[22] when the market is stable rather than unstable. Moreover, the delay necessary to obtain reliable data on prices and quantities matters, as well as its nature. For example, professional associations sometimes publish information on prices, productions or capacity utilization rates. It first matters whether this information is about aggregate or individual data, since in the latter case it is easier to identify a deviant firm.[23] The time lag elapsed between the pricing period and the publication period is also important. Even detailed information may not help to sustain collusion if it is available only after a long delay.

*Demand growth*

Collusion is easier to sustain in growing markets, where today's profits are small compared with tomorrow's. Conversely, collusion is more difficult to sustain in declining markets, where tomorrow's profits (with or without retaliation) will be small anyway – in the limiting case where the market is on the verge of collapsing, there is almost no 'future' and therefore no possibility of inducing firms to abide to a collusive conduct.

*Business cycles and demand fluctuations*

These hinder collusion. This is a corollary of the above impact of demand growth or decline. The idea, formally captured by Rotenberg and Saloner (1986) and Haltiwanger and Harrington (1991), is that when the market is at a peak, short-term gains from a deviation are maximal while the potential cost of retaliation is at a minimum. Hence collusion is more difficult to sustain at those times.

To see this, suppose that demand fluctuates from one period to another and, to fix ideas, assume for the moment that demand shocks are independent and identically distributed across periods. In this hypothetical scenario, firms know that they face an uncertain future, but in each period the prospects are the same; the probability of benefiting from a good shock is for example the same in each future period, and likewise for the probability of bad shocks. This in turn implies that the amount of future retaliation to which a firm exposes itself in each period remains the same over time. However, in periods where demand is higher than average, the short-term benefits from a deviation are themselves higher than average. Therefore, in such a period, the firm must trade off higher-than-average gains from deviation against a constant (and thus 'average') level of punishment. Clearly, deviations are more tempting in such a period and, by the same token, collusion is more difficult to sustain than in the absence of demand fluctuations, where both the short-term gains from deviations and retaliation possibility would always remain at an average level.

As fluctuations gain in scale, collusion becomes more and more difficult to sustain, at least in those states where demand is especially high. Firms are then obliged to collude 'less' (by lowering the collusive price) or even abandon any collusion when demand is high. A similar analysis applies to more deterministic fluctuations, as for example in the case of seasonal or business cycles. There again, undercutting rivals is more tempting when demand is high, and the perceived cost of future price wars is lower when the cycle is currently at its top, since retaliation will only occur later, thus in periods of lower demand.

*Product differentiation*

This factor can have a more ambiguous impact on collusion, since it affects both the incentives to undercut the rivals and their ability to retaliate. This is particularly the case when product differentiation consists of offering different combinations of characteristics, possibly at comparable prices but targeted at different types of customers; this corresponds to what economists refer to as 'horizontal differentiation'. Such differentiation aims at segmenting customers, in order to gaining market power over specific segments by creating customer loyalty: a customer may then be reluctant to switch away from a favourite brand, even if s/he would benefit from a small price reduction by turning to an alternative brand. This segmentation strategy affects the scope for collusion in two ways. First, it limits the short-term gains from undercutting rivals, since it becomes more difficult to attract their customers. Second, it also limits the severity of price wars and thus the firms' ability to punish a potential deviation. Overall, therefore, the impact of horizontal differentiation appears quite ambiguous.

Indeed, the economic work on this issue has shown that collusion may become easier or more difficult, depending on the exact nature of the competitive situation (for example, competition in prices versus competition in quantity).[24] Raith (1996) notes, however, that product differentiation may exacerbate informational problems in non-transparent markets. That is, even if firms do not observe their rivals' prices or quantities, they may still be able to infer the relevant information from their own prices and quantities. But such inference may be easier to achieve when all firms offer the same goods than when they offer highly differentiated products. This may be one reason why antitrust authorities usually interpret product homogeneity as facilitating collusion.

Product differentiation has a less ambiguous impact on collusion, however, when one firm has a 'better product' than the others (what economists refer to as 'vertical differentiation'). In essence, firms are then in an asymmetric situation and, as already noted above, the analysis is then similar to that of asymmetric costs. A firm that has a better quality (possibly adjusted for the cost) is in a situation somewhat similar to that of a firm that would offer the same quality as the others, but at a lower cost. This firm would have more to gain from cheating on a collusive path (put another way, it may require setting a price that does not fully reflect the increase in quality), and it has less to fear from a possible retaliation from the other firms.

To see this more precisely, consider a duopoly with an inelastic demand where one firm offers a better quality (at the same cost $c$, for the sake of presentation), which translates into a monetary bonus $b$ for its customers. That is, consumers are willing to pay $r$ for the lower quality and $r + b$ for the higher quality. Collusive conducts must then maintain a price differential of $b$ between the two firms (otherwise, one firm would take over the entire market). For example, the firms could try to maintain a price of $r$ for the low-quality good and of $r + b$ for the high-quality good, whereas price competition would have the high-quality firm sell the entire market but at a lower price $c + b$.

This situation is formally equivalent to the one, already discussed, where the two firms offer the same quality but face different costs.[25] The conclusions of the previous discussion thus readily apply. The high-quality firm is more tempted to cut prices, because it enjoys a higher margin and thus gains more from stealing any additional customer away from its rival. To ease that firm's incentive constraint, collusive conducts can give a bigger market share to the high-quality firm, but collusion will nevertheless remain more difficult to sustain: when firms offer different cost-adjusted qualities, collusion is more difficult, the larger the competitive advantage of the high-quality firm.

*Demand elasticity*

It is often perceived that low demand elasticity should exacerbate collusion concerns. The elasticity of the demand has in fact no clear impact on the sustainability of collusive prices. For example, in the case of an oligopolistic industry where $n$ identical firms produce the same good and repeatedly face the same demand $q = D(p)$, it can be shown that collusion is sustainable when the discount factor of the firms lies above a threshold equal to $1 - 1/n$; this threshold does not depend on the shape of the demand $D(p)$, which is therefore irrelevant in this case for assessing the sustainability of collusion: if the firms have a discount factor lower than $1 - 1/n$, no collusion is sustainable and, whatever the shape of the demand, only equilibrium yields the competitive price $p^c = c$. Conversely, if the discount factor of the firm exceeds $1 - 1/n$, the firms can sustain any collusive price, even the monopoly price, whatever the shape of this demand function. Thus, demand elasticity has indeed no impact here on the sustainability of collusion. This comes from the fact that demand elasticity (and more generally, the shape of consumer demand) affects in the same way both the short-term gains from undercutting rivals and the long-term cost of foregoing future collusion.

Collusion is, however, more profitable when demand elasticity is low. When picking a collusive price, the firms must trade off the increased margins generated by higher prices with the reduction in sales that these higher prices would trigger. The industry's ideal collusive price is the monopoly price, $p^M$, which maximizes the joint profit of the firms, $\pi^{\text{joint}} = (p - c)D(p)$. It is well known that this price is higher when the demand elasticity is lower.[26] This reflects the fact that, when demand is highly elastic, firms would lose too many sales if they tried to impose high prices. Conversely, if demand is low, then the firms can afford to maintain high prices without losing too many customers; the trade-off between sales and margins is then best solved for relatively high prices. Therefore, for a given market size, the firms have more to gain from sustaining the monopoly price when demand elasticity is low. In that sense, demand elasticity may constitute a relevant factor, although of a different nature than the factors listed above.[27] In addition, collusion is a larger concern for consumers when demand is inelastic than when it is elastic. This is both because the potential for a large profitable increase in prices above the 'normal' level decreases when demand becomes less elastic, and because consumers are hurt more by a given price increase when they have few alternatives.[28]

*Buying power*

A related factor concerns the countervailing buying power of the customers. If buyers are powerful, even a complete monopolist may find it difficult to impose high prices. The profitability of collusion is similarly reduced. In addition, Snyder (1996) notes that large buyers can successfully break collusion by concentrating their orders, to make firms' interaction less frequent and to increase the short-term gains from undercutting rivals; more generally, large buyers can design procurement schemes that reduce the scope for collusion.

### *Multi-market contact*

It is well recognized that firms can sustain collusion more easily when they are present on several markets.[29] First, multi-market contact increases the frequency of the interaction between the firms. Second, it may allow softening asymmetries that arise in individual markets. For example, one firm may have a competitive advantage in one market and its rival can have its own competitive advantage in another market. While a market-level analysis may then suggest that collusion is difficult to sustain, multi-market contact restores in such a case an overall symmetry that facilitates collusion. Third, multi-market contact may allow the firms to sustain collusion in markets where the industry characteristics alone would not allow such collusion.

For example, recall that for a symmetric oligopoly with $n$ firms, collusion is sustainable when the discount factor of the firms is larger than $1 - 1/n$. Suppose that two firms are in a duopoly situation in one market and face one more competitor in another market. The firms could sustain collusion in the first market if their discount factor were higher than 1/2, but could not *a priori* collude in the second market if their discount factor were below 2/3. Yet they can actually sustain collusion in both markets. The idea is that they can give a higher market share to the competitor in the second market, in order to induce it to collude, and rely on their interaction in the first market to discipline them. Overall, therefore, multi-market contact facilitates collusion.

### *Bidding markets*

The principles reviewed above apply as well to bidding markets. For example, collusion is easier when there are fewer bidders that repeatedly participate in the same bidding markets, when the frequency of these markets is high (for example, daily markets), and so forth. In addition, however, bidding markets can be designed in ways that either hinder or facilitate collusion. For example, sealed bid auctions generate less information (that is, except if the auctioneer reveals the details of all the bids afterwards) than public descending procurement auctions, where sellers observe at each moment who is still bidding at the current price. Therefore, a close look at the organization of the bidding markets may be necessary to assess the likelihood of collusion.[30]

### *Club or network effects*

Some markets are subject to club or network effects, where consumers benefit from being in the same 'club' by using the same software or the same keyboard layout, subscribing to the same operator, and so forth.[31] Club effects have several implications. They tilt the market in favour of a single participant, thereby creating a 'winner-take-all' type of competition which is not prone to collusion. In addition, club effects create lock-in effects that reinforce the position of the market leader and thus increase the benefits derived from such a position. By undercutting its rivals a firm can trigger snowballing effects that could easily tilt the market in its favour and thus obtain a durable leadership position. Club effects therefore exacerbate the gains from undercutting the rivals and, at the same time, lock-in effects limit retaliation possibilities. Both factors contribute to make collusion less likely.

*Note: Collusion in other dimensions than prices*

The above discussion also applies to other forms of competition. Where for example firms compete in quantity or production capacity, collusion consists in reducing the production levels below competitive levels and retaliation can either take the simple form of reverting to 'normal' competition, with higher output levels, or involve temporary large increases of competitors' outputs, in order to further depress prices to punish the deviating firm. While the nature of competition is different and often less intense under quantity competition than under price competition, it does not follow that the scope for collusion is larger or smaller, since retaliation possibilities are affected as well as the short-run gains of deviations from collusion: increasing one's production level is less profitable, since prices will adjust to sell out the competitors' output, but at the same time, retaliation is somewhat more difficult since the firm can always adapt its output level.

## 3. WHAT CAN COMPETITION AUTHORITIES DO?

Assessing the scope for collusion is a difficult issue. In particular, the above analysis underlines relevant factors but does not allow a definite conclusion as to whether collusion takes place or not. For one thing, tacit collusion relies to a large extent on firms' conjectures about the future reactions of their competitors to their pricing or production decisions. For this reason there is not one but many potential forms that it may take for a given market structure, defined in terms of firms, products, costs and demand.[32] Whether a particular form of coordination occurs or not depends on factors that affect managers' perception of the industry and that can be to some extent subjective or manipulated by specific participants. In particular, even in situations where collusion is indeed sustainable, firms may well end up 'competing' in each and every period if they expect their rivals to do so: the fact that firms *could* sustain collusion does not mean that they actually succeed in doing so, and firms may well compete in each period as if it were the last one, even if there exists another equilibrium in which they could maintain monopoly pricing forever.[33] While there is a good understanding of the mechanisms underlying tacit collusion in general, as well as of many factors that hinder or enhance the ability to co-ordinate, this is not so for the conditions under which any particular form of tacit collusion emerges at a specific point in time.

Even assessing the likelihood of collusion is a tricky issue. We have seen that the sustainability of tacit collusion depends on many factors – and only some of them can be quantified with a reasonable degree of precision. In addition, it is rarely the case that all factors go in the same direction in a given industry. Evaluating the likelihood of collusion thus often involves a delicate assessment of the net result of possibly conflicting effects. The exercise is made all the more difficult in that we do not have good models that incorporate the various effects; incorporating all the relevant dimensions into a single integrated model would in most cases be unmanageable and unlikely to yield clear-cut predictions.

That being said, how can antitrust fight collusion? There are several possible courses of action:

- *Ex ante*, competition authorities can prevent the emergence of an industry structure that is prone to collusion, by taking into account this concern when examining proposed mergers or joint venture agreements. They can also ban facilitating practices.

- *Ex post*, competition authorities can take steps to fight collusion *per se*, for example by uncovering evidence of explicit coordination or by attacking specific agreements that again facilitate collusion.

I discuss these approaches below.

### 3.1 Merger Control

As just mentioned, the same market situation can give rise to many alternative equilibria. It is thus impossible to rely on theoretical analysis alone to determine whether collusion is actually taking place. While the analysis of the history of the industry may help determine whether collusion occurred in the past, it provides limited help for evaluating whether it will occur in the future, and even more so if a merger takes place.[34]

Thus, it will not be possible to reach a definite conclusion from available market data on whether tacit collusion will actually occur as a consequence of the merger or not. The merger control office can, however, address a different and yet relevant question: will the merger create a situation where collusion becomes more likely, that is, will collusion be significantly easier to sustain in the post-merger situation?

A merger often affects many of the factors that are relevant for the sustainability of collusion and it can affect them in ways that tend to offset each other. The impact of the merger on collusion can thus involve a difficult assessment of possibly conflicting effects. Ideally, this could be done by building a 'meta-model' encompassing all relevant characteristics, but such a 'global model' would probably not be tractable and thus be quite useless.

The above discussion provides, however, a basis for prioritizing the relevant factors. It also advocates a structural analysis. Rather than a pure 'check-list' of relevant factors, it seems indeed more appropriate to develop a clear understanding of why each dimension is relevant, as well as of how it affects collusion – and is affected by a merger. This helps to prioritize the relevant factors, with an emphasis on the necessary ingredients (high entry barriers, frequent interactions and little role for innovation) and on the most important factors (number of market participants, their degree of symmetry, and so forth). But this structural analysis also makes it easier to build an overall assessment when several factors have a role and push in different directions.

For example, a merger reduces the number of competitors, which tends to facilitate collusion, but it can make the remaining competitors more asymmetric, which tends to hinder collusion. Economic modelling of these two aspects provides an analytical framework for assessing how they affect the effectiveness of retaliation conducts, and thus how these retaliation possibilities are modified by a merger. This, in turn, helps to assess the net impact of the merger on the sustainability of collusion.

We should keep in mind, however, that evaluating the impact of a merger on collusion is by nature more difficult than the analysis of single dominance. This increased difficulty is also reflected in the more limited help offered by quantitative or econometric approaches. In particular, while some successful efforts have been made to evaluate *ex post* the likelihood of collusion in a particular industry, predicting the impact of a merger on the future likelihood of collusion appears substantially more challenging. If firms were not tacitly colluding in the pre-merger phase, past market data and econometric studies can help in assessing key structural parameters but will not provide direct information on potential collusive behaviour. But even if there is some evidence on past collusive conducts, one has to account for the fact that firms will adapt their conduct to accommodate the new environment created by the merger, which again requires some prospective analysis.

### 3.2 Antitrust Enforcement against Collusion

Fighting collusion is not an easy task. To be sure, antitrust authorities can attack explicit forms of collusion where, say, managers meet, exchange information and conclude agreements on prices or market shares. The main difficulty in that case is to establish the existence of such explicit agreements and get hard evidence that would stand in court. Antitrust enforcement can launch detailed investigations and down-raids – and rely on indirect or informal evidence to target likely suspects.

Another possibility is to encourage informed parties to provide the needed evidence. The interest of this approach is exemplified by the development of leniency programmes which have already encountered substantial successes, first in the USA and then in Europe.[35] Leniency programmes vary in their design and scope: some apply to companies and others to individuals; some provide full protection while others only offer limited leniency; some apply to first informants only while some also apply, to a lesser extent, to additional informants; and so forth. The performance of these leniency programmes also varies,[36] which provides some ground for enhancing and fine-tuning their design.[37]

In the absence of any hard evidence of explicit agreements, which could be caught in the EU under Article 81 and in the USA under Section 1 of the Sherman Act, it is difficult if not impossible to directly fight collusion *per se*. There might actually be a debate as to whether antitrust authorities should take actions against purely tacit collusion, where by definition firms set prices non-cooperatively. Such actions would come close to regulating prices, something that competition authorities and courts are generally reluctant to do. In the USA, Section 2 of the Sherman Act condemns for example monopolization but not the exploitation of market power through high prices. In the EU, charging an excessive price can constitute an abuse of dominant position. Article 82 thus provides a basis on which tacit collusion could be attacked. However, building a case on the abuse of a dominant position requires establishing dominance, which in this case would amount to proving that collusion indeed occurred. This raises large difficulties on both legal and economic grounds.

First recall that the above discussion only provides an analytical framework for assessing the sustainability of collusion; it does not allow us to conclude whether collusion will effectively occur. Of course, the past history of prices and market shares does bring additional information, which can reinforce the suspicion of collusion, but it will not in general provide a definite conclusion. Even the most advanced econometric models, assuming that the required data are available, typically provide only probability estimates. Second, from a legal perspective, establishing collusion would involve a standard of proof similar to the high standard of proof established by the European Court for Article 81 cases.[38] Overall, it is unlikely that an abuse of collective dominance cases could rely solely on the past realization of prices or market shares.[39]

Competition authorities can still use the past history of prices, say, to alert industry 'supervisors' about abnormally high prices. They can for example provide such information to consumer associations, in order to increase customers' awareness of the problem. In regulated industries, competition authorities can also alert regulators or point to deficiencies in the regulatory environment.

Short of fighting collusion directly, competition authorities can attack those practices that facilitate it. Thus, for example, antitrust authorities may want to block Resale Price Maintenance when it facilitates collusion by generating more uniform prices. In the same vein, competition agencies may want to have a close look at marketing practices that tend to make the market more 'transparent', in the sense that they allow the market participants to infer more easily the pricing strategies of their competitors.

## 4. THE TELECOMMUNICATIONS INDUSTRY

The general principles discussed above apply to the telecommunications industry. Several specific aspects deserve some additional comments, however. These concern the role of regulators, the need for co-operation in interconnection agreements and the importance of innovation and growth. I also briefly discuss some issues associated with the development of e-commerce.

### 4.1 Regulation, Competition and Collusion

Fighting collusion directly comes close to regulating prices, and here regulators can thus complement the action of competition agencies. While antitrust authorities generally assess the lawfulness of conducts, regulators have more extensive powers. They can set constraints on wholesale and retail prices, either directly or through cost-of-service regulation schemes. Also, regulators can operate *ex ante* and not only *ex post*; while antitrust authorities assess conduct after the fact, regulators define pricing rules *ex ante*. Concomitantly, the regulatory process must be more expedient. The necessity not to halt progress puts pressure on regulators to take rapid decisions. In contrast, the *ex post* nature of antitrust intervention does not call for a similar expediency. As a result, regulators are in practice in a better position to adopt corrective measures when needed. Another advantage is that regulators have superior expertise. They are involved in a long-term relationship with the industry and have larger staffs for monitoring firms' accounts on a continuous basis; they can also insist on specific accounting principles (such as accounting separation) and disclosure rules.

To be sure, regulatory environments have their own caveats. *Ex ante* intervention and ongoing relationships may foster collusion between the industry and the overseer. The need for expertise also exacerbates revolving-doors problems: it imposes some constraints on recruitment and, at the same time, because of their deeper knowledge of a particular industry, regulators' natural career evolution is more likely to involve close links with this industry. Superior expertise may also exacerbate ratchet effects that arise from limited commitment powers, where regulators (or politicians) are tempted to confiscate the efficiency gains that firms may try to realize. Similarly, excessive attention may inhibit firms' initiative and distort their technology choices.[40] Regulators are also constrained by their institutional environment. While they may lack accountability and follow their own agenda, they face constraints – partly motivated by this accountability problem – on their means and powers. They are also subject to industry lobbying and political pressures, which can reflect various interest group pressures and be sensitive to alternative motivations (such as favouring domestic or public operators). Last, national regulators have limited incentive and opportunity to develop international benchmarking.

For all these reasons, there are natural complementarities between the role of national regulators and that of a community-wide antitrust supervision. First, having multiple instances may reduce the scope for capture.[41] For example, the antitrust process may provide a forum for entrants when the regulator is excessively lenient with the incumbent. Second, the separation of tasks may create a more focused mission to each overseer.[42] Third, the *ex ante* nature of regulation and the *ex post* nature of antitrust creates a natural division of labour between decision-maker and assessor and reduce incentives for cover-ups. Overall, antitrust authorities have an important watchdog role. Last, the community dimension of antitrust supervision facilitates the development of international comparisons, which can take advantage of the substantial heterogeneity observed in national environments and promote 'best practices'.

Collective dominance and tacit collusion may play an increasingly important role in that respect. Historically, the industry was dominated by a monopolistic operator. Accordingly, in the first phase of liberalization the main concern was to strike an appropriate balance between promoting the entry of efficient competitors and protecting the incumbent's incentive to maintain and develop its own infrastructure – which in many cases was and still is the only infrastructure used by all operators. We are now in a second phase, at least in some sectors (for example mobile telecommunications and Internet backbones), where several operators can compete head to head; it may well be the case that no operator enjoys a dominant position any more, and even when one operator is still dominant, the scope for abuses of such single dominance may no longer be the only concern or even the main one. The prior experience of the Commission with collective dominance cases further contributes to give it a natural role. The Commission can take advantage of this experience and help regulators adapt to the new issues arising in this second phase of liberalization; the above discussion may also help to identify the situations in which operators may be particularly tempted and able to maintain cosy relationships and soften competition.

## 4.2 Co-operation and Collusion

An important characteristic of the telecommunications industry is that operators often need to co-operate in order to provide service and thus be in a position to compete effectively. Telecoms customers want 'universal connectivity', which requires interconnection agreements, the setting of standards, and so forth.

This need for connectivity is of course exemplified in the hotly debated disputes between incumbents and new entrants, such as the famous Clear–Telecom battle in New Zealand. Incumbents can indeed threaten the successful development of new competitors by refusing to interconnect or, more realistically, by insisting on high interconnection charges or by offering *de facto* degraded connectivity; conversely, entrants can take advantage of asymmetric regulatory environments to impose costly demands on the incumbents.

However, the need for mutual interconnection among competitors and the required co-operation that these interconnection agreements involve also create some danger for the effectiveness of competition between mature operators. First, they provide one more way in which competitors interact, which creates more scope for retaliation and thus for collusion.

Second, incumbents may collude in the terms they propose to new entrants, in order to protect their incumbency advantages.[43] Third, and perhaps more importantly, interconnection agreements have a direct impact on the operators' perceived costs, and thus exert a direct influence on their pricing decisions, as well as on their decisions concerning new capacity investments, technological choices, and so forth. But then, operators can design the agreements so as to soften competition.

This point can be illustrated in a simple way in the context of bilateral interconnection agreements between two operators that, at the same time, compete in (linear) prices for subscribers.[44] Suppose, to fix ideas, that two operators negotiate a bilateral access charge for terminating each other's calls. Whenever one operator's subscriber calls one of the other operator's subscribers, the former operator must pay that access charge to the latter one. Thus, if in equilibrium each operator gets half of the market, and assuming that calling patterns are uniform across on-net and off-net calls, each operator will have to pay the termination charge over about half of its subscribers' calls. Therefore, denoting by $c$ the actual cost of communication and by $m$ the termination mark-up that the access charge adds to the actual cost of termination, each operator's perceived marginal cost of communications is equal to $c' = c + m/2$. That is, an increase in the access mark-up $m$ will inflate the cost of all off-net calls, and thus inflate in turn the marginal cost of attracting or stealing one more subscriber, as well as the marginal cost of inducing subscribers to call more often.[45] Both factors contribute to discouraging the operators from pricing aggressively and give them instead an incentive to maintain high prices. Note that the termination charge affects operators' marginal cost, and thus their prices, even if in equilibrium the traffic is balanced and payments thus offset each other. Operators can take advantage of this feature and agree to set the access charge at a level that eliminates *de facto* any effective competition between them.[46]

To be sure, the extent to which operators can manipulate these interconnection agreements so as to soften competition depends largely on the exact nature of retail competition as well as on regulatory supervision.[47] The point remains, however, that the negotiation of interconnection agreements gives the operators a natural avenue for altering the incentives to compete and thus deserves particular attention from regulators and antitrust authorities.

A similar remark applies to infrastructure-sharing agreements, a particularly relevant issue for the development of third-generation mobile communications networks. There again, such agreements have an impact on the effectiveness of competition, by enlarging the scope for retaliation but also by directly affecting the operators' pricing and other strategic decisions. This potential impact on competition is, however, smaller when operators simply share towers (and the associated fixed costs), which leaves many technology choices open, than when they also co-operate on network design and on technology choices that condition the nature of services that can be offered.

## 4.3 A Fast-Growing Industry

Despite the troubles associated with the UMTS technology and the financial market bubbles, the telecommunications industry is and remains a fast-growing industry. As we have seen, market growth is a relevant factor when assessing the likelihood or sustainability of collusion. Let me, however, emphasize again here that, as such, demand growth tends to facilitate collusion. More precisely, for a fixed number of market participants, collusion is easier to sustain in growing markets: collusion is easier to sustain when the short-term gains from a deviation are small compared with the cost of future retaliation, and this is indeed more likely to be the case in growing markets, where today's profits are small compared with tomorrow's.[48]

This conclusion appears somewhat at odds with some court cases and opinions expressed by the European Commission in guidelines.[49] Indeed, demand growth is in practice often interpreted as a factor hindering collusion. One possible reason for this apparent discrepancy is that the above reasoning assumes that the number of market participants remains fixed despite market growth, while in practice, entry may be easier in growing markets.[50] It is the prospect of future entry that then hinders the ability to collude.

However, it may be useful to disentangle the intrinsic effect of market growth discussed above from the impact of entry and other factors, so as to assess their relative strengths. In markets with low entry barriers, market growth is indeed likely to generate entry, and the overall impact may well be detrimental to collusion. However, in those markets where entry barriers are high (for example, because of spectrum licences, as in mobile telecommunications), the intrinsic impact of market growth may prevail and facilitate collusion. This is not to say that collusion is necessarily likely, since other factors – and notably the role of innovation – also play an important role. But again, distinguishing the role of each such factor can contribute to providing a more accurate overall assessment of the scope for collusion.

## 4.4 Network Effects

As already mentioned, club effects tend to destabilize collusion by generating tipping effects that both exacerbate the gains from undercutting the rivals and limit retaliation possibilities. Is it not clear, however, that such club effects appear in 'network' industries such as the telecommunications industry, where all networks are interconnected. Because of universal connectivity, a subscriber from one network can communicate with subscribers from any other network, which evacuates the scope for club effects.

Club effects may, however, become relevant when networks develop proprietary standards and offer network-specific services. Indirect club effects can also arise from network-based price discrimination. Both types of practices may generate problems. For example, proprietary standards degrade the quality of the 'connectivity' that is *de facto* offered to subscribers, while termination-based discrimination can be used strategically to exacerbate installed base advantages. Both types of practices may, however, also contribute to making collusion less likely.

## 4.5 E-commerce

Let me conclude with a digression on the impact of e-commerce on collusion in various goods and services markets. Of course, at this stage many uncertainties remain about the development of electronic wholesale and retail trade; the remarks that follow are thus highly speculative.

A first source of concern relates to the development of electronic procurement platforms. These platforms can involve some close co-operation between competitors and may enhance the information that each of them has about his rivals' production and development plans.

'Co-operation' on the buying side may thus have some adverse consequences on the 'competition' on the selling side.

At the retail level, many observers predict instead that the development of electronic commerce will foster competition and contribute to lower prices. Setting up a web site certainly involves much lower fixed costs than in the case of traditional brick-and-mortar retail outlets. In addition, regulations that limit the development of large stores are likely to have little bite on electronic commerce. Furthermore, producers may themselves decide to bypass distributors and sell directly 'on-line'. Last, consumers can more easily compare prices – by 'surfing' on the net, or by using 'shop bots' that do the research for them.

The reduction in search costs and the increase in the overall number of (electronic or traditional) retailers suggest that, indeed, prices will fall, particularly for homogeneous goods for which price comparison is the most effective, as well as for those goods that do not require inspection or trial prior to purchase.[51]

It can, however, be stressed that electronic commerce also facilitates price-matching strategies. For example, Barnes & Noble.com can react within a couple of hours to a change in the prices that Amazon.com charges for best-sellers – even though its traditional bookstores do not change their prices. The possibility of such quasi-instantaneous price matching (thanks to the sellers' own search engines) can foster collusion.

Electronic trade also makes new strategies available to retailers. For example, Books.com offers a 'Price Compare' option which requires some time but looks for the prices of Amazon.com, Barnes & Noble.com and Borders.com, and aligns its own price on the cheapest one. Such an 'automatic' price-matching service resembles 'most-favoured nations' clauses and a commitment power whose anticompetitive effects have long been recognized.[52]

There again, it is premature to predict the future evolution of these strategies. This will for example depend on suppliers' ability to control the electronic commerce of their product – and in particular, to protect their traditional retailers – as well as on distributors' ability to refuse access to these 'shop bots' – as some have tried against Bargain Finder.

## NOTES

* This chapter builds on joint work with many colleagues, particularly Bruno Jullien, Paul Seabright and Jean Tirole, but does not necessarily reflect their views. All errors are mine.

1. Selten (1984) points out that tough price competition may discourage entry; fighting collusion may thus 'backfire' by reducing the number of market participants – a similar observation applies to investments and other endogenous sunk costs, as emphasized by Sutton (1991, 1998). As noted by D'Aspremont and Motta (1994), however, intensifying competition selects the most efficient competitors.

   Using a panel of UK manufacturing industries, Symeonidis (1999) finds that the UK cartel laws from the late 1950s triggered tougher price competition and had a strong effect on the structure of previously cartelized markets, but little impact on firms' profits.
2. At a given demand level, the benefits from a small price cut increase as the price-cost margin or the own-price elasticity of demand increase.
3. See for instance the work of Porter (1983) on the Joint Executive Committee for the railroads industry in the 1880s.
4. For example, in *Compagnie Maritime Belge* (case C-395/96P) shipping companies were alleged to charter 'fighting ships' specifically designed to compete head to head against the ships of a targeted company.
5. The exact meaning of market transparency in the analysis of tacit collusion is a delicate issue that will have to be discussed in detail.
6. A discount rate $R$ means that the firm weights the profits in period $T$ with a multiplicative discount factor

$\delta = 1/(1 + R)^T$. If the firm faces no risk and can freely access the credit market, the discount rate corresponds to the market interest rate.

7. Of course, other industry characteristics such as market transparency, which is discussed below, also affect the length of time before retaliation effectively occurs. But the point here is that retaliation will not even be feasible in the absence of frequent interaction.
8. The idea that co-ordination is more difficult in larger groups is intuitive but there is little economic literature on this issue. See, for example, Compte and Jehiel (2001).
9. This insight is valid when holding all other factors constant. The number of firms is, however, endogenous and reflects other structural factors such as barriers to entry and product differentiation.
10. For example, in the case of an oligopolistic industry where $n$ identical firms produce the same good, it can be shown that collusion is sustainable when the discount factor of the firms lies above a threshold equal to $1 - 1/n$. This threshold increases by 33 per cent (from 1/2 to 2/3) when adding a third competitor to a duopoly, whereas it increases by only 12.5 per cent (from 2/3 to 3/4) when adding a fourth competitor.
11. See Bain (1948) for an early discussion. Gertner (1994) validates this insight for environments with 'immediate responses' where collusion is otherwise straightforward to achieve through simple price-matching strategies, even in the absence of repeated interaction.
12. It is for example well known that the monopoly price is an increasing function of the industry's marginal cost.
13. Side-transfers need not be monetary, however. They may for example consist of in-kind compensations or, when the same firms are active in several markets, of concessions made in one of these other markets. Still, such collusion schemes are not very plausible in the absence of any explicit agreement, and thus go beyond the scope of this report. For a discussion of these issues, see Osborne and Pitchik (1983) and Schmalensee (1987).
14. Mason et al. (1992) note in experimental duopoly games that co-operation is more likely when players face symmetric production costs.
15. See, for example, Abreu (1986) for a symmetric Cournot context and Brock and Scheinkman (1985) for a first analysis of a symmetric Bertrand context, later extended by Lambson (1987).
16. Brock and Scheinkman (1985) show for example in a linear model that, with exogenously given symmetric capacity constraints, the highest sustainable per capita profit varies non-monotonically with the number of firms.
17. Lambson (1994) provides a first partial characterization of optimal collusion schemes in this context. Lambson (1995) shows further that introducing a slight asymmetry in capacities hinders tacit collusion; and Davidson and Deneckere (1984, 1990) and Pénard (1997) show that asymmetric capacities make collusion more difficult in duopolies, using particular forms of collusive strategies.
18. A new entrant can also appear to destabilize a pre-entry collusion during a transition period, until a new collusive situation is reached. This is a rather different scenario, where the temporary absence of collusion simply reflects a tâtonnement process for reaching a new focal point.
19. A firm that uses a different production technique than others will be affected by the price of different inputs, or the labour cost may fluctuate in a different manner.
20. See Harrington (1989) for an analysis of collusion between firms that have different discount factors.
21. Martin (1995) provides a detailed analysis of this issue.
22. For example, in the above hypothetical industry, in the absence of any demand shock firms could perfectly detect any deviation by their rivals by simply looking at their own sales.
23. See, for example, Kühn (2001).
24. See, for example, Ross (1992) and Martin (1993).
25. More precisely, the situation is formally the same as if consumers were willing to pay $r$ for any of the products, but one firm faces a high cost $c_H = c$ whereas the other faces a low cost $c_L = c - b$.
26. More precisely, the monopoly price is such that the Lerner index is inversely proportional to the demand elasticity: $L = (p - c)/p = 1/\varepsilon(p)$, where the elasticity is given by $\varepsilon(p) = pD'(p)/D(p)$.
27. The profitability of collusion can in turn influence the firms' willingness to design and implement practices that facilitate the implementation of a collusive action. It can also induce firms to engage in more explicit collusion, at the risk of being caught by antitrust enforcement.
28. The potential harm to consumers is thus the larger, the less elastic is the demand. The impact on total welfare, however, is more ambiguous. The reason is that price increases generate fewer distortions when demand is inelastic (see, for example, Tirole, 1988 for a discussion of this issue).
29. The classic reference is Bernheim and Whinston (1990). See also Parker and Röller (1997) and Evans and Kessides (1994) for empirical evidence.
30. See, for example, Klemperer (2001).
31. One important issue concerns the 'compatibility' of rival clubs or networks. Club effects are fully internalized – and thus become irrelevant – when rival networks are fully compatible. This is, for example, the case in the telecommunications industry, where all operators are interconnected, so that subscribing to one or the

other network does not affect who someone can communicate with. However, compatibility can be imperfect (for example, some services can be proprietary) and pricing policies can also induce indirect club effects (for example, when it is cheaper to call subscribers of the same operator).

32. This translates formally into the multiplicity of possible equilibria. Game theorists have long recognized that repeated games (so-called 'super-games') easily generate many equilibria – infinitely many, actually. Theory identifies the possible equilibria, including collusive ones, but so far it does not predict which of these equilibria will emerge.
33. That is, the mere repetition of the 'static' or 'non-collusive' equilibrium is always an equilibrium of the repeated game (it is even 'subgame-perfect', that is, it satisfies an additional credibility criterion); therefore, any collusive pricing equilibrium comes in addition to the standard static equilibrium.
34. Past behaviour can, however, provide some information about specific characteristics of the market participants, which can for example be useful to identify whether firms are prone to collusion or of a 'maverick' type.
35. In the USA, firms bringing information before an investigation is opened have benefited from such a leniency programme since 1978. The EU adopted a leniency programme in 1996, which allows firms that bring information to benefit from reduced fines. The Office of Fair Trading in the UK and the recent competition bill in France have also introduced leniency programmes.
36. In the USA, for example, it is only after the reform of 1993 extending leniency to firms that bring information after the investigation has been opened (as long as the Department of Justice has not yet been able to prove collusion) that the leniency programmes became effective. Thanks to this reform, on average two cartels are now disclosed every month, and the fines often exceed $100 million (not to mention jail for some managers). In 1999 only the Antitrust Division secured more than $1 billion in fines, which is more than the total sum of fines imposed under the Sherman Act since its adoption more than a century ago. The EU has also recently amended its leniency programme to improve its performance.
37. These programmes have also triggered a body of theoretical work. See, for example, Spagnolo (2000a, 2000b) and Motta and Polo (2000).
38. See the *Woodpulp* Judgment of 31 March 1993. The Court basically required to prove that no other behaviour than collusion could explain the observed realization of prices, a rather insurmountable task.
39. This suggests that a 'pure' Article 82 case is unlikely for collective dominance. However, cases could be built – and have already been so; see, for example *Compagnie Maritime Belge* – on both Articles 81 and 82.
40. See, for example, Crémer (1995) and Aghion and Tirole (1997).
41. This benefit from having multiple 'principals' has been extensively studied in the contract theory literature. See, for example, Laffont and Martimort (1999).
42. On the desirability of 'advocates', see Dewatripont and Tirole (1999).
43. In addition, incumbents that collude – or take advantage of their own bilateral agreements to maintain soft competition, as stressed below – have actually higher incentives to keep entrants out, since their arrival could endanger collusion and restore more effective competition.
44. See Armstrong (1998) and Laffont et al. (1998a).
45. The reasoning assumes here that operators cannot engage in termination-based price discrimination. If instead they can charge higher prices for off-net calls, the access charge would not affect their incentives to induce more on-net calls. Termination-based price discrimination, however, creates indirect forms of externalities (friends and relatives, for example, are then induced to subscribe to the same network), and are the source of other problems. See Laffont et al. (1998b) and Jeon et al. (2001).
46. That is, the equilibrium price is increasing with the level of the reciprocal access charge. In particular, there exists a level of this access charge for which the 'normal' competition between the operators leads them to charge retail prices at the monopoly level, even in the absence of any 'repeated interaction', simply because they perceive a high cost for each additional piece of traffic they generate.
47. If for example the operators compete in 'two-part tariffs' (subscription fees plus usage prices), the termination charge still affects their incentives with respect to usage prices but has less direct impact on the competition in subscription prices, which are not usage-sensitive. In addition, termination mark-ups introduce pricing distortions that hurt the operators' profitability by reducing the net 'quality' of the services they offer (high usage prices means that subscribers call less often than would be efficient).
48. If for example demand grows steadily at a constant rate $g$, the situation is analogous to that of a stagnating market where firms' discount factors are multiplied by the same constant $g$, since future periods weigh more when the market grows. Thus, in a growing market collusion can be sustained for lower values of the firms' actual discount factor.
49. See for instance the recent guidelines for market definition in electronic communication markets.
50. Market growth may also be a sign of a lack of maturity, or of a highly innovative market.
51. Brynjolfsson and Smith (1999) find that books are 9 per cent to 16 per cent cheaper on the Net, shipping costs included. Prior studies had, however, found smaller differences – and even, sometimes, higher prices

on the Net. Brokerage commissions for stock exchanges or plane tickets are also cheaper on-line, even though traditional brokers and travel agencies have started to align their own tariffs.
52. The commitment to match any price cut reduces the competitors' incentives to price aggressively.

# REFERENCES


Abreu, D. (1986), 'Extremal Equilibria of Oligopolistic Supergames', *Journal of Economic Theory*, **39**, pp. 191–223.

Abreu, D., D. Pearce and E. Stachetti (1985), 'Optimal Cartel Equilibria with Imperfect Monitoring', *Journal of Economic Theory*, **39**, pp. 251–69.

Aghion, P. and J. Tirole (1997), 'Formal and Real Authority in Organizations', *Journal of Political Economy*, **105**, pp. 1–29.

Armstrong, M. (1998), 'Network Interconnection', *Economic Journal*, **108**, pp. 545–64.

Bain, J. (1948), 'Output Quotas in Imperfect Cartels', *Quarterly Journal of Economics*, **62**, pp. 617–22.

Bernheim, D. and M. Whinston (1990), 'Multimarket Contact and Collusive Behavior', *Rand Journal of Economics*, **21**, pp. 1–26.

Brock, W.A. and J. Scheinkman (1985), 'Price Setting Supergames with Capacity Constraints', *Review of Economic Studies*, **52**, pp. 371–82.

Brynjolfsson, E. and M. Smith (1999), 'Frictionless Commerce? A Comparison of Internet and Conventional Retailers', MIT Sloan School of Management.

Compte, O., F. Jenny and P. Rey (2002), 'Capacity Constraints, Mergers and Collusion', *European Economic Review*, **46**(1), pp. 1–29.

Compte, O. and P. Jehiel (2001), 'Multi-party Negotiations', mimeo, CERAS.

Crémer, J. (1995), 'Arm's-length Relationships', *Quarterly Journal of Economics*, **2**, pp. 275–95.

D'Aspremont, C. and M. Motta (1994), 'Tougher Price Competition or Lower Concentration: A Trade-off for Antitrust Authorities', Université Catholique de Louvain, Discussion Paper No. 9415.

Davidson, C. and R.J. Deneckere (1984), 'Horizontal Mergers and Collusive Behavior', *International Journal of Industrial Organization*, **2**, pp. 117–32.

Davidson, C. and R.J. Deneckere (1990), 'Excess Capacity and Collusion', *International Economic Review*, **31**, pp. 521–41.

Dewatripont, M. and J. Tirole (1999), 'Advocates', *Journal of Political Economy*, **107**(1), pp. 1–39.

Evans, W.N. and I.N. Kessides (1994), 'Living by the "Golden Rule": Multimarket Contact in the U.S. Airline Industry', *Quarterly Journal of Economics*, **109**, pp. 341–66.

Gertner, R. (1994), 'Tacit Collusion with Immediate Responses: The Role of Asymmetries', mimeo, University of Chicago.

Green, E. and R. Porter (1984), 'Non-Cooperative Collusion under Imperfect Price Information', *Econometrica*, **52**, pp. 87–100.

Haltiwanger, J. and J. Harrington (1991), 'The impact of cyclical demand movements on collusive behavior', *Rand Journal of Economics*, **22**, pp. 89–106.

Harrington, J. (1989), 'Collusion Among Asymmetric Firms: The Case of Different Discount Factors', *International Journal of Industrial Organization*, **7**, pp. 289–307.

Jeon, D.-S., J.-J. Laffont and J. Tirole (2001), 'On the Receiver Pays Principle', mimeo, University of Toulouse.

Jullien, B. and P. Rey (2002), 'Resale Price Maintenance and Collusion', mimeo, University of Toulouse.

Kühn, K.U. (2001), 'Fighting Collusion by Regulating Communication between Firms', *Economic Policy*, **32**, pp. 169–204.

Laffont, J.-J. and D. Martimort (1999), 'Separation of Regulators against Collusive Behavior', *Rand Journal of Economics*, **30**(2), pp. 232–62.

Laffont, J.-J., P. Rey and J. Tirole (1998a), 'Network Competition: I. Overview and Nondiscriminatory pricing', *Rand Journal of Economics*, **29**(1), pp. 1–37.

Laffont, J.-J., P. Rey and J. Tirole (1998b), 'Network Competition: II. Price Discrimination' (with Jean-Jacques Laffont and Jean Tirole), *Rand Journal of Economics*, **29**(1), pp. 37–56.

Lambson, V.E. (1987), 'Optimal Penal Codes in Price-Setting Supergames with Capacity Constraints', *Review of Economic Studies*, **54**, pp. 385–97.
Lambson, V.E. (1994), 'Some Results on Optimal Penal Codes in Asymmetric Bertrand Supergames', *Journal of Economic Theory*, **62**, pp. 444–68.
Lambson, V.E. (1995), 'Optimal Penal Codes in Nearly Symmetric Bertrand Supergames with Capacity Constraints', *Journal of Mathematical Economy*, **24**(1), pp. 1–22.
Martin, S. (1993), 'Endogenous Firm Efficiency in a Cournot Principal–Agent Model', *Journal of Economic Theory*, **59**, pp. 445–50.
Martin, S. (1995), 'R&D Joint Ventures and Tacit Product Market Collusion', *European Journal of Political Economy*, **24**, pp. 357–79.
Mason, C.F., O.R. Phillips and C. Nowell (1992), 'Duopoly Behavior in Asymmetric Markets: an Experimental Evaluation', *Review of Economics and Statistics*, **74**(4), pp. 662–70.
Motta, M. and M. Polo (2000), 'Leniency Programs and Cartel Prosecution', mimeo, available at http://www.iue.it/Personal/Motta/.
Osborne, M. and C. Pitchik (1983), 'Price Competition in a Capacity-Constrained Duopoly', *Journal of Economic Theory*, **38**, pp. 238–60.
Parker, P.M. and L.-H. Röller (1997), 'Oligopoly and the Incentive for Horizontal Merger', *American Economic Review*, **75**, pp. 219–27.
Pénard, T. (1997), 'Choix de Capacités et Comportements Stratégiques: une Approche par les Jeux Répétés', *Annales d'Economie et de Statistique*, **46**, pp. 203–24.
Porter, R. (1983), 'A Study of Cartel Stability: The Joint Executive Committee, 1880–1886', *Bell Journal of Economics*, **14**, pp. 301–14.
Raith, M. (1996), 'Product Differentiation, Uncertainty and the Stability of Collusion', London School of Economics–STICERD Discussion Paper Series EI/16:49.
Ross, T.W. (1992), 'Cartel Stability and Product Differentiation', *International Journal of Industrial Organization*, **10**, pp. 1–13.
Rotemberg, J. and G. Saloner (1986), 'A Supergame-Theoretic Model of Business Cycles and Price Wars during Booms', *American Economic Review*, **76**, pp. 390–407.
Schmalensee, R. (1987), 'Competitive Advantage and Collusive Optima', *International Journal of Industrial Organization*, **5**, pp. 351–68.
Selten, R. (1984), 'Are Cartel Laws Bad for Business?', in H. Hauptmann, W. Krelle and K.C. Mosler (eds), *Operations Research and Economic Theory*, Berlin/Heidelberg: Springer-Verlag.
Snyder, C.M. (1996), 'A Dynamic Theory of Countervailing Power', *Rand Journal of Economics*, **27**(4), pp. 747–69.
Spagnolo, G. (2000a), 'Optimal Leniency Programs', mimeo, Stockholm School of Economics.
Spagnolo, G. (2000b), 'Self-Defeating Antitrust Laws: How Leniency Programs Solve Bertrand's Paradox and Enforce Collusion in Auctions', mimeo, Stockholm School of Economics.
Stigler, G. (1964), 'A Theory of Oligopoly', *Journal of Political Economy*, **72**, pp. 44–61.
Sutton, J. (1991), *Sunk Cost and Market Structure*, Cambridge, MA: MIT Press.
Sutton, J. (1998), *Technology and Market Structure*, Cambridge, MA: MIT Press.
Symeonidis, G. (1999), 'Are Cartel Laws bad for Business? Evidence from the UK', mimeo, University of Essex.
Tirole, J. (1988), *The Theory of Industrial Organization*, Cambridge, MA: MIT Press.

# 7. Some thoughts on collective dominance from a lawyer's perspective

**Lambros Papadias***

## 1. INTRODUCTION

The notion of collective or oligopolistic dominance has been widely discussed in Europe and elsewhere over recent years by competition lawyers and economist alike.[1] Its application under the EC competition rules (Article 82) and especially under the Merger Control Regulation (MCR)[2] has aroused particular interest since neither the relevant Treaty provision (Article 82) nor the MCR gives a meaningful definition of the term 'collective dominance'. Moreover, the recent jurisprudence of the Court of Justice and of the Court of First Instance has contributed in raising more questions than providing needed answers.[3] In particular, the entry into force in 1990 of the MCR forced the Community judge to start thinking about an articulated and perspective structural antitrust analysis and the effective treatment under the Community rules of the so-called 'oligopoly problem',[4] a problem dealt in the past exclusively under Article 81 of the Treaty and the 'cartel' theory.

The current discussions, however rich and thought-provoking they may be, have not yet succeeded in clarifying certain core issues of the theory of collective dominance. There is still a certain degree of confusion as to what collective dominance means or should mean for the purposes of applying EC competition law. This is due, among other things, to the fact that the 'oligopoly problem' dealt under the notion of collective dominance and especially the concept of tacit collusion which lies at its heart has a different connotation in Europe than in the USA. Economists – and to a lesser extent lawyers – often fail to make this critical distinction between the two legal systems and as a result their reasoning appears to be a mixture of various aspects of antitrust analysis under Articles 81 and 82 of the EC Treaty and Sections 1 and 2 of the Sherman Act. However, this is an important distinction because ultimately antitrust enforcement against dominant oligopolies or 'collectively' dominant undertakings must comply with existing case-law. Any theory which fails to take into account the hierarchy of norms, the context and the objectives of the EC Treaty provisions is destined to be, one way or another, rejected by the Community judge, the latter being the ultimate arbitrator as far as antitrust enforcement is concerned.

Following the recent adoption of a new regulatory framework for electronic communications markets, the notion of collective dominance has also permeated the area of *ex ante* sector-specific regulation, thus becoming one of the most powerful tools in the hands of national regulators dealing with present or expected market failures in concentrated electronic communications markets. In particular, under the new regulatory framework, national regulatory authorities (NRAs), when applying in an *ex ante* environment the notion of collective or oligopolistic dominance, will have to comply with the principles of EC competition law and the relevant jurisprudence of the Court of Justice.[5] It follows that the way collective dominance is applied under Article 82 or under the MCR will have a direct impact on the corresponding market analysis carried out by NRAs under the new framework.

The purpose of this chapter is not to discuss or develop new theories of collective or oligopolistic dominance. This has been done exhaustively by others, especially by economists. Rather, the goal here is to try to clarify the legal context within which the notion of collective dominance has evolved and to shed some more light on the jurisprudence of the Community Courts (Court of Justice, Court of First Instance). In doing so, reference will also be made to the Commission's own practice in order to examine the evolving interrelation between the Commission, in its capacity as a decision-making body, and the Court as the final interpreter of the legislator's will.

## 2. THE LEGAL CONTEXT

### 2.1 Oligopolies, Concerted Practices and Conscious Parallel Behaviour

Collective dominance, as such, is not a new concept in EC competition law. Article 82 of the EC Treaty explicitly provides that a dominant position can be held by one or more undertakings. However, for more than 30 years since the entry into force of the Treaty, collective dominance remained a mere theoretical possibility for the Commission and the Court alike. This was due mainly to the fact that collectively dominant undertakings were dealt *de facto* under Article 81 of the Treaty which prohibits anticompetitive agreements and concerted practices between undertakings as well as decisions by an association of undertakings. Members of cartels were treated as parties to anticompetitive agreements rather than as parties to a collectively dominant entity. By prohibiting not only explicit or implicit agreements, but also 'concerted practices', Article 81 enabled the Commission to pursue further more elaborated forms of collusion within oligopolistic markets, thus rendering reliance on Article 82 and the notion of collective dominance unnecessary.

As far as cartels or oligopolistic markets are concerned, the main issue has always been to prove that the undertakings concerned had colluded either in fixing prices, restricting output, sharing markets and/or customers, and/or foreclosing entry to other competitors. Collusion or co-operation between such undertakings could be either explicit or tacit. However, growing awareness and familiarity with the competition rules meant that, in practice, the undertakings were 'intelligent enough' to avoid entering into any kind of explicit anticompetitive agreements. Instead, more sophisticated means of co-ordination and co-operation were deployed in order to avoid antitrust enforcement. As a result, the notion of 'concerted practices' became a progressively useful and often critical tool in the hands of the Commission when dealing with tacit co-operation or collusion between members of a cartel or an oligopoly.

As early as 1972, the Court had ruled that 'a concerted practice refers to a form of coordination between undertakings which, without having been taken to the stage where an agreement properly so-called has been concluded, knowingly substitutes for the risks of competition practical co-operation between them'.[6]

The problem was, however, that the notion of concerted practices as defined by the court could not encompass all possible forms of anticompetitive outcome. Oligopolists faced with an Article 81(1) prohibition had always argued that the anticompetitive outcome observed in the market was the result of oligopolistic interdependence or of a conscious parallel behaviour alone, and not the result of a tacit collusion or concerted practices. In the *Dyestuff* case,[7] which involved the legality of simultaneous and uniform price increases, dyestuff producers argued that the Commission had wrongly identified the concept of concerted practices with the conscious parallelism of members of an oligopoly, the incriminated conduct being due to the independent decisions of each undertaking, determined by objective business needs and in particular by the need to increase the unsatisfactorily low rate of profit on the production of dyestuffs. In addition it was argued that the identical price increases within the oligopoly were attributed to a price leadership of one of the undertakings concerned.[8]

The Court accepted that

> although parallel behaviour may not by itself be identified with a concerted practice, it may however amount to strong evidence of such a practice if it leads to conditions of competition which do not correspond to the normal conditions of the market, having regard to the nature of the products, the size and number of the undertakings, and the volume of the said market. This is especially the case if the parallel conduct is such as to enable those concerned to attempt to stabilise prices at a level different from that to which competition would have led, and to consolidate established positions to the detriment of effective freedom of movement of the products in the common market and of the freedom of consumers to choose their suppliers.[9]

Although the Court's ruling in *Dyestuffs* made it clear that parallel conscious behaviour was not considered sufficient by itself to trigger an Article 81(1) prohibition, the question of how to deal with oligopolies where few undertakings could easily follow simultaneously and uniformly a pattern of behaviour which is not considered optimal and which does not either result from any kind of explicit or tacit agreement remained unanswered.[10] Was there any scope in Article 81(1) for interpreting the notion of 'concerted practices' in a way that would enable control of anticompetitive oligopolistic behaviour? How far could an undertaking go in adapting intelligently to the existing and anticipated conduct of its competitors?

The issue was raised again in *Woodpulp II*,[11] where the main question was whether the parallel conduct consisting essentially in simultaneous quarterly price announcements by pulp producers and the fact that the announced prices were identical was a proof of a concerted practice.[12] Since the Commission had failed to provide documentary evidence to establish the existence of concertation between the producers of woodpulp, the Court had to examine whether the system of quarterly price announcements, the simultaneity or near-simultaneity of the price announcements and the parallelism of price announcements constituted as such 'a firm, precise and consistent body of evidence of prior concertation'. In other words, the Court sought to establish whether the existence of what appeared to be a co-ordinated oligopoly implied collusion (within the meaning of Article 81) or whether the price parallelism observed in the market was due to the oligopolistic structure of the market.

The Court recalled that

> parallel conduct cannot be regarded as furnishing proof of concertation unless concertation constitutes the only plausible explanation for such conduct. It is necessary to bear in mind that, although Article 8[1] of the Treaty prohibits any form of collusion which distorts competition, it does not deprive economic operators of the right to adapt themselves intelligently to the existing and anticipated conduct of their competitors. Accordingly, it is necessary in this case to ascertain whether the parallel conduct alleged by the Commission cannot, taking account of the nature of

> the products, the size and the number of the undertakings and the volume of the market in question, be explained otherwise than by concertation.[13]

On the basis of an 'expert report' the Court concluded that

> the organisation of the market, in conjunction with its very high degree of transparency, leads in the short term to a situation where prices are slow to react. The producers know that, if they were to increase their prices, their competitors would no doubt refrain from following suit and thus lure their customers away. Similarly, they would be reluctant to reduce their prices in the knowledge that, if they did so, the other producers would follow suit, assuming that they had spare production capacity. Such a fall in prices would be all the less desirable in that it would be detrimental to the sector as a whole: since overall demand for pulp is inelastic, the loss of revenue resulting from the reduction in prices could not be offset by the profits made as a result of the increased sales and there would be a decline in the producers' overall profits.

For the Court, the oligopolistic characteristics of the market and its very high degree of transparency were some of the main reasons for the price stability and the price parallelism observed.[14] The Court also put emphasis on the existence of potential competitors. Given that pulp production by outsiders was at 40 per cent of total consumption of pulp in the Community, the Court concluded that 'it would have been difficult for a cartel to operate only as between the undertakings found to have committed an infringement by engaging in concertation'.[15]

Having carried out a thorough market analysis the Court noted that

> concertation is not the only plausible explanation for the parallel conduct. To begin with, the system of price announcements may be regarded as constituting a rational response to the fact that the pulp market constituted a long-term market and to the need felt by both buyers and sellers to limit commercial risks. Further, the similarity in the dates of price announcements may be regarded as a direct result of the high degree of market transparency, which does not have to be described as artificial. Finally, the parallelism of prices and the price trends may be satisfactorily explained by the oligopolistic tendencies of the market and by the specific circumstances prevailing in certain periods. Accordingly, the parallel conduct established by the Commission does not constitute evidence of concertation.[16]

*Woodpulp II* was the culmination of a series of important judgments which resisted legal attempts to deviate from the core nature and language of Article's 81(1) prohibition which applies only to agreements or 'practices' which are characterized by a minimum of co-operation and/or collusion between independent undertakings. For Article 81(1) to apply within an oligopolistic market, there should be no other explanation for the anticompetitive outcome (conscious parallel behaviour) than concerted practices or collusion.

What *Woodpulp II* amply demonstrated was that the cartel theory enshrined in Article 81 could not be relied upon to address issues related to anticompetitive outcomes within (tight) oligopolistic markets. If an oligopoly performed in an anticompetitive manner either because of price leadership or because of oligopolistic interdependence, so be it. It was clear that unless Article 82 and the notion of collective dominance were to be given a purposive or teleological interpretation, there would be no possibility – similarly to the situation prevailing in US antitrust law – to deal *ex post* with conscious parallelism in oligopolistic markets. Alternatively, such markets could be dealt with under the *ex ante*, prospective analysis required by the MCR were a structural change in the market to trigger its application.

## 2.2 Oligopolies and the Emerging Concept of Collective Dominance

Initially the Court did not consider that the notion of collective dominance could apply to an oligopoly. Its early case-law suggested that Article 82 could apply to abusive conduct by a number of undertakings only if the latter belonged to the same group or economic entity.[17] It was for the first time in 1998, within the context of the *Alsatel* case, that the Commission suggested that the Court consider whether parallel behaviour on the part of several independent undertakings, which does not leave their customers any possibility of negotiating the terms of their contracts, could place those undertakings collectively in a dominant position within the meaning of Article 82 of the Treaty, a suggestion which, however, remained unanswered since it was not raised formally in the preliminary reference by the French judge.[18]

Three years later, in 1991, the Commission published its 'Guidelines on the application of EEC competition rules in the telecommunications sector',[19] where it stated that although at the time both its own practice and the case-law of the Court were still developing, it would consider two or more undertakings to be in a collective dominant position when they had substantially the same position *vis-à-vis* their customers and competitors as a single company has in a dominant position, provided that no effective competition existed between them. The lack of competition could be due, in practice, to the existence of certain links between those companies without, however, the existence of such links being a prerequisite for a finding of collective dominance.[20]

From 1991 onwards, the concept of collective dominance will be tested in a number of decisions taken by the Commission under Regulation 17 and under the MCR, decisions which will ultimately be brought before the Court of First Instance and the Court of Justice for judicial review. It is not a coincidence that this new wave of jurisprudence coincided with the entry into force in 1990 of the MCR and the requirement for a prospective, structural analysis of mergers, and with the creation, at the same period, of the Court of First Instance with jurisdiction to hear all antitrust cases and conduct thorough fact-finding assessments. As will be shown below, the advances in legal and economic theory under the MCR will have a direct impact on the corresponding interpretation of Article 82. The Court will be forced to fill in the legal gap left open by *Woodpulp II* by *de facto* revisiting its case-law under Article 82 as far as conscious parallel behaviour or oligopolistic interdependence is concerned.

## 2.3 Oligopolies, Collective Dominance and the Jurisprudence of the Court

One of the first things that the recent case-law clarified was that the expression 'one or more undertakings' in Article 82 of the EC Treaty means that a dominant position may be held by two or more economic entities which are legally and economically independent of each other.[21] The previous (wrong) suggestion that a collective dominant position could only be held by entities belonging to the same group or economic entity was thus definitively abandoned.[22]

The notion of collective dominance under Article 82 was first discussed in some detail in '*Flat Glass*',[23] a case decided in 1992 when the Court of First Instance (CFI) was still in its early years of functioning and clearly unaware of the magnitude or dimension of the issues at stake. It is not surprising therefore that later on *Flat Glass* would be effectively rewritten and implicitly corrected by *Gencor* (see below).

In *Flat Glass* the CFI considered that

> there is nothing, in principle, to prevent two or more independent economic entities from being, on a specific market, united by such economic links that, by virtue of that fact, together they hold a dominant position vis-à-vis the other operators on the same market. This could be the case, for example, where two or more independent undertakings jointly have, through agreements or licences, a technological lead affording them the power to behave to an appreciable extent independently of their competitors, their customers and ultimately of their consumers.[24]

In referring to 'agreements' or 'licences' the CFI may have been influenced by the Commission itself, which, during the court proceedings, had stated that it did not intend to apply the concept of collective dominant position to the undertakings in question solely on the ground that they form part of a tight oligopoly controlling more than 80 per cent of the Italian market in flat glass, but mainly because the three producers in question presented themselves on the market as a single entity and not as individuals.[25] Giving no rational explanation, the CFI also rejected the thesis following which a collective dominant position could be created by virtue alone of agreements or concerted practices between undertakings and the mere aggregation of their respective market shares. What the CFI considered 'absolutely essential' for an assessment of the question of dominance was the extent to which the degree of competition was weakened by the allegedly unitary presence of the three producers on the market.[26]

*Flat Glass* seemed to imply that a finding of collective dominance would have to be based on proving the existence of economic links, in the sense of structural links, or other factors which could give rise to a 'connection' between the undertakings concerned.[27] The CFI's insistence on structural links could be explained by the fact that *Flat Glass*, being an Article 82 *ex post* case, the whole analysis was inspired by and based upon the traditional cartel theory, whereby undertakings with a commonality of interests, explicitly or implicitly, co-operate and collude with each other. As a result, the CFI did not address the issue of the 'oligopoly problem' and that of 'conscious parallelism'.

The question of whether collective dominance could also apply to an oligopolistic market, in the absence of any kind of links among the undertakings present in such a market, would, however, be raised for the first time, four years later, in *Gencor*.[28] The case concerned the legality of a decision adopted by the Commission under the MCR prohibiting the notified transaction on the grounds that it would lead to the creation of a duopoly market conducive to a situation of oligopolistic dominance.[29] Before the CFI, the parties argued that the Commission had failed to prove the existence of 'links' between the members of the alledged duopoly within the meaning of *Flat Glass*.

In a landmark judgment the CFI dismissed the application by stating, *inter alia*, that there was no legal precedent suggesting that the notion of 'economic links' was restricted to the notion of structural links between the undertakings concerned:

> there is no reason whatsoever in legal or economic terms to exclude from the notion of economic links the relationship of interdependence existing between the parties to a tight oligopoly within which, in a market with the appropriate characteristics, in particular in terms of market concentration, transparency and product homogeneity, those parties are in a position to anticipate one another's behaviour and are therefore strongly encouraged to align their conduct in the market, in particular in such a way as to maximise their joint profits by restricting production with a view to increasing prices. In such a context, each trader is aware that highly competitive action on its part

designed to increase its market share (for example a price cut) would provoke identical action by the others, so that it would derive no benefit from its initiative. All the traders would thus be affected by the reduction in price levels.[30]

As the Court pointed out, market conditions may be such that 'each undertaking may become aware of common interests and, in particular, cause prices to increase without having to enter into an agreement or resort to concerted practice'.[31]

Given that the notion of collective dominance under the MCR and Article 82 is fundamentally the same, it follows that *Gencor* implicitly overruled and, in an elegant manner, corrected *Flat Glass* by making it clear that oligopolistic interdependence within a 'tight oligopoly' was a sufficient condition for the creation of a collective dominance. In fact, with a single and decisive stroke, the CFI brought back within the boundaries of antitrust law liability, the notion of parallel (conscious) behaviour and oligopolistic interdependence that Article 81 had failed to capture. It is not clear whether the Court was actually cognizant of the wider implications of its pronouncement in *Gencor*.

The CFI's ruling in *Gencor* was later endorsed by the Court of Justice in the appealed *Compagnie Maritime Belge* ('*CMB*') case.[32] It is important to stress that *CMB*, as an Article 82 case, when first tried by the CFI, did not raise any definitional issues of oligopolistic dominance. *CMB* was a traditional collective dominance case in the sense that the undertakings concerned were united by agreements and other structural links. However, when the case came on an appeal before the Court, *Gencor* was already out and the question was whether the Court would seize the opportunity to reject or endorse the 'new' theory of oligopolistic dominance already embraced by the CFI. The Court could have chosen not to expand on the theory of oligopolistic dominance, first because this was not the issue in *CMB*, and second, on the ground that *CMB* was an Article 82 case where the market analysis was not prospective, as in *Gencor*, but based on an *ex post* assessment of past (abusive) behaviour. The Court decided, however, that the time had come to take a closer look at its own previous case-law and, in a manner which is characteristic of seminal judgments, to spell out an articulated, rigorous and methodological approach for making a finding of collective dominance, be it for the purposes of the MCR or of Article 82.

According to the Court, in order to show that two or more undertakings hold a collective dominant position, it is necessary to consider whether the undertakings concerned together constitute a collective entity *vis-à-vis* their competitors, their trading partners and their consumers on a particular market.[33] This will be the case when (i) there is no effective competition among the undertakings in question; and (ii) the said undertakings adopt a uniform conduct or common policy in the relevant market.[34] Only when that question is answered in the affirmative is it appropriate to consider whether the collective entity actually holds a dominant position.[35] In particular, it is necessary to ascertain whether economic links exist between the undertakings concerned which enable them to act independently of their competitors, customers and consumers. The Court recognized that an implemented agreement, decision or concerted practice (whether or not covered by an exemption under Article 81(3) of the Treaty) may undoubtedly result in the undertakings concerned being linked in a such way that their conduct on a particular market on which they are active results in them being perceived as a collective entity *vis-à-vis* their competitors, or the consumers.[36]

The mere fact, however, that two or more undertakings are linked by an agreement, a decision of associations of undertakings or a concerted practice within the meaning of

Article 81(1) of the Treaty, does not, of itself, constitute a necessary basis for such a finding. As the Court emphatically stated, 'a finding of a collective dominant position may also be based on other connecting factors and would depend on an economic assessment and, in particular, on an assessment of the structure of the market in question'.[37]

It follows from *Gencor* and *Compagnie Maritime Belge* judgments that although the existence of structural links can be relied upon to support a finding of a collective dominant position, such a finding can also be made in relation to an oligopolistic or highly concentrated market whose structure *alone* is conducive to conscious parallel behaviour or co-ordinated effects on the relevant market.[38]

## 2.4 Has '*Airtours*' Changed Anything?

By decision of 22 September 1999, the Commission declared that the acquisition by the tour operator Airtours of one of its main competitors, First Choice, was incompatible with the common market on the ground that it would give Airtours/First Choice and the other two large tour operators in the UK short-haul foreign package holiday market (Thomson and Thomas Cook) a collective dominant position leading to higher prices and increased profits.[39]

In its decision the Commission maintained that Airtours' acquisition of First Choice would result in increased concentration and that the three operators would then be able to co-ordinate their conduct tacitly (without needing to enter into an agreement or a cartel) and to impose, by adopting a common policy, higher prices than those dictated by effective competition.

In a judgment rendered on 6 June 2002, the CFI considered that the Commission failed to prove to the requisite legal standard that the three leading tour operators would have an incentive to cease competing with each other and that the three main conditions for a finding of collective dominance were not met. It therefore annulled the Commission's decision.

It is beyond the scope of this Chapter to comment in detail on the judgement of the CFI. This has been done elsewhere by others.[40] What needs to be underlined is that, in our view, *Airtours* has not altered the existing case-law on oligopolistic dominance.[41] In *Airtours* the CFI tried to refine and/or discuss further certain aspects of the notion of oligopolistic dominance as applied in the Commission's decision. In carrying out its assessment, the CFI disagreed at the end with most of the findings of the Commission. It did not, however, disagree with the essence of the legal principles applied by the Commission. If it is true that the CFI has the last word on the facts of a case, this is subject, however, to the authority of the Court of Justice as far as the interpretation of legal issues is concerned. Accordingly, it will be premature to consider that *Airtours* has settled all legal questions concerning oligopolistic dominance.[42] Ultimately, it will be for the Court of Justice to decide in the future whether the methodological approach laid down in *Airtours* by the CFI was fully compatible with existing case-law.[43]

In *Airtours*, the CFI seems to have been heavily influenced by the fact that both the applicants and the Commission were in agreement as to the three main conditions for a finding of a collective dominance. The first condition concerns the required transparency of the 'oligopolistic' market: 'there must, therefore, be sufficient market transparency for all members of the dominant oligopoly to be aware, sufficiently precisely and quickly, of the way in which the other members' market conduct is evolving'.[44] The second condition (retaliation) results from the first, in the sense that transparency allows the monitoring of the competitors' behaviour. Should an oligopolist choose to deviate from a tacit conduct, 'retaliation' will follow. It is important to stress that according to the CFI, retaliation is not conceived as a punishment mechanism. Rather

> there must be adequate deterrents to ensure that there is a long-term incentive in not departing from the common policy, which means that each member of the dominant oligopoly must be aware that highly competitive action on its part designed to increase its market share would provoke identical action by the others, so that it would derive no benefit from its initiative.[45]

The third condition (the destabilizing factor) encompasses current and future competitors as well as consumers, which would 'jeopardise the results from the common policy'.[46]

The Court's reasoning in *Airtours* as exposed in the introductory part of the judgment entitled 'General considerations',[47] although in line with the main features of the concept of oligopolistic dominance, is not entirely satisfactory or as comprehensive as expected. It would have been better had the Court first defined the notion of 'tight oligopoly', then examined the circumstances under which tacit collusion or parallel conduct[48] is likely to occur before concluding as to the minimum requirements necessary for a finding of oligopolistic dominance. Having followed the reverse process, it is in the assessment of the parties' main arguments that the conditions of a tight oligopoly and of tacit collusion are discussed for the first time.[49] This explains why two of the most important conditions for collective dominance according to the Court of Justice, that is absence of effective competition[50] and ability to implement a common policy as a single entity,[51] are only mentioned *en passant* by the CFI as part of the Commission's burden of proof.[52]

To sum up, *Airtours* should be seen more as a judgment dealing essentially with the nature and the degree of the required burden of proof placed on the Commission than with the relevant legal principles.

## 3. THE COMMISSION'S DECISION-MAKING PRACTICE AND ANNEX II OF THE FRAMEWORK DIRECTIVE

The Commission has applied the concept of collective dominance in relation to oligopolistic markets the structure of which was considered conducive to co-ordinated effects or conscious parallel behaviour in a number of decisions adopted under the Merger Control Regulation.[53]

The approach by the Commission consists, in essence, in a two-step examination:

(a) whether the characteristics of the market makes it conducive to tacit co-ordination, and
(b) whether such form of co-ordination is sustainable, that is, (i) whether any of the oligopolists has the ability and incentive to deviate from the co-ordinated outcome, considering the ability and incentives of the non-deviators to retaliate and (ii) whether buyers/fringe competitors/potential entrants have the ability and incentive to challenge any anticompetitive co-ordinated outcome.[54]

Most of the criteria relied upon by the Commission in applying the notion of collective dominance under the MCR have also been summarized in Annex II of the recently adopted Framework Directive for electronic communications services and networks.[55] According to this Annex, two or more undertakings can be found to be in a joint dominant position within the meaning of Article 14 if, even in the absence of structural or other links between them, they operate in a market, the structure of which is considered to be conducive to co-ordinated effects.[56] This is likely to be the case where the market satisfies a number of appropriate characteristics, in particular in terms of market concentration, transparency and other characteristics mentioned below:

- mature market
- stagnant or moderate growth on the demand side
- low elasticity of demand
- homogeneous product
- similar cost structures
- similar market shares
- lack of technical innovation, mature technology
- absence of excess capacity
- high barriers to entry
- lack of countervailing buying power
- lack of potential competition
- various kind of informal or other links between the undertakings concerned
- retaliatory mechanisms
- lack or reduced scope for price competition.

As Annex II of the Framework Directive also states, the above is not an exhaustive list, nor are the criteria cumulative. Rather, the list is intended to illustrate the sorts of evidence that could be used to support assertions concerning the existence of a collective (oligopolistic) dominance.[57]

The above list, which reflects the present state of law under the EC competition rules, shows that in the absence of structural links, in order to establish whether a market is conducive to collective dominance, it is necessary to consider a number of characteristics of the market.[58] While these characteristics are often presented in the form of the above-mentioned list, this is not a cumulative 'checklist'. In particular, depending on the circumstances of the case, the fact that one or another of the structural elements usually associated with collective dominance may not be clearly established is not in itself decisive for excluding the likelihood of a co-ordinated outcome.[59]

In an oligopolistic market where most, if not all, of the above-mentioned criteria are met, what does need to be established is that market operators have a strong incentive to converge to a co-ordinated market outcome and refrain from reliance on competitive conduct. This will be the case where the long-term benefits of an anticompetitive conduct outweigh any short-term gains resulting from resort to competitive behaviour.[60]

It must be stressed that a mere finding that a market is concentrated does not necessarily warrant a finding that its structure is conducive to oligopolistic dominance.[61]

Ultimately, in applying the notion of collective/oligopolistic dominance, the criteria which will carry the most sway will be those that are critical to a co-ordinated outcome in the specific market under consideration. For instance, in Case No. COMP/M.2499, *Norske Skog/Parenco/Walsum*, the Commission came to the conclusion that even if the markets for newsprint and wood-containing magazine paper were concentrated, the products were homogeneous, demand was highly inelastic, buyer power was limited and barriers to entry were high, none the less the limited stability of market shares, the lack of symmetry in costs structures, the lack of transparency of investments decisions and the absence of a credible retaliation mechanism rendered unlikely and unsustainable any possibility of tacit co-ordination among the oligopolists.[62]

### 3.1 Collective Dominance and the Electronic Communications Sector

The concept of collective dominance has also been applied to a few cases involving the electronic communications sector. For instance, in *MCI WorldCom/Sprint*, the Commission examined whether the merged entity together with Concert Alliance could be found to enjoy a collective dominant position on the market for global telecommunications services (GTS). Given that operators on that market competed on a bid basis where providers were selected essentially in the first instances of the bidding process on the basis of their ability to offer high-quality, tailor-made sophisticated services, and not on the basis of prices, the Commission's investigation was focused on the incentives for market participants to engage in parallel behaviour as to who wins what bid.[63] After having examined in depth the structure of the market (homogeneous product, high barriers of entry, customers countervailing power, and so on), the Commission concluded that it was not able to show absence of competitive constraints from actual competitors, a key factor in examining whether parallel behaviour can be sustained, and thus decided not to pursue further its objections in relation to that market.[64]

In *BT/Esat*,[65] one of the issues examined by the Commission was whether market conditions in the Irish market for dial-up Internet access lent themselves to the emergence of a duopoly consisting of the incumbent operator, Eircom, and the merged entity. The Commission concluded that this was not the case for the following reasons. First, market shares were not stable; second, demand was doubling every six months; third, Internet access products were not considered homogeneous; and finally, technological developments were one of the main characteristics of the market.[66] In *Vodafone/Airtouch*,[67] the Commission found that the merged entity would have joint control of two of the four mobile operators present on the German mobile market (namely D2 and E-Plus, the other two being T-Mobile and VIAG Interkom). Given that entry into the market was highly regulated, in the sense that licences were limited by reference to the amount of available radio frequencies, and that market conditions were transparent, it could not be ruled out that such factors could lead to the emergence of a duopoly conducive to co-ordinated effects.[68]

Finally, in *France Telecom/Orange* the Commission found that, prior to the entry of Orange into the Belgian mobile market, the two existing players, Proximus and Mobistar, were in a position to exercise joint dominance. As the Commission noted, for the four years preceding Orange's entry, both operators had almost similar and transparent pricing, their prices following exactly the same trends.[69] In the same decision the Commission further dismissed claims by third parties as to the risk of a collective dominant position of Vodafone and France Telecom in the market for the provision of pan-European mobile services to internationally mobile customers. Other than significant asymmetries between the market shares of the two operators, the market was considered to be emerging, characterized by an increasing demand and many types of different services on offer and on price.[70]

## 4. CONCLUDING THOUGHTS

Without doubt, given today's tendency towards further consolidation in most sectors of the economy, collective dominance will play an ever-increasing role in the application of EC competition rules. Looking into the Court's relevant jurisprudence, it is rather reassuring to see that the Community judge, through a purposive, teleological interpretation, has managed to offer Article 82 of the Treaty and the MCR the required flexibility and ammunition to adapt to the dynamic and constant evolution of markets and the strategic behaviour and interaction of players with market power. Of course, as in every case of purposive interpretation, the question is how far the language of a given provision can go, and where one sets the limits beyond which it is for the legislator to intervene, not for the judge.

The case-law of the Court also shows that there if there is still an intellectual gap between the perception that a judge has of the notion of collective dominance and that most economists have, none the less, as *Gencor*, *CMB* and *Airtours* demonstrate, that gap is gradually closing. This being said, divergences between these two schools of thought will always persist since the Court performs a much wider role, that of ensuring the respect and maintenance of the 'rule of law'. It is the absence of knowledge of the main features of the Community legal edifice that often leads to unnecessary misunderstandings as to how the Treaty provisions and secondary Community law should apply.

What is important from a legal point of view is to ensure that the judicial interpretation and application of texts does not in the end become counterproductive and is not overtaken by the constant developments in the functioning and organization of markets and society. Ultimately, it is incumbent on the Commission to ensure, subject to the judicial review of the Court, that through its decision-making powers the notion of oligopolistic dominance remains a workable tool in the maintenance of undistorted competition within the single market.

## NOTES

* Former *Référendaire*, Court of First Instance of the EC. All views expressed are strictly personal.

1. See, in particular, M. Schrödermeier, 'Collective dominance revisited: An analysis of the EC Commission's new concept of oligopoly control', European Competition Law Review (1990), p. 28; A. Briones, 'Oligopolistic dominance: is there a common approach in different jurisdictions? A review of the decisions adopted by the Commission under the Merger Regulation', European Competition Law Review (1995), p. 334; T. Soames, 'An analysis of the principles of concerted practice and collective dominance: A distinction without a difference?', European Competition Law Review, p. 24 (1996); J. Venit, 'Two steps forward and no steps back: Economic analysis and oligopolistic dominance after KALI&SALZ', Common Market Law Review (1998), 1101; R. Whish, 'Collective Dominance', in D. O'Keefe and A. Bavasso (eds), *Judicial Review in European Union Law, Liber Amicorum in Honour of Lord Slynn of Hadley*, Vol. 1 (Kluwer, 2000); T. Penard, '*Collusion et comportements dynamiques en oligopole: une synthèse*', (CREREG, Université de Rennes 1, 1997). Kai-Uwe Kühn, 'Closing the Pandora's Box? Joint dominance after the "Airtours" judgment', University of Michigan, John M. Olin Center for Law and Economics, Paper 02-013; Kai-Uwe Kühn, 'An economist's guide through the Joint Dominance', University of Michigan, John M. Olin Center for Law and Economics, Paper 02-014; Massimo Motta, 'EC merger policy and the Airtours case', *European Competition Law Review* (2000), p. 197; John Temple Lang 'Oligopolies and joint dominance in community antitrust law', in B. Hawk (ed.), 2001 Annual Proceedings of the Fordham Corporate Law Institute, (2002), p. 269; Frederic Jenny, 'Collective Dominance and the EC Merger Regulation', in B. Hawk (ed.), 2001 Annual Proceedings of the Fordham Corporate Law Institute, (2002), p. 361; Giorgio Monti, 'The scope of Collective Dominance under Article 82 EC', *Common Market Law Review* (2001), p. 131; Heiko Haupt, 'Collective Dominance under Article 82 EC and EC Merger Control in the light of the Airtours Judgment', *European*

*Competition Law Review* (2002), p. 434; Juan Briones and Atilano Jorge Padilla, 'The complex landscape of oligopolies under EU Competition Policy', *World Competition* (2001), p. 307; Borris Etter, 'The assessment of mergers in the EC under the concept of Collective Dominance. An analysis of the recent decisions and judgments – by an economic approach', *World Competition* (2000), p. 103; Sigrid Stroux, 'Is oligopoly control outgrowing its infancy?', *World Competition* (2000), p. 50; Marcel Canoy and Sander Onderstal, 'Tight oligopolies. In search of proportionate remedies', CPB Netherlands Bureau for Economic policy analysis, CPB Document No 29, (February 2003); NERA, 'Merger appraisal in oligopolistic markets', OFT Research Paper 19, (November 1999); Europe Economics, 'Study on assessment criteria for distinguishing between competitive and dominant oligopolies in merger control', Report for the European Commission Enterprise Directorate General (May 2001).

2. Council Regulation (EEC) No. 4064/89 of 21 December 1989 on the control of concentrations between undertakings (OJ L 395, 30.12.1989, p. 1), as amended by Council Regulation (EC) No. 1310/97 of 30 June 1997, OJ L 180, 9.7.1997, p. 1.
3. See in particular, Joined Cases C-68/94 and C-30/95 *France and Others v. commission* [1998] ECR I-1375, Joined Cases C-395/96 P *Compagnie Maritime Belge des transports and Others v. Commission*, [2000] ECR I-1365, Case C-497/99 P [2001] I-5333, Case C-309/99 *Wouters and Others* [2002] ECR I-1577, Case T-342/99, *Airtours v. Commission* 2002 [ECR] II-2585, Case T-102/69 [1999] ECR II-753, and Case T-374/00 *Verband der freien Rohrwerke eV and Others v. Commission* [2003] ECR II-0000.
4. See, in particular, G.J. Stigler, 'A theory of oligopoly', *Journal of Political Economy*, (1964) p. 44; C. Shapiro, 'Theories of Oligopoly behaviour', in R. Schmalensee and R. Willig (eds), *Handbook of Industrial Organisation*, Amsterdam: North-Holland, p. 329; L. Phlips, 'Parallélisme de comportements et pratiques concertées', *Revue d'Économie Industrielle*, (1993), p. 63. W.T. Stanbury and G.B. Reschenthaler, 'Oligopoly and conscious parallelism: Theory, policy and the Canadian cases', *Osgoode Hall Law Journal*, (1977), p. 617; R.B. Heflebower, 'Conscious parallelism and administered prices', in Almarin Phillips (ed.), *Perspectives on Antitrust Policy*, Princeton, NJ: Princeton University Press, Chapter 5, p. 86, (1965); Fritz Machlup, *The Economics of Sellers' Competition: model Analysis of Sellers' Conduct*, Baltimore, MD: The Johns Hopkins Press (1952), pp. 432–46; Richard Posner, 'Oligopolistic pricing suits, the Sherman Act and economic welfare. A reply to Professor Markovits', *Stanford Law Review* (1976), p. 903.
5. References to the Court of Justice also cover the Court of First Instance (CFI), the latter being institutionally an organic part of the former.
6. Case 48/69, *ICI* v. *Commission* [1972] ECR 619.
7. Case 48/69, *ICI* v. *Commission* [1972] ECR 619. As the Court has also ruled in *Züchner*, 'parallel conduct in the debiting of a uniform bank charge on transfers by banks from one member state to another of sums from their customers' funds amounts to a concerted practice prohibited by Article 85(1) of the Treaty if it is established by the national court that such parallel conduct exhibits the features of co-ordination and co-operation characteristic of such a practice and if that practice is capable of significantly affecting conditions of competition in the market for the services connected with such transfers', Case 172/80, *Züchner* v. *Bayerische Vereinsbank* [1981] ECR 2021, para. 22.
8. In fact, the term 'concerted practice' is synonymous with the terms 'collusion' and 'co-operation'. If undertakings co-operate in pricing their products, they are engaged in collusion, and in the absence of a written or tacit agreement, they are also engaged in concerted practices. See also Fritz Machlup, 'The forms and degrees of collusion', in *The Economics of Sellers' Competition* (Baltimore, MD: Johns Hopkins Press, 1952) pp. 432–46.
9. *ICI* v. *Commission* [1972] ECR 619, paras 66–68. In particular, 'the criteria of co-ordination and co-operation laid down by the case-law on restrictive practices, far from requiring the working out of an actual plan, should be understood in the light of the concept inherent in the provisions of the Treaty relating to competition that each economic operator must determine independently the policy which he intends to adopt on the common market. Although it is correct to say that that requirement of independence does not deprive economic operators of the right to adapt intelligently to the existing and anticipated conduct of their competitors, it does however strictly preclude any direct or indirect contact between such operators, the object or effect whereof is either to influence the conduct on the market of an actual or potential competitor or to disclose to such a competitor the course of conduct which they themselves have decided to adopt or contemplate adopting on the market'. See Joined Cases 40/73 to 48/73, 50/73, 54/73, 55/73, 56/73, 111/73, 113/73 and 114/73 *Suiker Unie and Others* v. *Commission* [1975] ECR 1663, para. 174, Joined Cases T-202/98, T-204/98 and T-207/98, *Tate & Lyle plc, British Sugar plc and Napier Brown & Co. Ltd* v. *Commission* [2001] ECR 11-2035. In *Suiker Unie* the Court clearly stated that 'the fact that a vendor aligns his price on the highest price charged by a competitor is not necessarily evidence of a concerted practice but may be explained by an attempt to obtain the maximum profit'.
10. Distinguishing between tacit agreement or collusion and parallel oligopolistic behaviour has been one of the most debated issues in the USA under the Sherman Act, Section 1, since in the absence of an agreement or conspiracy, Section 1 could not apply to an oligopolistic dominance ('the shared monopoly debate'). As far

as Section 2 of the Sherman Act is concerned, US courts that have squarely addressed the issue have determined that Section 2 of the Sherman Act applies to the conduct of single firms only, rather than to the conduct of a small number of firms engaged in tacit collusion, as in cases involving oligopoly, shared monopoly, or duopoly. See *Rebel Oil Co., Inc. v. Atl. Richfield Co.*, 51 F.3d 1421, 1443 (9th Cir. 1995), stating that the 'fact that competitors may see proper to follow the prices of another manufacturer does not violate the Sherman Act. To pose a threat of monopolization, one firm alone must have the power to control market output and exclude competition', and that 'an oligopolist lacks this unilateral power. By definition, oligopolists are interdependent. An oligopolist can increase market price, but only if the others go along'; *Indiana Grocery, Inc. v. Super Valu Stores, Inc.*, 864 F.2d 1409, 1416 (7th Cir. 1989). See also, H.L. Hayden, 672 F. Supp. at 741 (S.D.N.Y. 1987) expressing 'considerable discomfort' with shared monopoly theory and noting that the 'notion that two competitors could conspire to monopolize is, seemingly, antithetical'. In the same vein, Consolidated Terminal Sys., Inc. v. ITT World Communications, Inc., 535 F. Supp. 225, 229 (S.D.N.Y. 1982), stating that an oligopoly, or a shared monopoly, does not in itself violate Section 2 of the Sherman Act. For a more recent case, see *ID Security Systems Canada, Inc., v. Checkpoint Systems, Inc., District Court for the Eastern District of Pennsylvania*, 249 F. Supp. 2d 622; 2003 U.S. Dist. LEXIS 4807; 2003-1 Trade Cas. (CCH) P74, 004, (decided March 28, 2003) and *Santana Prods. v. Sylvester & Assocs.*, 2000 U.S. Dist. LEXIS 20213, 9-10. For a contrary, but clearly a minority view, see JTC Petroleum Co. v. Piasa Motor Fuels, Inc., 190 F.3d 775, 780 (7th Cir. 1999), noting that oligopolistic interdependence may be actionable under Section 2 where an agreement is proven. As a result, inferring from circumstantial evidence the existence of a tacit agreement or of a conspiracy is the only way to deal with oligopolistic dominance under the Sherman Act, Section 1. See also Michael K. Vaska, 'Conscious Parallelism and Price Fixing: Defining the Boundary', *University of Chicago Law Review*, **52**, p. 508. As commented in the past, the real problem is that to try to challenge pure oligopolistic behaviour (that is, pricing) would be tantamount to challenging the industry structure *per se*, which is not an appropriate application of the concept of conspiracy or tacit agreement, D. Turner, 'The definition of agreement under the Sherman Act: Conscious Parallelism and Refusals to deal', *Harvard Law Review* (1962), p. 665. For an excellent counter-argument, see R. Posner, 'Oligopoly and the Antitrust Laws: A Suggested Approach', *Stanford Law Review* (1969), p. 1562; L. Kestenbaum, 'What is "price signalling" and does it violate the law?', *Antitrust Law Journal* (1980), p. 683. Needless to say, the influence of this debate on the early case-law of the Court of Justice and in particular on its *Woodpulp II* case (see below) is more than obvious. The main difference with the US approach is that Article 82 and the MCR do cover so called 'shared monopoly' situations.

11. Joined Cases C-89/85, C-104/85, C-114/85, C-116/85, C-117/85 and C-125/85 to C-129/85, *Ahlström Osakeyhtiö and Others* v. *Commission* [1993] ECR 1-1307, para. 42, hereafter referred to as '*Woodpulp II*'.
12. See also G. Van Gerven and E. Navarro Varona, 'The Wood Pulp case and the future of concerted practices', *Common Market Law Review* (1994), p. 575.
13. Ibid., at para. 70.
14. According to the Court, 'the transparency of the market could be responsible for certain overall price increases recorded in the short term: when demand exceeds supply, producers who are aware – as was the case on the pulp market – that the level of their competitors' stocks is low and that their production capacity utilisation rate is high would not be afraid to increase their prices. There would then be a serious likelihood of their being followed by their competitors.' *Woodpulp II*, at para. 105.
15. Ibid., para. 116.
16. Ibid., para. 126. In *Binon/AMP*, the Court found that 'parallel conduct on the part of different publishers vis-à-vis the approval of retail outlets is to be regarded as a "concerted practice" within the meaning of article 81 of the EEC Treaty where ... the parallel conduct occurs within a contractual framework'. The fact that all publishers of foreign newspapers distributed their newspapers in Belgium through a single distribution agency and refused to supply retailers directly resulted not from a parallel conduct but from a prior set of contracts, namely the individual exclusive agreements made between the publishers and the distribution agency and those made between the latter and retail sellers. See Case 243/83, [1985] ECR 2015, paras 15–16. As noted above, *Woodpulp II* is fully in line with US antitrust laws and jurisprudence. Under Section 1 of the Sherman Act, an anticompetitive agreement may be inferred from parallel conduct if the undertakings involved cannot show legitimate, independent business justifications for engaging in such practices, see *Weit* v. *Continental Ill. Nat'l Bank & Trust Co.*, 641 F.2d 457, 463 (7th Cir. 1981) holding that parallel conduct plus opportunity to conspire did not justify conclusion of conspiracy when rebutted by evidence that the adoption of the parallel conduct was supported by independent business justifications, cert. denied, 455 US 988 (1982).
17. Case 78/70, *Deutsche Grammophon* v. *Metro* [1971] ECR 487, *Suiker Unie* v. *Commission*, op. cit.
18. Case 247/86 *Alsatel* v. *Novosam* [1988] ECR 5987, paras 21–2.
19. OJ C 233, 6.9.1991, p. 2.
20. Ibid., para. 79.
21. See Joined Cases T-68/89, T-77/89 and T-78/89, *SIV and Others* v. *Commission* [1992] ECR 11-1403, paras

357–8, and Joined Cases C-395/96 P and C-396/96 P, *Compagnie Maritime Belge and Others* v. *Commission* [2000] ECR I-1365.

22. See *Suiker Unie* and the Opinion of Advocate General Mayras; Case 85/76, *Hoffmann–La Roche* v. *Commission* [1979] ECR 461, para. 39; *Züchner*, para. 10, and the Opinion of Advocate General Slynn, p. 2039; Case 293/83, *CICEE* v. *Commission* [1985] ECR 1105, and the Opinion of Advocate General Lenz, p. 1110; Case 75/84 *Metro* v. *Commission* [1986] ECR 3021, para. 84; and Case 247/86, *Alsatel* [1988] ECR 5987, paras 20–22.
23. Joined Cases T-68/89, T-77/89 and T-78/89, *SIV and Others* v. *Commission* [1992] ECR II-1403.
24. *Flat Glass*, para. 358.
25. In its defence the Commission argued that the three Italian producers constituted or acted as a single entity not because of the structure of the oligopoly but because of the agreements and concerted practices which led them to create structural links amongst themselves, such as, in particular, the systematic exchanges of products, at para. 354.
26. Ibid., para. 363. In particular, the Court found that the Commission had not attempted to gather the information necessary to weigh up the economic power of the three producers against that of Fiat (a monopsony buyer in the 'automotive market'), which could cancel each other out, ibid., at para. 366.
27. *Flat Glass*, para. 358. This line of reasoning was also taken up later by the Court of Justice, Case C-393/92, *Almelo* [1994] ECR I-1477, para. 43; Case C-96/94, *Centro Servizi Spediporto* [1995] ECR I-2883, para. 33; Joined Cases C-140/94, 141/94 and C-142/94, *DIP* [1995] ECR I-3257, para. 62; Case C-70/95, *Sodemare* [1997] ECR I-3395, para. 46; and Joined Cases C-68/94 and C-30/95, *France and Others* v. *Commission* [1998] ECR I-1375, para. 221.
28. Case T-102/96, *Gencor* v. *Commission* [1999] ECR II-753.
29. Case No. IV/M.619, *Gencor Lonhro* (OJ 1997, L 11, p. 30).
30. *Gencor* v. *Commission*, at para. 276.
31. Ibid., at para. 277.
32. Joined Cases C395/96 P and C-39/96 P, *Compagnie Maritime Belge and Others* v. *Commission* [2000] ECR I-1365. See also Sigrid Stroux, 'Case note', *Common Market Law Review* (2000), p. 1249.
33. Ibid., at para. 39.
34. See in particular, *France and Others* v. *Commission*, para. 221.
35. *Compagnie Maritime Belge*, at para. 39.
36. Ibid., at para. 44. In this respect, *CMB* seems to throw into doubt *Flat Glass*, where the CFI ruled that it is not 'sufficient to recycle the facts constituting an infringement of Article 81, deducing from them the finding that the parties to an agreement or to an unlawful practice jointly hold a substantial share of the market, that by virtue of that fact alone they hold a collective dominant position, and that their unlawful behaviour constitutes an abuse of that collective dominant position', *Flat Glass*, para. 360.
37. Ibid., at para. 45. With this last *dictum* the Court implicitly endorsed and incorporated *Gencor* and the theory of oligopolistic interdependence into its own case-law. See also Recital 26 of Directive 2002/21/EC on a common regulatory framework for electronic communications networks and services (OJ 2002 L 108, p. 33): 'Two or more undertakings can be found to enjoy a joint dominant position not only where there exist structural or other links between them but also where the structure of the relevant market is conducive to co-ordinated effects, that is, it encourages parallel or aligned anti-competitive behaviour on the market'.
38. The use of the term 'co-ordinated effects' is no different from the term 'parallel anticompetitive behaviour' also used in the Commission's decisions applying the concept of collective (oligopolistic) dominance.
39. Case IV/M.1524 *Airtours/First Choice* (OJ 2000 L 93, p. 1).
40. For a critical assessment, see, in particular, Andrew Scott, 'An immovable feast?: Tacit collusion and collective dominance in Merger Control after Airtours', CCR Working Paper CCR 02-6 and the references therein. Stroux, Sigrid, 'Collective dominance under the Merger Regulation: a serious evidentiary reprimand for the Commission, *European Law Review* (2000), p. 736.
41. In fact the reasoning of the CFI starts in para. 59 by making references to the principles laid down first by the Court of Justice in *Kali und Salz* (Joined Cases C-68/94 and C-30/95, *France and Others* v. *Commission* [1998] ECR I-1375) and in *Gencor*. The marked absence of any reference to *CMB* may be an indication that the CFI intends to distinguish cases under Article 82 from those falling under the MCR. However, this would not be in line with the jurisprudence of the Court of Justice. *CMB* makes explicit reference to *Kali und Saltz*; see para. 41 of *CMB*.
42. It is rather surprising, given the issues at stake, that the case was not heard by the plenary of the CFI but by one of its chambers.
43. For instance, according to the CFI, for the purposes of applying the MCR and the notion of collective dominant position, the Commission 'must ascertain whether the concentration would have the *direct* and *immediate* effect of creating or strengthening a position of that kind, which is such as significantly and lastingly to impede competition in the relevant market' (our emphasis). Nowhere in the case-law of the Court of Justice is there a requirement to ascertain the 'direct and immediate' effect of a merger on creating or

strengthening a collective dominance. The CFI makes reference to para. 94 of its judgment in *Gencor*. However, para. 94 of *Gencor* only refers to the fact that the application of the MCR is justified under public international law when it is foreseeable that a proposed concentration will have an immediate and substantial effect in the Community. Is it realistic to require from the Commission, as the CFI seems to suggest, that under a prospective analysis the Commission must prove that the day following the consummation of a merger the latter will 'immediately and directly' lead to the creation of a collective dominant position? From a strict legal point of view, it might be rather difficult to reconcile such a requirement with the repeated interaction of oligopolists which ultimately but not 'immediately' can lead to an oligopolistic interdependence and thus to an oligopolistic dominance?

44. *Airtours* v. *Commission*, para. 62.
45. Ibid. The Court is careful enough to avoid speaking about a retaliation mechanism *per se*: 'the Commission must not necessarily prove that there is a specific retaliation mechanism involving a degree of severity, but it must none the less establish that deterrents exist, which are such that it is not worth the while of any member of the dominant oligopoly to depart from the common course of conduct to the detriment of the other oligopolists'. This is an important aspect of the Court's reasoning. If oligopolistic dominance is assessed not under the MCR but *ex post* under Article 82, it will be, by definition, impossible to find proof of a retaliation mechanism *stricto sensu*. The incentives not to depart from a parallel conduct should come from the realisation that the deviation will be more costly in the long term than the (short term) gains expected from a deviation. Therefore, under an Article 82 assessment, and in the absence of any prior evidence of 'a retaliation mechanism', the sustainability of the tacit co-ordination will have to be inferred from the structure and characteristics of the market. If one can adduce evidence of an agreed punishment mechanism, then it is more likely that the oligopolists have already formed a cartel in which case Article 81 might apply.
46. The Court's dictum covers both the so-called 'mavericks' and the countervailing power that consumers or users may be able to exercise. Jonathan B. Baker has rightly suggested reconstructing the antitrust analysis of the loss of a firm whether through a merger or exclusion around the role played by a maverick in constraining tacit co-ordination: 'A focus on identifying the industry maverick can increase the precision of the antitrust analysis of mergers and exclusion, allowing courts to understand better when the loss of a firm is likely to promote tacit collusion and when it may not'; Jonathan B. Baker, 'Mavericks, mergers and exclusion: proving coordinated competitive effects under the Antitrust law', *New York University Law Review*, (2000), p. 135. Although the CFI's reference to *actual* competitors (fringe competitors, mavericks) makes sense and calls for a verifiable assessment, it is difficult to identify potential competitors and to prove to the 'requisite legal standard' that their mere existence suffices to undermine the sustainability of tacit collusion raises. The burden of proof is too high and quite arbitrary. Rather, the focus should be on the existence or absence of entry barriers and the overall contestability of the market. This is at the end the conclusion reached by the Court in para. 269 of its judgment, where it concludes that the Commission failed to take into account the 'fact that the lack of barriers to market entry is likely to allow potential competitors to gain access to and offer their products on the relevant market and therefore to take fast and effective action'.
47. *Airtours*, paras 56–65.
48. In paras 62 and 192 of the judgment the Court refers to the obligation on the Commission to examine whether any parallel anticompetitive conduct to which the notified transaction might give rise is sustainable over time. 'Common policy', 'tacit collusion' and 'parallel conduct' are used invariably by the Court to describe the existence of an oligopolistic interdependence.
49. Ibid., paras 111 (stability of historic market shares), 133 and 139 (mature market, stable demand), 156 (sufficient transparency) 195 (effective and quick deterrents/retaliatory measures) 210 (ability of actual/ potential competitors to counteract the creation of a collective dominant position) 274 (consumers' buyer power) 285 (structural/strong economic links).
50. 'The evidence must concern in particular factors playing a significant role in the assessment of whether a situation of collective dominance exists, such as for example, *the lack of effective competition* between the operators alleged to be members of the dominant oligopoly and the weakness of any competitive pressure that might be exerted by other operators' (our emphasis). *Airtours*, para. 63.
51. *CMB*, para. 39. As the Court ruled in *CMB*, it is only where the question concerning the existence of a single entity is answered in the affirmative that it is appropriate to consider whether that collective entity actually holds a dominant position. The only plausible explanation for *Airtours* having downplayed the requirement of a 'single entity' may be that if under Article 86 and an *ex post* assessment one may be able to prove that the oligopolists acted as if they were a single entity, the same assessment is meaningless within an *ex ante*, prospective assessment, where what matters most is to establish the likelihood of parallel behaviour or a common policy. The 'single entity' requirement appears to be somewhat more demanding than proving the likelihood of 'conscious parallelism' *post merger*. In our view, the Court ought to have clarified these issues.
52. *Airtours*, paras 76–7.
53. See in particular, Case No. COMP/M.2553, *BP/E.ON* (OJ L 276, 12.10.2002); Cases No. COMP/M.2498,

*UPM–Kymmene/Haindl* and COMP/M.2499, *Norske Skog/Parenco/Walsum*; Case No. COMP/M.2201, *MAN/Auwärter*; Case No. COMP/M.2389, *Shell/DEA* (OJ L 15, 21.01.2003, p. 35); Case No. IV/2824, *Ernst & Young/Andersen Germany*; Case No. COMP/M. 2097, *SCA/Matsä Tissue*; Case No. COMP/M. *1882, Pirelli/BICC*; Case No. IV/M.1527, *OTTO Versand/Freemans*; Case No. COMP/M.1741, *MCI WorldCom/Sprint*; Case No. IV/M.1524, *Airtours/First Choice* (OJ L 93, 13.4.2000, p. 1); Case No. IV/M.1383, *Exxon/Mobil*; Case No. IV/M.1313, *Danish Crown/Vestjyske Slagterier* (OJ L 20, 25.1.2000, p. 1); Case No. IV/M.1225, *Enso/Stora* (OJ L 254, 29.9.1999, p. 9); Case No. IV/M.1517, *Rhodia/Donau Chemie/Albright & Wilson*, Case No. IV/M.1016, *Price Waterhouse/Coopers & Lybrand* (OJ L 50, 26.2.1999, p. 27); Case No. IV/M.619, *Gencor/Lonrho*; Case No. IV/M.308, *Kali + Salz/MdK/Treuhand* (OJ L 186, 21.7.1994, p. 38); and Case No. IV/M.190, *Nestlé/Perrier* (OJ L 356, 5.12.1992, p. 1).

54. This is in essence the type of analysis carried out by the Commission in past decisions related to collective dominance. See for instance, Case No. IV/M.190, *Nestlé/Perrier*; *Gencor/Lonrho*, Case No. IV/M.1383, *Exxon/Mobil*, para. 259; and Case No. COMP/M.2499, *Norske Skog/Parenco/Walsum*, para. 76.
55. Directive 2002/21/EC of the European Parliament and of the Council of 7 March 2002 on a common regulatory framework for electronic communications networks and services (Framework Directive) OJ L 108, 24.4.2002, p. 33.
56. See also Recital 26 of the Framework Directive: 'two or more undertakings can be found to enjoy a joint dominant position not only where there exist structural or other links between them but also where the structure of the relevant market is conducive to co-ordinated effects, that is, it encourages parallel or aligned anticompetitive behaviour on the market'.
57. See also Case No. COMP/M.2498, *UPM–Kymmene/Haindl*, and Case No. COMP/M.2499, *Norske Skog/Parenco/Walsum*, at para. 77.
58. For a recent example where the Commission found that most of the necessary conditions for a finding of collective dominance were met, see Cases COMP/M.2389, *Shell/DEA* and COMP/M.2533, BP/E.ON. These two cases concerned the acquisition of the German oil and petrochemicals company DEA by Royal Dutch/Shell and the combination of the petrochemicals businesses of BP and Veba, a subsidiary of the E.ON group. The Commission found that the combination of the respective petrochemical activities of Shell and DEA, on the one hand, and of BP and E.ON, on the other, would result in the creation of a joint dominant position on the market for the supply of ethylene on the pipeline network called 'ARG' which links various production sites, sea terminals and ethylene consumers in Belgium, the Netherlands and Western Germany. Both transactions' major impact was the elimination of the only downstream non- integrated ethylene producers from the market, which were also the most important suppliers to the merchant market. This would have left independent ethylene buyers only with suppliers which compete with their customers in the downstream markets. Both merged entities would control the biggest part of the ethylene market, with a unique position (control) of the ARG pipeline. The transaction was finally cleared after the parties offered commitments to ensure effective access to the pipeline. See in particular, paras 116–25 of *Shell/DEA* for a detailed assessment of the retaliation mechanism and the incentives for deviation.
59. See for instance, Case No. COMP/M.2097, *SCA/Metsä Tissue*.
60. See also Thierry Penard, 'Structures de marché et pratiques collusives: une approche par la théorie des jeux repétés', forthcoming in *Economie Rurale* (2003).
61. For instance, in Case No. COMP/M.2201, *MAN/Auwärter*, despite the fact that two of the parties present in the German city-bus market in Germany, MAN/Auwärter and EvoBus, would each supply just under half of that market, the Commission concluded that there was no risk of joint dominance. In particular, the Commission found that any tacit division of the market between EvoBus and MAN/Auwärter was not likely as there would be no viable co-ordination mechanism. Second, significant disparities between EvoBus and MAN/Auwärter, such as different cost structures, would make it likely that the companies would compete rather than collude. Likewise, in the *Alcoa/British Aluminium* case, the Commission found that despite the fact that two of the parties present in the relevant market accounted for almost 80 per cent of the sales, the market could not be said to be conducive to oligopolistic dominance since (i) market shares were volatile and unstable, and (ii) demand was quite irregular, making it difficult for the parties to be able to respond to each other's action in order to tacitly co-ordinate their behaviour. Furthermore, the market was not transparent in relation to prices and purchasers had significant countervailing power. The Commission's conclusions were further reinforced by the absence of any credible retaliation mechanism likely to sustain any tacit co-ordination and the fact that competition in the market was not only based on prices but depended to a large extent on technological innovation and after-sales follow-up. Case No. COMP/M.2111, *Alcoa/British Aluminium*.
62. Likewise, in Case No. COMP/M.2348, *Outokumpu/Norzink*, the Commission found that even if the zinc market was composed of few players, entry barriers were high and demand growth perspectives low, the likelihood of the emergence of a market structure conducive to co-ordinated outcome was unlikely if it could be shown that (i) parties could not manipulate the formation of prices, (ii) producers had asymmetric cost structures and there was no credible retaliation mechanism in place in the markets examined in that case, the

possibility of parallel behaviour resulting in a co-ordinated outcome depended mainly on the transparency of investments decisions and the availability of a device that would allow firms to co-ordinate, *Norske Skog/Parenco/Walsum*, para. 127. Even if in some respects the parties could engage in tacit co-ordination (restriction of 'output downtimes'), the existence of fringe players capable of increasing capacities made any parallel strategy of co-ordination on 'downtimes' unsustainable. Ibid., para. 146.

63. See Case No. COMP/M.1741, *MCI WorldCom/Sprint*, para. 263.
64. Ibid., paras 257–302.
65. Case No. COMP/M.1838, *BT/Esat*.
66. Ibid., paras 10–14.
67. Case No. IV/M.1430, *Vodafone/Airtouch*.
68. Ibid., at para. 28. The likely emergence of a duopolistic market concerned only the three largest mobile operators, that is D2 and E-Plus, on the one hand, and T-Mobil on the other, given that VIAG Interkom's market share was below 5 per cent. The Commission's concerns were finally allayed after the parties proposed to divest Vodafone's entire stake in E-Plus.
69. Case No. COMP/M.2016, *France Telecom/Orange*, at para. 26.
70. Ibid., at paras 39–40.

# Section 4

# Economic Aspects of Access to Networks

# 8. Access to telecommunications networks

**Marcel Canoy, Paul de Bijl and Ron Kemp**

There are things known, there are things unknown, in between are doors.

Jim Morrison

## 1. INTRODUCTION

Access has been a key issue for regulators right from the start of telecommunications market liberalization. Despite some notable submarket successes, even by 2002 – more than ten years after the first steps towards liberalization – incumbent operators still maintain a strong hold on many aspects of the sector, and infrastructure competition has not matured across the industry. Most notably, there has been little infrastructure competition within the local loop, while its unbundling progresses slowly (Buigues, 2002). As a result, incumbent operators with local loops are able to engage in anticompetitive behaviour, for example by providing access against unfavourable terms. Because *ex post* application of the competition law can have only a limited effect in such situations, regulators are tasked with setting the terms and conditions of access charges.

Although the legitimacy of involving regulators in access charges can be easily understood, things are a little trickier when it comes to determining just what that involvement should entail. Tasking them with setting access charges poses particular problems for a number of reasons. First, the economic theory on determining optimal methods of setting access charges is not always clear-cut. Second, even in cases where the theory is unequivocal, the practice is not as easy as theory may suggest. The 'difficulties of the daily regulation business' is not just the usual cliché. Admittedly, the usual problems do exist here: required data are often simply not available and time is lacking. But on top of that, access regulation often serves too many goals. Access charges are used to allocate scarce capacity, to provide incentives for productive efficiency, to promote entry, to promote investments by incumbents as well as by entrants – and all taking into consideration the fair allocation of common costs while also meeting equity goals such as a universal service obligation. This is quite a burden for an access charge, which is, in its simplest form, a one-dimensional price. Goals can so easily conflict. Low access charges can be good for stimulating service competition, but by the same token they might hamper infrastructural investment. The very least any regulator has to do is to determine just which goals it wants to reach with an access charge, and, in case of conflicts, which goals take priority. Sometimes 'smart regulation' is possible in attaining seemingly conflicting goals: an access charge which starts low but rises over time can promote services competition in the short term (enhancing static efficiency), while also providing incentives for entrants and incumbents to invest in infrastructure (enhancing dynamic efficiency). But more often than not, choices have to be made and tradeoffs emerge.

This chapter provides an overview of access charge theory, mindful of the policy perspective. We draw a distinction between one- and two-way access.[1] One-way access refers to an entrant who needs access to an infrastructure owned by an incumbent, though the reverse is not true. Examples are unbundled local loop and carrier select. Two-way access refers to network interconnection, where operators need mutual access in order to terminate calls on each other's networks. Policy implications can be quite different in one- and two-way access.

Determining access prices implies taking into consideration both the short and the long term. Particularly in the early days, both economic theory and regulation practices focused strongly on static efficiency (that is, the short term). However, dynamic considerations are important and should not be neglected for a number of reasons, one of which being that if infrastructure's competition matures, less regulation is needed. For this reason this chapter devotes extra attention to dynamic considerations.

An interesting issue, briefly addressed here, is why the European Commission is involved in access policy. Typically, few worthwhile cross-border trade or arbitrage opportunities exist, and neither is there a 'policy-race-to-the-bottom' argument for centralization. Thus there would seem to be a strong case here for decentralization. But there seems little point in having the national regulators constantly reinventing the wheel. Access problems in Finland appear very similar to those in Greece. The problems are highly complex, so that when it comes to knowledge, achieving economies of scale is an important reason for the European Commission's involvement.

The main sources used for this chapter are surveys of the economic literature on telecommunications by Armstrong (2001) and Laffont and Tirole (2000), and policy-oriented research by Cave et al. (2001) and (to a lesser extent) Bennett et al. (2001), as well as a number of policy documents, most notably from the European Commission and Oftel.

This chapter is organized as follows. Section 2 discusses one-way access. Section 3 discusses two-way access. Section 4 analyses four special topics, namely fixed–mobile termination, margin squeeze, the allocation of common costs, and static versus dynamic efficiency. Section 5 contains the conclusion.

## 2. ONE-WAY ACCESS

### 2.1 Introduction

One-way access is characterized by a vertically integrated incumbent, usually the former state-owned monopolist, with a local access network. One or more entrants will not have such a local network, nor will they be able to build one in the reasonably short term. Entrants may have their own long-distance network, but this is not necessary.[2] However, in order to compete with the incumbent in the retail voice telephony market and to match the incumbent's pattern of offers, they need access to the incumbent's local access network.

One-way access is common in competition's early phases, when entrants have not yet been able to roll out (some of) their own local connections; but it is also relevant in the market's mature stages. An important example of one-way access is unbundled local loop

(ULL): an entrant gets access to the incumbent's local network copper-cable pairs, and has control over the use of (some of) the copper-cable pairs' total frequency spectrum. The main implication of this is that it enables new network operators to offer high bandwidth directly to consumers, thereby fostering competition for high-bandwidth services.[3]

Why is so much policy attention devoted to ULL? In the first place, ULL allows for a direct and comprehensive relationship with the end-consumer without requiring a full rolled-out network. It also puts pressure on operators to offer consumers a whole new range of services using (some of) the total bandwidth, such as fast Internet, receiving richer content such as video on demand, and so on. Although there might be alternative technologies, which can provide broadband services, such as cable and UMTS, ULL seems to be deployed most rapidly.

At the moment, the roll-out of broadband services (for example ADSL) is relatively limited (Buigues, 2002). This might be related to the difficulties new entrants face: behaviour of the incumbents such as excessive pricing, delays in delivery, predatory pricing or price squeezes and refusals to supply the necessary information or space in the incumbent's locations. All these aspects deserve the regulators' attention.

Carrier select is another example of one-way access: an entrant has originating and terminating access to the incumbent's local network. By dialling a prefix (usually consisting of four digits), consumers can indicate that they want their call to be carried by the entrant instead of by the incumbent. Such access generally works well after some initial problems and other matters needing attention (such as margin squeeze, or scarcity).

Because of the asymmetry between the bargaining positions of incumbents and entrants, as well as between their interests, it is highly unlikely they will agree on the access price level. Specifically, the incumbent has a strong incentive to raise the access price as high as possible, perhaps even to foreclose entry, whereas entrants prefer cheap access.[4] Their interests are thus strongly opposed, necessitating access price regulation. The most notable central issue is that of optimal access policy.

## 2.2 Access Pricing

### 2.2.1 First-best solution

The 'first-best' solution, in other words the theoretical solution where there are no imperfections, is to make the access price equal the marginal access cost. Marginal cost here refers to traffic-dependent cost, which is practically zero if one applies a strict definition (roughly speaking it would equal the electricity cost per time unit of transmitting a signal). Given that the sales/marketing departments making pricing decisions typically attribute various fixed costs to telephony traffic, it may be advisable to apply a somewhat broader definition of marginal cost. Furthermore, where there is a substantial fixed cost in serving an additional consumer, a traffic-independent 'access price' may also be needed, such as the monthly lease price for a local line in the case of an unbundled local loop. This discussion covers traffic-dependent access prices, however. With an access price based on marginal costs, there are no retail price distortions, and entrants receive all the right signals: they can make positive profits only if they are more efficient than the incumbent. This access price is also fair and non-discriminatory, because it is the same for all entrants, and is not usage-based. In this first-best world, the incumbent's fixed network cost is covered by the state making a lump-sum payment to the incumbent.

The first-best concept can rarely (if ever) be implemented in practice. An obvious problem with this solution is that lump-sum transfers from the government are usually not feasible, certainly not without creating large distortions elsewhere in the economy while seriously impeding the incumbent's incentive structure. Therefore, one thus has to look for an access price above marginal cost, that which would help to cover the incumbent's fixed network cost.

### 2.2.2 Ramsey pricing

The practical problems involved in setting access prices equal to marginal costs bring about the necessity of introducing mark-ups to enable the incumbent to recover its network costs. On a more general level, one can try to set not only the incumbent's access prices, but also its retail prices to such an extent that welfare is maximized, subject to the constraint that the incumbent recoups its fixed cost. Note that welfare, defined as the sum of consumers' and producers' surplus, refers here to static efficiency. The solution to this problem is known as Ramsey pricing. Expressed in another way, the problem is to find the price structure for a multi-product (that is, access and retail services) regulated incumbent, which will maximize welfare, given that the incumbent has to break even overall. *A priori*, Ramsey pricing implies that there should be mark-ups in the incumbent's access prices as well as in its retail prices. The central idea is that to maximize welfare, all the incumbent's prices (both wholesale and retail) should play a part in the recovery of fixed costs. In particular, the optimal access price will be equal to the marginal cost of access plus a 'Ramsey term' (that is, a specific mark-up).

In their construction, Ramsey prices (wholesale and retail) reflect underlying costs as well as demand characteristics (see Laffont and Tirole, 2000, section 2.2). Hence, Ramsey prices are both cost-based and usage-based. This implies that they are, in principle, compatible with a firm's standard pricing practices, since these also take into account costs and demand. More precisely, it can be shown that welfare-maximizing prices result if the firm maximizes its profits under the proviso that it offers a certain minimum level of surplus to its customers (the 'Ramsey–Boiteux' social welfare level). Note that the regulator is supposed to have full information covering cost and demand characteristics. Intuitively, if the price elasticity for a specific service is relatively low, then the mark-up in the price for this service can be relatively high. This is optimal for welfare, since the prices of services with higher elasticities can be reduced while still satisfying the incumbent's cost recovery constraints. For instance, it may be optimal to increase the access price above marginal cost in order to reduce retail prices. In general, mark-ups can be higher when they lead to less distortion in the optimal allocation.

It is important to note the potential problems attached to Ramsey pricing (see Laffont and Tirole, 2000, section 3.4). First, Ramsey pricing's informational requirements are very high; that is, they require the regulator to have knowledge on cost levels and demand elasticity – information the regulator generally doesn't have. However, one should not rule out the possibility of obtaining reasonable estimates of these data. Failing this, the regulator can delegate pricing decisions to the better-informed firm, within the constraints imposed by Ramsey pricing. In observing this, it is usually apparent that unregulated firms have access to fine-tuned and sophisticated pricing tactics, suggesting that they have access to far more information than the regulators. In particular, pricing behaviour tends to reflect cost levels, elasticities, competitive pressure, and more. It can be shown that the structure (not the levels) of an unregulated monopolist's prices is the same as the Ramsey price structure. Thus a regulator may determine the Ramsey pricing structure by imposing a global price cap so that (i) access is treated as a final item and included in the price cap, and (ii) the weighting of the price cap is determined externally based on forecast quantities. If the weighting is set appropriately, then the operator internalizes net consumer surpluses in maximizing its profits. Although the principles for setting the optimal weighting are fairly straightforward, the information required (on forecast quantities) depends on actual costs and demand elasticities, and may therefore be difficult to obtain (see Laffont and Tirole, 2000, section 4.7). Nevertheless, this is much less demanding than obtaining the information needed to set an optimal partial price cap.

Second, usage-based access prices are discriminatory access prices, and would therefore appear to contradict the non-discrimination policy principle. More precisely, Ramsey pricing prescribes that the charge paid by an entrant must depend on its use of the service. The access price typically depends on the incumbent's price-cost margin in the relevant retail market, on demand-side substitution possibilities, on supply-side bypass possibilities, and on the elasticity of access demand. For instance, customers of services which are not very price-sensitive contribute more to cost recovery. From an economic perspective, this dependence simply maximizes welfare: access prices should be higher when they are used for services for which demand is less elastic. However, applying Ramsey prices can imply a (very) skewed distribution of prices and hence it often has a dubious political feasibility.

Overall, from an economic perspective the problems mentioned above are not necessarily vexatious for policy-makers. By construction, Ramsey pricing is the best way to set access and retail prices simultaneously; it is the 'least-bad' departure from first-best prices. At worst, if application turns out to be prohibitively difficult in practice or too unattractive politically, Ramsey pricing can (or should) serve as a useful benchmark for access regulation.

### 2.2.3 ECPR

Whereas Ramsey pricing aims at choosing optimal access and retail prices, it may happen that the access regulation problem is separated from retail pricing. In such an event, the efficient component pricing rule (ECPR) is a popular pricing rule, since it provides a link between the access and retail pricing.

ECPR, which is also known as the Baumol–Willig rule, has its background in the theory of contestable markets.[5] A retail market is said to be 'contestable' if there is potential hit-and-run entry, which constrains the incumbent's retail price – in other words where the threat of quick entry disciplines the incumbent's pricing behaviour. An underlying assumption is that potential entrants take the incumbent's price as given. Thus the theory is that for a fixed price charged by the incumbent, a more efficient firm can enter and capture the market; there is no 'in-market' competition.

Another ECPR underlying assumption is that access price regulation is separated from price setting in the retail market. In particular, retail prices are fixed in advance by the regulator. Thus ECPR's prescription is to choose the access price which maximizes welfare given the incumbent's retail prices. The regulator is not concerned with overall welfare maximization, as in the case of Ramsey pricing, but aims instead at cost recovery and productive efficiency.

Assuming a homogeneity in final products (which seems reasonable for voice telephony) and that the market is contestable (a 'heroic' assumption, as discussed above), ECPR prescribes that the access price should not be higher than the incumbent's opportunity cost, which is equal to its marginal cost of access plus its forfeited retail mark-up. One could equally state that the access price should not be higher than the incumbent's retail price minus its cost in the competitive retail activity.

An attractive feature of ECPR is that entrants receive the right signals; that is, they enter only if they have a cost advantage. Another positive feature sometimes put forward is that entry is revenue-neutral for the incumbent. But it is not clear why one should want to achieve revenue-neutrality. If the incumbent's profits are excessive, they will also remain so under ECPR if market entry is possible. ECPR and Ramsey pricing do not generally coincide, although this divergence can possibly be restored if one makes specific assumptions about symmetry and the absence of entrants' market power. An important difference is that ECPR ignores the relationship between the wholesale and retail markets. ECPR is a partial rule, in contrast to Ramsey pricing. If retail prices are regulated at Ramsey levels, then ECPR is optimal.

## 2.3 Dynamic Pricing Rules

The access pricing rules discussed in the previous subsection were based on the implicit assumption that firms do not invest or develop new technologies. Their purpose was simply to compensate the incumbent for providing entrants with access to its network, such that its fixed investments could be recouped. Since telecommunications markets change very rapidly precisely because of investments and technological progress, dynamic considerations should also play a role in access regulation. In particular, it is important to note that any access price affects operators' (potential) profits, and hence also their incentives to enter the market, to invest in new technologies, to roll out networks, to maintain and upgrade existing networks, and so on. To deal with such dynamic issues, regulators have developed specific access pricing rules, discussed below. More generally, when assessing the effects of access regulation on firms' incentives to invest, one must distinguish between:

- the effects on entrants' incentives to roll out networks themselves (versus using an incumbent's existing network), and
- the effects on an incumbent's incentives to maintain and upgrade its existing network.

Among other things, these effects depend on both current and anticipated access prices. For instance, if the access price is low and firms anticipate that it will remain low, then the incumbent is not very eager to invest in its existing network, while entrants feel no urge to roll out their own networks. On the other hand, short-term competition will be intense, as entrants can easily compete by using the incumbent's network. In the short term this is good for consumers, but it is less certain if this also holds true in the longer term. If the access price is high and firms expect it to remain thus, then it makes more sense for entrants to start rolling out their own networks, which in turn imposes a discipline on the incumbent to invest as well, in order to remain competitive. But in the short term it is expensive for entrants to start building up market share (for example, by starting with carrier select services while their networks are not yet ready), thus negatively affecting consumers over the longer period. The following subsections discuss the dynamic consequences of access rules.

### 2.3.1 Backward-looking access pricing rules

Backward-looking cost-based access pricing rules are based on the incumbent's actual or historical costs (also called embedded costs). An example is embedded direct costs (EDC). A backward-looking rule can be called fair in the sense that the incumbent is compensated for its actual network investments. But on an ongoing basis it may provide weak incentives for reducing costs, since the implicit message is that any cost will be reimbursed. One would also intuitively expect backward-looking access prices to be relatively high, compared with forward-looking prices (see below), making it harder for entrants without networks to compete with an incumbent. Recent experience with local loop unbundling has shown that this is not always true, however (see also the next section).

### 2.3.2 Forward-looking access pricing rules

Forward-looking cost-based access pricing rules are based on state-of-the-art, currently available technology. Hence they explicitly take technological progress into account. If a new network can be rolled out at half the cost of the incumbent's network, then the cost of the new technology serves as the relevant benchmark. Accordingly, forward-looking rules incorporate cost efficiency; they can be used to correct for an incumbent's possible inefficiencies. Arguably, they can be used to mimic competition, which does not yet exist. It is even possible that an incumbent's access service, using an existing or obsolete technology, becomes a loss-making activity because of the downward pressure on access prices. Hence, in theory, forward-looking rules give the incumbent an incentive to keep up with technological progress, and to keep investing in its network. A downside of forward-looking rules is that they tend to neglect so-called 'stranded assets'. If an investment has a fifty–fifty chance of success, then the revenue needed if the investment is successful should take into account the possibility that it might have failed. If access rules do not take this into consideration, the result may be under-investment and risk-averse behaviour.

The main example of a forward-looking rule is long-run incremental cost (LRIC). LRIC prices can be substantially lower than the incumbent's actual cost levels. However, it should be noted that LRIC can also lead to substantially higher prices (for example because of increasing labour costs, or better but more expensive technology), especially when it comes to access to the local loop. LRIC does not appear to be well suited to incorporating corrections for quality differences between old and new technologies. It may thus be more appropriate for interconnection fees (that is, two-way access prices; see the next section) than for pricing local loop access.[6]

Despite its intuitive appeal, LRIC has further drawbacks (see also Laffont and Tirole, 2000; Leo et al., 2001). For instance, it should be noted that the cheapest technology available cannot be derived from standard accounting systems. Hence the regulator must determine just how efficient the incumbent should be (that is, the regulator does not determine the desired efficiency level, but can put a lower boundary on it). In addition, LRIC may not allow the incumbent to make a profit margin on access. If this happens, the incumbent has strong incentives to deny access to entrants by using non-price anticompetitive practices, such as refusals to deal and delays in interconnection. Accordingly, a possible consequence of LRIC (and other forward-looking rules) is that the regulator has to continue to play a key role in managing entry. In other words, LRIC may imply that regulation remains heavy-handed, in contrast with the intention of gradually withdrawing regulation as competition matures.

LRIC's disadvantages originate in the implicit assumption that markets are contestable. Since telecommunication markets are typically not contestable, LRIC fails to take into account that assets can become stranded and that operators need risk premiums to cover that eventuality. It follows that applying LRIC without taking these dynamic considerations into account could be fairly disastrous.

To summarize: forward-looking rules such as LRIC have an appeal for interconnection. For access to the local loop there are serious dangers, in particular if used without taking stranded assets and common costs into consideration.

Overall, one cannot draw a clear-cut conclusion on the dynamic considerations. The problem is that access prices, especially in situations of one-way access as described above, often have to perform too many simultaneous tasks. Those tasks may conflict, as indicated above. The section on policy (section 4) will consider possible solutions to this problem.

## 3. TWO-WAY ACCESS

### 3.1 Introduction

Two-way access refers to a situation where there are two or more operators, each with their own infrastructure (of long-distance and local access networks), needing mutual links so that any consumer can call anyone else. These operators compete for network subscribers and need each other in order to offer their customers maximum network benefits (interoperability). The typical two-way access situation is network interconnection.

Another important two-way access example is fixed–mobile terminating access: a network operator can charge a consumer for making use of the network (for example a two-part tariff). The network operator can also charge other parties to reach their subscribers. There is competition for subscribers, but no competition for reaching the subscribers of a network. The characteristics of fixed–mobile call termination are that there is competition for mobile subscribers, but no competition for providing access to mobile customers once they have subscribed to a particular mobile network. The main difference with network interconnection is that the services are provided to non-competing operators.[7] For example, in fixed–mobile termination, the mobile operator needs to access the fixed network and vice versa, though to a certain extent they are not exactly competing head to head. Much of the theory discussed in this section applies to fixed–mobile termination. For this reason the present authors have opted to deviate from the categorization introduced by Armstrong (2001), who offers a separate discussion on 'competitive bottlenecks' (of which fixed–mobile termination is considered to be the main example). There are indeed a number of interesting and separate policy questions related to fixed–mobile termination. For example, access to the incumbent's fixed lines has had greater regulation than access to mobile. This creates some policy questions but does not necessarily warrant new theory. The policy aspects of fixed–mobile termination will be discussed in section 4.

Another two-way access situation concerns international network interconnection, which was relevant even before the start of liberalization. The crucial difference with the examples above (and the reason we do not consider this to be an example of two-way access) is that there is no competition for subscribers. This creates its own problems, which will not be discussed here. The same applies to international roaming. If a subscriber roams on another network to enable his or her mobile to function abroad, the visited network operator bills the home network operator for this call and vice versa. Since international roaming is a competitive service in principle, and since possible related problems are beyond the access discussions of this section, it will not be pursued here.

As with one-way access, the central question is the extent of the access price, or in this case the terminating access price. A difference here is that, by definition, there is always more than one access price – each operator charges its own terminating access price. Hence an additional issue is whether terminating access prices should be reciprocal, that is, the same for all operators. A third issue concerns the so-called 'missing price'. Because of the 'calling party pays' principle (CPP), there is no fee for receiving a call. This fact will have great relevance in the discussion.

### 3.2 Access Pricing

Two important policy questions govern terminating access prices. The first is the optimal level of access prices: which levels of access prices maximize welfare (or, depending on the regulator's objective, consumers' surplus) in such a way that total industry profits are sufficient to cover fixed costs? The second issue concerns the discretion of firms in setting access prices. Should operators be allowed to choose access prices, for instance by negotiation? Or should these prices be regulated?

The first economic theory articles on access pricing in telecommunications focused mainly on symmetric, mature markets – in other words, operators are assumed to be identical in terms of networks, costs and market coverage, with the possible exception of an asymmetry introduced by assuming horizontal product differentiation (see Laffont et al., 1998; Armstrong, 1998; and Carter and Wright, 1999). Consider a duopoly, and suppose that access prices are set before the operators start to compete in the retail market. The operators compete by setting linear prices (that is, they charge customers a per-minute price only and not a subscription fee). Then, if access charges are chosen non-co-operatively by the operators, not surprisingly access prices turn out to be symmetric (and therefore also reciprocal) in equilibrium. Moreover, the operators have incentives to set the access price above the marginal access cost. This is because they benefit from raising their rival's cost, which generates higher profits without actually colluding in the retail market. By way of explanation, note that a high access price increases the cost of a unilateral retail price cut, because such a price cut increases the net outflow of calls to the rival operator, which is costly because of the access mark-up. In the absence of a joint and common cost, welfare is not maximized by these above-marginal cost access charges. Quite the opposite is in fact the case: welfare is maximized by setting the access price below marginal cost in order to compensate for the mark-up generated by the imperfect retail market competition (see Laffont and Tirole, 2000, ch. 5). In the presence of a joint and common cost, the optimal access price can be below or above marginal cost, because a mark-up on access contributes to cost recovery.

If operators compete in two-part tariffs (consisting of a subscription fee and a per-minute price), they have no incentive to charge an access mark-up. The reason is that, although a higher per-minute price (induced by an access mark-up) increases the per-customer profit, competition in subscription fees becomes more intense as a result of an access mark-up. In equilibrium, operators optimally set per-minute prices equal to marginal call costs, and recoup fixed costs on a per-customer basis through subscription fees. Welfare is maximized by setting the access price equal to the marginal cost of access (Laffont and Tirole, 2000, ch. 5).

An interesting observation in the literature is that, at least in certain circumstances, terminating access prices in a mature and symmetric market can be used in support of tacit collusion. On the other hand, this result depends heavily on the assumptions of the model (for a more extensive discussion, see Armstrong, 2001, and Laffont and Tirole, 2000, ch. 5). Overall, the risk of tacit collusion should not be neglected in a mature market, but it is not necessarily the most pressing problem facing policy-makers in infant markets such as the current one for fixed telecommunications, given the assumptions under which the result is derived (in particular, the twin assumptions of symmetric operators and linear pricing). In the mobile telephony market, where operators are more symmetric and linear pricing schemes may be observed in practice, tacit collusion may be a policy concern.

In the fixed telephony market's current transition, most entrants are still relatively small when measured against the incumbents. A newcomer in the market, starting from scratch and slowly building up its market share, finds itself in a much more difficult position with regard to the incumbent than vice versa. In such a situation, asymmetric access regulation can be useful in the short term to protect small entrants as well as to prevent a reduction in consumer surpluses (see de Bijl and Peitz, 2002a, 2002b).[8] In particular, if an entrant is initially disadvantaged against an incumbent (for example in terms of market shares while there is customer lock-in, or because it lacks a quality track record and a reputation for quality), as is apparent in immature markets, it may be optimal to regulate the incumbent's access price at the marginal cost level (possibly allowing for a mark-up that yields a reasonable return), while allowing the entrant to charge a substantial access mark-up. Such an asymmetric access mark-up increases the entrant's profits as well as consumers' surpluses. Welfare is barely affected, but the incumbent's profits are obviously reduced. This latter effect is undesirable in principle and should therefore not loom too large, since otherwise the incumbent may lose its incentives to invest. But within safe margins it is outweighed by the positive effects on consumer surpluses and the entrant's profits. Thus non-reciprocal access prices may be temporarily useful to stimulate competition in an immature (and asymmetric) market. In the longer term, when the entrant has built up market share, cost-based access prices for both operators are optimal for consumer surpluses and welfare.

In summary: optimal policy with regard to terminating access prices is fundamentally different in infant and mature markets, a difference which is especially important in the light of the current fixed voice telephony transition phase. In an infant market, generally characterized by asymmetry between the incumbent and entrants, asymmetric access prices which favour entrants stimulate competition and consumer surpluses. In a mature and symmetric market, however, asymmetric access regulation should be lifted, while oversight may be necessary to prevent tacit collusion through the access price.

The preceding discussion assumes that the 'calling party pays' principle (CPP) applies. And indeed CPP does apply in all European countries. Recent literature (Laffont et al., 2001) has shown that most of the problems mentioned above may disappear when the CPP principle is abolished. The reason is that a price is 'missing' under CPP, namely a price for receiving a call. As a consequence, all costs have to be incurred by the caller, while technically the costs should somehow be split. One of the consequences is that operators have monopoly power over their incoming calls, which generates problems, for example in the fixed–mobile termination traffic (see Section 4.1). Laffont et al. (2001) show that interconnection charges will be set at the competitive level if the missing price is recovered, that is, when mobile operators are able to charge their customers for receiving calls. In contrast with the CPP case, perfect retail competition can exist and is viable. In the case of perfect competition, the access charge serves to determine how the cost of communication is split between the caller and the recipient, with the operators demonstrating indifference (since they achieve zero profits anyway). However, whenever there is imperfect competition (market power, albeit limited), operators' favoured access charges need not coincide with the socially optimal ones; unfortunately, the nature of the operators' bias is not that easy to predict.

### 3.3 Dynamic Considerations

There are two issues relevant to the discussion between the short and the long term. First, since it has been pointed out that asymmetric access regulation is optimal in an infant market, an important issue queries the point at which the market can be considered to be sufficiently mature for asymmetric regulation to be lifted. Second, as with one-way access situations, any access price affects operators' profit levels, and hence their incentives to invest.

With regard to the first issue, if the regulator imposes asymmetric access prices to stimulate competition in the short term, a crucial question is: when does the transition phase end – in other words, when are entrants sufficiently large so that there is effective competition? For policy purposes, this is largely an empirical question. Clearly, the 'stopping rule' or 'sunset clause' should be based on observable market outcomes. In practice, policy-makers can look at various indicators, such as the growth of entrants' market shares, reductions in retail prices, quality/price ratios and variety for end-users. However, there is a risk involved in making asymmetric regulation lifting dependent on market outcomes.[9] For instance, a criterion based on market share furnishes both the incumbent and entrants with incentives to compete less aggressively by increasing retail prices. By way of illustration, consider that the incumbent wants the entrant to gain market share more quickly so that it no longer incurs the access mark-up charged by the entrant, while the entrant wants to keep enjoying the access mark-up for as long as possible. To this end various regulators have been investigating the criteria for sunset clauses (for example Oftel, 1999). However, it is not yet clear what all these dynamic regulation methods have yielded.

On the second issue, note that the policy prescription of asymmetric access regulation (as discussed above) does not conflict with the possibility that the regulator may want to introduce a bias towards facilities-based entry. Nevertheless, if the regulator explicitly wants to stimulate entrants into rolling out networks themselves, it is more direct to make other types of entry (Carrier Select, unbundled access to the local loop) increasingly unattractive. This may be sufficient to induce entrants who don't have their own local loops, but who do have experience and market share, to start building them.

## 4. POLICY ISSUES

This section covers four policy issues in more detail. These issues are fixed–mobile termination, margin squeeze, the allocation of common costs, and static versus dynamic efficiency.

The issues are highly relevant for regulators at this stage of the telecoms market's development. Considerable policy attention concerns the issues of fixed–mobile termination and margin squeeze. The allocation of common costs, and static versus dynamic efficiency issues, are both important general considerations in regulation.

This section is structured as follows. Each policy issue is preceded by a discussion of the policy's relevance, followed by an overview of what economic theory has to say. Occasionally this entails a summary from earlier sections. The policy issues proceed with practical experiences and end with some policy conclusions which combine insights from the earlier sections with the practical experiences.

## 4.1 Fixed–Mobile Termination

### 4.1.1 Policy relevance

Fixed–mobile termination garners a great deal of policy attention. It is an issue that is the subject of a number of fixed network operators' complaints, while politicians also discuss the high fixed–mobile termination tariffs. In practice, one can concede that these access prices are generally very high throughout Europe, compared to the cost of access. There is also a price asymmetry; that is, the cost of being called is higher than that of making a call (although they place similar demands on the network). When multiple mobile operators entered the market, the retail cost of making a mobile call decreased whereas the terminating access price decreased much more slowly, and price asymmetry increased. One might conclude that the fixed networks are subsidizing the mobile networks. OPTA (2002) calculates that the revenue of fixed–mobile termination for 2000 and 2001 was about €270 million above revenue based on cost orientation, that is, a cross-subsidy of about €270 million from the fixed networks to mobile networks. Furthermore, termination revenues are highly significant. For Telefonica, 75 per cent of mobile revenue originated from termination and roaming services (OECD, 2000). This might give cause for regulators to intervene.

What intervention options are available? Regulators can decide not to intervene. One can expect the situation to continue unchanged because no alternatives for mobile termination are anticipated in the short term. Fixed subscribers subsidize mobile subscribers.

Regulators may decide to intervene to treat fixed–mobile termination as a competitive bottleneck and to regulate the terminating access tariffs. Different pricing principles can be applied to calculate cost-oriented tariffs (see Section 2). These principles have to allocate the common costs across the different services, as the common costs need to be recovered somewhere. Ramsey pricing and LRIC are the two polar cases. When Ramsey pricing is applied, the problem may not be solved in a way that satisfies the regulators (excessive prices).[10] Common costs will be allocated based on the demand (elasticities). Demand in the retail market is more elastic than in the fixed–mobile terminating market. This implies that the largest segment of the common costs will be recovered in the terminating market, which itself implies that the current price asymmetry might be in line with Ramsey pricing. However, the low elasticity of the terminating access service would appear to be the result of market power. Fixed subscribers generally have no choice when they want to dial a mobile customer, so that the mobile operator on which these calls terminate possesses substantial market power. The regulator may wish to protect consumers from mobile operator market power abuse by regulating the terminating access prices (see, for example, Armstrong, 2001, section three). In other words, with given levels of market power and the opportunities for abusing market power, Ramsey pricing is a useful welfare notion. However, the regulator may have an interest in changing market power, which would also change elasticities and thereby Ramsey prices. The alternative to Ramsey is LRIC, commonly used by regulators in this situation. To recover the common costs, a fair share of these costs is often added to the price based on LRIC. The calculation of this fair share is somewhat arbitrary (see also Section 4.3.2 on common costs).

### 4.1.2 Economic theory

When fixed subscribers want to reach mobile users, they have no choice but to pay the high access fee (as part of the overall per-minute price). The level of the fixed–mobile termination access price does not directly influence the level of the usage-based mobile retail prices, but forms part of the fixed cost related to the mobile subscribers (for example lower subscription fee or subsidies on handsets). In essence, high access mark-ups imply that fixed customers subsidize mobile users. Mobile users benefit from high fixed–mobile access charges to the extent that retail prices are subsidized.

Standard assumptions can show that welfare is maximized by making the terminating access prices equal to marginal costs (see Armstrong, 2001). However, it appears that no effective competitive pressure exists to set terminating prices at marginal costs. The policy discussion focuses on the extent to which different mechanisms restrict the mobile network operator's (MNO's) ability to behave independently of its competitors, customers and ultimate consumers to an appreciable extent. In other words, does the MNO enjoy a dominant position or not? If the MNO is dominant and (also) sets excessive terminating access prices, intervention would seem to be called for.

How much do mobile consumers care about what it costs others to call them? One might argue that mobile users could also be contacted in other ways, for instance via a fixed connection. However, this argument misses the point, which is that one generally tries to reach mobile users at their mobile phone when they are mobile, that is, away from home or office where there is a fixed line with a known phone number (or away from other contact alternatives). Even where there are substitution possibilities, they may not be significant enough to result in substantial downward access charges pressure.

One might also argue that market forces already exert sufficient discipline on mobile operators. For example, suppose that mobile customers derive benefit from being called (which is undoubtedly true in many cases). If consumers want to receive more calls, they may care about the termination access price charged by their operator, and hence they may take the costs of inbound calls into account when choosing a mobile subscription. As a consequence, the presence of 'call externalities' gives mobile operators an incentive to reduce fixed–mobile termination access prices, although it is unclear how strong this effect is in practice. Armstrong (2001) actually shows that if mobile subscribers derive benefit from incoming calls, then welfare is maximized by setting termination charges *below* the marginal cost of access. This is because access below cost encourages fixed subscribers to make calls to mobile subscribers. Another way in which consumers may take the costs of inbound calls into account occurs if one assumes that they care about the costs incurred by people calling them. In theory, if mobile customers allow for the welfare of people who will call them when choosing a mobile operator, this again provides the incentive for mobile operators to reduce termination fees (Armstrong, 2001). This scenario does have some grounding in reality, since people most often receive calls from family members or direct colleagues.

Ultimately, it is empirical as to how much mobile customers care about the costs of incoming calls. Existing evidence (for example Oftel, 2001) together with casual observation (the rates are indeed very high throughout Europe) suggests that it is most likely that mobile customers don't take much notice of the costs incurred by people calling them when choosing a subscription.

As the current system of 'calling party pays' (CPP) and customer pressure does not appear to exert enough downward pressure on the terminating access prices, one might choose to change the payment principle. The alternative is to let mobile operators charge their customers for incoming calls. In other words, call recipients pay a per-minute price to receive calls from fixed subscribers (instead of the CPP principle). This remedy removes the bottleneck problem (given that the mobile operator no longer charges the fixed operator for the termination), since customers who have to pay for incoming calls will be more sensitive to the prices levied for that service.[11] Changing the system may create a distortion; that is, subscribers pay too little for calls to mobile users if the costs between callers and recipients are split for fixed–mobile but not for fixed–fixed. It is not clear how serious this distortion would be, but to solve it one might also consider introducing the split for other services.

The conclusion is that policy intervention appears to be necessary. There might indeed be some mechanisms available to restrict MNO behaviour, but the effect is too small to conclude that there is effective competition in the fixed–mobile termination market. In the short term, most research indicates that these mechanisms will not increase sufficiently in strength to create an effective competition situation (see, for example, Armstrong, 2001).

### 4.1.3 Practical experiences

Central to the debate on fixed–mobile termination is whether fixed–mobile call termination is a 'bottleneck' and whether MNOs can exercise market power. There is significant policy discussion between regulators and market players on this very point. The principle that the 'calling party pays' is central to this discussion. Arguments can be made as to why there is no (or a limited) effect on the extent of the terminating access prices versus arguments as to why there is enough pressure on MNOs to compete on the terminating access prices (see, for example, Oftel, 2001; IRG, 2002). Despite the results of this discussion, Armstrong showed that in most cases the pressure from these mechanisms is too limited to restrict MNO behaviour in setting their termination access prices. The high terminating access prices appear to support this argument. Most regulators also follow this line of reasoning, and intervene.

Oftel (2001) discusses different mechanisms that might constrain MNO behaviour. The first mechanism is that mobile subscribers might care about the costs of calling them and take the termination prices into account when choosing an MNO. The conclusion was that mobile callers pay little attention to terminating prices (this might be different for different groups). Second, the effect of closed user groups[12] generates a limited pressure on termination prices. In addition, Oftel concludes that there are no real alternatives and that buyers have no countervailing power. The overall conclusion is that there is insufficient competitive pressure to constrain the terminating prices, and that pressure is expected to remain insufficient.

OPTA uses more or less the same arguments as Oftel, and reaches the conclusion that mobile call termination is a bottleneck facility. OPTA decided to intervene and to set a timeframe to implement cost-oriented terminating access tariffs in 2003. In the meantime, OPTA suggested what it considered to be fair terminating prices.

In calculating terminating access prices, regulators use the LRIC pricing principle, or are planning to use it in the near future. LRIC is preferred over Ramsey pricing because Ramsey pricing might be difficult to calculate and might result in an unfair allocation of common cost (cross-subsidization and/or major price asymmetries). Common costs also seem to be a relatively small problem (see Section 4.3 on common costs).

### 4.1.4 Policy conclusions

The policy conclusions combine insights from the earlier sections with the practical experiences from the previous one.

- Fixed–mobile termination is a hot topic in the telecoms sector and is the focus of much regulator attention. So far, the conclusion of most regulators is that fixed–mobile termination is (virtually) a bottleneck facility. There are too few competitive pressures to restrain MNO behaviour; hence regulation is considered to be necessary. However, it is important to review the development of potential mechanisms which might bring about effective competition. For example, it is unclear how closed user groups will develop and to what extent they might restrict MNO behaviour.
- Another important issue is the potential of alternatives to the 'calling party pays' (CPP) principle. CPP is common practice in Europe. CPP could be exchanged for 'receiving party pays' (RPP) or a combination of both. RPP can put effective pressure on the call termination prices because the individual who chooses the network operator is charged for both outgoing and incoming calls. The individual has the option of switching to a different network operator offering better charges. Countries with CPP have significantly higher terminating access prices than countries with RPP (OECD, 2000, p. 50). Thus RPP might bring about lower terminating access prices. On the other hand, RPP might also have disadvantages. Prepaid could become less attractive. The subscriber has to pay not only for outgoing calls, but for incoming ones as well – calls over which he or she has no control (other than disconnecting the mobile). It will be more difficult for a prepaid subscriber to keep a rein on telephone expenses. It is also argued that CPP exhibits more potential for helping the mobile market to grow (OECD, 2000). Under CPP it becomes increasingly desirable to join a network, because receiving calls is 'free', while prepaid cards are more attractive in markets with CPP. Finally, with the new proposed 'open network provision' (ONP), it will be easier to designate MNOs with significant market power in narrower markets (for example mobile termination) and not to intervene in the more competitive retail market.
- Competitive pressures might also be introduced by offering consumers more options. For instance, it might be possible to offer consumers the possibility of choosing a different network for receiving calls than the one used for making calls. The mobile subscriber could select the network operator offering the lowest originating charge and another with the lowest terminating charge. This implies the creation of two separate markets. It is not clear to what extent operators focusing on a separate market (for example only terminating) can compete with operators providing a package of originating and terminating services. For this option, it would also be a requirement that SIM cards could communicate with all mobile networks. To be effective, it would also be necessary for the mobile subscriber to take the costs of the calling party into account.

- The calling party might also be able to call a mobile subscriber by using different MNOs. The mobile subscriber might have multiple mobile phones (not very realistic) or the mobile subscriber could be reached via various MNOs. This offers MNOs an incentive to compete on terminating access prices. This last option also assumes that the SIM cards could communicate across different mobile networks. It implies that the mobile subscriber owns the SIM card instead of the MNO, and that the networks need to be highly standardized. However it is unclear whether this could be either possible, or desirable, in the future.
- Increasing mobile subscribers' awareness of terminating access costs might increase their willingness to take these terminating costs into account. Transparency on the pricing structures is a tool to this end. To be effective, fixed network operators need to define their retail tariffs for the various MNOs.

## 4.2 Margin Squeeze

### 4.2.1 Policy relevance

Telecoms operators or independent service providers may rely on the key input (upstream market) of the vertically integrated operator, while at the same time competing with that operator in the downstream market. If the vertically integrated operator is dominant in the upstream market, there could be scope for it to leverage its dominant position into the downstream market, resulting in a margin or price squeeze.

A margin squeeze occurs if the retail prices charged by the dominant operator are so close to the wholesale prices of services offered to competitors that even a more efficient competitor cannot enter into or survive in that market. The vertically integrated operator can subject its competitors to high access prices (raising the cost of the key input) and/or lower its prices in the downstream market. The vertically integrated operator's total revenue may remain unchanged. In extreme cases, the access price may even be higher than the incumbents' retail prices. In a margin squeeze situation, there is too little room for competitors to compete. Thus a margin squeeze can harm the development of effective competition over the long term (relatively high prices, low quality, and so on).

The lack of (efficient) entrants or competitors does not necessarily reduce welfare. If the incumbent operates more efficiently than its (potential) competitors and the customers reap the benefits, the lack of entrants/competitors is not an issue. If this is not the case, then a margin squeeze is indeed a problem, because efficient competitors do not enter into or stay in the market and competitive forces cannot do their job.

More formally, a margin squeeze can be defined thus (given that access and retail services are strictly comparable):[13]

$$a + c > p$$

where:

$a$ is access price charged by incumbent to entrants;
$p$ is incumbent's corresponding retail price;
$c$ is incumbent's cost of delivering the retail service, on top of costs already included in access price $a$.

This implies that: the incumbent would make losses if it bought access at price $a$, given retail price $p$; $p - a$ does not reflect the incumbent's own downstream cost $c$; the 'notional downstream margin' on the incumbent's retail activities is negative, $p - (a + c) < 0$.

Is the possibility of a margin squeeze sufficient reason to intervene as a regulator? Is a margin squeeze part of the process of becoming a competitive industry? Is a margin squeeze the result of normal business practices or anticompetitive behaviour? To assess the need for (potential) intervention, it is worth looking at what economic theory has to say about margin squeeze.

### 4.2.2 Economic theory

The critical point in a margin-squeeze case is that an efficient (not vertically integrated) competitor cannot enter into or stay in the market and earn a normal profit because of the pricing behaviour of a (dominant) vertically integrated incumbent. In order to understand any margin-squeeze case and the potential remedies, one should first consider its causes. There can be three different instances which may give rise to a margin squeeze:

- access price is too high (and competitive retail price); risk of exploitative abuse;
- retail price is too low (and cost-based access price); possibly predatory pricing;
- margin squeeze between the prices above; possibly cross-subsidization.

These issues will be discussed in greater detail.

*Access price is too high: risk of exploitative abuse*

When it comes to excessive access prices, it is vital to consider the reason for the level of the access price and its relation to the underlying costs. Excessive access prices are characterized by the absence of a reasonable relationship between the price and the economic value of the product or service supplied. If there is a large gap between the price and underlying costs, profits are higher than might be anticipated in a competitive setting. There may be various reasons for excessive access prices. Access prices may be too high as the result of a dominant position. From an access perspective, this is a one-way access situation (see Section 2). On the other hand, access prices may be too high because of conflicting regulation practice objectives.

In theory, excessive access prices on the upstream market are not a problem if this attracts new entrants into the network market. This depends on this market's entry barriers. Particularly in the local loop, entry barriers are high and no disciplinary forces can be expected in the short term. Regulators might thus opt to intervene.

Access prices may also be too high even though this is not an intentional abuse by the incumbent. Price regulation rules set by the regulators could result in excessive access prices (in relation to retail prices). This could result from conflicting objectives and/or different pricing principles for the access and retail markets. The objective of efficient investment (access price based on cost) might contradict the objective of creating effective entry and competition at the retail level (retail minus). In such a situation, the regulator has to set priorities: not all objectives can be achieved simultaneously and some are more important than others.

*Retail price is too low: possibly predatory pricing*

Various causes can result in retail prices being too low. Retail prices might be too low because of different pricing rules charged by regulators on the wholesale and retail market (compared to the access price). A uniform national retail price might also be too low compared to cost-based access prices in a rural area. On the other hand, low retail prices can also result from an intentional abuse by the incumbent, that is, a special case of predatory pricing.

A margin squeeze as a particular form of predatory pricing implies that an efficient entrant or competitor cannot redeem its downstream costs, and that it will not enter or leave the market. The incumbent opts for a short-term loss in order to make long-term extra profits thanks to a dominant position.

Predatory pricing implies that there is a price reduction which is profitable only because of the added market power the predator gains from eliminating, disciplining or otherwise inhibiting the competitive conduct of a rival or potential rival (Bolton et al., 2000). In the short term customers may benefit from lower prices, but over a longer period weakened competition will lead to higher prices, lower quality or less choice. The fact that an activity is being run at a loss is not sufficient to establish a case of predatory pricing; the question is whether it has an anticompetitive effect. In order to prove the anticompetitive effect of predatory pricing, Bolton et al. (2000) propose a five-criteria rule:

1. a facilitating market structure,
2. a scheme of predation and supporting evidence,
3. probable recoupment,
4. price below cost, and
5. the absence of efficiencies or business justification defence.

Subnote (1): The market structure must make predation a feasible strategy. A company must have the power to raise prices (or to otherwise exploit consumers or suppliers) over some significant period of time (dominant firm or small group of jointly acting firms, entry and re-entry barriers).

Subnote (2): Predation pricing and recoupment require that predation is plausible *ex ante* (that is, based on prediction and extrapolation) and probable *ex post* (that is, retrospectively). This means there must be a predatory scheme *ex ante* under which the predator can expect to recoup its initial losses. Using the tools of applied game theory can assist in identifying economic conditions under which predation is rational profit-seeking conduct on the part of a dominant firm. *Ex post* probability is shown by the subsequent exclusion of rivals and post-predation market conditions making future recoupment likely.

Subnote (3): At the very least, losses incurred in a predation strategy must be recouped somehow. Should the operator be unable to recoup the losses, because of competition from existing or potential competitors, the predation strategy is not viable. Recoupment is only possible if there is an exclusionary effect on (potential) rivals, or through disciplining the rival's competitive conduct. The most common and straightforward recoupment occurs when prices rise above the predatory market's competitive level. In more complex settings, recoupment can occur through other channels, for example by raising the prices of complementary or closely related services. It is essential that these latter price increases should be explained unambiguously by the earlier predatory pricing (see also Cabral and Riordan, 1997).

Subnote (4): In the predatory period, prices should be below average variable cost, although prices which are also above average variable cost but below average total cost might be predatory and injurious to competition.[14] The most-used cost standards are average total cost (ATC) and average variable cost (AVC) (OFT, 1999) or long-run average incremental cost (LRAIC) as a substitute for ATC and average avoidable cost as a substitute for AVC (Bolton et al., 2000). If prices are above ATC, there is no problem. If prices are below AVC, predation can be assumed. A price between ATC and AVC is either presumptively or conclusively legal. If the price is presumptively legal, there is a need for evidence that the operator intends to eliminate or to discipline a competitor.

Subnote (5): Finally, cases can arise where below-cost pricing by a dominant operator might be efficiency-enhancing rather than predatory. However, in such cases one has to consider very carefully whether the efficiency enhancement is also to the benefit of consumers in the long term. Otherwise the argument could be abused to foreclose a market on the grounds that it is 'efficient' to do so.

The five-criteria rule provides a clear procedure for handling a potential predatory pricing case. However, predatory pricing might be difficult to prove, particularly the intentional aspect. The section on policy conclusions (Section 4.2.4) will include a more detailed discussion of this rule's practical value.

*Margin squeeze between access price and retail price: possibly cross-subsidization*

Cross-subsidization occurs where an operator uses revenues from one market to subsidize losses in another. If the revenues derive from a market in which the operator has a dominant position and are used to subsidize losses in a competitive market, this hampers competition and might be illegal. In the case of cross-subsidization, losses in one market are redeemed by another market. Cross-subsidization is particularly viable if the market is characterized by economies of both scale and scope, that is, scale and scope economies broaden the opportunities for incumbents to use part of their business to cross-subsidize other parts. For example, causally attributable and avoidable service costs can be labelled as common costs resulting in too low incremental costs. If the price in such a situation is set according to incremental cost, then a case of cross-subsidization exists (see also Section 4.3 on common costs).

There is cross-subsidization if revenues are expected not to cover the cost of the associated activities over their economic lifetime. If the costs are indeed covered over the service's economic lifetime, relatively low prices when introducing the service could be a normal business strategy (penetration strategy), and thus do not constitute anticompetitive behaviour. For example, if the production of a service is characterized by economies of scale or learning effects, an appropriate business strategy might be to offer the initial service below cost price (for initial customers). As demand grows significantly because of the low price, this will yield a sharp decrease in production costs. This option might yield a higher profit over the service's economic lifetime (low price, large volumes and substantially lower costs) than the alternative of a high price and a slower drop in costs.

Assessing cross-subsidization relates to predatory pricing assessment. If revenues exceed the long-term incremental cost (including the cost of capital), the service is sustainable in the long term and there is no cross-subsidization (Oftel, 2000). If services share common costs, the evaluation is more complex (see also Section 4.4). In this instance, total costs consist of the long-term incremental costs of all services and the common costs. In the case of cross-subsidization, a dominant operator might charge a higher price to cover its common costs in the market in which it enjoys dominance, using a price that equals the incremental costs in the competitive segment. Although this is not illegal in terms of predation, it might hamper competition. A competitor who does not offer all services available from the dominant operator might be at a cost disadvantage in the competitive market (that is, it cannot recover the common cost from another market).

Economic literature focuses on two tests for cross-subsidization: the incremental cost test and the stand-alone test (Braeutigam, 1989, pp. 1337–42; Sharkey, 1982; Faulhaber, 1975, as cited in Nicolaides and Polmans, 1999). The incremental cost test requires that the revenues from each subset *S* at least cover the increment to the total costs, which occurs when *S* is produced, as opposed to not being produced at all. If the revenues of *S* do not cover the incremental cost of *S*, then service *S* is subsidized. The incremental cost test sets the lower boundary on the subsidy-free price range. The stand-alone test argues that if the revenues of *S* exceed the costs of providing service *S* alone, then users of *S* are subsidizing other services. The stand-alone test sets an upper level on the revenues generated by *S*. The tests must be done for all possible subsets[15] (often referred to as the combinatorial cost test). If the price is between the average incremental cost and the average stand-alone cost, then the activity is neither a subsidy source nor recipient.

Economic theory yields important insights into margin squeeze, its causes and ways of assessing different forms. How can such theoretical insights be interpreted in real-life cases?

### 4.2.3 Practical experiences

Until recently, regulation and intervention by regulators focused on preventing excessive prices (for example price caps). Over the past two to three years, regulators have devoted more attention to the potential harmful effects of low prices. As a result of complaints from entrants and new competitors, it became more obvious that low competitive prices have a hampering effect. However, the distinction must be drawn between low prices resulting from predatory behaviour and low prices resulting from legitimate competitive behaviour. In case law, Dutch manufacturer *Akzo* v. *The European Commission* is the leading example (ECJ, 1991). The European Court of Justice argued that 'prices below average variable costs (that is those which vary depending on the quantities produced) by means of which a dominant undertaking seeks to eliminate a competitor, must be regarded as abusive' (para. 71), and 'Moreover, prices below average total cost, that is to say, fixed cost plus variable costs, but above average costs, must be regarded as abusive if they are determined as part of a plan to eliminate a competitor' (para. 72). In the first situation, the ECJ found the conduct of a dominant company to be an abuse without explicitly considering whether recouping losses would be feasible. Thus if the accused company is dominant, a case of predation does not require explicit analysis of the predatory scheme (see ad 2 and 3, p. 31).

Interest in margin squeeze increased as a result of the potential problems incurred in local loop unbundling (European Commission, 2001). It must be emphasized that margin squeeze does not necessarily imply intentional abuse. It can be the result of an incomplete price rebalancing. If prices are not completely rebalanced on a cost-oriented basis, such prices do not adequately represent the underlying costs. This may result in overcharging or underpricing. The Commission can launch infringement proceedings against Member States to counter this problem.

Accounting inconsistencies may also result in margin squeeze. Certain pricing principles focus on static efficiency (for example Ramsey pricing), whereas others focus on dynamic efficiency (for example LRIC). National uniform retail prices could conflict with differences in the regional costs of providing access. Pricing principles may conflict and may result in margin squeeze.

Most regulators studied the effects of price squeezing and suggested price-squeeze tests when approving the incumbent's interconnection tariffs (for example Oftel, 2000; OPTA and NMa, 2001). This *ex ante* regulation can be interpreted as part of the regulators' cost orientation check. In a price-squeeze test, the costs of the incumbent are studied, more specifically the costs that the incumbent incurs to provide the service internally. The extra costs the incumbent incurs to deliver the access service (for example billing) should be omitted from the test. In determining the access tariffs, most regulators use LRIC, or plan to use LRIC in the near future.

A problem could arise when using LRIC in a sector with high common costs – the telecommunications sector being just such a one. A multi-product supplier may choose the market in which to recover these common costs, provided it does not abuse its dominant position. If all services are priced at LRIC, the common costs are not recovered. To test that there is no case of predation, Oftel tests whether the service prices are at least at LRIC level plus a fair share of the common costs. If a dominant operator is pricing below LRIC, predation is presumed, unless the operator provides solid arguments to the contrary. In addition, all the relevant services should be able to recover all common costs together (combinatorial test, Oftel, 2000). If the dominant operator prices above LRIC, but overall revenue fails to cover total costs and the dominant operator has the intention of eliminating a competitor, predatory pricing is presumed.

OPTA and NMa (2001) published price-squeeze guidelines. These guidelines explain how the price-squeeze test is implemented. The test could be applied to all the dominant operator's retail products, and will be applied to the various retail price components (set-up cost, peak, off-peak, and so on). It will be difficult for entrants to cherry-pick as a result of this detailed approach.

Another issue needing attention is the allocation of costs (costs related to access versus those related to retail activities; see also Section 4.3 on common costs). An incumbent might allocate costs related to retail activities to access activities. As a result, access costs are too high and retail costs are too low. If access prices are cost-oriented based on these (high) access costs, then the entrant is paying too much (that is, part of the incumbent's retail costs). Thus when setting access prices based on an incumbent's information, the cost relevance must be addressed.

In conclusion: most regulators recognize the price-squeeze problem. They have developed or are planning to develop a price-squeeze test. This instrument is used as part of the cost orientation check. As this instrument is relatively new, it is too soon to draw conclusions on the effectiveness of price-squeeze tests.

### 4.2.4 Policy conclusions

Theoretical insights require a practical interpretation. A margin squeeze can be difficult to assess, proper data are difficult to obtain and interventions can also come at a cost. Mistaking competitive pricing for a margin squeeze will tend to distort the market. On the other hand, mistaking predation for competition may result in less competition and higher prices over the longer term.

It is vital to look at the margin-squeeze cause in order to minimize these mistakes in practice. Depending on the cause and the effect on competition, regulators might decide to intervene. Thus the regulators should first determine the actual cause of the margin squeeze. The discussion in the theoretical section can be an important starting-point. Some issues do need special attention. These are: the time dimension, allocation of (common) costs, information asymmetry, and *ex ante* regulation versus *ex post* intervention.

Time is an important issue in margin-squeeze cases caused by predatory pricing and cross-subsidization. In a predatory pricing case, time refers to the period required for recoupment. The period in which recoupment is possible is likely to be short, particularly in the telecoms sector with its rapid technological development. On the other hand, path dependence might imply that once a company has missed a stage, it will be difficult to make up the disadvantage. Entry barriers for new technologies might also increase the period for recoupment. In a cross-subsidization case, time is important to evaluate whether certain business practices constitute anticompetitive behaviour or not. For normal business practice, revenues over the economic lifetime of the service should cover the costs. If the costs are covered, the operator uses a penetration price to generate a large custom base quickly. If costs are not covered over the service's economic lifetime, there might be a case of cross-subsidization.[16]

A margin-squeeze case can be very complex if common costs are involved. If the revenues of a service are higher than the LRIC plus the common costs, then there might be a case of cross-subsidization. By using the incremental cost test and the stand-alone test one can identify whether there is indeed such a case. Predatory pricing is also more difficult to prove when common costs are involved. Part of the problem is related to the allocation of the common costs across the relevant services.

It is clear that a margin-squeeze case caused by predation or cross-subsidization is difficult to prove. It is difficult to obtain all the necessary cost information, and to allocate the costs to the relevant activities. If a regulator cannot perfectly oversee the incumbent's cost levels, the incumbent may try to hide such a margin squeeze by allocating costs to its wholesale activities that are actually incurred as a result of its retail activities. Accordingly, in practice, the risk of a margin squeeze is closely linked to information asymmetry. Second, it is difficult to prove operators' intentions. The likelihood of a specific case satisfying all criteria is limited.

Finally, there is a tradeoff between *ex ante* regulation and *ex post* intervention. An *ex ante* margin-squeeze test clearly helps to make regulation transparent, and creates clarity for market parties. It can stimulate more market competition. On the other hand, it can also result in over-regulation or needless regulation, particularly if the price-squeeze test is very detailed. Margin-squeeze cases can also be handled *ex post*, with the risk that intervention is too late (complaints procedures take a very long time). Regulators have to make a tradeoff between possibly needless regulation and the risk of intervention, which is too late.

## 4.3 The Allocation of Common Costs

### 4.3.1 Policy relevance

One of the toughest regulation problems concerns the allocation of common costs. Technically speaking, entrants only use part of the infrastructure when leasing a line or using only parts of the spectrum, for example. However, access would be of no use if the other parts were not there (that is, the parts of the infrastructure it does not directly use). In other words, parts of the infrastructure are 'used' for various services, but it is unclear in just what proportion the costs of those parts should be allocated across those services. A related but slightly different issue occurs when the costs do not depend on the amount of the bandwidth actually used or the number of services which run on it. Finally, there are also often provisions for public services to be shared among the various services. An intervening regulator has to come up with an answer by deciding on some method of allocating the common costs. Misallocating these common costs can distort competition.

### 4.3.2 Economic theory

The necessity of financing part of the common costs is a problem reasonably well covered in the economic literature (see, for example, Armstrong 2001 or Laffont and Tirole, 2000). At the same time, it is acknowledged there is no easy way of doing it. Determining access charges consists typically of two stages. In stage one, the regulator determines the causally attributable costs (see Section 2 for various methods). Stage two then tackles the common costs. The failure to allocate appropriate costs in stage one implies that the residual will also contain costs which might otherwise have been allocated as causally attributable costs, but which will be allocated as if they were common costs. The question then becomes: how does one allocate these common costs?

Theoretically there is an optimal way (Ramsey pricing, see Section 2.2.2). In short, common costs are allocated inversely proportional to the service's price elasticity; that is, the service with the lowest price elasticity carries the highest burden. It will be recalled that there are three reasons why Ramsey prices are rarely used in practice. The first is political: Ramsey prices often involve a (very) skewed distribution of prices across services and may imply sky-high prices for some services, which may be politically unattractive. The second is practical: to calculate Ramsey prices, one may need very detailed information, which is often unavailable. The third is that Ramsey prices do not allow for clear comparisons between countries or operators. Such comparisons could be useful for benchmarking (see, for example, Jullien, 2001).

Because of these practical difficulties, regulators had to devise alternative approaches (see DTe, 2000; Europe Economics, 2001; Oftel, 2001). Four are detailed below:

*Equal proportionate mark-up (EPMU)*

The mark-up for common costs is proportional to the incremental costs of the service provided. It differs from Ramsey prices in that the willingness to pay does not count – only the costs. Thus this method neglects demand-side factors and scores worse than Ramsey pricing on allocative efficiency. In cases where common costs are small relative to incremental costs, this distortion is modest.

*Pro-rata appointment*

The mark-up on common costs is decided as a fixed proportion between various services. The ratio can be decided by the regulator. This method's advantage is that it is very simple to implement, and that it can easily be tailored to a specific goal by the regulator. However, unless the regulator comes up with plausible arguments, the proportions are set in an arbitrary manner.

*Incremental and stand-alone costs*

This method is based on the calculation of two extremes. The boundaries of the costs to be allocated to any service are determined by:

- a floor, the incremental cost – the cost which would be avoided were that service not provided, and
- a ceiling, the stand-alone cost – the cost that would be incurred if that service were provided in isolation.

The difference between these two boundaries represents the level of common costs. With pricing at incremental cost, the service concerned makes no contribution to common costs; with pricing at stand-alone cost, the service concerned bears the totality of common costs. At intermediate prices, the service makes some contribution towards common costs. This method is particularly useful in casting light on what causes costs. The method provides an estimate of the magnitude of common costs, which is independent of the accounting data and thus could be applied as a cross-check in identifying attributable and common costs. As a method of determining some intermediate level, the same criticism is applicable as to the pro-rata method.

*Commercial negotiation*

The regulator sets prices consistent with a hypothetical commercial negotiation between two independent parties in advance of the expenditure being incurred (Europe Economics, 2001). The advantage of this method is that it takes demand-side considerations into account, but is not likely to lead to the extreme Ramsey outcomes, nor does it require such a high information burden. The downside is that the outcome of this method is not clear *a priori*. At worst, the method is as arbitrary as pro-rata appointment.

### 4.3.3 Practical experiences

Before considering the main telecommunications experiences, it should be observed that other utilities face quite similar problems. Worthy of mention below are the examples of an airport and electricity supply.

*Airports UK*

In a number of European airports, regulators are considering a so-called 'dual-till' system. Under a dual till, airport activities would be divided into 'aeronautical' and 'non-aeronautical' activities, and only costs and revenues in the former category, termed the 'regulatory till', would be taken into account when setting the price control for regulated charges (Europe Economics, 2001). The UK has the longest and best-documented experience with this, hence the UK focus here.

Europe Economics (2001), in advising the regulator on matters such as common costs, rejects Ramsey pricing for practical and political reasons, such as those discussed above. It rejects the EPMU, because it would cause profitable activities to cease (or prevent others from being introduced). Here the absence of demand-side considerations takes its toll. As a compromise, Europe Economics (2001) suggests the commercial-negotiation principle. In practice, the method boils down to a variant of EPMU, but corrected for demand-side factors.

*Electricity in the Netherlands*

The Dutch regulator considers 'the local electricity loop' as regional natural monopolies and regulates access to the local loop via yardstick competition. Naturally the regional structure of the electricity networks helps the regulator in applying such yardstick competition. But even TenneT, the national grid company, is benchmarked against best-practice grids in the USA and Europe (DTe, 2000). Yardstick competition implies that there is no relationship between access charges and the costs of the network. Access charges are set based on the best-practice performance in the country (or elsewhere). Yardstick competition obviously avoids any discussion on common costs. Its applicability to telecommunications depends on just how comparable accounting and regulation principles are between European countries. If they are indeed comparable, then yardstick competition could be a useful alternative to other methods of price-cap regulation. A recent example is the local loop unbundling in Ireland, where tariffs are based on international comparisons (Office of the Director of Telecommunication Regulation Ireland, 2001).

We move on now to the two main telecommunications applications.

*Unbundled local loop/shared access in the UK*

Unbundled local loop (ULL) enables competing operators to lease the fixed-line incumbents' local loop and to install equipment at their exchange sites. The main implication is that it enables these operators to offer high bandwidth directly to consumers and thus fosters competition for high-bandwidth services. ULL is a typical one-way access situation. Section 2 has already discussed access charges with ULL, while leaving the common-cost aspect until this section. One way of reducing additional consumer costs for hiring these new operators with their high bandwidth is to introduce shared access (see, for example, EC Commission Recommendation C2000 1059, Oftel, 2000b). An issue specific to shared access is how to split the common costs between high and low frequencies.

Oftel has suggested a method for sharing common costs in the case of shared access (Oftel, 2000b). It first notes that shared loops differ from fully unbundled local loops, in that the competing firms only lease a portion of the local loop – the high-frequency one. So after applying the attributable costs in step one (LRIC is used in this instance), in step two it must be determined just what to do with the costs of the loop which are invariant with the bandwidth used.

Oftel rejects Ramsey pricing for the usual reason: the information burden is too high. It goes on to state that its objective is to adopt a method which ensures that the take-up of higher-bandwidth services is neither discouraged nor deterred, that is, allocative efficiency in DSL services appears to be the main goal. Oftel concludes that the easiest way to achieve these goals is to let the local loop provider (LLP) set the access charges itself so long as there is non-discrimination; that is, the LLP cannot charge competitors a different amount to that which it itself charges. Since the LLPs do not charge any of the common costs to their own downstream higher-bandwidth business, Oftel suggests that the total of common costs will be allocated to voice.

*Fixed–mobile termination*

It has been concluded from Sections 3 and 4.1 that fixed to mobile termination is a service which should be regulated since mobile network operators (MNOs) have a near-monopoly position in this service. Indeed, a number of regulators have adopted RPI–X regulation to prevent MNOs from exploiting their near-monopoly position (see, for example, Oftel, 2000b).

Oftel opts for the EPMU system for the common costs (Koboldt and Maldoom, 2001). It endorses this choice since the recovery of these costs is a relatively minor issue in the context of mobile termination, because a mobile network's common costs are relatively small. In Oftel's cost model, common costs amount to only about 3 per cent to 5 per cent of the total costs of a 900 MHz and 1800 MHz network. The common costs comprise the cost of a network management system plus the fixed costs of coverage, that is, the cost of acquiring, renting and operating the number of sites required to provide coverage. The characteristics of a 1800 MHz spectrum mean that more base stations are required to provide the same degree of coverage, which is why common costs are a slightly higher proportion of total costs than in a 900 MHz network. As a consequence, EPMU distortion is modest. Since EPMU is easy to implement, the choice is clear.

### 4.3.4 Policy conclusions

- Ramsey pricing, though theoretically appealing, is often discarded as being impractical or politically unfeasible. None the less, Ramsey pricing's message is that demand-side factors can be important, in particular in situations where the common costs are fairly high relative to incremental costs.
- When common costs are high, Ramsey prices, or more generally demand-side factors, should not be readily dismissed. Difficult data requirements can sometimes be resolved. Imperfect demand estimations can be better than no demand-side estimations at all. Political (income distribution) problems can also often be resolved.
- In cases where Ramsey pricing is not feasible, the regulators will find themselves at a crossroads since none of the options can be deemed to be superior to the other, *a priori*.
- The two leading questions in determining the best available alternative are: (i) to what type and level of distortion does the method lead? (ii) which are the main objectives the regulator wants to achieve with the access charges?
- The first question demonstrates sensitivity to the relative damage that the method inflicts: in the case of fixed–mobile termination, for example, we saw that the com-

mon costs were relatively small, implying that an alternative such as EPMU only causes a small distortion.

- The second question gives weight to the distortions: one distortion is not as bad as the other. In the airline example, we saw that demand-side considerations were considered very important, which made EPMU unappealing.
- A totally different perspective emanated from the electricity market. Indeed, one can wonder how international comparison could be a feasible future option in the telecommunications field, whether by benchmarking or even by the more formal companion to benchmarking, yardstick competition.

## 4.4 Static versus Dynamic Efficiency

### 4.4.1 Policy relevance

Sections 2 and 3 devoted space to the dynamic consequences of access prices. Because of the policy relevance in virtually all access problems, and the lack of a theoretical consensus in general, as well as the lack of a well-developed analysis of the linkage between access pricing and dynamic efficiency, this section reverts to the issue in more detail. The notions of static and dynamic efficiency are used throughout this section; see Box 8.1. This enables the possible tradeoff between short- and long-term policy choice implications to be addressed.[17]

### BOX 8.1 STATIC AND DYNAMIC EFFICIENCY

Roughly speaking, static efficiency is high if competition is sufficiently intense, if there is downward pressure on prices, if consumers can choose between several suppliers, and if they get good value for money. Some static efficiency drivers are: low entry barriers, no collusion and no substantial consumer switching costs. Unfortunately one cannot measure the static efficiency of a market, but in practice it can be roughly assessed by consulting industry experts, antitrust economists and consumer organizations which may each have specific views on certain indicators related to static efficiency drivers.

Dynamic efficiency is high if there are low entry barriers for new technologies developed by rival firms, and if firms have incentives to invest in R&D and innovation. Some dynamic efficiency drivers are: low entry barriers for players with new technologies, the possibility of recouping R&D investments, little regulatory uncertainty, and effective standardization. Again, it is very difficult to measure dynamic efficiency, but in practice one can look at the levels of investment in product and process innovation, R&D budgets, the number of patents, the prospects for new players introducing new technologies, and consumer take-up of new products. Such indicators may provide a good approximation of dynamic efficiency. It is important to stress that a high level of innovation does not always mean that dynamic efficiency is high. There can also be too much innovation in a market, for example in a situation of planned obsolescence.

Why do regulators struggle with this issue? There are several explanations. First, theory is well developed for static access rules, not for dynamic ones. Second, while the emphasis was on developing competition for services in the early stages of liberalization, attention has shifted (at least to a certain extent) to competition for infrastructures. Third, there could be tradeoffs between static and dynamic efficiency. Regulating access prices, which foster service competition, does not automatically foster infrastructure competition. The well-known example here is low one-way access prices: entrants who can 'free-ride' on the incumbent's network do not have incentives to build their own networks. Fourth, partly because of the low market sentiments for telecommunication, investments in new infrastructure are not as booming as expected (with some notable exceptions). Fifth, there is considerable political pressure on regulators to go easy on telecommunications operators because tight regulation is allegedly harmful to investment, particularly given current market sentiments. Finally, the emphasis on dynamic considerations cannot be overstated. The telecommunications industry is one that is driven largely by innovation and investments in new infrastructure. So putting right the investment and innovation incentives can easily override any of the static considerations. This notion is further reinforced when considering a regulatory time path. Bennett et al. (2001) showed that if investment and/or innovation can create a breakthrough in market structure, creating a market-driven transition from low efficiency to high efficiency, this route should not be obstructed (see also Leo et al., 2002). Since market-driven routes are often preferable to regulation-driven ones, the likelihood of this route should be assessed first before considering some form of regulation.

### 4.4.2 Economic theory

Summarizing the theoretical insights of Sections 2–3, six conclusions can be drawn:

1. Forward-looking access rules (for example LRIC) may offer better incentives for investment than backward-looking ones, although the empirical evidence is somewhat mixed (Cave et al., 2001).
2. However, by implicitly assuming that markets are contestable, LRIC rules can have adverse dynamic consequences, which can completely offset the positive incentives mentioned above. Typically, telecommunication markets are not contestable: there are substantial non-recoverable costs as well as many unsuccessful investments (stranded assets). Failing to take this into account can hamper investments and innovation.
3. Providing *ex ante* clarity on the leading regulation principles reduces uncertainty and thereby contributes to dynamic efficiency.
4. Uncertainty can further be reduced by defining 'sunset clauses'; that is, regulators predetermine conditions under which regulation can be lifted or softened.
5. Dynamic efficiency is not one-dimensional: access charges which serve well as investment incentives for incumbents are not always as good for entrants, and investment is not equivalent to innovation.
6. A simple focus on incentives for investments by the entrants is too short-sighted: investments can be wasteful. It is only useful to provide incentives for entrants if the incumbent's infrastructure is replicable.

### 4.4.3 Practical experiences

There are a number of practical experiences with access in a dynamic context. While any number of practical experiences could be listed here, the focus has been laid instead on a couple of instances where the dynamic nature was most apparent. Much of this subsection on practical experience is taken from Cave et al. (2001) and Bennett et al. (2001).

*Country experiences*

Three country studies can be mentioned, taken from Cave et al. (2001). The fourth is taken from Bennett et al. (2001).

In the UK, Oftel realized in 1996 that its interconnection regime should be revised, because insufficient attention had been given to dynamic considerations. The new regime was characterized by low (forward-looking) interconnection charges for entrants. Entrants obtained a special status (relevant connectable systems) which allowed them these low charges. At the same time, Oftel made it clear that operators could not assume they would maintain that status unless they were prepared to invest.

For the USA, there is certainly empirical evidence that access charges influence investment levels, to the extent that lower access charges have promoted a greater deployment of digital technology among US incumbent local-exchange carriers.

For the Netherlands, empirical evidence suggests that interconnection policies have influenced the level and structure of investment, most notably in relation to ULL. A new group of DSL providers has entered the market, boosting investment in fibre-optic networks.

There has been a Dutch policy debate as to whether cable companies should be forced to provide access to their cable infrastructure. Mandated cable access is good for the television market, where cable companies hold dominance. It can also be good for static efficiency in the Internet market. The effects on dynamic efficiency are unclear, since it may provide disincentives for cable companies to invest. Potential harm to dynamic efficiency can be reduced by time-dependent access regulation, taking into consideration the stranded assets discussion (see Section 2.3.2).

The conclusion drawn from these national scenarios is that various countries do certainly have experience with the dynamic consequences of access charges. While these consequences are considered important, it is not yet clear what lessons to draw from these experiences.

*Unbundling the local loop*

The discussion on ULL also yields interesting results for the discussion on static versus dynamic efficiency. Consider various types of entry, where entrants adopt different strategies or where the maturity of the market differs (quoted from Cave et al., 2001; we have added some notes in italics).

> First, a cable operator like UPC [*a Dutch cable operator*] will already have replicated some aspects of the local loop, although further investments are required both for telephony and for the development of Internet services. In this respect, local access pricing makes little difference. The cable operator has to buy call termination from the incumbent, probably in respect of the majority of calls. This service is wholly non-replicable. But reciprocal pricing between operator and incumbent neutralises this factor. Finally, the cable operator needs access to long-distance convey-

ance, which is replicable, either by the cable operator or by other entrants and which should, with time, be competitively priced.

Second, consider the case of Tele2 [*the largest Dutch Carrier select player*], whose strategy consists of targeting a mass market, involving marketing and advertising expenditure, on the basis of – initially at least – a minimal investment in infrastructure. As time passes, Tele2 makes further investments in switching and conveyance at the national level, but its investments are limited (possibly confined to marketing costs).

Third, consider the case of non-cable entry into the high-bandwidth markets. This is a new market, in which the incumbent has no historic market share, although clearly it has the advantage of providing the related service of basic telecommunications to the vast preponderance of domestic and business customers. Incumbents and entrants are under the same necessity to make the appropriate investments in servers. In this instance, the key non-replicable resort for entrants is the local loop, or access to part of the bandwidth provided by the loop. Setting on one side wireless technology, which is essentially untried, unbundling of the loop is a necessity.

This raises the key question of how the unbundled loop should be priced [*see also Section 2*]. Commitment to a low regulated rental for the unbundled loop would clearly encourage a competitor's complementary investment. If, however, the regulator were concerned about the entrant's short-term cash position, it might choose to alleviate its difficulties in the early phase of entry by a lower rental charge, which would then rise to above cost. The choice of a final level on the price curve at which the price would be stabilised would be influenced by the regulators' preference for network duplication. If the preference were slight, then a level-pricing policy would be preferable (or one which levelled off at cost). If the preference for duplication was strong, then this could be reflected in a pricing strategy, which rose to above cost. A time limitation on mandated access to the local loop might also be appropriate. But this last policy, which has been adopted in certain areas in Canada, appears to carry the risk that entrants' fears about what might happen towards the end of the mandated access period might discourage investment.

This all leads to the conclusion that regulators have two instruments with which to influence investment decisions. The first relates to whether incumbents' assets are replicable or not. Clearly, entrants are not sensitive about access charges in parts of the network they cannot replicate. Second, access charges can have a dynamic nature. Most entrants choose a gradual investment strategy, leasing lines from the incumbent in the initial stages and meanwhile investing in their own networks. A dynamic access policy takes this into consideration: access charges can rise over time, or the level can be made contingent on the level of investments.

### 4.4.4 Policy conclusions

A number of policy conclusions emerge from the above.

*Universally high access prices are not smart*

A possible implication of static analysis is that the best way to stimulate infrastructure investment is to have universally high access prices, that is, access prices which are high and which remain high over time. The argument is that high access prices provide an investment incentive for incumbents (risk premiums, stranded asset discussion) as well as for entrants (who can avoid high access prices by investing themselves). This is not necessarily true when taking a dynamic view. There is neither theoretical nor empirical evidence to support high-cost access pricing as a means of encouraging infrastructure competition. This is because future profit opportunities are more important than current price levels when contemplating an investment decision. Admittedly, if entrants are certain that future access levels will be at such low levels that it will always be cheaper to obtain access than to invest, then (future) high access levels are problematic. But there are many reasons why that picture may look different. The investments can yield other types of returns (unrelated to access) and there can be a belief that regulators will reward investments in infrastructure.

*Eligibility and replicability*

Access charges, which explicitly take entrants' investments and investment possibilities into consideration, do well on dynamic efficiency. Think of Oftel's strategy of conditioning charges, or the status of operators on investment decisions. The same applies to replicability: on which parts of the network is it reasonable to provide investment incentives? This is a tough question for regulators. A few years ago local loop competition seemed to be the name of the game, with wireless, cable and Internet coming up. Nowadays, over-investment predominates throughout Europe. Perhaps tomorrow – who knows – everyone will quietly forget it when fibre to the house is installed.

*Sunset clauses and other dynamic regulation*

Regulatory uncertainty can be lowered by pre-specifying sunset clauses, not only (or rather: not necessarily) purely the date, but perhaps the market condition under which some specific regulation can be removed. Access charges which rise over time or which contain some other temporal element enable the fulfilment of both static and dynamic goals.

*A dynamic focus can reduce the need for future regulation*

Setting appropriate access charges is difficult. Stimulating investments in new infrastructure provides the basis for removing access regulation and replacing it with bilateral agreements on interconnection charges.

## 5. CONCLUSIONS

This chapter provided an overview of the literature on access regulation and noted some lessons for regulation practice.

A common element in all this chapter's discussions is that access prices can never be neutral by definition. Access prices affect competition – or more precisely, operators' profits, market shares and retail prices. Much can be gained by acknowledging this and by using economic theory to analyse the various links between markets and prices.

A second element is that access charges often perform too many tasks. Access charges are used to stimulate static as well as dynamic efficiency while at the same time serving equity goals such as universal service obligation. This is a bit much for an access charge which, in its simplest form, is a one-dimensional price. Several of these goals can easily conflict, so choices have to be made and tradeoffs persist.

A third element is that the focus has shifted from competition in service to competition in infrastructure, or, more generally, attention has shifted from static access prices to dynamic notions. The emphasis on dynamic considerations cannot be overstated. The telecommunications industry is largely driven by innovation and investment in new infrastructure. So putting right the incentives for investments and innovation can easily dwarf any of the static considerations. Successful competition for infrastructure can reduce the need for access regulation and does not preclude competition for services taking place at the same time. Dynamic access policies, for example access charges that rise over time, enable the fulfilment of both static and dynamic goals. It is important to reduce uncertainty by providing regulatory transparency on these time-related issues.

A fourth element concerns the various methods of determining access charges. No one method is superior to another, *a priori*. It is both the relative levels and weighting of distortions that determine which method is best. Some methods distort dynamic efficiency, but if the focus is on static efficiency, such a method could be adequate. Other methods ignore demand-side considerations, but if the distortion is small, such a method could indeed be used. Ramsey methods require a high information burden, but sometimes this burden can and should be overcome. LRIC, when used too simplistically, can produce economically perverse outcomes, reducing incentives to invest instead of achieving the intended incentive boost.

A final element is the potential of an alternative to the 'calling party pays' (CPP) principle. CPP is common practice in Europe (with the notable exception of international roaming). The CPP principle implies that a receiver obtains a service (receiving a call) but doesn't actually pay for it. Some distortions take place as a result, for example in interconnection tariffs and fixed–mobile termination, while Internet charges may be set at suboptimal levels – where a balanced cost allocation would lead to incentives to set the charges at competitive levels (Laffont et al., 2001), thus reducing the need to regulate services which are currently under regulator scrutiny.

## NOTES

1. Armstrong (2001) also mentions competitive bottlenecks as a separate category, but we prefer to discuss competitive bottlenecks in the policy section.
2. Entrants without their own backbone may lease long-distance capacity from the incumbent. Since the incumbent may have spare capacity, this may be in its own interest.
3. Only the leasing part of ULL is one-way access. The interconnection part is two-way access. It is the leasing aspect which is the most attractive here.
4. There need not be a conflict of interest if the incumbent wants to increase traffic on its network (compare with virtual mobile operators).
5. See Tirole (1988), ch. 8.
6. Applying LRIC to certain services (for example, wholesale services) and backward-looking prices to others (for example, retail services) could lead to inconsistencies.
7. Or at least less directly competing operators.
8. Armstrong (1998) and Laffont et al. (1998) also address asymmetric competition between an incumbent and an entrant, but focusing on different issues, for example the entrant's optimal network coverage.
9. This is pointed out by de Bijl and Peitz (2002b).
10. It is not clear, however, whether this would constitute a problem in welfare terms.
11. However, it does not solve the *originating* access from fixed to mobile. These charges should still be regulated.
12. Closed user groups are groups which have an interest in how much it costs to call each other and therefore all subscribe to the same network (for example family members or a company). Special offers are often made for these groups.
13. If different retail services use the same access service, the allocation of common costs to the different retail services will come into play; see Section 4.3.
14. If prices are below cost but stay at that level 'for ever', then there is no predation. This can be a viable

business strategy in industries characterized by network externalities or learning by doing. The period of below-cost pricing extends no longer than necessary to achieve the network economics, or until the cumulative production experiences result in lower costs. See also the next section on cross-subsidization.
15. This also solves the discussion as to which subset to start with.
16. A practical issue is how to determine the economic lifetime. This is an accounting matter.
17. See also Bennett et al. (2001) and Leo et al. (2002), in which the notions of static and dynamic efficiency play a central role in addressing various telecommunications market policy issues.

## REFERENCES

Armstrong, M. (1998), 'Network interconnection in telecommunications', *Economic Journal*, **108**, pp. 545–64.
Armstrong, M. (2001), 'The theory of access pricing and interconnection', in M. Cave, S. Majumdar and I. Vogelsang (eds), *Handbook of Telecommunications Economics*, Amsterdam: North-Holland. http://www.nuff.ox.ac.uk/economics/people/armstrong/acc-int-num.pdf
Bennett, M., P. de Bijl and M. Canoy (2001), 'Future policy in telecommunications: an analytical framework', *CPB Document* (005), CPB Netherlands Bureau for Economic Policy Analysis, The Hague. http://www.cpb.nl/eng/pub/document/5/doc5.pdf
Bolton, P., J.F. Brodley and M.H. Riordan (2000), 'Predatory pricing: Strategic theory and legal policy', *The Georgetown Law Journal*, **88**, pp. 2239–330.
Braeutigam, R.R. (1989), 'Optimal policies for natural monopolies', in R. Schmalensee and R.D. Willig (eds), *Handbook of Industrial Organization, Volume II*, Amsterdam: Elsevier Science Publishers.
Buigues, Pierre (2002), Latest progress in LLU as reflected by the sector inquiry, LLU hearing 8 July 2002, downloadable from EC web site.
Cabral, Luis M.B. and Michael H. Riordan (1997), 'The Learning Curve, Predation, Antitrust, and Welfare', *Journal of Industrial Economics*, **45** (2), pp. 155–69.
Carter, M. and J. Wright (1999), 'Interconnection in network industries', *Review of Industrial Organization*, **14**, pp. 1–15.
Cave, M., S. Majumdar, H. Rood, T. Valletti and I. Vogelsang (2001), 'The relationship between access pricing regulation and infrastructure competition', Report to OPTA and DG Telecommunications and Post, Brunel University.
de Bijl, P. and M. Peitz (2002a), 'New competition in telecommunications markets: regulatory pricing principles' (2002), forthcoming in *IFO Studies*, also published as *CESifo Working Paper* No. 678 (9), CESifo, München.
de Bijl, P. and M. Peitz (2002b), *Regulation and Entry into Telecommunications Markets*, Cambridge: Cambridge University Press.
DTe (2000), *Guidelines for price cap regulation of the Dutch electricity sector*, (http://www.nma-dte.nl/pdf/guidelines9mei.PDF)
European Commission (2001), *Pricing issues in relation to unbundled access to the local loop*, ONPCOM 01-17.
Europe Economics (2001), *Airport Cost Allocation Report for the CAA*.
IRG (2002), Working paper on market definition for mobile termination, February, IRGs Significant Market Power – Working Group.
Koboldt, Ch. and D. Maldoom (2001), 'Optimal fixed-to-mobile interconnection charges', paper presented at the 12th European regional ITS conference, Dublin, 2–3 September.
Laffont, J.-J. and J. Tirole (2000), *Competition in Telecommunications*, Cambridge, MA: MIT Press.
Laffont, J.-J., P. Rey and J. Tirole (1998), 'Network competition: I. Overview and non-discriminatory pricing', *RAND Journal of Economics*, **29**, pp. 1–37.
Laffont, J.-J., S. Marcus, P. Rey and J. Tirole (2001), 'Internet Peering', *American Economic Review*, **91** (2), pp. 287–91.
Leo, H., M. Pfaffenmayr and G. Schwarz (2002), *Innovation und Regulierung im Telecom-sektor*, (in German) WIFO publication, Wien.
Nicolaides, P. and R. Polmans (1999), 'Competition in EC Telecommunications: Cross-subsidisation, access and predatory pricing', *World Competition*, **22** (4), pp. 21–43.

OECD (2000), *Cellular mobile pricing structures and trends*, Paris: OECD, DSTI/ICCP/TISP(99)11/final.
Office of the Director of Telecommunication Regulation Ireland (2001), 'Local Loop Unbundling – Eircom's Access Reference Offer (ARO)', *Decision Notice* D8/01, September.
OFT (1999), *Assessment of individual agreements and conduct*, OFT 414.
Oftel (1999), *Ideas for relevant markets/Sunset criteria for access market*. http://www.OFTEL.gov.uk/ind_info/international/irgtf99.htm
Oftel (2000a), *The Application of the Competition Act in the Telecommunications Sector*, http://www.OFTEL.gov.uk/publications/ind_guidelines/cact0100.htm
Oftel (2000b), *Statement on the Implementation of Shared Access to the Local Loop in the UK*, www.oftel.gov.uk/publications/broadband/llu/shac1200.htm
OPTA and NMa (2001), *Richtsnoeren prijssqueeze*.
OPTA (2002), *Beleidsregels inzake de regulering van mobiele terminating tarieven*, 28 maart.
Sharkey, W.W. (1982), *The theory of natural monopoly*, Cambridge: Cambridge University Press.
Tirole, J. (1988), *The Theory of Industrial Organization*, Cambridge, MA: MIT Press.

# 9. The price of access: the unbundling of the local loop in the EU

**Juan Delgado, Jérôme Fehrenbach and Robert Klotz***

## 1. INTRODUCTION

This chapter attempts to show why interest in local loop unbundling is now waning in Europe, and why the whole process, while initially heralded as a major breakthrough, may turn out to be nothing more than a wasteful regulatory experiment with little benefit for competition. As the main threat to the success of local loop unbundling, we identify the pricing policies implemented by the incumbent telecommunications operators and in some cases facilitated or even mandated by the national regulatory authorities.

Section 2 briefly reviews the main theoretical issues behind local loop unbundling. Section 3 describes the European policy approach, and Section 4 provides an overview of the current situation in Europe and of the problems detected by access seekers. Section 5 analyses in more detail the access pricing policies adopted in the EU and suggests possible causes of the slow take-up.

## 2. THEORETICAL FRAMEWORK

### 2.1 Why Unbundle the Local Loop?

The local loop is the part of a telecommunications network that links the consumer premises to the first exchange of the telecommunications operator. It is no more than a copper pair linking the end-user to the operators' networks.

The local loop provides access to the final customer and is unique for each end-user line; that is, it is not shared by several customers as is the case for the rest of the network. Before the liberalization of the telecommunications sector the local loop was protected by a legal monopoly. This fact and the fact that the duplication of the local loop is uneconomic (and wasteful from a social welfare perspective unless the alternative infrastructure implies cost savings and increases consumer surplus) confer on it a bottleneck nature.

The bottleneck nature of the local loop and the existence of network externalities mean that granting to any firm the freedom to deploy its own networks and to provide services is not sufficient to guarantee the existence of effective competition. For competition to develop, what is also required is to guarantee the interconnection and inter-operability of the incumbents' networks with the networks of the new entrants and mandate the provision of access by the incumbent to its competitors.

The provision of access to the local network of the incumbent can follow different patterns. New entrants can interconnect their networks to the incumbent's local network and simply 'resell' the local services on their own or by adding extra services (such as long-distance transit or Internet access). Another policy option is to provide unbundled access to the local loop. Through unbundled access, new entrants take full control of the local loop and offer their own services through it directly to the end-user. The first option leaves little room for new entrants to modify the service provided by the incumbent while local loop unbundling allows competitors to fully configure their own services to the end-customer.

As we shall see in the next section, European Regulation imposed on the incumbents the obligation to provide access through interconnection in the first 'liberalization wave'[1] during the 1990s. Only in 2000, with the aim of accelerating and promoting competition in local telephony services and, more importantly, in the provision of broadband services, did the European Commission mandate the provision of unbundled access to the local loop.[2]

Access policies usually have two main dimensions: ownership of the facility to which access is mandated and access price determination. On the ownership dimension, the two ends of the segment either keep the local loop the property of the incumbent or impose local infrastructure divestiture. The price determination dimension is delimited, at one end, by the determination of prices through free commercial negotiations, and, at the other, by the orientation of the price to its cost. There are several options in between, including different degrees of formal separation[3] and different forms of price regulation.

Keeping ownership of the local loop in the hands of the incumbent may induce the incumbent to foreclose access, while divestiture may reduce incentives to innovate and imply high transition costs and efficiency losses. Under free commercial negotiations the incumbent may use its higher bargaining power to set high access prices and keep monopoly rents. On the other hand, cost orientation may encourage the bottleneck owner to deny or foreclose access (since no benefit is obtained from granting it) and may deter investment in alternative infrastructures.

The option chosen by most countries mandating local loop unbundling is price regulation, while keeping the ownership of the local loop in the hands of the former monopolist.[4] In Europe, for example, there is no obligation to separate the provision of access from other activities.[5] Incumbent operators are obliged to provide access to the local loop 'on the basis of cost orientation',[6] at a price that ensures that 'the local loop provider is able to cover its appropriate costs … plus a reasonable return'.[7]

## 2.2 Pricing the Local Loop

When pricing the local loop, regulators face a well-known dilemma: too high a price prevents competitors from renting the local loop and then reduces service-based competition. On the contrary, too low a price promotes service competition but deters investment in alternative infrastructures, or 'short term may kill long term'.[8,9] The European Regulation is aware of this issue when noting that 'pricing rules for local loops should foster fair and sustainable competition, bearing in mind the need for investment in alternative infrastructures';[10] however, it gives little indication of how to translate this into reality.

The apparent conflict between static and dynamic efficiency is an issue that is not trivial to solve: perhaps solution is a highly ambitious aim to be achieved only with the use of a single static tariff. As it is a dynamic objective, the promotion of short-term and long-term competition might require a more complex mechanism than a fixed price over time.[11]

Economic theory states that the efficient pricing of an essential facility should be equivalent to marginal cost plus a mark-up to cover fixed and common costs.[12] In order to allow for dynamic efficiency, it may also include a mark-up to compensate for innovation and investment risks.[13] The concept of marginal cost is of little use in telecommunications. Since most costs are fixed, the cost of increasing production marginally is negligible (in the absence of congestion). Therefore the concept used in practice is that of incremental cost,[14] which is the average cost (that is, the cost per unit of output) over a specific period of time of increasing output by a substantial arbitrary amount.

The fact that a time factor is included in the cost calculation implies that some short-term fixed costs become variable in the long term and then the mark-up to cover fixed costs should become lower. In addition, since the bottleneck situation does not result from innovation but from a previous situation of legal monopoly, the return on innovation (that could make sense in the case of a patent made available to competitors) is senseless in this case.

How can this cost be measured? One possibility is to use the incumbent's accounting books to calculate this cost (which is known as the historic cost approach). The problem with this approach is that part of the equipment used by the incumbent may be depreciated and obsolete and its cost may not reflect the current cost of increasing production. Replacement costs differ from historic costs. As pointed out by Cave (1997), 'prices based on historic costs might discourage entry by more efficient suppliers of access services'.

Forward-looking costs try to mimic the current and future costs of the relevant network elements. Therefore, the network is evaluated at its actual replacement cost rather than at its past deployment cost, taking into account, in particular, technological and economic changes. In this way, the right signals are given to competitors in order to encourage efficient entry.

The forward-looking incremental cost approach (commonly known also as long-run incremental cost, or LRIC) is not exempt from difficulties. Indeed, cost allocation is still subject to a high degree of arbitrariness. The choice of depreciation methodologies and periods is equally somewhat arbitrary, and the rule of depreciation may depart from the real technical lifetime of the asset. Finally, technological evolution is uncertain and thus the future evolution of the costs is difficult to predict. LRIC is, however, proposed by the European Commission in the ONP Directive[15] and in the USA by the 1996 Telecommunications Act.[16]

How does this translate in the case of the local loop?[17] The local loop is not a traffic-sensitive network element; that is, neither its cost not its dimensioning depend on the traffic carried. Therefore, in contrast to other network elements located at a higher level of the network hierarchy, the traffic is not the relevant unit of increment in the case of the local loop.

In principle, the basis for the calculation of LRIC of the local loop should be the local loop itself, that is, the cost incurred by unbundling plus the average annualized costs of maintaining and upgrading the loop and other associated facilities. The cost should normally be structured in two elements: a one-off fee that should cover the necessary works to transfer the loop to the new entrant, and a recurring fee that should cover two kinds of costs:

- the cost of local core network elements which have to be re-dimensioned because of an increase in the number of loops;

- the cost of maintaining and upgrading the local loop itself (taking into consideration the expected life of a local loop given the technical state of the network, and the necessity of replacing obsolete or deteriorated equipment).

## 3. THE EU APPROACH TO LOCAL LOOP UNBUNDLING

In early 2000, the Lisbon Summit concluded that it was necessary to accelerate Europe's transition to the new information society, in particular through the development of cheap Internet services. To meet this challenge, competition was seen as a key factor to encourage innovation and maximize consumer welfare.

The USA had opened the path to local and Internet access competition with their 1996 Telecommunications Act and the hands-on approach of the Federal Communications Commission (FCC). The EU followed suit, at a time when the quick emergence of high-speed Internet services made it all the more necessary to handle the bottleneck issue alluded to above. The European Commission took several initiatives in early 2000, with a view to achieving both by means of its competition tools and through sector-specific legislation what is technically known as local loop unbundling.

In April 2000 the Commission released a recommendation on local loop unbundling[18] and a communication[19] on the same subject. As these two instruments were not legally binding, but belonged to what is conventionally called 'soft law', which might have undermined their effectiveness, the Commission decided to go a step further and in July 2000 it proposed a regulation to the Parliament and the Council, which was adopted in December 2000.[20] In parallel, the Commission launched a sector inquiry on local loop unbundling on the basis of its competition powers under Regulation 17/1962 to monitor the evolution of the sector and possibly detect anticompetitive practices.[21]

The philosophy of the unbundling regulation (hereafter 'the Regulation') was to devolve large powers to national regulatory authorities (NRAs) to ensure quick and efficient enforcement. It imposed on incumbents obligations relatively similar to those already imposed in the context of interconnection, with, in particular, the obligation to publish *ex ante* a public reference offer covering the items listed in the annex to the Regulation (technical and economic conditions for local loop unbundling and related facilities). It gave the power to NRAs to modify these reference offers when they did not meet the criteria set out in the Regulation. The Regulation covers both the provision of full, unbundled access and the provision of shared access to the local loop whereby the access seeker only exploits the high part of the bandwidth spectrum.

The Regulation was perceived as an efficient tool to enforce rapid unbundling on a large scale.

## 4. THE POOR RESULTS OF LLU IN EUROPE AFTER TWO YEARS OF ENFORCEMENT

### 4.1 Facts and Figures

At the end of May 2002, only 900 000 lines, that is, 0.5 per cent of all EU fixed lines, had been unbundled. The number of lines subject to shared access, a few thousands across Europe, is simply negligible. It is clear that in a majority of Member States, unbundling had not gone beyond a purely experimental stage nearly 18 months after the entry into force of the Regulation. In France, Spain, Sweden, Portugal, Belgium and the UK, the unbundling process had no statistic visibility, with only a few hundred lines concerned. In late spring 2002, the process was slowly beginning to gain momentum in Austria, Italy and the Netherlands, although the proportion of unbundled lines did not exceed 0.3 per cent. Even in those countries where unbundling had been on offer since 1997 or 1998 (Denmark, Finland, Germany), the proportion of unbundled lines was below 2 per cent.

In the various Member States, collocation conditions, that is, the conditions offered by the incumbents to new entrants to install equipment on their premises, resulted in concentrating a high degree of complexity and in creating a wide range of difficulties. Although in the spring of 2002, more than 20 per cent of all main distribution frame (MDF) sites connecting 45 per cent of all fixed lines actually offered collocation facilities, the lightest – and least discriminatory solutions – such as co-mingling and virtual collocation, whereby the access seeker is treated on the same footing as the incumbent, were still extremely marginal, with less than 5 per cent of the sites concerned.

This slow progress of unbundling is all the more worrying as in the meantime the landscape of the broadband market is rapidly taking shape. In the first half of 2002, the progress of incumbents throughout Europe was spectacular in this regard. The number of new DSL lines opened every week by the EU incumbents totalled 65 000 units, while no more than 6000 lines were unbundled weekly. Given that cable providers are generally less well positioned in Europe than in the USA, with the exception of a few Member States, incumbent telecoms operators are set to dominate very largely the supply of broadband solutions, especially in the residential market.

### 4.2 Grievances Expressed by Access Seekers

The sector inquiry carried out by the Commission allowed us to glean a large amount of market information which provides interesting feedback on the perception by incumbents and by alternative telecoms operators of the problems associated with local loop unbundling.

Access seekers complain about a so-called delay–degrade strategy on the part of the incumbents. They allege discriminatory treatment, some refusals to supply and unjustifiable delays. The core of the problems, however, seems to revolve around costs and prices. In nearly all Member States the new entrants complain about margin squeezes and excessive prices for certain components of the incumbents' reference offers.

The situation may improve slightly in the future with some regulatory or competition problems gradually overcome. Most reference unbundling offers are now 'cleaner' than at earlier stages of the process, with the most egregious anticompetitive provisions usually removed.

However, we believe that local loop unbundling will never exceed a very limited scale under the current regime if present trends continue. And there is no reason for these trends not to continue if the EU approach to unbundling remains unchanged.

The reason why we are not especially sanguine about the future of unbundling in Europe is that in our view the economic equation of unbundling, at least as has been conceived so far, is essentially flawed.

## 5. ACCESS PRICING POLICIES

### 5.1 A Conflict of Objectives

#### 5.1.1 Benefits from existing economies of scale

The objective of local loop unbundling when it was mandated in 2000 was to give access to third parties at reasonable cost to an economic asset that had some features of a natural monopoly, with considerable sunk costs that would have made uneconomic and wasteful the duplication of the existing infrastructure. The idea was that a nation-wide copper network is not reasonably replicable in a sensible timeframe on profitable or at least acceptable financial terms. It is not defensible in terms of allocation of resources and economic welfare unless the new infrastructure creates value for the consumer by offering, for example, new services that cannot be offered through the existing infrastructure. The local loop infrastructure essentially implies sunk costs, that is, capital costs that cannot be recovered if the investment fails. Mandated access to the last mile of telecoms networks was thus designed to make alternative telecoms operators benefit from economies of scale acquired by the incumbents in their local access network, the latter often deployed in recent decades at the expense of the taxpayer.

#### 5.1.2 Concerns about too low unbundling tariffs

This commendable objective was soon influenced by debates on the right pricing of access to the local loop and the associated facilities (provision of tie-cables, collocation, and so on). It was argued, when the LLU Regulation was drafted, that the pricing methodology should give preference to a long-run incremental costs approach. The prevailing fear at that time in the academic world was that pricing of the local loops below a reasonable level would not only be unfair, but also economically not sensible.[22] The idea (as honourable as the principle of LLU itself) was that an artificially low price for local access would undermine investment in alternative platforms such as cable networks, wireless local loop and fibre-optic. More generally, the fear expressed by some economists was that pricing access to the copper pair at too low levels would amount to an implicit subsidy to those firms that purchased unbundled network elements. The negative implications for efficiency, competition and innovation were likewise underscored many times. An excessively low price for access to the copper pair was doomed to deter investments in the necessary network upgrades and would, more or less, paradoxically lead to a slow demise of the whole existing infrastructure and stifle attempts to upgrade it. The very process of mandatory local loop unbundling was sometimes denounced as a valuable option given to competitive telecoms carriers to take advantage as free-riders of an investment made at the incumbent's own risk. Any investment in the network would in any event rest on the shoulders of the incumbents, while the competitors would be at any time in a position to withdraw from the process and thus avoid losing large amounts of capital.

These concerns were translated as follows in a somewhat contorted recital of the LLU Regulation:[23]

> Pricing rules should ensure that the local loop provider is able to cover its appropriate costs …, in order to ensure the long term development and upgrade of local access infrastructure. Pricing rules for local loops should foster fair and sustainable competition, bearing in mind the need for investment in alternative infrastructure …

With now nearly two years of experience of the unbundling process at the EU level, it is clear, however, that the concerns about the risks involved in too low pricing of the network elements at stake were largely overstated. The assumption as to the nefarious effects of low prices rested on intellectual premises which, though intellectually correct, did not fit properly with the economic and market reality of the sector in Europe.

First, as regards the local loop infrastructure itself, one may wonder whether the risk of investment disincentive, argued by many economists and by incumbents, was not a fallacy. Some erroneous economic assessments would certainly have been avoided by a closer look at the economic and technical nature of the infrastructure at stake. The historic cost of the copper pair is minimal. Most of the infrastructure is depreciated. The proportion of the local network needing repair or replacement due to physical obsolescence or accidental deterioration is insignificant, and certainly below what the application of a theoretical accounting depreciation rate to the replacement costs of the entire network would give. The cost of maintaining and operating this infrastructure is limited; the capital outlay necessary to update network intelligence, though significant, is equally limited within sensible bounds. It is certainly true that incumbents have to invest heavily in equipment to develop new services such as those belonging to the DSL family. They have to purchase and install DSLAMs in their local exchanges, upgrade their ATM networks, and install broadband access servers. Nobody would deny that these various components do not represent sizeable amounts, often hundreds of million euro. However, there is no link between the price of access to the copper pair and the cost of these elements, precisely because none of these elements is included in the price of access to the local loop. The infrastructure we are talking about here is not a complex, technology-rich and sophisticated one. It is a mere bare, raw twisted metallic pair. Some components may be added to these network elements to activate the whole bandwidth spectrum offered by the copper pair and thus provide new services, but the Regulation does not cover these elements. Therefore, it is inaccurate to claim that the pricing of local loop unbundling influences the propensity of incumbents to invest in the development of broadband technologies such as ADSL. On the contrary, local loop unbundling was designed to enable competitive telecoms companies to deploy their own equipment on top of the raw copper network, and thus to relieve incumbents of part of the investment risk involved in the nationwide roll-out of these relatively capital-intensive new services.

Second, the fear of deterring investment in alternative infrastructures should have been evaluated and qualified in view of the European situation. This concern was largely imported from the USA,[24] where cable infrastructure is widespread (more than 75 per cent of the population) and sometimes more appropriate for the delivery of broadband services than telephone lines, given the relatively high average distance between local exchanges and customers' premises. In Europe, the overall situation is clearly different, and cable networks do not generally supply a valid substitute for local telephone wirelines. For instance, while cable penetration is very high in a limited number of Member States (over 90 per cent in Belgium and the Netherlands), it reaches roughly half the households in the UK and Germany, but cable networks do not pass more than 30 per cent of the population in large countries such as France, Spain and Italy.[25] The state of these cable networks differs widely from one country to another, and even in countries where cable penetration is significant, cable networks have not everywhere undergone the necessary upgrades to accommodate broadband services. Wireless local loop is so far deemed too expensive to be able to offer affordable prices to the mass residential market. Fibre-optic can be a substitute to the local loop, but only in densely populated areas with a significant proportion of corporate customers. The provision of broadband through satellite is still an almost unexplored realm.

As a result, while in a long-range time horizon (for example 20 years) a low price for local loop facilities may to some extent impair investment incentives, in the short run there is little risk of distortion in the allocation of resource, since the various platforms at the time of writing obviously do not stand within the same area of technical and economic solutions to reach the consumer. An 'incorrect' pricing of unbundled loops (for example 10 per cent too low or 10 per cent too high) would influence the incentives to invest in alternative platforms if the technical features and the respective costs of these various solutions were relatively similar. It is still widely thought, however, that the costs of these solutions differ too much to allow them to be considered as real alternatives among which an investor would exercise an arbitrage decision. The assumption that infrastructure competition should not be deterred is flawed, since there is no real alternative to the copper pair in many instances and in many large areas. The idea of infrastructure competition only makes sense in areas where it is reasonably achievable. In the long run, it does constitute an understandable policy objective. But the rapid development of broadband markets does not allow reasoning over a too long period of time. To the Commission's knowledge, there is no instance of an attempt at national level to adapt the pricing of the local loop to the circumstances of time and space to take into account the actual degree of competition pressure by alternative infrastructure. Lastly, it is no longer possible to reason in an abstract world with unlimited capital resources. In the context of the financial downturn, it is clear that the scarcity of funds is a strong constraint, and that the incentives to invest in any alternative technological platform have considerably shrunk. In this context the debate on investment disincentives seems somewhat surreal, at least rather theoretical.

### 5.1.3 Excessive unbundling tariffs

In practice, it is worth noting that the cautious wording of the Regulation was nearly universally construed as an implicit incentive to resort to a long-run incremental cost approach, and that the notion of LRIC was in turn interpreted in a way which very much coincided with the incumbents' strategic views.

Contrary to what was expected, the application of LRIC methodologies did not result in low unbundling prices, but in high tariffs as compared to current retail fees, which make it totally uneconomic for alternative telecommunications operators to enter the market. The regulatory preference for LRIC has induced incumbents to develop (and NRAs to endorse) heavy cost models, which are not based on the actual costs incurred by the incumbents, but on the intellectual re-creation of what the network would cost if it were to be rebuilt entirely now.

The costs presented by the incumbents are in fact in many instances purely notional costs, and bear little or no relation to the real cash outflow generated for the incumbent by running the local access network. The fees charged to new entrants for the rental of existing space within the premises of the incumbents in the framework of collocation arrangement is a case in point. These fees are often based on references to the real-estate market. However, the buildings at stake have often long been depreciated and could not seriously be converted to a different use than the accommodation of technical equipment and thus rented out on commercial terms. The fees for collocation then represent 'windfall revenues' bearing little or no relation to the actual expenditures necessary for the maintenance of the building.

Moreover, most of the costs alleged by incumbents are difficult to challenge, despite the insight of the regulators, due to a high degree of information asymmetry. In addition, the notion of 'incremental costs' being rather unclear, many incumbents took on board not only avoidable costs, but also costs that would have in any case been incurred, irrespective of whether local loop unbundling had been mandated or not. The decision taken in France in April 2000 to put the fees for shared access clearly in line with the pure incremental costs associated with line sharing (filters, costs of maintenance and repair specifically caused by line sharing), which led to a very significant reduction in the price, is still unfortunately a rather isolated example across Europe.

Not only did incumbents claim purely notional costs and interpret the notion of incremental costs to their own advantage. In many instances they also managed to apply different methodologies for the pricing of the retail line rental on the one hand, and the pricing of the local loop on the other, whereas both services rely on the very same infrastructure and should be cost-oriented according to Community sector-specific law. This methodological inconsistency frequently resulted in unjustifiable situations, where the retail tariff for telephone line rental was set significantly below the wholesale price for access to the unbundled copper pair, while it should logically have been the contrary.

The inherent contradiction with local loop unbundling soon became obvious: whereas (1) the very principle of unbundling derives from the idea that new entrants should be spared the obligation to replicate the local access network, (2) access to the local loop is in most countries priced exactly as if the infrastructure had to be entirely replicated on a green field. Therefore, the pricing of local access does not allow new entrants to benefit from the economies of scale of the incumbents, which contradicts its very objective and annihilates the purpose of this policy.

## 5.2 Importance of a Coherent Tariff Structure

### 5.2.1 Application of the margin-squeeze test

When analysing the conditions for market entry in the local access market, particular attention must be given to the issue of margin squeeze. A margin-squeeze test must ensure that the wholesale price for input products is not such that a vertically integrated operator would sell retail services at a loss if it were to apply internally to its downstream arm the wholesale prices it charges to its competitors.

It follows from the Commission's decision-making practice that if a dominant undertaking retains a margin between the wholesale price charged for input to its competitors in downstream markets and its own retail prices in those downstream markets which is insufficient to reflect that dominant undertaking's own costs in providing the relevant services in these downstream markets, with the result that competition in the derived product is restricted, this constitutes an abuse of dominant position.[26]

Regarding more particularly local loop unbundling, it can be deemed that a margin squeeze occurs when the combined amount of the monthly and one-off fees to be paid to the incumbents for wholesale access, as well as the other necessary cost factors, forces alternative operators to charge their own customers more than the incumbents charge their retail customers. If the wholesale fees are higher than retail fees, it is obvious that the incumbents' competitors cannot make any profits, however efficient they may be. The same holds true if the competitors are at least as efficient as the incumbents, given that they have to cover extra costs on top of the wholesale fees, for example for marketing, billing, debt collection and so on.

By fixing the tariffs for wholesale local loop access above the retail subscription tariffs, the incumbents prevent new entrants from competing for the offer of retail local access services, in addition to voice telephony services. The incumbents thus force new entrants, who would be interested in ordering unbundled local loops to offer retail services to their own customers, to equally subsidize their loss-making access services with call revenues. However, the competitors often have no possibility of using revenues from calls for cross-subsidizing the loss-making local access market. An insufficient spread between the two prices may amount to anticompetitive conduct if it excludes competition from the alternative carriers which are at least as efficient as the incumbent on the downstream market.[27]

Comparability of wholesale and retail access services is a key issue in conducting a margin-squeeze test. The effect of an insufficient spread between an upstream access service price and a specific downstream retail service price might be justified if the upstream access service allows competitors to provide other retail services at the same time. The margin-squeeze test therefore requires determining whether the retail services and the wholesale services of the incumbent operator can be compared to each other in respect of technical features enabling the provision of comparable services at both levels.[28]

In principle, the incumbent's wholesale tariffs for unbundled access to the local loop cannot be directly compared with the retail subscription fees, because wholesale access allows the incumbent's competitors to offer to their customers a wider array of retail services than merely the basic retail subscription limited to analogue voice telephony. There are, however, two ways in which the incumbent's wholesale and retail services can be compared. The first comparison takes into account all forms of retail local access (analogue, digital, narrow and broadband) in an aggregated manner, based on the number of each type of line the incumbents have rented out to their end-customers. The second comparison is limited to the market for local access to high-speed data services and thus applies to line-sharing and digital broadband services separately.

By reaching full access to the local loop, alternative operators are enabled to offer a range of retail services, such as analogue narrow-band access, digital narrow-band access, or broadband access in the form of DSL services. In such a case, one single wholesale service has to be compared to an array of different retail services. The easiest way of taking into account the different services at wholesale and retail level would be to have different wholesale fees reflecting the different retail services and to apply the retail-minus approach on each of the wholesale tariffs. Such differentiated wholesale tariffs exist in a number of EU Member States,[29] while in others one single wholesale tariff was fixed by the regulators, whatever downstream service the competitor provides with the rented line.

In the case of full local loop unbundling, the tariffs to be taken into account for the comparison of wholesale and retail services must therefore cover the full range of retail services which competitors are enabled to offer. Consequently, comparable tariffs are those for full unbundling of a basic analogue copper pair (the wholesale service) on the one hand and for basic telephone subscription, ISDN as well as ADSL (the retail services) on the other. In order to determine the incumbent's price for the full range of retail local access services, a weighted approach has to be applied, which consists in calculating the incumbent's average price per retail line, based on the quantities of each variety of subscriber lines effectively held by the incumbents and the respective prices for those lines.

This method is based on the principle that the incumbent's tariff structure must enable competitors to compete effectively, and at least to replicate the customer pattern of the incumbent. One cannot proceed on the assumption that the customer structure and the basket of services of the competitors will be more profitable than those of the incumbent. Although the digital broadband market seems to be the most profitable, or at least the most promising one among those to which access is being offered via the local loop, this is not the only market the competitors are interested in entering. Apart from some operators specializing from the outset in DSL services to frequent users and business customers, most competitors therefore not only tried to gain customers for digital broadband, but also for analogue and ISDN services. The main reason for such a business strategy was to enter the mass market of telephone customers with analogue lines. However, even in those Member States where local loop unbundling started earlier than 2001, market penetration by the competitors did not take place more extensively, because of the high wholesale access fees. When digital services started to be rolled out more broadly, the competitors' interest in analogue lines remained high because of the attractive prospect of upgrading their own customers with higher-value lines and services.

The spread between the incumbent's retail and wholesale tariffs can be negative or positive. A negative spread, with the wholesale fees being higher than the retail fees,[30] is evidence of a margin squeeze without having to look at the downstream costs. If the spread is positive, with the wholesale fees below the retail fees,[31] a margin squeeze can be demonstrated if this spread is insufficient for the incumbent to cover the downstream costs for the provision of its own retail services to end-customers.[32]

### 5.2.2 Possible policy dilemmas

The discussions held on 8 July 2002 on the occasion of the public hearing organized by the Commission on competition in local access illustrated the paradox and dangers of the current approach to LLU. During the discussions, the confrontation of views between Telecom Italia and Wind was a case in point. Both companies presented with the keenest earnestness their respective assessment of the same situation in the Italian marketplace. Telecom Italia (TI) compared the retail monthly fee with the monthly fee for access to the local loop, and concluded that there was no such thing as a price squeeze in Italy. Wind compared the retail fee not only with the wholesale fee for access to the copper pair, but with the whole range of costs (apportioned to an average telephone line) involved in local loop unbundling, including the upfront costs (one-off fees, collocation, and so on). Wind blamed high and artificial entry barriers and drew a conclusion opposed to that of TI, that a spectacular price squeeze still prevails in Italy. It is impossible to say who is wrong, since both firms were right, at least from their own perspective.

It seems that the reconciliation of the logics of TI (*the local loop must be priced at the cost of its replication*) and Wind (*a price squeeze must be avoided at any rate to safeguard possibilities of market entry*) would imply a significant cost for the consumer. It would mean that the retail price offered even by the incumbent should be very significantly raised to be in line with what new entrants have to pay for unbundled access, including the related facilities. Italy is not an isolated case. The feedback from the sector inquiry in all Member States shows that if the costs for collocation and the one-off fees were to be taken on board in a margin squeeze test (which would be perfectly legitimate from a new entrant's perspective), the retail price should be nearly doubled to avoid detrimental effects on new entrants.

This situation seems absurd, and may contradict the aim of the unbundling process to benefit the consumer. The current approach to the pricing of local access seems to either (1) preclude new entrants from addressing the residential market and confine them to less price-sensitive market segments, or (2) penalize the end-customer through artificial retail price increases.

The conclusion may be drawn that the very process of unbundling is meaningless, if not absurd. We take a different view. The principle of unbundling itself is not to blame. It is rather the current approach to the pricing of the local loop which creates difficulties. The only way to escape the traditional competition dilemma between favouring the competitors or favouring the consumer would be to review the pricing principles of the local loop in the light of economic reality and not of remote, potential and unmaterialized alleged economic risks. The pricing rules have so far too much accommodated the strategic interests of the incumbents, and totally overlooked the economies of scale and externalities that the unbundling process was supposed to bring to the market and in the first place to the consumer. In the absence of fair and non-discriminatory pricing strategies adopted by the incumbent operators in the local loop, it is now one of the major challenges for both regulators and competition authorities to ensure that coherent pricing principles are implemented, in order to achieve and secure sustainable competition in the telecommunications sector.

## NOTES

* The views expressed by the authors are purely personal and do not necessarily reflect the official position of the European Commission.

1. Directive 97/33/EC, on interconnection in telecommunications with regard to ensuring universal service and inter-operability through application of the Open Network Provision principles, of 11 June 1997.
2. Regulation 2887/2000 of 18/12/2000 on unbundled access to the local loop (OJ L 336/4, 30.12.2000).
3. See OECD (2001) for an analysis of different forms of separation and the current practices in regulated sectors in the OECD countries.
4. See, for example, the *Eighth Report from the European Commission on the Implementation of the Telecommunications Regulatory Package*, Brussels, 3 December 2002, pp. 24–8. http://europa.eu.int/information_society/topics/telecoms/implementation/annual_report/8threport/finalreport/com2002_0695en01.pdf
5. 'Cable' Directives 95/51/EC and 99/64/EC require, however, accounting and legal separation respectively between telecommunications services and cable television network for dominant operators.
6. See Article 3.3 of Regulation 2887/2000.
7. See Recital 11 of Regulation 2887/2000.
8. See Bourreau and Dogan (2002) and Hausman and Sidak (1999) on this debate.
9. This dilemma may not always exist. As pointed out by Woroch (2002), a new entrant may be interested in purchasing unbundled network elements 'in those markets where it is uneconomical to build a second network'. Local loop unbundling may also allow 'an embryonic carrier to quickly achieve a broad footprint as it builds out its network'. In those cases, local loop unbundling may become a complement and not a substitute to facility-based competition.

10. See Recital 11 of Regulation 2887/2000.
11. Bourreau and Dogan (2002) describe the pros and cons of a fixed rental price versus a time-dependent rental path.
12. See Rey et al. (2001).
13. Harris and Kraft (1997) and Hausman and Sidak (1999) put into question the bottleneck nature of the local loop at least in densely populated areas and for business customers. They claim that in those cases the deployment of alternative technologies may be feasible and profitable. Hausman and Sidak state that the offering of unbundled network elements at their cost not only deters investment in alternative technologies and discourages incumbents' investment in their own access networks but also introduces distortions in the market for access by inducing artificial demand at a price below the market price. Harris and Kraft defend the example of Canada, where urban loops were not considered bottlenecks and then were excluded from mandatory unbundling.
14. There is little theoretical support for the properties of this standard and its appropriateness.
15. See note 1.
16. Section 251 (c) Para. 3 of the Telecommunications Act of 1996, Pub. LA. No. 104-104, 110 Stat. 56 (1996).
17. We focus on the recurring tariff for full unbundled access, which is the most visible tariff item. However, it should be borne in mind that there are many other services associated with local loop unbundling, such as collocation, cabling, backhaul and so on, whose pricing is also highly relevant for the elaboration of the new entrants' business cases.
18. Commission Recommendation on unbundled access to the local loop enabling the competitive provision of a full range of electronic communications services including broadband multimedia and high-speed Internet, OJ L 156/44, 29.6.2000.
19. Communication from the Commission: Unbundled access to the local loop: enabling the competitive provision of a full range of electronic communication services, including broadband multimedia and high-speed Internet, OJ C 272/55 of 23.9.2000.
20. Regulation 2887/2000.
21. Commission Press Release IP/00/765 of 12.7.2000.
22. See Hausman and Sidak (1999) and Jorde et al. (2000).
23. Recital 11 of Regulation 2887/2000.
24. See Hausman et al. (2001).
25. In the last two countries cable penetration does not exceed 5 per cent.
26. Commission Decision 88/518/EEC of 18 July 1988, *Napier Brown/British Sugar*, OJ 19.10.1988 L 284/41, para. 66.
27. Access Notice: 'Notice on the application of the competition rules to access agreements in the telecommunications sector – Framework, relevant markets and principles', OJ C 265/2, 22.8.1998, paras 117–19.
28. ONP-Committee Document No. ONPCOM 01-17 of 25.6.2001.
29. In Belgium, Italy, Luxembourg, the Netherlands, Portugal and Finland, there are two different monthly wholesale fees, and in Sweden and Norway, there are three different monthly wholesale fees; see ONP-Committee Working Document No. ONPCOM 01-27 REV2 of 18.2.2002.
30. As of 31.12.2001, this was the case in Germany, Greece, Spain, France, Italy and the UK; see ibid.
31. As of 31.12.2001, this was the case in Belgium, Denmark, Ireland, Luxembourg, the Netherlands, Austria and Sweden; see ibid.
32. See for an example of both scenarios, the recent decision against Deutsche Telekom AG, OJ L 263/9, 14.10.2003.

## REFERENCES

Bourreau, M. and P. Dogan (2002), 'Service-based vs. Facility-based Competition in Local Access Networks', mimeo, ENST, Paris.

Cave, M. (1997), 'From Cost Plus Determination to a Network Price Cap', *Information Economics and Policy*, **9**, pp. 151–60.

Harris, R.G. and C.J. Kraft (1997), 'Meddling Through: Regulating Local Telephone Competition in the US', *Journal of Economic Perspectives*, **11**(4), pp. 93–112.

Hausman, J.A. and J.G. Sidak (1999), 'A Consumer-Welfare Approach to Mandatory Unbundling of Telecommunications Networks', *Yale Law Journal*, **109**(3), pp. 417–505.

Hausman, J.A., J.G. Sidak and H.J. Singer (2001), 'Cable Modem and DSL: Broadband Internet Access for Residential Customers', *AEA papers and proceedings*, pp. 302–7.

Jorde, T.M., J.G. Sidak and D.J. Teece (2000), 'Innovation, Investment and Unbundling', *Yale Journal on Regulation*, **17**(1), pp. 1–37.

OECD (2001), *Structural Separation in Regulated Industries*, Paris: OECD.

Rey, P., P. Seabright and J. Tirole (2001), 'The Activities of a Monopoly Firm in Adjacent Competitive Markets: Economic Consequences and Implications for Competition Policy', mimeo, IDEI, Toulouse.

Woroch, G.A. (2002), 'Local Network Competition', in M. Cave, S. Majumdar and I. Vogelsang (eds), *Handbook of Telecommunications Economics*, Amsterdam: Elsevier Science.

# Section 5

# Competition and Allocation of Scarce Resources: the Case of UMTS

# 10. Competition in mobile communications and the allocation of scarce resources: the case of UMTS

**Jörn Kruse**

## 1. INTRODUCTION

In the 1990s, mobile communication in Europe was a tremendous success story in terms of technical and economic developments. Prices declined and penetration rates grew surprisingly. To facilitate these socially desirable developments, private companies invested heavily in communication infrastructures.

These developments were accompanied by (and not independent of) a comparatively low level of regulation – with the very significant exception of spectrum licensing. Government licensing predominantly shaped industries and market structures in the individual countries.

Since spectrum is a scarce resource under the governance of the countries' authorities, its allocation is a very important element for the procedures and for the efficiency results of competition in mobile communications. Spectrum is not only a technically essential resource. The availability of more or less spectrum (and what kind of spectrum) is of primary importance for a mobile operator's economic success. Adequate spectrum allocation is a crucial factor for the competitiveness and efficiency of the mobile sector. Since national regulatory authorities (NRAs), or national governments respectively, were (and are) using different methods of spectrum allocation, this raises questions regarding competition issues on the European level.

Government licensing in mobile communications is (from an economic viewpoint) only justified by the relevance of spectrum allocation. But government decisions and influence were going much beyond that. They basically predetermined the number of market players, usually on a step-by-step basis over time, and they often chose the players themselves, by the visible hand of government committees. Also, they mostly decided in favour of a high degree of vertical integration, instead of leaving these issues to the market.

Government decisions seem to be responsible for some of the problems the mobile communications industry has had to deal with recently, especially with respect to UMTS. Although UMTS may be technically quite different from GSM, its economic and competitive developments depend heavily on earlier GSM market structures.

## 2. THE ECONOMICS OF CELLULAR NETWORKS AND SERVICES

### 2.1 Network Architecture in GSM and UMTS

Both GSM and UMTS networks consist of (1) core networks (CNs) and (2) radio access networks (RANs).

RANs provide the radio coverage of a specific geographic area in order to connect mobile users. In GSM, this is done by BTSs (base station transceivers) and BSCs (base station controllers), which are connected to MSCs (mobile switching centres). MSCs are doing the switching job, which means routing outgoing calls to their destination within fixed or mobile networks. In UMTS, the equivalent terms are Node B and RNC (radio network controller). The radio networks require about 75 per cent of the investments and they are the real mobile part of the communication system.

The core systems in GSM and in UMTS include network intelligence and operating management, home and visitor location registers, service platform, connection to ISDN and IN-networks and so on. The core systems are most crucial for mobile network operators' competitive strategies.

The most significant feature of UMTS is the mobile availability of general or specific services (like Multimedia Messaging Service [MMS] and Internet access) and content (news, games, music, video and so on). For these value-added services, which are supposed to generate most of the UMTS revenue in the longer run, third-party suppliers (from the media and from the e-commerce sector) are entering the picture, asking for about 90 per cent of their content revenue.

### 2.2 Scale Economies and the Significance of Spectrum Availability

Spectrum is an essential resource for radio-based transmission of data and voice from a transmitter to one or more receiving stations. The term spectrum is often used as a synonym for frequency or radio channel. It is defined by a specific range of MHz (band) in the overall electromagnetic spectrum in a specific area.

Spectrum is not only absolutely necessary for mobile network operators. The amount of available spectrum also significantly influences cellular operators' production costs. Let us assume that a certain number of GSM base stations is needed to secure radio coverage of a desired or regulatorily required percentage of population with a minimum number of traffic channels. If airtime minutes increase, the same base stations will be equipped with more transceiver stations, operating with a given stock of spectrum. Since incremental costs are low, the resulting average cost function $BAC_A$ in Figure 10.1 is strongly degressive.

If increasingly more traffic has to be handled by the network, the given stock of spectrum is the limiting factor. This will be first noticed in urban areas, since the blocking rate will increase, particularly in peak hours. To avoid quality deterioration, the network has to be remodelled by cell-splitting, which increases the reuse factor of the available spectrum, increases the number of base stations, and reduces the signal power and the areas of cells. This results in additional costs for investments in base stations, wired and wireless connections, and for additional operating costs. The resulting average incremental costs are shown by *AIC*.

If a specific number of frequencies (GSM channels) allows no more traffic than $Q_1$, $AIC_1$ shows the additional average costs of cell-splitting. If the network operator had more

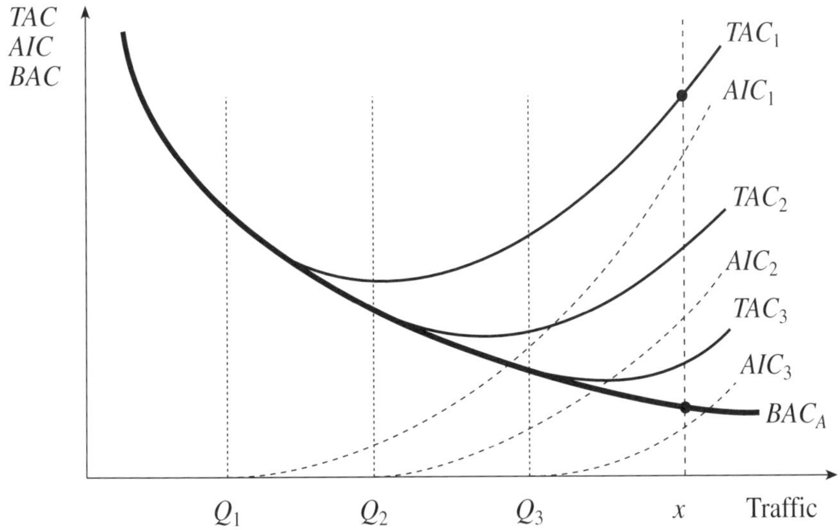


*Figure 10.1 Mobile operators' average costs with different available spectrum*

spectrum available, the respective cost functions would be $AIC_2$ or $AIC_3$. The total average cost functions $TAC_1$ in Figure 10.1 (sum of $BAC_A + AIC_1$) demonstrate that a network with more available spectrum has significant cost advantages, whereas a network with only few frequencies will be forced to limit its growth or will be subject to quality deterioration (higher blocking rate) or higher average cost.

For a UMTS system, the functional relationships are somewhat more complex, due to different technological methods of spectrum use and different responses to increasing traffic volume. If more $2 \times 5$ MHz spectrum packages are available for a UMTS operator, additional layers can be used to respond to increased traffic, depending on the specific area and demand characteristics. Holding quality (data rate) constant, basically the same relationships (as indicated in Figure 10.1) between spectrum availability, usage quantity, and costs apply.[1]

Thus the allocation of spectrum is significant for the competitiveness of a specific network operator and may also be relevant for the whole industry's competitiveness and efficiency (see Kruse, 1993 and 1996).

## 3. THE ECONOMICS AND METHODS OF SPECTRUM ALLOCATION

### 3.1 Intramodal Spectrum Allocation

In order to be able to use a specific frequency properly, there should not be any other transmitter using the same frequency within the same area, which would otherwise cause interference. Because of such interference, the use of spectrum is subject to consumption rivalry. For simplicity, we assume here that this is a direct rivalry (Kruse, 1995).

Even if spectrum is costless in the sense that no economic resources have to be used in order to make frequencies available, and even if contemporary use will not reduce the amount of spectrum in the future (spectrum is not exhaustible), the opportunity costs of spectrum (resulting from consumption rivalry) call for spectrum fees.

The concept of 'spectrum allocation' is used to refer to the allocation of property rights for the use of a specific part of the spectrum to a transmitting unit. The allocation of spectrum (dedicated to a specific service) to individual users (transmitters) is called intramodal spectrum allocation. In the USA, this is called spectrum assignment (Cramton, 2002).

Consider the networks' demand function for spectrum.[2] In Figure 10.2, $SD_2$ represents the aggregate demand function of network operators for a specific type of spectrum. Suppose the spectrum available is $Q^*$. The relevant supply function is $SS^*$, that is, perfectly price-inelastic. The market equilibrium is at $D$, with the efficient price $P_3$, which equals opportunity cost. The price $P_3$ would be the price in a competitive market.

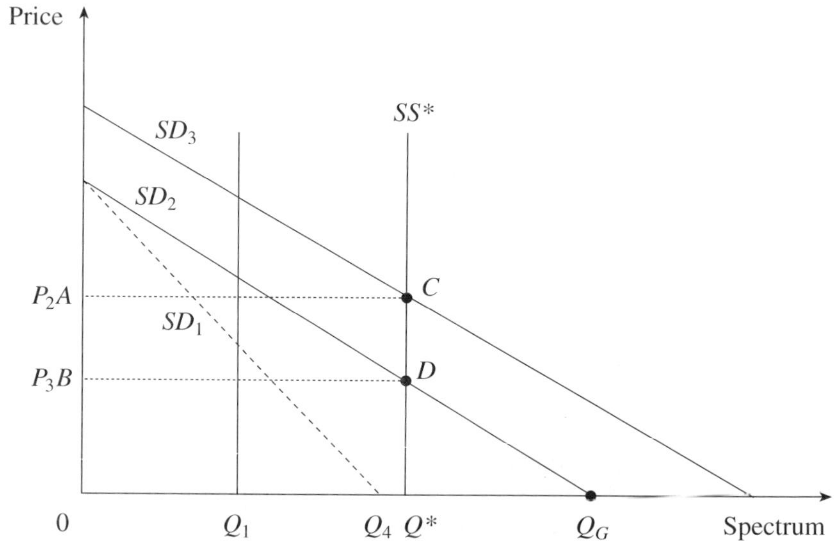


*Figure 10.2 Intramodal spectrum allocation (spectrum assignment)*

Under competitive conditions (free trade or auction) spectrum would be made available to those users with the highest willingness/ability to pay. In general, these are expected to be the most efficient operators. If spectrum is given away free of charge, spectrum will be wasted.

The price $P_3$ is a scarcity price. If an auction were held, the government would collect a scarcity rent of $Q^* \cdot P_3$ (rectangle $0Q^*DB$). Theoretically, its amount does not (contrary to public opinion) influence prices and quantities on the telecommunications service markets. If demand for spectrum were $SD_3$ instead, opportunity cost and prices would be $P_2$, and government revenue would be $0Q^*CA$.

Were, instead, the spectrum demand function $SD_1$, opportunity costs and prices would be zero. Spectrum would not be scarce in this case. The function $SD_1$ may result, if the operators that otherwise demand $SD_2$ invest in more efficient equipment so that they can serve customers' demanded airtime minutes with less spectrum. The optimal rate of substitution between spectrum and other factors of production (such as transmission and receiving stations) will be simultanously derived with efficient spectrum prices.

This creates a general economic theoretical case for a competitive market approach to spectrum allocation if a specific amount of spectrum is made available by the government. But it does not determine the specific procedure to apply under real-world conditions nor the number of licences to be issued. Nor does it say how much spectrum would be adequate for the specific services (see Section 3.2) if different services (at least potentially) compete for spectrum. Therefore, intermodal competition for spectrum is necessary.

## 3.2 Intermodal Spectrum Allocation

Intermodal spectrum allocation is the allocation of certain parts of spectrum to different uses and services (for example TV, cellular mobile communication, or military use).[3] There has never been a market for this allocation problem. Intermodal allocation always has been (and still is) decided on by administrative decision-making, often with international consultation and standard-setting.[4] This has caused some of the well-known problems that occur when market forces are not used for the allocation of resources.

Let us consider two services, *A* and *B* (say, television and cellular), which may use the same spectrum. In Figure 10.3, $0_D0_R$ denotes the quantity available for intermodal spectrum allocation decision-making. Let $SD_A$ be *A*'s demand function for spectrum (where quantity is shown in the usual way, starting from $0_D$ and going to the right). For service *B*, quantitatives are shown right to left, starting at $O_R$. This shows the direct consumption rivalry between the services.

If spectrum allocation between services *A* and *B* were performed by a competitive market, point *S* would indicate the equilibrium with price $P_S$ and quantity $0_DQ_S$ for *A* and $O_RQ_S$ for *B*. This would be the most efficient spectrum allocation maximizing welfare.

Let us assume that regulatory authorities were discretionarily granting the quantity $o_DQ^*$ to service *A*, leaving $O_RQ^*$ to *B*. At the borderline $Q^*$, the value of the frequency is $P_2$ for *A* and $P_1$ for *B* (with the difference of $P_2 - P_1$). The welfare loss equals the area of the triangle *DRS*. Efficiency could be enhanced if the 'border' were shifted to the right, until $Q_S$. Welfare would then be the sum of $0_DQ_SSE$ (for *A*) and $SQ_SO_RT$ (for *B*), instead of the sum of $0_DQ^*DE$ and $Q^*O_RTR$.[5]

Thus, if the government sells the spectrum quantity $0_DQ^*$ for *A* intramodally (for example for UMTS), this would result in $P_2$ with a revenue of $0_DQ^*DB$. But this will be inefficient in an intermodal allocation picture, since the spectrum price and the spectrum costs for the networks are too high compared to the optimum. The institutional problem then is not the *A* procedure itself, but the intermodal failure. Thus high prices in spectrum auctions, like those for UMTS, may also reflect the fact that governments did not allocate enough spectrum to UMTS.

This problem would be automatically solved by a general intermodal market for spectrum rights. This may be regarded in analogy to markets for real estate. In the longer run, this will be the superior solution.

The problem in the short run is that frequencies may not be totally fungible, but specifically defined by bands of MHz in a specific use of a standard. In a longer perspective, these problems will be solvable. To a large extent, the potential instruments are comparable to the conventional restrictions on the use of real estate, in order to manage a socially desirable outcome.

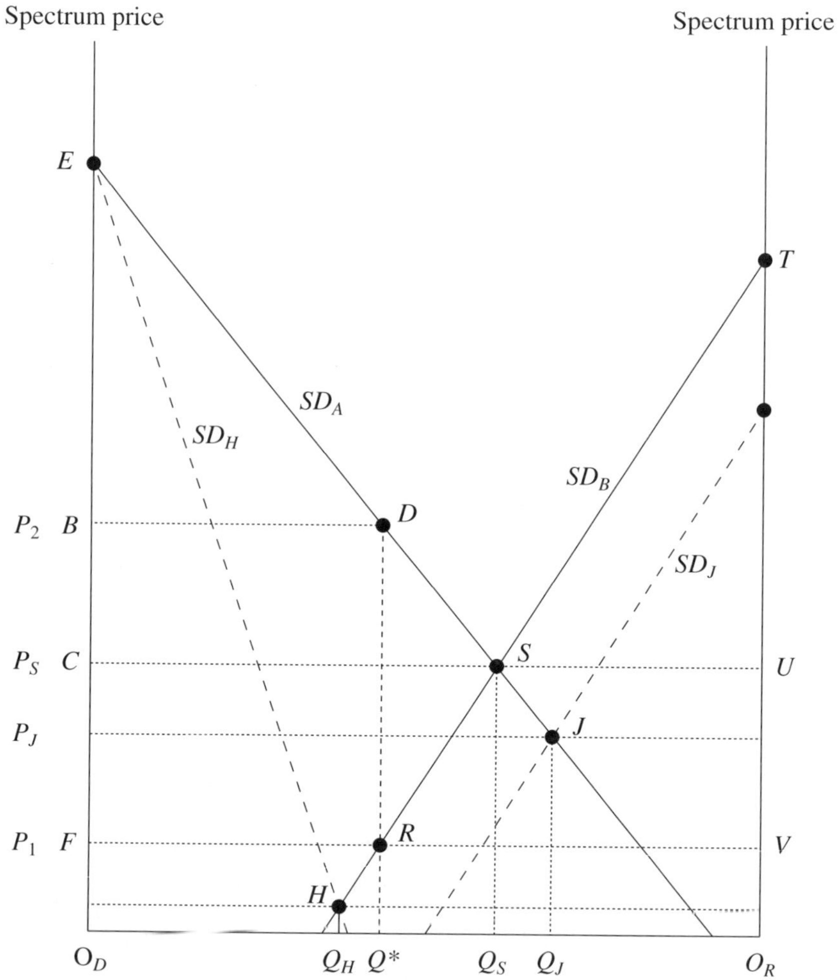


*Figure 10.3 Intermodal spectrum allocation (different services)*

## 3.3 Methods for the Allocation of Scarce Resources

Long before economists and industry representatives looked at licensing and spectrum allocation from an efficiency and competition point of view, governments had used different methods for allocating property rights for frequencies.

If we look at the methods used in European countries in GSM (see Figure 10.6) and UMTS (see Figure 10.8), we see that different methods apply in different countries, and for different bands of spectrum in a particular country. In the following, the main methods for allocating scarce resources are briefly outlined and summarized in Table 10.1. Of course, governments always can impose restrictions and specific rules that will change the assessments shown in the table.

*Table 10.1 Allocation methods*

| | Efficiency | Transaction cost time | Non-discriminating, transparent | Social and fiscal objectives | Government's influence (from their point of view) |
|---|---|---|---|---|---|
| FCFS | -- | + | -- | - | - Depends |
| Lottery | -- | ++ | ++ | -- | -- |
| Auction | ++ | + Depends | ++ | ++ | -- |
| Discretionary decisions | -- Depends | Depends | -- | Depends | ++ |
| Beauty contest (criteria contest) | + Depends | -- | - Depends | Depends | + |

*Notes*: The cells denote good (+), very good (++), poor (-) and very poor (--) fulfilment of the respective criteria.

**First come first served (FCFS)**

This is the most traditional method, especially in earlier times when spectrum was not yet scarce. In mobile communications, FCFS often meant that the incumbent monopolies got the available spectrum. This was basically relevant with respect to the analogue systems (first generation – 1G). Thus its market relevance vanishes with growing market shares of digital systems – except in some countries, where the first GSM licences were also issued on a first-come-first-served basis.

**Lotteries**

Lotteries have been used in a number of smaller US digital markets because this enabled the regulatory agency to issue the licences quickly and reliably, avoiding time-wasting legal fights. Lotteries are non-discriminating and transparent and incur only low transaction costs. But efficient spectrum use is not only not guaranteed, it is not even likely if we neglect the possibility of second-hand trade. However, the latter has mostly been ruled out by law (although the firms themselves may be traded). Societies' scarcity rents are granted like gifts to lottery winners.

**Auctions**

Auctions for spectrum that open a new market are generally non-discriminatory and transparent. They may incur relatively low transaction costs, but this will depend on auction design. The efficient use of spectrum is supposed to be most likely. Efficient spectrum prices foster efficient substitution between spectrum use and other factors of production. Auctions do not leave much discretionary control to the governments or influence to the lobbyists. In some cases, they may generate significant contributions to countries' budgets. Additional provisions will be necessary to avoid monopolization and to reduce the risk of collusion, which is a matter of auction design.

**Discretionary decisions**

This category includes any procedure where the authority has large discretionary powers to grant spectrum to whom it wants (lack of strictly binding criteria). The results depend on the incentive structure of the (mostly political) decision-makers. They may pick a licensee who may be an efficient user of the spectrum or not. At one extreme, they may feel that they do it in the best interest of their country. At the other, they may give the licence to partisans or family members or to those who gave the highest donations to the ruling party. This method is the most discriminating and untransparent one. From the point of view of most economists it is heavily criticized and regarded as unacceptable.

**Beauty contest (or criteria contest)**

These terms cover a whole range of institutional methods, depending on the specific procedures and criteria. Some beauty contests include strict and transparent economic criteria. They may include auction elements (spectrum fees, population coverage and so on). Some of them are performed by independent commissions on the basis of objective and transparent criteria and valuation. Some may look more like a black-box procedure, which could be closer to discretionary decisions.

In general, economists strongly advocate auctions. But in practical terms, this very much depends on the specific institutional framework and auction design.

## 3.4 Auction Methods

The specific institutional settings and the rules for auctions may be very different. Economic textbooks distinguish between four basic auction methods.

The English auction is an ascending auction starting from a low price, where a bidder may overbid the actually prevailing bid in each round. The auction stops when no overbidding occurs and the remaining highest bidder gets the licence, paying his bid price. This very simple and intuitive procedure may become very complex, with dozens of rules, when a number of objects is auctioned simultaneously, for example for different frequencies (different bands or different regions), since there may be aggregation economies between them.

The Dutch auction is a descending auction starting from a high price, which is automatically lowered. The bidder who first stops the descending procedure gets the licence at the prevailing price.

In sealed-bid auctions every bidder submits a single (non-renewable) bid. The highest bidder wins the licence.

In the first-price sealed-bid auction (FPSB) the highest bidder pays the price of his bid. In multi-object auctions (like spectrum auctions with different frequencies) this is also called a discriminating auction, because different bidders may pay different prices, even for equivalent spectrum.

In the second-price sealed-bid auction (SPSB), called the Vickrey auction, the winner pays the amount of the second highest bid, which is the opportunity price. In multi-object auctions, each winner pays the same price, which in practice is often the bid of the margin winner.[6]

Although economic auction theory shows that – under some restrictive assumptions – expected prices will be the same for all four methods (equivalence theorem) and equal to opportunity costs, this may not hold under realistic conditions. Here, phenomena such as collusion, valuation uncertainty, risk aversion and so on may prevail.

There is a broad theoretical literature on different auction mechanisms under different assumptions with more or less relevance to real-world conditions in licence and spectrum auctions (for broader treatments and more references see McAfee and McMillan, 1987, 1994, 1996; McMillan, 1995; Klemperer, 1999; Cramton, 2002). These will not be summarized here.[7]

Under real-world conditions, it is generally recognized that auctions are subject to the risk of explicit or implicit collusion. But this very much depends on the specific auction method, basically because the incentives to cheat vary (if there should be any collusion).

In English auctions, incentives for cheating are almost non-existent. Thus there is a comparatively high risk of collusion (see Table 10.2). This may be surprising since almost all spectrum auctions have been English auctions.[8]

The same applies with regard to a criterion that may be relevant in a market where additional spectrum is auctioned, while incumbents who already supply the market will have incentives to ensure that additional companies will not enter the market. Their chances to overbid any newcomer's bid in case of an entry threat, without committing to overpaying, are best in English auctions. Therefore the newcomer's incentives to enter via spectrum auctions are comparatively low.

Table 10.2 roughly summarizes the basic methods with these two criteria and some others that may be relevant. The information for bidders depends on the specific information policy

*Table 10.2 Auctions and criteria*

| | Low risk of collusion | Low risk of entry block | Endogenous information for the regulator | High bidder information and low risk aversion | Suitability for multiple-object cases with complex cost and/or demand efficiencies |
|---|---|---|---|---|---|
| 1 | 2 | 3 | 4 | 5 | 6 |
| English auction (Ascending *A*) | - - | - - | 0 | + | + + |
| Dutch auction (Descending *A*) | + + | + + | - | - | - - |
| First-price SB (Discriminating *A*) | + + | + + | + | - | - |
| Second-price SB (Vickrey *A*) | - | - | + + | - | - - |

which may be relevant for forthcoming auctions. The endogenous information for the regulator will be valuable for intermodal spectrum allocation and for intramodal allocation of additional spectrum, if available. The informational advantages of the second-price sealed-bid auction relate to the fact that any bidder has incentives truthfully to reveal his valuation (if he actually knows it) in his bid.

## 4. SPECTRUM ALLOCATION IN EUROPEAN MOBILE COMMUNICATION

### 4.1 GSM Spectrum and Licence Allocation in European Countries

The analogue mobile telephone systems NMT, AMPS, TACS, C-450 and so on, which are nowadays included in the term 'first generation' (1G), were the beginning of individual mobile communications on a large scale (former systems had only very small user groups), though their subscriber numbers were still limited when compared to figures for the time of writing.

In most countries, these systems were run by the traditional telecommunication incumbents on a monopoly basis. The duopolistic competition between Racal Vodaphone and BT Cellnet has been an exception rather than the rule. Spectrum, mostly in the 450 MHz (and later in 900 MHz) area, was allocated by traditional administrative procedures (first come first served). Table 10.3 shows for (1) different countries, and (2) different areas of mobile spectrum (1G: analogue; 900: GSM-900 MHz band, even if temporarily used for analogue; 1800: DCS/GSM-1800) how much spectrum has been allocated by the three different methods (3) to (5).[9] The figures denote the quantities (in MHz) for uplink as well as downlink.

The licencing of GSM operators and the allocation of the original GSM spectrum of $2 \times 25$ MHz offered a unique opportunity to introduce competition and to apply market-based institutional mechanisms. It was the beginning of a new market or, at least, of new market dimensions in mobile telephony.

But disappointingly, quite a number of countries in Europe were still operating on a monopoly basis, even after the beginning of GSM, starting 1992 (see Figure 10.4). Others decided to start with two operators (UK, Germany, Finland, France). It was not really questioned at that time whether more than two licences in GSM (and only 8 MHz or less spectrum for each operator) could have been economically viable.

Spectrum was practically treated as a free (in the sense of costless) resource, although definitely scarce. Only administrative fees that did not reflect economic values were applied. This can be viewed as a foregone opportunity to introduce market forces and to determine the economic value of spectrum. The introduction of spectrum auctions would have been much easier at the starting-point when the firms' technical and market conditions were relatively more symmetrical than three or five years later.

When GSM markets started (and even before), it was acknowledged that there was not enough spectrum available in the 900 MHz band to serve the expected large numbers of subscribers. Even more important, some regulatory agencies found it desirable to have more network operators in order to intensify competition.

Additional spectrum in the 1800 MHz band was allocated to cellular use. This raised more problems because of several market asymmetries between incumbents and newcomers. One is the cost disadvantage of 1800 MHz operators, which is significant at the

*Table 10.3 Spectrum allocation methods in European countries*

| Country (1) | Spectrum (2) | FCFS, Extensions (3) | Beauty Contest (4) | Auction (5) |
|---|---|---|---|---|
| | 1G | 11 | | |
| Austria | 900 | 16 | 8 | |
| | 1800 | 10 | 31.6 | 19.2 |
| | 1G | | | |
| Belgium | 900 | 12 | 12 | |
| | 1800 | 4.5 | 15 | |
| | 1G | 10.3 | | |
| Denmark | 900 | 8.8 | 8.8 | |
| | 1800 | | 50.4 | |
| | 1G | 6.6 | | |
| Finland | 900 | R 31.4 | | |
| | 1800 | 28.8 | | |
| | 1G | | | |
| France | 900 | 21.6 | 3.1 | |
| | 1800 | 26.4 | 23.2 | |
| | 1G | 5.6 | | |
| Germany | 900 | 12.5 | 12.5 | |
| | 1800 | | 45 | 10 |
| | 1G | | | |
| Greece | 900 | | 20 | |
| | 1800 | 25 | | |
| | 1G | 10 | | |
| Ireland | 900 | 7.5 | 12.3 | |
| | 1800 | 28.8 | 14.4 | |
| | 1G | 11.8 | | |
| Italy | 900 | R 10.5 | R 10.2 | |
| | 1800 | 9.6 | R 24.4 | |
| | 1G | | | |
| Luxembourg | 900 | | 23.2 | |
| | 1800 | | 19.6 | |
| | 1G | | | |
| Netherlands | 900 | 16 | 7.4 | 6.2 |
| | 1800 | | | 77.4 |
| | 1G | | | |
| Portugal | 900 | 8 | 15.8 | |
| | 1800 | 18 | | |
| | 1G | 16.6 | | |
| Spain | 900 | 11.8 | 11.8 | |
| | 1800 | | 40.2 | |
| | 1G | 6.4 | | |
| Sweden | 900 | 7.2 | 14.4 | |
| | 1800 | | 31.8 | |
| | 1G | | 16.3 | |
| UK | 900 | 11.6 | 24.4 | |
| | 1800 | | 60 | |

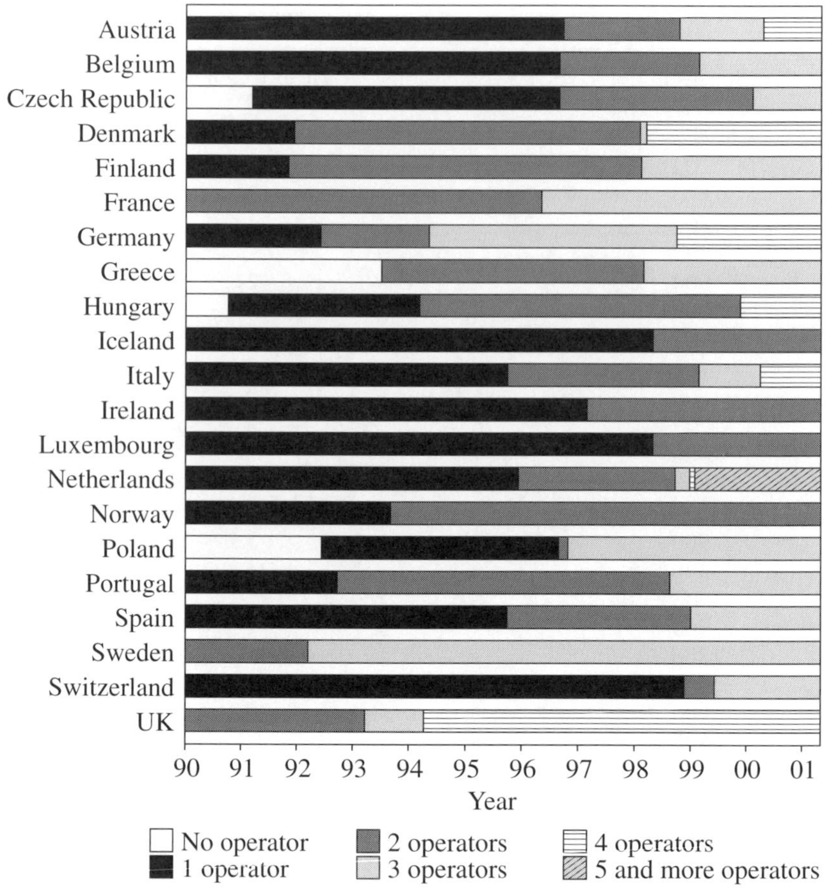


*Figure 10.4 Number of GSM operators in European countries, 1990–2001*

beginning and tends to fade if more traffic forces GSM-900 operators to split cells and to increase the number of base stations.[10] Another is that the price-inelastic customers with large monthly bills had already subscribed to the GSM-900 operators.

A specific issue was the allocation of the remaining frequencies out of the 1800 MHz band. The GSM networks usually had a strong interest in getting 1800 MHz spectrum, because they had capacity problems in city areas (and some other hot spots) and they intended to build a 1800 MHz infrastructure and to use this spectrum only in these areas. Because of their capacity problems and to avoid further cell-splitting, they valued this spectrum most highly. Some NRAs chose to auction at least parts of the remaining 1800 MHz spectrum.

### 4.2 UMTS Licensing in Europe

UMTS licensing was another chance to introduce market forces into mobile spectrum allocation – on the assumption that mobile markets would begin again. But this assumption is not totally justified, since 2G operators with already existing infrastructure and customers will have significant advantages over total newcomers. Anyway, economists were much more satisfied with the UMTS licence procedures, in which some NRAs used auctions for spectrum allocation (see Table 10.4).

A number of countries (Finland, Sweden, Norway, Spain, Portugal and France) still relied on 'beauty contests' for UMTS licensing allocation. They mostly chose (or planned) to issue four UMTS licences with 2 × 15 MHz of paired spectrum and an additional 5 MHz of unpaired spectrum. Thus the number of offered UMTS licences mostly exceeded the number of incumbents by one, and at least one new entrant could be expected.

In France, only two applicants were willing to pay the fee the government had set discretionarily. After the procedure, and on request of the firms, the fees have been reduced by 86 per cent. In Spain, the original fees have even been revised twice, at first more than eight times up and later three times down, with a resulting increase of 280 per cent. In Poland, a procedure including beauty contest as well as auction elements was announced. During a somewhat opaque administrative process some potential newcomers withdrew and the licences were granted to the three incumbents.

Other countries used auctions. Most of them (in Germany, Italy, Netherlands, Austria and Switzerland) were simultaneous English ascending auctions. This basically also applied to the UK (and was also planned for Belgium and Greece), where sealed-bid elements have been added. This was called an Anglo-Dutch auction (Klemperer, 2001b). Denmark used a second-price sealed-bid auction.

The NRAs mostly decided, before the auction, to offer four (Switzerland, Belgium, Denmark, Greece) or five (UK, Italy, Netherlands) UMTS licences. These numbers mostly exceeded the number of incumbents in these markets by one. Therefore, it could be expected that at least one new entrant would show up in the country's mobile market,[11] and that the incumbent GSM network operators would not have to fight too hard in order to get a UMTS licence. Anyway, in Italy and Denmark, one incumbent failed to receive a 3G licence. In Belgium and Greece only the three incuments were interested, leaving the fourth licence unsold.

Two other countries (Germany and Austria) used an auction design which did not determine the number of licences in advance, but determined the number of licences endogenously (in the range of four to six). In both cases, the result was that six licences were issued, including two newcomers to the respective markets.[12]

As a result of UMTS licensing, the number of actual or prospective mobile operators was increased by one (UK, Spain, Portugal, Denmark) or two (Germany, Austria, Italy, Sweden). But, given the industry's perspectives as of mid-2002, it remains uncertain whether every UMTS licensee that is a newcomer in the national market will actually roll out a network. In Germany, Austria and Switzerland it is almost certain that the number of UMTS networks entering the market will be smaller than the number of licences.

The UMTS spectrum fees, which have either been set by NRAs or determined by auctions, are very different among the countries. These fees, in relation to the individual countries' populations, are listed in column 7 of Table 10.4.[13] These numbers are used by most authors and statistics to represent the 'price for spectrum' in those countries, and they are used here to rank the countries.

However, this may not be entirely adequate since the number of licences and the average amount of spectrum for each operator varies. If the number of licences is smaller, each

*Table 10.4 UMTS spectrum allocation in Europe*

| Country | Month | Method | Number of GSM incumbents | Number of UMTS licences offered/ issued | Number of GSM + UMTS licences | Fixed spectrum fees (euro per population) | Cost+value index of licence and spectrum (per pop.) |
|---|---|---|---|---|---|---|---|
| 1 | 2 | 3 | 4 | 5 | 6 | 7 | 8 |
| UK | April 2000 | Auction Engl. + SB | 4 | 5 / 5 | 5 | 648 | 32.40 |
| Germany | Aug 2000 | Auction Engl. | 4 | 4–6 / 6 | 6 | 610 | 36.60 |
| Italy | Oct 2000 | Auction Engl. | 4 | 5 / 5 | 6 | 212 | 10.60 |
| Netherlands | July 2000 | Auction Engl. | 5 | 5 / 5 | 5 | 171 | 8.55 |
| Austria | Nov 2000 | Auction Engl. | 4 | 4–6 / 6 | 6 | 103 + PoR | 6.18 |
| Denmark | Sept 2001 | Auction SPSB | 4 | 4 / 4 | 5 | 95 | 3.80 |
| Poland | Dec 2000 | Administrative | 3 | 3 / 3 | 3 | 51 | 1.51 |
| Greece | July 2001 | Auction Engl. + SB | 3 | 4 / 3 | 3 | 45 + PoR | 1.35 |
| Belgium | March 2001 | Auction Engl. + SB | 3 | 4 / 3 | 3 | 44 | 1.32 |
| Portugal | Dec 2000 | Beauty contest | 3 | 4 / 4 | 4 | 40 | 1.59 |
| Spain | March 2000 | Beauty contest | 3 | 4 / 4 | 4 | 13 / 37 + PoR | 1.48 |
| France | Febr 2001 | Beauty contest | 3 | 4 / 2 | 3 | 169 / 23 + PoR | 0.47 |
| Switzerland | Dec 2000 | Auction Engl. | 3 | 4 / 4 | 4 | 19 | 0.76 |
| Norway | Dec 2000 | Beauty contest | 2 | 4 / 4 | 4 | 11 | 0.44 |
| Sweden | Dec 2000 | Beauty contest | 3 | 4 / 4 | 5 | 0.0 + PoR | 0.00 |
| Finland | March 1999 | Beauty contest | 3 | 4 / 4 | 4 | Free | – |

*Note*: PoR = percentage of revenues.

network operator can expect a larger market share (and enjoy more cost degression for its own services and content) and, in some cases, less competition from the upstream (content provision) and downstream (customers) markets. If he has more spectrum available (that is, 2 × 15 MHz instead of 2 × 10 MHz), he can take advantage of scale economies (see Section 2.2) and better quality. Thus the 'cost and value index' in column 8 may be used to control for different numbers of UMTS licences.

The numbers in Table 10.4 do not include additional variable payments. In some countries (Sweden, France, Spain, Austria, Greece), operators have to pay a percentage of their 3G revenues (PoR) to the government. These additional fees vary from 0.08 per cent to 2 per cent of UMTS revenues (McKinsey, 2002, p. 19).

The numbers also do not control for additional requirements, such as regional coverage and roll-out plans,[14] or regulatory benefits, such as infrastructure sharing, roaming, or MVNO, which also vary significantly among countries.

### 4.3 Comments on the UMTS Auctions

If we look at the spectrum fees, it is remarkable that they are extremely different among countries (especially among those with auctions). This cannot be explained by different average income of the respective countries. It was, even at that time, significant that the first auctions, in the UK in April 2000 and in Germany in August 2000, resulted in incredibly high fees, which totalled €37 billion in the UK and €50 billion in Germany.

Some theorists (especially Klemperer, 2001b, and others who did not consider the specific industry in detail) claim that most revenue differences can be explained by auction design (Börgers and Dustman, 2002; Grimm et al., 2001a, 2001b). But the contribution of auction theory to explaining real-world 3G spectrum auctions' outcomes still remains questionable (Melody, 2001; Van Damme, 2002; Wolfstetter, 1996, 2001).

Industry-specific, time-specific, country-specific, or other real-world factors and unique conditions may be even more relevant. This does not mean that auction design does not matter, of course. In the following nine points, some comments are made on influencial factors for the results of the UMTS auctions and some other features of UMTS licensing:

(1) Objective function: revenues

An important aspect is the auction's objective function, which is supposed to be the most relevant criterion for assessments and comments. Most assessments are mainly or solely based on total revenue per capita in the specific country. Whether this is an adequate criterion is very questionable, even if it is difficult to create alternatives which are quantifiable at the time of writing.

Usually the expected maximization of revenue is the primary reason for holding an auction, instead of selling the good through other institutions. Therefore it is adequate for theoretical treatments to use revenue as the objective function for the majority of auctions.

However, this does not apply totally to spectrum auctions. Here, the main reason for regulatory agencies to use auctions instead of beauty contests or other administrative procedures is efficiency in the mobile markets (plus non-discrimination, transparency and minimal transaction costs, respectively). The proper functioning and efficient outcome of the mobile markets are supposed to be the main objectives for the spectrum and licensing auctions.

For these 'main' markets, spectrum markets are only auxiliary pre-markets. The firms' willingness to pay on spectrum markets is supposed to be a good indicator of their efficiency in the mobile market,[15] if these two markets actually interacted. This would be the case if spectrum were tradable during the years of network operation. However, in UMTS the auctions took place many years before efficiency differences would show, even for industry experts. Thus the relationship between the auctions and the markets, where efficiency really counts, is pretty loose.

Fiscal revenues should only be a side benefit for the government, but not a principal concern (at least under the prevailing conditions of intermodal spectrum allocation inefficiencies). And this has also been the case in reality, at least for the first auctions. For example in Germany, the ministry of finance was not involved in auction design. And the German officials were not even aware that the UMTS auction revenues could be substantial for the federal budget, before the UK auction in April 2000 produced surprisingly high results.

Almost the only reason to use revenues for comparison is the lack of alternatives, since the individual 3G markets' efficiency is unlikely to be apparent before 2010. And revenue is a very simple and well-known measure.

(2) Competition and efficiency

An alternative (but indirect) method to approach the issue of efficiency on the UMTS markets is the intensity of competition, which may be increased by the number of networks. This suggests that five or six networks are better than three or four. But, as in other markets, the 'optimal' number of producers also depends on scale economies, market volume, product variation and differentiation, technological and commercial developments, growth rates and so on.

Thus any predetermination of the number of licences by the regulatory agency, far ahead of real operations, includes the risk of being wrong. This risk of 'regulatory failure' could be reduced if the number of licences were determined endogenously by the bidders within the auction.[16] This was the case in the spectrum auctions in Germany and Austria, which came up with six UMTS licences each. However, if regarded in the light of recent developments and problems in the mobile industry, this outcome raises questions, too. To what extent were these country-specific (see point 6)?

Since competitive forces in markets with high entry barriers can easily deal with 'too many' networks, but not with 'too few', markets with more mobile licences (such as Germany, Austria, Italy, the UK and Denmark) appear to be more promising from an efficiency perspective. The firms' view, of course, will be different.

(3) Auctions too early

The timing of the auctions was crucial for several reasons. First, all the auctions (and all other UMTS licensing) might be regarded as 'too early' in the process of technical and market developments of UMTS. Product and service characteristics, market opportunities and demand as well as technical specifications and restrictions, potential problems, costs, cost structures (scale economies), availability of handsets and third-party services, competitive relationships to the GSM based GPRS (General Packet Radio Service) and W-LAN (Wireless Local Area Network) have been rather unclear when the first auctions started in the year 2000.

Thus assessing the value of a UMTS licence (and the value of a specific amount of spectrum) was rather difficult in general – and particularly for bidders who were newcomers – or even those in another national market.

Therefore UMTS bidders were particularly subject to the risk of the winner's curse of 'paying too much'. On the other hand, it is well known that the lack of information among bidders may lead to downward biasing in order to avoid or reduce the winner's curse. The bidding in 'high-revenue countries' (for example UK and Germany), which was surprising for most experts, may be explained as a winner's curse, as a result of holding the auctions too early.

The lower revenues in other countries can be interpreted as a result of downward biasing. This is especially convincing, because the financial results of the UMTS auctions in the UK in April and Germany in August 2000 caused a shock in the European telecommunications industry. Avoiding these extraordinary spectrum fees, which could be ruinous for their business plans, was a widely and intensively discussed topic in the industry in the respective European countries, where the next auctions were held only some weeks or months later.

(4) The decline of the Internet and mobile hype

A second point where timing matters is the telecommunications industry's overall developments during 2000 and 2001. For the explanation of different auctions' outcomes in different countries it is relevant to consider the timing of individual auctions in this context, as shown in Figure 10.5.[17] This gives the Dow Jones European telecoms stock price index for 2000 and 2001. The vertical lines represent the ends of the countries' UMTS auctions. The filled dots indicate revenues per capita in those countries. The crosses show the cost-and-value index for spectrum from Table 10.4, column 8.

The first auctions (in the UK and Germany) took place almost at the peak of (or slightly after) a decade of a mobile success story with ever-increasing penetration rates (and even increasing growth rates). In addition, a general Internet-optimistic feeling was prevalent,

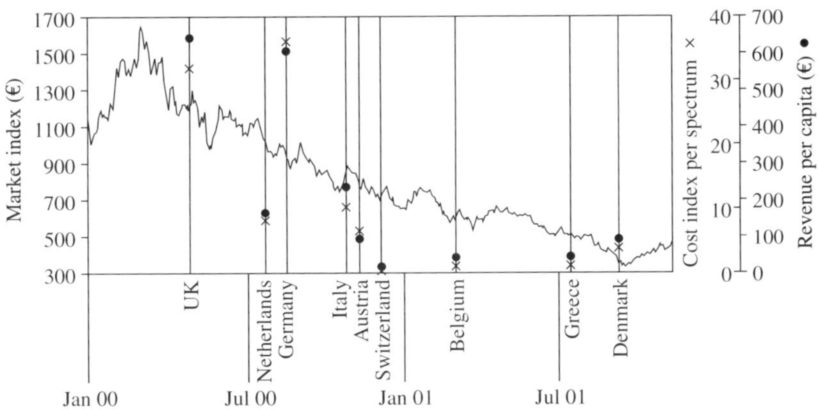


*Source*: Based on Klemperer (2002).

*Figure 10.5 Telecommunications stock prices and UMTS auction fees in European countries, 2000–2001*

which was later referred to as the 'Internet hype' or bubble. This is especially relevant with respect to capital markets and investors' behaviour. Investment money for mobile and Internet projects was abundant.

When the Internet and mobile hype and over-optimism vanished, and telecoms stock prices fell, and the industry's assessments of UMTS licences became more realistic – or even too pessimistic.

This may explain part of the low-revenue results (that is, in Belgium, Switzerland, Austria and even the Netherlands). I do not share the view that these auctions have been disasters (as said above) and that this is mostly due to poor auction design (see below). Those revenues may adequately reflect UMTS licence values at the time of the auctions.

If it is true that the industry's perspectives have become more realistic during 2000 and 2001 and/or if the high revenues in the UK and in Germany could be interpreted as a winner's curse, it would no longer be justified to use them as yardsticks for other countries' auctions, as Klemperer and others do.

(5) Attracting newcomers

It has often been mentioned that it is important to attract newcomers to the auction. In general, this is obviously true. But the incentives to enter a market are often not primarily (and never solely) in the hand of the auction designers.

In the early phases of UMTS concept and development as well as in those of legal planning for licensing procedures, the assumption prevailed that UMTS would mostly create a 'new market' (with services sufficiently new, enhanced and different from the 'old' mobile telephony). However, it became more and more clear that (1) most of the revenues would come from the same services that already exist in GSM mobile markets,[18] at least for quite a number of years, and especially, if the GSM enhancement of GPRS is considered. Additionally, (2) a UMTS operator would not be able to survive on a stand-alone basis for a long time, but needs a GSM network (or at least access to a GSM network) to secure regional coverage and so on. Also, an existing customer base and the possibility of upgrading them from GSM to UMTS is very important.

In some countries, the newcomers' problems could be expected to be even larger, if four or more mobile networks already supply the market (as in the Netherlands, the UK, Denmark, Italy, Germany and Austria) and/or if the latest GSM entrants still have severe problems.

As a result, the 'natural economic differences' between incumbents and newcomers are very significant – and increasingly considered to be crucial. Therefore the economic value of a UMTS licence is much higher for an incumbent mobile operator in a specific country than it is for a newcomer.[19] The differences are not only firm-specific. They may, at least to a certain extent, also reflect economic efficiency issues from society's point of view. In such circumstances, it will generally be difficult to attract serious newcomers to participate in a licence auction.

Also, the GSM market structure in a specific country (see Figure 10.4 and Table 10.4) is very relevant for the UMTS auctions. The individual companies' position and strategy is most important. In general, every GSM incumbent could be expected to fight for a UMTS licence in order to protect its business and its corporate future. Any potential newcomer will calculate its chances of getting a reasonable licence with regard to the number of incumbent operators, the number of licences offered and the auction design. If the number of incumbents and licences are equal, it is relatively easy for the incumbents with the above-mentioned advantages to block entrants in English auctions. And entrants will anticipate that.

In UMTS, the number of licences ($n + 1$) often exceeded the number of incumbents ($n$) by one (see Table 10.4). Potential newcomers are encouraged to participate in the auction, since their rivals are newcomers also. When the number of licences is endogenous (Germany, Austria), they may expect that the incumbents will not invest into a costly fight for a market structure that would not really change the competitive intensity in the mobile market. But it is still risky (see point 6).

(6) The UK and Germany

Figure 10.5 shows that the most significant variation in spectrum prices is between the UK and Germany versus the rest of the countries. This has (besides the above-mentioned reasons) also to do with the countries' sizes and their importance for a pan-European mobile market (if anything like that is expected to exist). The UK and Germany were considered two out of a few European core countries, where you have to be present if you want to be a European (or even world) mobile and telecommunications player. And most of the other larger countries (except Italy) in Europe used beauty contests instead of auctions.

Both countries showed some consideration for newcomers to participate. The UK did this explicitly by reserving the best spectrum package for an entrant. This provision might be regarded as a subsidy for a new entrant, which could be disputable.

The design in Germany even allowed for two newcomer licences, but did not guarantee it, since the number of licences was endogenous to the auction. This has already been criticized in advance, mostly because of the risk that the four incumbents could keep newcomers away because of the English ascending design (Jehiel and Moldovanu, 2000). Even if these considerations were theoretically convincing, it simply did not happen.

A potential explanation for the extraordinarily high revenues refers to the fact that the UK and Germany were among the first auctions. The individual companies might have had incentives to demonstrate their intended leading role in the mobile services industry in Europe and to gain a 'reputation for being tough'. Thus their high bids could have been signalling. They may have reflected their expectations of a 'first winner advantage' for the countries to follow. This expectation may have been relevant with regard to following auctions as well as to countries with beauty contests, where the licences were comparatively cheap.

According to the conventional measures (which I do not totally accept, as outlined above), 'total revenues' and the 'number of new mobile licence holders', the German auction was the most successful in Europe. Why?

Besides the arguments already made under points in subsections 4.3.3 to 4.3.5, the line between strong and weak bidders was not simply between incumbents and outsiders. The fourth incumbent, VIAG Interkom (now $O_2$), was considered pretty weak, whereas one of the newcomers, Mobilcom, was (backed by) France Telecom (with a pan-European strategy and financially strong at that time). Another was a consortium of Spain's Telefonica and Finland's Sonera, and a third was Debitel, owned by the financially sound Swisscom.

The potential newcomers could expect that the incumbents would not invest in a costly fight for a market structure that could not really change the intensity of competition in the mobile market. Vodafone and T-Mobile (the GSM market leaders in Germany) have been blamed for their attempt to keep newcomers out of the market. It is questionable whether their bidding was motivated by market structure objectives. This seems not very likely if we consider their positions in the market and those of the potential newcomers. More relevant may have been the intention to get the full package of 2 × 15 MHz of spectrum considered essential for service quality and/or cost efficiency.

Even more important for the outcome, and somewhat surprising, is the bidding behaviour (€8.5 billion for only 2 × 10 MHz of UMTS spectrum and with five competitors) of the remaining newcomers in the final rounds (France Telecom/Mobilcom and Telefonica/Sonera). They knew that they would have to develop networks and customer bases almost from scratch (except a part of Mobilcom's service provider's customers). This can (probably only) be explained by the hype situation and the ambitious strategies to become European players.

(7) Auction design

Of course, auction design matters. The question is merely whether auction design features (or auction theory in general) can explain most of the real-world outcomes, as Klemperer and other auction theorists suggest, or whether other factors (like those mentioned above), country-specific factors and so on, are even more important.

Klemperer (2001b) suggests that the UK auction, where he acted as the 'principal economic advisor' (Binmore and Klemperer, 2002), has been almost the only successful one, because of optimal auction design. The auctions in other countries, which failed to use Klemperer's wisdom, suffered from (more or less) inadequate design, with the sole exception of Denmark.

He attempts to explain every single auction's outcome by its design characteristics and other reasons (like timing and so on). Most of the individual arguments are quite persuasive as such. But they appear very much as *ex post* explanations and case-specific.

The extraordinary spectrum fees in the UK fostered incentives in the following countries to collude (among other motives to downward bidding plans). Therefore it is somewhat surprising that most of the other regulatory agencies (with the exception of Denmark) still relied on English ascending designs where some sealed-bid elements might have been advisable – at least from the perspective of the governments' short-run budgets. Whether overall efficiency would have been improved remains doubtful.

One strategy has been to merge and/or to build consortia (joint enterprises) before the auction starts. This has been facilitated by the rule that any potential bidder had to be listed well in advance. No bidder could show up unexpected, which would have made collusion much riskier and less desirable. This made it easier to merge (pre-auction collusion) and/or to decide on entry-blocking strategies.

(8) Fees

It has been argued that the very different spectrum fees in different European countries constitute competitive disadvantages to companies in some Member States. With the assumption of efficient capital markets, this argument is not convincing, since UMTS markets are basically national markets. But under imperfect capital markets and companies operating in different countries, these concerns may indeed be relevant.

There has also been concern that high licence fees will lead to high mobile communications prices for the customers. This is not convincing either, since licence fees are fixed costs (for exceptions see Table 10.4, column 7) for the operators (and also sunk). Thus they are not a part of marginal costs. Since licence fees are fixed and sunk, high fees provide incentives to (even aggressively) increase the number of customers and also their usage quantity, in particular by reducing prices below those of other firms, by advertising and so on. Thus high fees may intensify competition. The enormous licence fees in Germany and the UK are mostly at the expense of the shareholders of the licence-winning companies.

But if licence fees are high enough to question the financial stability of the companies (and potentially the whole industry) and to lower its capital market ratings, it may reduce further investments and innovation. Therefore 'too high' licence fees (as in the UK and Germany) actually do have an impact on mobile markets.

(9) The European Union
The main problem for the 'European competition level' was the fact that some countries were still using beauty contests with rather opaque criteria. These procedures have always been under the suspicion that national governments are using them to discriminate against foreign companies. The results from the UMTS licensing shows that this concern was (and is) actually justified, since 'comparative bids tended to favor national players, whereas auctions appear to have favored multinational players' (McKinsey, 2002, p. 16). Of 37 licences awarded by auctions, 68 per cent are held by multinational operators. In beauty contests 28 per cent of the licences were granted to multinational operators.[20]

## 5. COMPETITION IN EUROPEAN MOBILE COMMUNICATION

### 5.1 Fierce Competition or Collusion?

As a result of licensing and spectrum allocation in GSM in Europe, is there fierce competition or collusion in European mobile communication markets? Let us have a short look at the structural factors.

There is no straightforward economic theory telling us whether there will be collusion and collective dominance in a specific few-players market. But there are several structural factors that make collusion more or less likely (see also Chapter 6 by Rey, in this volume).

In Table 10.5, eight structural criteria are applied to European mobile communication markets and roughly assessed for the mid-1990s and for 2002 (see Kruse, 1997, 2002b), and accompanied by a forecast for the future (when UMTS markets will have started).

Since mobile communication markets are mostly national markets which differ to a certain extent, this requires a somewhat more general overview of several countries at the same time, not allowing judgments on specific national markets.

(1) Number of operators and concentration
If there are only two (instead of more) firms on the supply side, collusion is easier, more favourable and thus more likely (although not securely predictable), with competition intensity being reduced. More firms make fierce competition more likely. The same can be expected if market concentration is reduced.

In the beginning of GSM, there were mostly only two networks (plus several service providers in some countries) which have often been structurally dominant in the mid-1990s. When more networks came in (see Figure 10.4), concentration was reduced and competition

*Table 10.5 Factors of competition in mobile communications*

| | Structural feature/factor | 1995 | | 2002 | | Future | |
|---|---|---|---|---|---|---|---|
| | | Assess. | Relevance for high (+) and low (–) competition intensity | Assess. | Relevance for high (+) and low (–) competition intensity | Assess. | Relevance for high (+) and low (–) competition intensity |
| | a | b | c | d | e | f | g |
| 1 | Number of operators | Small | – | Relatively small | + | Medium | + |
| | Concentration | Very high | – | High | 0 | | |
| 2 | Entry barriers | High | – – | High | – | High | 0 |
| 3 | Fixed costs | Very high | + + | Very high | + + | Very high | + + |
| 4 | Sunk costs | Very high | 0 | Very high | + | Very high | + |
| 5 | Elasticity of market demand, substitution | Low | – | High | + | Very high | + + |
| 6 | Homogeneity and transparency | High | + + | High | + + | High | + + |
| 7 | Switching costs | Medium | 0 | Medium | – | Low | + |
| 8 | Technical and economic dynamics | High | – | Medium | + | Low | ++ |
| 9 | Total | | Moderate | | Relatively high | | High |

was intensified. At the time of writing we see that the number of suppliers has significantly increased and concentration has decreased. Including UMTS, mostly four or more operators are licensed (Table 10.4, column 6).

(2) Entry barriers

The existence of significantly high entry barriers is a necessary, but not a sufficient, condition for collusion. In mobile communication, entry barriers are very significant because of the scarcity of spectrum and the licensing of the regulatory agencies. But the relevance for competition intensity has decreased because of the increased number of networks and changes in other factors.

(3) Fixed costs

The cost structure is a very important element for competition. The higher the fixed costs (as compared to variable costs), the more profit is gained by extra sales (since marginal costs are low), thus generating incentives for price reductions. In mobile communication, a large amount of costs is fixed, because of network investments. In UMTS, in some countries, high licence fees added to extraordinary cost fixity.

(4) Sunk costs

Most of the investments (and most licence fees in UMTS) are also sunk, resulting in high exit barriers. Thus a small company losing market share would reduce prices rather than leaving the market. And if a company failed, this would not reduce the number of players, but the networks would be sold for a price that allows making profit from mobile operation.

(5) Elasticity of market demand

For collusion to be successful, the relevant part of the demand function has to be inelastic.[21] This actually was the case when GSM started, due to wealthy and business customers. The more mobile telephony has grown into a mass market, the more elastic market demand has been. In addition, intermodal substitution is becoming relevant (GSM, UMTS, Wireless LAN).

(6) Homogeneity and vertical market transparency

GSM telephony has mostly been a nearly homogeneous product (with the exception of newly started networks with smaller regional coverage, especially in the 1800 MHz band). Vertical market transparency (which is transparency for the consumers) has mostly been rather high due to extensive media coverage. Although in some cases, horizontal market transparency (for the other suppliers) may foster collusion, vertical transparency for the consumers intensifies competition in most cases.

(7) Switching costs

If customers' switching costs are high, this may reduce competition. This has probably gained some relevance with growing penetration rates over the years. But switching costs are (or will be) reduced by the introduction of MNP (mobile number portability).

(8) Technical and market stability

If technological and market conditions are stable, markets are more likely to be collusive. Technology and markets were basically stable for the first years of GSM. Now, technology, supply and demand are more uncertain for network operators. This holds especially for UMTS, since the vertical relationships with third-party service and content providers still need to be established.

(9) Conclusion
Summing up the economic factors, Table 10.5 shows that several values have changed towards increased competition. The prospects for UMTS are such that competition intensity in mobile markets is likely to be further increased.

The main problem for UMTS in European markets is that there will not be enough customers and services, so that revenues will not increase quickly enough to encourage further service developments. Indeed, because of the large outlays for UMTS networks (and sometimes also for licence fees), network operators will have to struggle for customers and revenues.

The theoretical picture is supported by actual developments in European mobile markets. Penetration of mobile services has increased tremendously in a few years, especially at the end of the 1990s with the introduction of pre-paid services without monthly fees. Average prices for consumers have been constantly reduced over the years.

International comparisons of mobile prices and their reduction over the years are somewhat complicated, because of very different tariff structures and handset subsidies. This can be seen in Table 10.6, which relies on an Oftel benchmark study. The ranking of the five countries (and the respective index values) with and without handset prices (left side versus right side of the table) shows that handset subsidies are very significant for consumers' 'effective prices'. Competition concentrates on different parameters in different countries, due to different consumer preferences and behaviour (reaction elasticity toward different elements of the tariff structure).

*Table 10.6 Price index for mobile services (international benchmark by Oftel)*[22]

| | Handsets included | | | | | Handsets not included | | | | |
|---|---|---|---|---|---|---|---|---|---|---|
| | Baskets/Minutes | | | | Average | Baskets/Minutes | | | | Average |
| | 245 | 468 | 1133 | 3812 | | 245 | 468 | 1133 | 3812 | |
| Italy | 128 | 130 | 125 | 127 | 122 | 120 | 129 | 124 | 110 | 116 |
| Sweden | 99 | 98 | 82 | 104 | 108 | 126 | 118 | 102 | 101 | 144 |
| UK | 100 | 100 | 100 | 100 | 100 | 100 | 100 | 100 | 100 | 100 |
| France | 102 | 97 | 99 | 89 | 94 | 145 | 148 | 118 | 86 | 141 |
| Germany | 85 | 94 | 95 | 104 | 91 | 172 | 164 | 136 | 105 | 146 |

## 5.2 The Case for UMTS Infrastructure Sharing

One of the regulatory problems in UMTS is 'infrastructure sharing'. In fact, this has a great deal to do with the specific licensing policies of different regulatory agencies.

For GSM, the prevailing philosophy was the 'fully integrated networks and services' idea. Since future mobile subcribers, traffic volume, scale economies and so on were still unknown at the time of first GSM licensing, the regulatory agencies acted very cautiously by issuing only two licences in the 900 MHz band. They tried to avoid 'too small frequency-bands' for each network and were worried about ruinous competition among them.When GSM was successful, they adopted a step-by-step procedure, providing the third and fourth operator with 1800 MHz frequencies which were less valuable in the beginning of the system (see Table 10.3 and Figure 10.4). Thus GSM market asymmetries were, to a certain extent, regulation-made.

With UMTS, the NRAs changed their procedural approach and issued all licences simultaneously. This implied that they had to decide about the number of licences[23] (and thus the market structure) from the very beginning. Since UMTS is technologically quite different from GSM and because non-voice demand is rather uncertain, NRAs do not know much more about scale economies and traffic volume than was the case for GSM ten years earlier. Whether their decisions have been adequate will only become clear in five or ten years.

Another problem is that for most infrastructure-based services (in utilities, transportation and communications), as well as in other industries, scale economies are different for each production level. Some may be natural monopolies, some may allow a small number of operators and some may allow hundreds of competitors (especially in services). If the sector is fully competitive, the vertical structure (and the horizontal market structure as well) will almost automatically be adjusted to these conditions. In regulated industries this remains a problem.

In UMTS, the 'service and content level' allows for hundreds of players. The core networks exhibit moderate-scale economies, such that six networks (which is the largest number in any European country) may not be a problem. This also has to do with the fact that several players use part of their GPRS core network also for UMTS.

For the radio access networks (RANs), four parallel UMTS networks may already be a waste of resources. If we only look at the 'voice market', four GSM RANs plus four UMTS RANs add up to eight (partly parallel) networks. If the number of four (or even six) UMTS networks is a waste of resources, this is quantitatively significant, since the RANs require about 75 per cent of total investment for UMTS.

Since the regulatory agencies (probably) did not really consider these issues, they should adjust their policies by allowing infrastructure sharing of the RAN. The advantages of UMTS infrastructure sharing are most significant in out-of city areas and in the first years of roll-out and market development. Infrastructure sharing can make roll-out quicker and cheaper and allows a wider regional coverage (Kruse, 2002a).

As a result, the services will have a broader market, providing incentives for services and content developments, and marketing. This is especially relevant for those services that are subject to non-rivalry in consumption, inferring zero marginal costs and significant degression of average costs.

The question for NRAs and antitrust authorities will be: are there significant competition risks from infrastructure sharing? The answer is no.

In the beginning of UMTS, the networks will vigorously compete for subscribers (for reasons outlined above). The shared part is the most expensive one, but not the most significant for markets' efficiency. The 'intelligent' part of the network, which is embedded in the core networks, will be operated separately. And services and content will not only be offered by the network operators themselves, but also by numerous third parties (from the media and IT sector and so on).

## NOTES

1. I am grateful to Peter Leuthold (ETH Zürich), who provided some valuable information on these technical and economic issues. Of course, any remaining errors are my responsibility. See also Leuthold and Kemper (2002) and UMTS-Forum (1998).
2. The networks' willingness to pay for spectrum can be derived from the mobile communications costs and demand functions (Kruse, 1993).
3. Since it is generally possible to use a specific part of spectrum for mobile communication or other services, the question arises, whether GSM and UMTS were supplied with enough (or too much?) spectrum.
4. International standards (for example GSM-900) define the amount of MHz needed for a single channel and, therefore, the number of channels for a given amount of MHz.
5. The demand function $SD_H$, instead of $SD_A$, denotes the case that more efficient equipment may allow to produce the same services with less spectrum, but with additional costs elsewhere. The operator will then compare the incremental cost of equipment and spectrum. $SD_J$, instead of $SD_B$, characterizes the possibility that service *B* may as well use other spectrum or even other transmission methods (cable and so on). Incentives to do so will increase with spectrum prices.
6. The opportunity price in homogeneous multi-object auctions would be the bid of the margin loser, who is the next highest bidder. Such a price would be theoretically more convincing.
7. In Section 4.3 of this chapter, some comments are made on the European UMTS auctions.
8. The main exception is the UMTS auction in Denmark, where a second-price sealed-bid approach has been used.
9. Column (3) includes FCFS procedures, which are mostly to the advantage of incumbent firms, and extensions of other spectrum rights (for example 1800 MHz spectrum for existent GSM-900 operators. R denotes cases where spectrum rights are not country-wide, but regionally confined.
10. For technical reasons of wave propagation they need more base stations in order to cover a whole country (approximately 2.5 times more).
11. The UK explicitly reserved the most valuable frequency package for a newcomer.
12. Austria even has three new licensees, if we count Mannesmann 3G separately, after Vodafone had sold Tele.ring to Western Wireless.
13. The figures show the amount of euro the licensees together had to pay per capita in the countries. This does not include extra payments as a percentage of revenues (PoR) or extra burden such as mandated regional coverage.
14. On average, countries which use beauty contests have higher coverage requirements for the first years, whereas those using auctions imposed higher coverages for the years after 2005. See McKinsey (2002), p. 19.
15. This assumes that mobile markets are (and will be) competitive. Otherwise, willingness-to-pay could also reflect monopoly rents.
16. This argument is theoretically convincing, but it still depends on very good information of the bidders, which did not hold in the UMTS case (see point 3).
17. Figure 10.5 is drawn from Klemperer (2002). The numbers for a 'cost-and-value index for spectrum' (×), which controls for the economic effects of different numbers of licences and different average sizes of the licensees' spectrum availability (see Table 10.4) have been added to the original exhibit.
18. It is expected that in the beginning of UMTS most revenues will be generated by voice, with the share of data revenues increasing over the years.
19. National roaming, mobile virtual network operating, or other regulatory measures will not be able to fill the gap.
20. According to McKinsey (2002), p. 16, a 'national player is defined as an operator owned for more than 50 per cent by companies by this member state and a multinational operator is a player having operations in the member state considered but also in other member states and is owned for less than 50 percent by companies of the member state under consideration'.
21. If market demand were elastic, higher prices would not increase profits very much, since quantities sold would be reduced. Additionally, the incentives for individual cheating would be higher, and collusion would be less likely to occur or be stable.
22. Cf. Oftel (2001). Prices for August 2001, UK = 100. The baskets include a varying number of minutes, voice-mail, and SMS. The averages in the table are for all Oftel baskets. Details on the methodology of the Oftel study and for the different baskets can be found in http://www.oftel.gov.uk/publications/research/benc1200.htm.
23. Except Germany and Austria, where this was left to be determined by the auction process endogenously.

# REFERENCES

Binmore, Ken and P. Klemperer (2002), 'The Biggest Auction Ever: The Sale of the British 3G Telecom Licenses', *The Economic Journal*, **112**, pp. C74–C96.

Börgers, Tilman and C. Dustman (2002), 'Rationalizing the UMTS Spectrum Bids: The Case of the UK Auction', IfO-Studien, **48**, pp. 77–109.

Cramton, Peter (2002), Spectrum Auctions, in Martin E. Cave, S.K. Majumdar and I. Vogelsang, *Handbook of Telecommunications Economics*, vol. 1, Amsterdam: Elsevier.

Grimm, Veronika, F. Riedel and E. Wolfstetter (2001a), 'Low Price Equilibrium in Multi-Unit Auctions: The GSM Spectrum Auction in Germany', Discussion Paper.

Grimm,Veronika, F. Riedel and E. Wolfstetter (2001b), 'The Third-Generation (UMTS) Spectrum License auction in Germany', IfO-Studien, **48**, pp. 123–44.

Jehiel, Philippe and B. Moldovanu (2000), 'A Critique of the Planned Rules for the German UMTS/IMT-2000 License Auction', working paper, University of Mannheim and University College London.

Klemperer, Paul (1999), 'Auction Theory: A Guide to the Literature', *Journal of Economic Surveys*, **13**, pp. 227–86.

Klemperer, Paul (2001a), 'What really matters in auction design', paper, Nuffield College, Oxford.

Klemperer, Paul (2001b), 'How (not) to run Auctions: The European 3G Telecom Auctions', draft paper.

Klemperer, Paul (2002), 'Some Observations on the German 3G Telecom Auction: Comments on Grimm, Riedel and Wolfstetter', *IfO-Studien*, **48**, pp. 145–56.

Kruse, Jörn (1993), *Lizenzierung und Wettbewerb im Mobilfunk*, Heidelberg: Springer Verlag.

Kruse, Jörn (1995), 'Zugang zum Frequenzspektrum', in Ernst-Joachim Mestmäcker (Hrsg), *Kommunikation ohne Monopole II*, Baden-Baden: Nomos Verlag, pp. 449–72.

Kruse, Jörn (1996), Frequenzvergabe im digitalen zellularen Mobilfunk in der Bundesrepublik Deutschland, Studie für das Bundesministerium für Post und Telekommunikation, Bonn, also as discussion paper 174, Wissenschaftliches Institut für Kommunikationsdienste, Bad Honnef.

Kruse, Jörn (1997), 'Marktbeherrschung auf dem deutschen Mobilfunkmarkt', Discussion paper 159, Economics Institute, University of Hohenheim, December.

Kruse, Jörn (2002a), 'Ein gemeinsames Funknetz für UMTS in der Schweiz. Ökonomische Aspekte', hrsg. v. Avenir Suisse, Zürich, March, available at: www.avenir-suisse.ch.

Kruse, Jörn (2002b), 'Regulierung der Verbindungsnetzbetreiberauswahl im Mobilfunk', Discussion paper, University of the Armed Force, Hamburg, available at: www.unibw-hamburg.de/WWEB/vwl/kruse; a shorter version in MultiMedia und Recht, January 2003, pp. 29–35.

Leuthold, Peter E. and J. Kemper (2002), 'Ein gemeinsames Funknetz für UMTS in der Schweiz. Technische Aspekte', hrsg. v. Avenir Suisse, Zürich, March, available at: www.avenir-suisse.ch

McKinsey (2002), 'Comparative Assessment of the Licensing Regimes for 3G Mobile Communications in the European Union and their Impact on the Mobile Communications Sector', final report for the European Commission, 25 June.

McAfee, R.P. and J. McMillan (1987), 'Auctions and Bidding', *Journal of Economic Literature*, **25**, pp. 699–738.

McAfee, R.P. and J. McMillan (1994), 'Selling Spectrum Rights', *Journal of Economic Perspectives*, **8**, pp. 145–62.

McAfee, R.P. and J. McMillan (1996), 'Analyzing the Airwaves Auction', *Journal of Economic Perspectives*, **10**, pp. 159–75.

McMillan, John (1995), 'Why auction the spectrum', *Telecommunications Policy*, **19**, pp. 191–99.

Melody, William (2001), 'Spectrum auctions and efficient resource allocation: Learning from the 3G experience in Europe', *Info*, **3**(1), pp. 5–10.

Oftel (2001), 'International benchmarking study of mobile services', issued by the Director General of Telecommunications, London, 7 November.

UMTS-Forum (1998), 'Minimum Spectrum demand per Public Terrestrial UMTS Operator in the Initial Phase', Report No. 5, 8 September.

Van Damme, Eric (2002), 'The Dutch UMTS-Auction', IfO-Studien, **48**, pp. 175–200.

Wolfstetter, Elmar (1996), 'Auctions: An Introduction', *Journal of Economic Surveys*, **10**, pp. 367–420.
Wolfstetter, Elmar (2001), 'The Swiss UMTS Spectrum Auction Flop: Bad Luck or Bad Design', CESifo-Diskussionspapier.

# 11. The allocation of scarce resources, spectrum assignment and competition in mobile communications: the case of UMTS

**David Gabathuler and Wolf Sauter***

## 1. INTRODUCTION

While man-made bottlenecks, such as the fixed local loop or mobile call termination, currently raise particular challenges in the telecommunications sector, ultimately the scarcity of a finite natural resource – radio frequency spectrum – has the potential to become a determining element in shaping the future competitive landscape of a seamless wireline/ wireless environment. For the mobile communications sector that relies on electromagnetic spectrum as its bearer medium, the assignment of spectrum[1] is already of primary significance and continued technological advances are only likely to increase demand pressures. As a result, ensuring the efficient assignment of spectrum is becoming a political priority.

Until recently the spectrum allocation and assignment process had escaped any significant degree of harmonization at EU level, and was jealously guarded by the Member States invoking sovereignty rights. However, the uneven results of the recent 3G/UMTS[2] licence award procedures and the slump in the sector that followed shortly thereafter have placed both closer co-ordination of spectrum policy and improving the efficiency of frequency assignment on the political agenda. At the same time, the new EU sector-specific framework – in addition to the competition rules – seeks to actively promote effective competition in the use by undertakings of frequencies for communications purposes.

This chapter examines, first, the present and future regulatory framework for frequency allocation and assignment in the EU, in particular in relation to its impact on mobile communications, and discusses recent developments towards introducing spectrum trading. Next, it considers the regulatory and competition policy implications of the introduction of 3G/UMTS, with a focus on network sharing and MVNO access.

## 2. SPECTRUM ALLOCATION

Technically usable radio spectrum is a scarce natural resource.[3] That is, due to the physical nature of spectrum and the constraints inherent in its use (notably the restraints necessary to avoid electromagnetic interference), the amount of spectrum available is limited, and consequently demand cannot always be met. At the same time, the development of capacity-hungry new communications technologies adds to the demand for spectrum, further increasing its scarcity.

Assigning spectrum for use by individual operators has traditionally been the prerogative of individual national administrations following co-ordination on the general allocation of spectrum for particular uses at supranational level in the context of the European Conference of Postal and Telecommunications Administrations (CEPT) and the International Telecommunications Union (ITU).[4] Spectrum assignment has historically taken place as an *ad hoc* process, with large swathes of spectrum reserved for military and other public purposes, with the remainder generally based on a 'first come first served' principle that favoured incumbent operators. Evidently, this did not favour an optimal assignment and limited the scope for new technologies to emerge. The result has been a relatively inefficient assignment of spectrum, leading to further pressure on supply.

More recently, in order to develop a pan-European market for digital mobile communications, the EU Member States have started to co-ordinate frequency allocation for specific technologies at Community level. With the introduction of competition in mobile communications requiring licence assignment on objective, transparent and non-discriminatory terms, the Member States have also introduced spectrum pricing by means of mechanisms such as administrative fees and auctions in order to ensure more efficient assignment. In the award procedure for 3G licences, frequency auctions and competitive bids (or 'beauty parades') have been used in most Member States. However, the unco-ordinated use of different procedures has led to widely diverging outcomes in terms of the number of licences awarded and the proceeds derived from the process, and has given rise to widespread dissatisfaction and demands for co-ordination to create a more level playing field. At the same time, there is increasing interest in the introduction of further market-based mechanisms such as 'spectrum trading' that can help to ensure sustained efficiency in a constantly rearranged second-order distribution of frequency rights.

Although for the foreseeable future the allocation of spectrum to particular uses and its assignment to particular operators are likely to continue to be primarily determined at a national level, both the existing and new regulatory frameworks pursue further harmonization to promote effective spectrum management within the EU. There also appears to be growing recognition of the need for radio spectrum policy issues to be co-ordinated more closely at EU level in order to introduce more consistency in spectrum allocation and licensing.

## 2.1 Spectrum Allocation and Assignment under the Existing EU Framework

EU involvement in promoting the GSM family of standards pre-dates the liberalization of the telecommunications sector. It started in 1987 with a directive that required the Member States to reserve the relevant 900 MHz frequency bands for GSM, following a Member State-led effort to harmonize frequency use based on CEPT mandates aimed at facilitating network deployment and services adoption.[5] Another branch of early Community involvement was the promotion and development of technical standards through mandates to ETSI, the European Telecommunications Standards Institute that was created in 1989. Subsequently, the EU became involved in the joint liberalization of the telecommunications sector and the harmonization of the regulatory framework.

### 2.1.1 Liberalization

The liberalization of mobile communications networks and services occurred through a 1996 amendment of the Services Directive, a liberalization directive adopted by the Commission under its own authority based on Article 90 (now Article 86) of the EC Treaty,[6] that had abolished monopoly rights for fixed telecommunications services. In addition to abolishing the monopoly rights of the incumbent operators that had generally been awarded the only GSM licence available per Member State, the amended Services Directive set out certain principles concerning the award of new mobile licences. In particular, it required that where frequencies were available, licences should be awarded based on open, non-discriminatory and transparent procedures.

The Services Directive subsequently formed the legal basis of enforcement action by the Commission to ensure that equitable conditions consistent with the promotion of competition were imposed when the new mobile telephony licences for GSM telephony were awarded.[7]

### 2.1.2 Harmonization of licensing

Community-wide or regional licensing could have been one of the prime incentives for the emergence of transnational markets in the EU. However, the Member States have traditionally been jealous of their prerogatives in the area of the allocation and use of frequency rights that they associated not only with safeguarding against electromagnetic interference but also with the exercise of sovereignty over their national territory and with issues of national security. This attitude persisted even when the general principle of full liberalization of telecommunications markets by 1 January 1998 – and therefore open market entry – was accepted. Consequently, the Commission has only with difficulty succeeded in introducing elements of a harmonized approach concerning licensing, and its limited success in this area has contributed to some of the recent difficulties experienced in the sector.

The main landmark of efforts to harmonize the award of frequency rights in the EU in the context of the 1998 liberalization drive was the 1997 European Parliament and Council Directive on a common framework for general authorizations and individual licences in the field of telecommunications services (Licensing Directive).[8] This harmonization directive mirrored the key provisions of the earlier Commission liberalization directive (the above-mentioned Services Directive) in providing that licences should be awarded in accordance with principles of non-discrimination, proportionality and transparency, and that conditions attached to such licences should be limited to a set of essential requirements.[9]

Reflecting the Member States' sovereignty claims, both the Licensing and Services Directives sanctioned a continuation of the pre-existing situation whereby the licensing authority remained located at the national level, thus ensuring that economic activity in the sector remained fragmented along national lines. Although Article 13 of the Licensing Directive introduced a 'one-stop-shopping' procedure for obtaining licences or authorizations to operate in multiple Member States, this procedure has never been a success. Individual licensing was accepted for the provision of public voice telephony and public telecommunications networks, as well as any networks involving the use of radio frequencies. Limitation of the number of licences awarded, however, was allowed exclusively to the extent necessary to ensure the efficient use of scarce resources, that is, radio frequencies and numbering rights.

The EU directives required such licences to be granted through open, non-discriminatory and transparent procedures, based on objective criteria defined in advance. However, they did not specify the type of procedure to be used, leaving open the choice between a competitive (such as an auction) and a comparative (sometimes called a 'beauty parade') bidding process. Although the Licensing Directive sought to limit the charging of fees for individual licences to the administrative costs incurred in the issue of the rights, its Article 11 allowed the Member States to impose charges to ensure the efficient use of the frequency resources concerned. It is this article that was subsequently used to justify the use of frequency auctions that raised considerable fees from the award of mobile licences.

### 2.1.3 Frequency co-ordination

In the EU, as mentioned above, co-ordination of the use of frequency bands for specific technologies pre-dated liberalization and can be dated back to the GSM Directive of 1987. A further example was the 1997 Decision on the co-ordination of authorizations for satellite personal communications services.[10] Concerning 3G services, the co-ordination of the use of frequencies for this purpose was required by the Decision on the co-ordinated introduction of a third-generation mobile and wireless communication system (UMTS) in the EU, adopted at the end of 1998.[11] The UMTS Decision defined the capabilities of the new 3G services, called for harmonized spectrum use based on CEPT mandates and set out a timeframe for Member States to have their authorization systems in place by the start of 2000. Unlike the GSM Directive, which itself had set out the frequencies to be allocated to GSM use directly, the UMTS Decision authorized the Commission, assisted by the Licensing Committee,[12] to grant mandates to the CEPT to harmonize frequency use. The objective of the UMTS Decision was to enable operators to enter the market from 1 January 2002 at the latest. Due to a combination of unanticipated financial problems and commercial setbacks, this timeframe has proven illusory.

### 2.1.4 Frequency trading

Spectrum licences are term-limited, allowing spectrum to be re-farmed when it reverts back to the state after expiration of the licence period, or after the licence ceases to be valid for other reasons (for example failure to use the frequencies, to meet coverage obligations or other reasons generally set out in the award conditions). However, once spectrum has been awarded there is under the existing regulatory framework at EU level no provision for trading in the right to use this spectrum. In itself this omission does not exclude national rules that enable spectrum trading (provided that competition is not as a result distorted to a degree that affects trade between the Member States, for example by the creation or strengthening of a dominant position), but so far such rules do not exist. In principle at present transfers of rights to use spectrum can therefore only result from mergers and acquisitions, or when the licences revert to the responsible administration as cited above. The latter could subsequently reissue the frequencies concerned under a new licence pursuant to the applicable selection procedures.

## 2.2 Spectrum Allocation and Assignment under the New Regulatory Framework

On 25 July 2003, the new EU regulatory framework ('the new framework') for electronic communications, as implemented in national legislation, will enter into force. Although the new framework is not specifically directed at 3G/UMTS, its introduction will create a new regulatory environment. In particular, the new framework seeks to introduce a lighter touch of regulation and to limit regulation to specific cases where markets are failing. At the same time it provides for increased co-ordination between Member States.

### 2.2.1 Licensing and spectrum trading

An important aspect of this more restrained approach to regulation is that under the new Authorisation Directive 'general authorisations' – which require nothing more than a notification to a registry declaring that certain minimum conditions are met – will in future be the norm for electronic communications services and networks, with the possibility of requiring individual rights of use strictly limited to scarce radio frequencies and numbering resources.

Under the Authorisation Directive, as under the present regime, individual rights of use can be awarded either by means of competitive or comparative selection procedures.[13] The main change as regards the assignment of radio frequency rights is that spectrum trading – or 'the transfer of rights to use radio frequencies to other undertakings' – is now explicitly permitted for the first time.[14] However, allowing spectrum trading is not mandatory. In the event that a Member State does implement spectrum trading, according to Article 9 of the Framework Directive, a number of conditions must be fulfilled:

- the resulting transfer of rights to use must be notified to the responsible National Regulatory Authority ('NRA');
- it must take place according to procedures previously established by the NRA;
- it must be made public;
- it must not lead to a change of use of frequencies harmonized either by the Radio Spectrum Decision (see further below), or by earlier Community measures; and
- the NRA must ensure that competition is not distorted as a result of any such transaction.

The 'one-stop-shopping' procedure for pan-European licences or authorizations of the Licensing Directive based on centralized administrative co-ordination of the application procedures has been abandoned under the new Authorisation Directive. Instead a general legal basis for the co-ordinated award of rights of use has been introduced. In the context of the harmonization of the use of radio spectrum, Article 8 of the Authorisation Directive provides for the harmonized assignment of radio frequencies to individual undertakings based on access conditions and procedures agreed at EU level. Notwithstanding that no common EU procedure for awarding licences is foreseen, this provides for the first time a clear legal basis for the application of common procedures and a common selection process in particular cases. This is further promoted by harmonization in the context of the new radio spectrum policy framework.

### 2.2.2 The radio spectrum policy framework

The harmonized award of frequencies is at the core of the European Parliament and Council Decision on a regulatory framework for radio spectrum policy in the European Community (Radio Spectrum Decision).[15] The aim of the Radio Spectrum Decision is to establish procedures to facilitate harmonization of the use of radio spectrum, enable the effective implementation of radio spectrum policy, and ensure the co-ordination of Community interests in international negotiations where radio spectrum use affects policies of the EU.[16]

In this context both the harmonization of spectrum trading conditions and the timing of the introduction of licences ('sequencing') in different Member States will also be covered. Based on the experiences of the S-PCS (satellite personal-communications services) and UMTS Decisions, the relevant technical implementation measures and CEPT mandates will be established by the Commission under committee procedures, albeit with a new Radio Spectrum Committee of Member States' representatives. The Commission thus obtains a general competence regarding the harmonization of spectrum, while the central role of the CEPT is respected, ensuring compatibility within the ITU system.

In addition to the Radio Spectrum Committee, the Commission has created a Radio Spectrum Policy Group to advise on radio spectrum policy issues, on co-ordination of policy approaches and, where appropriate, on harmonized conditions with regard to the availability and efficient use of radio spectrum necessary for the establishment and functioning of the internal market. The Group will comprise high-level representatives from Member States and the Commission as well as observers, including from the European Parliament, EEA members and accession countries as well as representatives of CEPT and ETSI.[17]

With these measures, the basic instruments to prepare proposals towards the harmonization of spectrum use are in place.

## 2.3 Spectrum Trading and the Development of Secondary Markets

Spectrum trading refers to the development of a secondary market in frequency rights. By allowing the transfer of such rights to other interested parties who attribute a greater usefulness to them, it is potentially of great importance to the efficient distribution of frequency rights. Spectrum trading can vary by mode, duration and extent. *Mode* refers to how trading is organized to permit possibly not only changes of ownership but reconfiguration (that is, the dividing or aggregating of licences) and changes of use; *duration* to the time concerned, for example short- and long-term leases, or permanent transfers; and *extent* to the degree to which rights and obligations are transferred completely or partially, or continue to be held in common. The advantages of spectrum trading are found in terms of redistribution, where spectrum would be reassigned from low-value to high-value users achieving greater benefit; in terms of economic efficiency; and in terms of competition by means of allowing market entry.[18]

Given the development of different commercial uses for spectrum, interest in spectrum trading had increased markedly in recent years, following pioneering efforts in New Zealand in 1989, and the adoption of spectrum trading in Guatemala in 1996 and Australia in 1996. The US Federal Communications Commission adopted General Principles and a Notice of Proposed Rulemaking (NPRM) setting out a general framework to promote the development of spectrum trading in November 2000,[19] and in Canada spectrum trading has recently also been accepted at least in principle. In the EU the scope for spectrum trading is likewise increasing.

### 2.3.1 The emerging EU position

As mentioned above, the new Framework Directive for the first time specifically allows Member States to introduce spectrum trading, and the accompanying Radio Spectrum Decision ensures that the appropriate institutional structures will be in place to develop a co-ordinated European policy in this area. Although there is no specific mandate for or requirement of spectrum trading at EU level, it is likely that the Commission as well as the newly created Radio Spectrum Policy Group and the European Regulators Group will both be active in examining the feasibility of spectrum trading and drawing on other countries' experience in this area. Of the different Member States, the UK appears to be the furthest ahead and its experience may have a strong influence in shaping EU policy and also in determining whether other countries follow suit.

### 2.3.2 Developments in the UK

The UK government has considered for some time that the introduction of spectrum trading would result in a more efficient allocation of scarce (spectrum) resources by allowing the market-determined opportunity cost of the spectrum to be realized. The government also appears keen to raise additional revenue by releasing more spectrum on to the market, particularly by reassigning under-utilized frequencies. Introducing trading enhances the value of spectrum to investors and thereby also helps to generate additional revenue for the government. The impetus for spectrum trading in the UK resulted from work carried out by Professor Cave as part of his 'Independent Review of Radio Spectrum Management' published in March 2002.[20] Professor Cave's team carried out a wide-ranging review looking into all aspects and options of spectrum trading. The Review in its conclusions strongly supported the introduction of spectrum trading and made a number of detailed policy recommendations, including:[21]

- advocating the earliest and widest application of spectrum trading;
- allowing licensees to be as free as possible to determine the wireless service they provide and the technology they choose to deploy within the boundaries of the licences (to limit interference);
- trading should be introduced in a way that minimizes transaction costs, consistent with maintaining the integrity of spectrum management regime. Licensees should have the freedom to divide and partition their licences by frequency and geography for subsequent sale, and spectrum users should also be able to lease access to frequencies to others while retaining responsibility towards the regulator;
- spectrum trading should be subject to competition rules;
- *ex post* analysis of the impact of spectrum trading on competition in defined markets should be the primary safeguard but a more interventionist approach may be necessary where the input into the market is subject to sector-specific regulation, for example *ex ante* approval of specific trades.

As a follow-up, in July 2002 the UK published a Radiocommunications Agency consultation paper on 'Implementing Spectrum Trading', which reaffirms its commitment to spectrum trading and proposes that

> spectrum trading should be introduced progressively with the intention of being as flexible as possible to deliver maximum benefits as soon as possible. The greater the flexibility allowed to reconfigure geographical coverage and bandwidth and to change technology and applications, the greater the scope for trading to take place. There will need to be some limits to minimise risk of harmful interference and to ensure compliance with international obligations.[22]

It is likely that spectrum trading will be introduced in the UK shortly after the Communications Bill[23] becomes law but it will be on a more limited basis than envisaged by the Independent Review. In the consultation document, the UK government indicates that it expects some forms of trading to be implemented in some licences by the end of 2004. However, in the case of the five 3G licences, it is expected not to introduce trading earlier than three years after the first licensee's substantive launch of service within the UK, subject to further developments in the roll-out and market for 3G services.

### 2.3.3 Implications for 3G?

The traditional method of allocating spectrum licences has the advantage of limiting the risk of interference. However, the system's administrative nature may result in a suboptimal outcome because licensing authorities are unlikely to have sufficient information to allocate spectrum efficiently. The use of auctions allows a more economically efficient initial assignment of licences, but problems can arise where operators' original expectations are not met and as new technologies or commercial opportunities develop. The introduction of spectrum trading – subsequent to an initial assignment of spectrum on the basis of an auction to prevent windfall gains – should have positive efficiencies since it will make licensees consider the opportunity costs of the spectrum and also reduce barriers to market entry, thereby constraining market power.

For spectrum trading to have the greatest impact, the most flexible trading mode (change of ownership, reconfiguration and change of use) is necessary since it provides the most opportunities for trading and lessens the risks of cartelization and monopolization. It should also reduce the need for sector-specific regulation. However, in order to ensure efficient management of the spectrum so as to reduce the risk of interference, the most liberal mode may not be achievable in all sectors. Therefore a more interventionist regulatory framework based on *ex ante* rules may be necessary in certain sectors rather than simple reliance on *ex post* competition rules. In addition, public policy considerations such as the objective of ensuring media pluralism may necessitate supplementary *ex ante* rules.

3G operators are undoubtedly keen to unlock the value of their significant expenditure on UMTS licences and if necessary to sell off unused spectrum. Nevertheless, spectrum trading may not affect the existing market structure fundamentally: the mobile sector remains an oligopolistic market with very high barriers to entry and strong first-mover advantages. The high level of sunk costs means that the risks (and exit barriers) for new entrants are high and there is no opportunity of 'hit-and-run' entry. In addition, Article 9 of the Framework Directive excludes trading of frequencies whose use has been co-ordinated by Community measures for alternative uses.

Hence trading would in all likelihood involve existing mobile operators, and the likelihood of new entry based on spectrum trading appears limited. Nevertheless, because 3G's vastly increased bandwidth may make it necessary for operators to co-operate with innovative firms to provide content-rich services rather than rely on the limited imagination of their own vertically integrated downstream operations, entry via trading could come when such partners seek to expand across the value chain.

The introduction of spectrum trading during the initial 3G licence period would need to be considered carefully for reasons of legal certainty and legitimate expectations – as operators might have bid differently had they known trading would be possible. However, trading may favour existing operators where new entrants were granted more spectrum in their licences as an additional incentive to roll out costly networks in competition with established operators, and trading may allow this advantage to be lost. On the other hand, it would enable new entrants, if necessary, to sell on some of their under-used spectrum and might, in theory, even lead to new entry. Trading will clearly affect the competitiveness of the market, but should a market in spectrum trading develop, the application of the general competition rules should in most cases be sufficient to address 3G problem trades – and evidently they will apply if, in the absence of formal trading, mergers and acquisitions are relied on.

## 3. THE 3G ASSIGNMENT PROCESS IN THE EU

### 3.1 Licence Assignment

By the summer of 2002, all the Member States had completed their 3G licensing procedures, albeit with very differing results.[24] In total 62 licences were awarded across the EU, generating €109 billion in revenue, the vast majority (€103.9 billion) of which was raised through auctions. The bulk of this sum was raised in Germany (€50.8 billion) and the UK (€38.5 billion), the countries that were among the first to issue their licences. Market expectations for the subsequent awards were not met, resulting in reduced interest and lower licence fees. In some Member States (Belgium, France, Greece, Ireland and Luxembourg) not all available 3G licences were awarded due to a lack of interest. Consequently Member States lightened the regulatory burden: most notably in France (only two of the four available licences were initially awarded), where the government revised the cost of the licence downwards in order to attract a third 3G operator, increased the duration of the licence upward from 15 to 20 years and lightened the tax burden on the licensees.[25] Spain even saw two retroactive changes: the first to impose previously unannounced frequency reservation charges, the second to cap this regulatory burden.

The results of the 3G licensing process in the EU were mixed. On the one hand, the process saw the introduction of 14 new entrants in eight countries (that is, 48 out of 62 licences for 3G were awarded to operators already holding a digital licence). In most cases this was achieved by offering an additional licence compared to 2G (that is, $n + 1$), and sometimes encouraged by requiring existing operators to provide the new entrant with 2G roaming for an initial period. Overall the licensing process resulted in an initial 30 per cent increase in the number of 3G licences compared to 2G with at least four 3G licences in the majority of Member States.

On the other hand, the licensing process also strengthened the position of a small number of large pan-European operators such as Vodafone, Orange and T-Mobile. Moreover, the subsequent financing difficulties and the falling expectations in terms of roll-out and take-up of services in the sector may lead to further consolidation favouring the emergence of a handful of pan-European and regional groupings, and the withdrawal of marginal players – often entrants – in a number of markets (such as of Quam and Mobilcom in Germany). In most cases it is unlikely that where licences are returned to the awarding authorities, the resulting frequencies will be reissued in the short term.

### 3.2 The 3G Licence Legacy

Broadly speaking, the outcome of the 3G licensing process was positive for the market structure since it increased the number of licensees, allowed new entrants an opportunity to enter the market and gave pan-European operators a stronger position in national markets, thereby potentially strengthening competition. However, the combination of inflated market expectations, increased competition for licences and the 'one-off' character of the awards process (where all available 3G licences in any given Member State were awarded simultaneously) pushed bids up to unsustainable levels in the early awards procedures. Because it appeared that there would not be another opportunity to obtain a licence before technology had matured, incumbent 2G operators in particular, fearing that their existing business could come under threat if they did not expand into 3G, found themselves caught in a prisoner's dilemma, leading to inflated bids.[26] The high licence fees in the initial licence award procedures and the change in market circumstances made it difficult for operators to participate in further award procedures. This in turn led to distortions between Member States as fewer candidates appeared and spectrum valuation decreased over time.

The legacy of the auction process is to leave operators, in many cases, with very high start-up costs at a time when they face difficulties raising finance on the capital markets. Meanwhile, the uncertainty over the ability of operators to generate additional revenue from 3G services – combined with the existing high debt levels of some of the European operators – has led a small number of operators to postpone their planned 3G roll-out indefinitely. For example, Telefonica of Spain and Sonera of Finland have written off the cost of their German joint venture 'Quam' (€4.9 billion). Orange, controlled by France Telecom, has indicated that there will be at least a three-year delay in rolling out its network in Sweden, and is seeking to withdraw from its Mobilcom venture in Germany, which is on the verge of bankruptcy. Blu of Italy (which failed to obtain a 3G licence) returned its 2G licence to the Italian authorities and ceased trading.

It is clear that the launch of 3G is proving more difficult than initially envisaged by operators and governments alike. Only a handful of operators will have commercial 3G services available in 2003 and it is not yet certain whether the majority of operators will have their services up and running across their territories in 2004. In the end, the 3G landscape may not look very different from the existing 2G situation, with many markets concentrated in the hands of incumbents and fringe operators struggling to break even.

### 3.3 Lessons for the Future

Following the assignment of the licences, the Commission commissioned a study to draw lessons from the 3G licensing process in the EU and to consider what would be its likely current and future impact on the mobile communications sector.[27]

The study showed that the manner in which 3G services have been allocated and the licence fees and roll-out concessions extracted from the companies concerned led to an imbalance between the cost of supply and potential revenues. It also shows that this imbalance makes it unlikely that the investment costs in 3G will be recovered in a number of Member States.[28] Among the variables that can deliver cost reductions are network sharing and lower coverage roll-out obligations.

In order to develop and implement an adequate regulatory policy with regard to spectrum-based services, the study recommends that policy-makers and regulators take five guiding principles into consideration. These principles concern the market-clearing rules, the number of licences offered, and the roll-out and usage conditions.

Three principles focus on market entry:

- build spectrum assignment on the notion of 'sustainable market' – that is, based on an assessment of whether the market is ready to absorb additional spectrum, including in terms of the funding required for the costs of supply (roll-out and licence costs), keeping the costs of rolling out a new technology in line with revenue expectations from services based on this technology;
- allow for the gradual introduction of a new technology and/or capacity – for example by allowing market entry by new entrants first, followed by incumbents after a number of years, or with a model by which only one or two network operators are allowed, with MVNO access. This would also be compatible with allowing spectrum trading in order to correct spectrum assignment between operators;
- design the process for spectrum assignment so as to minimize distortions – that is, in particular avoid the 'now-or-never' notion by announcing further licensing rounds well in advance and, most importantly, by allowing spectrum trading. In addition, spectrum could be awarded simultaneously or within a limited timeframe between Member States to avoid the mutual influencing of bidding strategies in sequential award processes that can drive up licence fees.

Two principles address initial market functioning:

- align licence conditions and other regulatory levers to allow for financial stability – that is, the level of coverage obligations that determine up-front capital expenditures, the modalities of licence fee payment schedules (such as annual payments and licence fees expressed as a percentage of revenues), and the key determinants of the regulatory framework such as the regime for interconnection, roaming, and third-party access;
- support take-up of market demand, by stimulating R&D and best practice, with administrations becoming early users of technology.

In general this means that a substantial increase in public policy intervention in spectrum assignment is advocated, with a focus both on market entry and on initial market functioning in an attempt to nurture the development of new markets.

The principles advocated appear compatible with the new regulatory framework. However, in their implementation much will depend on the extent to which the Commission will be able to use the new telecommunications regulatory framework, in particular the Spectrum Decision, first to harmonize the spectrum assignment process, in terms of the assignment methods used, and second to promote co-ordination, for example in terms of the timing of licensing and licence payments.

## 4. REGULATORY AND COMPETITION IMPLICATIONS OF UMTS

There is at present considerable uncertainty surrounding the future of 3G. The initial optimistic forecasts of new entrants vying with incumbents to provide innovative services via the latest state-of-the-art technology to eager customers have all but disappeared. The reality is looking distinctly less attractive not only from a commercial viewpoint but also from a regulatory perspective. Even if the 3G market finally does take off, it may not be more dynamic than 2G given the exit of a number of new entrants and the delayed commercial roll-out of most operators. In any event the competitive outcome is likely to vary significantly from Member State to Member State, given the different market and licensing conditions existing across Europe.

However, the recent adoption of a new regulatory framework aligned on competition-law principles should ensure that regulatory measures are better targeted and that there is greater consistency across the EU. An issue that will require early consideration is the appropriateness of using regulatory measures as opposed to competition-law instruments to address market failings in 3G.

## 4.1 Competition versus regulation

### 4.1.1 Significant market power under the new framework

Under the current regulatory framework the two main forms of *ex ante* regulation (for example access or cost-orientation obligations) are limited to operators with significant market power in a particular telecommunications market. One of the main aspects of the ambition to lighten regulation is that under the new framework the relevant markets are no longer pre-defined in the directives but are defined according to competition-law principles and must have characteristics that may justify *ex ante* regulation. Moreover, the criterion of significant market power is assimilated with that of 'dominance' under the competition rules.

This means that in future, there is likely to be greater reliance on competition law to intervene where markets are not functioning properly as a result of restrictive arrangements or abuses of market power. Nevertheless regulation will still be required where a market is not sufficiently competitive and competition-law instruments are not effective.

There are three mechanisms to ensure co-ordination based on competition-policy principles. In the first place, whether markets are subject to *ex ante* intervention will depend on the applicability of three specific cumulative criteria set out in a Commission Recommendation:[29]

- whether a market is subject to high and non-transitory entry barriers, such as structural barriers where markets are characterized by substantial economies of scale, scope and density and high sunk cost, or legal and regulatory barriers;
- whether a market has characteristics such that it will tend over time towards effective competition, for instance where a sufficient number of undertakings with diverging costs structures and facing price-elastic demand are present;
- whether competition law by itself will be sufficient to remove entry barriers or restore effective competition.

Moreover, this Recommendation itself explicitly lists markets that appear to justify the imposition of *ex ante* regulatory obligations by NRAs, and is to be reviewed regularly. In such markets NRAs may impose obligations on operators of communications networks and services that are found to have 'significant market power' consistent with criteria set out in Annex II of the Framework Directive.

Second, the Commission issues Guidelines both on market definition and on the assessment of significant market power that are consistent with general competition-law analysis.[30] These Guidelines are to be taken into account by the NRAs in particular when they are analysing markets that are not covered by the Recommendation.

Finally, the Framework Directive provides for consultation by the NRAs on draft measures. Such consultation of the Commission and the other NRAs is mandatory in cases where an NRA imposes, maintains, amends, or withdraws obligations on undertakings with significant market power that would affect trade between Member States. In cases where an NRA notifies a decision defining markets that differ from those defined in the Recommendation, or a decision whether or not to designate an undertaking as having significant market power, the Commission may require the NRA concerned to withdraw the measure if it has serious doubts as to its compatibility with EU law. In all cases, the general competition rules will remain of parallel application.[31]

### 4.1.2 Market definition aspects of 3G

There is still no formal position as to the appropriate market definition for 3G. As regards the existing case law dealing with mobile telephony, most cases have considered that analogue and both GSM 900 and DCS 1800 (Digital Cellular System 1800) form part of the same market. However, the European Commission has also tested narrower market definitions to ensure that no dominant positions arose under any market definition.[32]

Initially the lack of sufficient reliable 3G coverage is likely to result in extensive reliance on both 2G and 3G networks by subscribers. In particular, as regards the provision of voice telephony services, there is likely to be overlap between 2G and 3G services. In addition, the extent to which there will be fixed/mobile network substitutability for the supply of higher bandwidth infrastructure and services remains open to discussion. However, because services and content available over 3G networks are expected to be considerably improved in relation to 2G both as regards data speeds and the range of available services, any substitutability between 2G and 3G is likely to be one-way. Hence, once 3G networks have been rolled out, the significant differences in the technology and of its uses are likely to make 3G into a separate market.

Applying competition-law analysis adopted in recent Commission thinking as reflected in the Commission Recommendation on relevant product and services markets may mean that 3G mobile markets will be broken further down *inter alia* into wholesale markets for call origination, call termination and international roaming markets, as well as into different retail markets (for example for voice and data, for residential users versus business and corporate customers, and so on).[33] It also appears possible that separate 3G wholesale markets may develop for data capacity and voice capacity.

So far most markets remain national in scope, with the exception of the emerging market for the provision of seamless pan-European mobile telecommunications services to internationally mobile customers that the Commission first identified in the Vodafone/Mannesmann case.[34] Geographical markets are likely to remain national due to the separate national regimes for the award and regulation of UMTS licences, and given the fact that pricing and services offers of mobile telephony are so far predominantly national.

However, Article 15(4) of the Framework Directive opens the possibility for the Commission to identify, by Decision subject to co-ordination with the Communications Committee, pan-European markets. These are markets that comprise the territory of more than one Member State. In the event that such markets are found not to be effectively competitive, and should competition policy not suffice, the NRAs concerned shall act jointly in imposing the requisite regulatory obligations.

### 4.1.3 Implications for 3G

It is too early to generalize on how competition problems arising in 3G will be dealt with, but it will have to be consistent with the third criterion for finding SMP set out in the above-mentioned Commission Recommendation on relevant product and services markets, that is, dependent on whether application of general competition law will suffice to remove entry barriers or restore effective competition. Hence it is likely that the Commission's current experience of applying the competition rules on the 2G markets for call termination and wholesale international roaming will help determine whether *ex ante* or *ex post* intervention is appropriate.

In addition, as is noted explicitly in recital 27 of the Framework Directive, and recital 14 of the Recommendation on market definition, newly emerging markets that are characterized by ongoing technological progress, but where the market leader has a substantial market share, should not be subject to 'inappropriate' *ex ante* regulation.[35]

This aspect is likely to be taken into account especially under the application of the second criterion for finding SMP, that is the question whether or not the market has dynamic characteristics such that it will naturally tend toward effective competition. Consequently intervention in the market – at least in the early stages – is likely to be limited to cases of clear and lasting market failure. Finally, the application of regulatory instruments – and by implication the choice of instruments used – has to be proportional to the gravity of the problem identified.[36]

## 4.2 Competition Issues Arising in 3G

It is relatively early to draw up a list of competition issues in 3G, as the markets involved are, in a hopeful view, at an emerging stage. Market structure and market characteristics will determine the competitive nature of the market but it is clear that the degree to which market shares in 2G markets will translate into 3G market share, and whether the high degree of vertical integration that characterizes the industry will be maintained, or will be replaced by more diffuse relationships between network operators, service providers, MVNOs and (premium) content providers will be key issues. However, it is possible to list likely topics of interest, including:

- Collaboration between network operators to share network infrastructure, including reliance on roaming, as well as closer collaboration within and across sectors to develop new services and content.
- Consolidation: costly investments combined with poor profit forecasts are widely believed to lead to consolidation in the industry around a few core pan-European and regional groups, and the exit of weaker local players (like Blu in Italy, and Quam and Mobilcom in Germany).
- State aid problems where national, regional or local governments step in to salvage floundering 3G operations.
- Network versus service competition, notably the degree to which collaboration at network level may be compensated by faster availability of services or increased services-level competition.
- The role of MVNOs, notably whether MVNOs have rights to access the network of

dominant operators, and under which conditions they have a right to obtain terminating interconnection with network operators.

Of these, network sharing is so far the issue that is most advanced in terms of practical relevance, but MVNOs are also likely to play an increasingly significant role in the 3G market.

### 4.2.1 3G network sharing

Network infrastructure sharing ('network sharing') on an *ad hoc* basis was common in 2G, but the need in 3G for greater network investment and the financial difficulties in the industry as well as environmental concerns surrounding mobile infrastructure have made it into an attractive cost-saving solution for operators. A number of network-sharing agreements have been entered into across the EU, for example, the mobile arms of mm$O_2$ ($O_2$) and Deutsche Telekom (T-Mobile) have agreed to share site infrastructure and engage in national roaming in both Germany and the UK. $O_2$ estimated that these forms of co-operation can reduce the build-out cost of a network by 20 to 40 per cent and in their case they expected to save at least £1.2 billion in capital expenditure over ten years from their various network-sharing arrangements.[37] Such estimates are supported by independent analysts.[38]

Network sharing is favoured by policy-makers both as a way of ensuring more rapid provision of 3G services and on environmental grounds, and on this basis the European Commission has repeatedly come out in favour of network-sharing arrangements subject to the caveat that competition rules are respected.[39] The above-mentioned network-sharing agreements between $O_2$ and T-Mobile in Germany and the UK were notified to the Commission for consideration. The Commission published its preliminary position which indicated that it was in favour of such forms of co-operation.[40] The final decisions provide guidance on the types of network sharing which are permitted under EC competition rules.[41] In addition, a number of national regulatory authorities have also produced guidance on the conditions under which network sharing is acceptable as a matter of national licence conditions, sector-specific rules, and competition law.[42]

*Types of network sharing*[43]

3G network sharing can take place at a number of different levels and involve varying degrees of co-operation. The degree of independence retained by an operator depends on which network elements are being shared and their remaining ability to install separate elements (planning freedom). Ranked by the increasing degree to which the network is shared, it is possible to distinguish between shared use of:

- sites, which ranges from sharing individual mast sites up to grid sharing (requiring a uniform layout of networks), and may include site support infrastructure, such as site support cabinets (SSC);
- base stations (Nodes B) and antenna;
- radio network controllers (RNCs);
- core networks, including mobile switching centres (MSCs) and various databases;
- frequencies.

National roaming, on the other hand, refers to a situation where the operators concerned do not share any network elements as such but simply use each other's network to provide services to their own customers.

In terms of the network, a key distinction can be drawn between the radio access network (RAN) and the core or backbone network (see Figure 11.1).

- *RAN*: this includes (i) mast/antenna sites, site support cabinets (SSC) and power supply, as well as antenna, combiners and transmission links; (ii) Nodes B, that is, the base stations that receive and send data across frequencies and control a particular network cell, and (iii) the radio network controllers (RNCs) that each control a number of such Nodes B and that are linked to the core network.
- *Core network*: this is the intelligent part of the network that consists of mobile switching centres (MSCs), various support nodes, services platforms, client home location registers and operation and maintenance centres. It is linked to the fixed ISDN (integrated services digital network) and Internet networks.

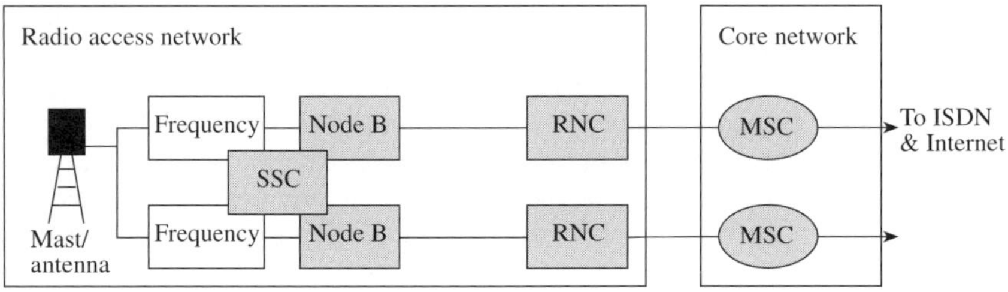


*Source*: European Commission.[40, 41]

*Figure 11.1 UMTS network*

*Competition issues*

Network-sharing agreements can raise significant horizontal as well as vertical competition concerns because they involve close co-operation between competitors in markets on which rivalry between firms is still not in all cases sufficiently effective to curtail anticompetitive outcomes. The principal restrictions on competition raised by such agreements are the following:

- reduction in network competition;
- risks of foreclosure on access to site sharing;
- resale restrictions on MVNO access;
- spillover effects.

REDUCTION IN NETWORK COMPETITION

It is expected that the costs saved as a result of network sharing constitute up to 30 per cent of the total network investment costs, which in turn accounts for a significant proportion of the costs of providing 3G services at retail level. If accompanied by the roll-out of similar networks as well as extensive mutual reliance on roaming, this could result in parties to a network-sharing agreement having similar cost structures, which may affect their ability to compete against each other in downstream markets (for example competition based on network capacity and quality as well as price). If the parties were to engage in RAN sharing, this would have an even greater impact as it would limit network-based competition even further.

However, the existence of roll-out obligations limits the effects on competition since parties to network-sharing agreements must deploy their own networks if they are to comply with national 3G licence conditions.

Whether the residual competition concerns are serious enough to warrant a prohibition, or may be exempted will depend largely on (i) the depth of co-operation between the parties and in particular the duration and extent of any roaming, (ii) the ability of parties to build out their own differentiated networks should market conditions develop, (iii) the scope of the roll-out obligations and (iv) the intensity of the remaining competition on the market.

In the case of the notified agreements in Germany and the UK, the conditions for an exemption on this count appear to be met. Although the agreements involve wide-ranging co-operation, the parties are required under licence conditions to cover a significant proportion of the population by 2005 in Germany and 2007 in the UK – therefore any roaming is limited in time and coverage. The parties also retained independent control over their networks and over network planning, as well as over service development, and their ability to compete with each other at the level of 3G retail services is not completely undermined. The operators in the UK and German 3G markets are likely to see significant competition from other operators and the agreements should also increase the parties' ability to compete with third parties.

#### Foreclosure

The existence of exclusivity provisions in network-sharing agreements limits the opportunity for other operators to share sites and potentially forecloses network competition. However, foreclosure may not be a problem provided there is widespread availability of sites. A problem could arise in relation to specific sites where planning restrictions limit the ability to build out or develop suitable sites. This is primarily a planning problem but regulatory intervention is available, albeit on a limited basis under both the existing and new EU Regulatory Framework.

#### MVNO access

Restrictions on re-selling to MVNOs raise particular competition concerns and are unlikely to be compatible with competition rules except in very limited circumstances. The German Agreement bars one party from re-selling the roaming capacity of the other operator to MVNOs unless the latter are acting as purely data MVNOs or voice SPs (service providers). This restriction is necessary to allow T-Mobile to keep control over who has access to its network, and to protect its investments, and it is exemptable as it promotes $O_2$'s position as a nationwide competitor on the German market leading to consumer benefits by enhanced competition.

SPILLOVER EFFECTS

Network-sharing agreements typically require close co-operation between competitors in oligopolistic markets where the risk of spillover into collusive behaviour is a concern. It is therefore necessary that any co-operation is narrowly delineated and that the exchange of commercially sensitive information is limited and tightly controlled: only information directly concerned with the roll-out plan and the related network infrastructure should be exchanged.

### 4.2.2 Mobile virtual network operators (MVNOs)

MVNOs remain an unknown quantity in the mobile market. Because mobile network operators are vertically integrated, comprising both network operation and retail services, they tend to view MVNOs with hostility as a threat to their own downstream operations. Regulators generally worry that intervention in favour of MVNOs who have not incurred significant investments in network infrastructure could upset legal certainty and would alter the structure of the market, with long-term effects for such infrastructure investment.

*The MVNO concept*

Although there is no common MVNO concept, these operators share the same basic characteristics. While they have a limited core (intelligent) network, they do not own frequency rights and consequently rely on the radio access network of a mobile network operator for a large part of the provision of service to their customers. MVNOs differentiate themselves from service providers (SPs) found in many EU Member States by the fact that unlike the latter they control basic elements of the core network, and may have their own number ranges and subscriber identity (issuing their own SIM cards).[44] This enables them to supply customers with their own tailored service offers (often a bundle of fixed and mobile services) by routing calls on the best possible terms through interconnection agreements with different operators.

The attractiveness of the MVNO model resides in the fact that it is less costly to operate since there is no ownership and operation of a radio access network (that is, the most expensive part of the network), while at the same time control over network intelligence enables it to be responsible for routing as well as billing. Because an MVNO controls the access to its own end-users, it is generally considered to be an operator of a public telecommunications network. This means that an MVNO is typically able to enter into interconnection agreements with other network operators on a 'cost-plus' basis, and to collect termination revenue, rather than merely obtaining wholesale airtime at less favourable 'retail-minus' prices like SPs.[45] It is therefore able to offer its own tailored retail packages more cheaply than SPs, and to offer network operators genuine price competition.

*The scope for MVNOs in 2G and 3G*

The development of MVNOs could clearly compensate at least to some degree for the limited opportunities for market entry due to spectrum scarcity, and would allow mobile network operators to make a more efficient use of their spectrum (as they could fill any excess capacity with MVNO traffic). For fear of undercutting their own retail operations, however, mobile operators have not been keen to grant MVNOs access to their network except where required by law on the same basis as SPs.[46] Consequently there have so far only been very few examples of commercial agreements between MVNOs and operators. Tele2 is the only high-profile operator active on a European-wide scale that has entered into MVNO agreements in a number of countries (including recently for 3G in Norway).

Regulatory involvement has so far also been limited, with significant developments only in Scandinavia. In the UK, Oftel published a position paper in 1999 in which it concluded that regulatory action to require the provision of services to MVNOs was not justified, in terms of their potential positive impact on competition and benefits for consumers.[47] Following a request by Tele2 France to be granted access to Orange's mobile network, the French regulator also declined to allow access to MVNOs but on the basis of a restrictive interpretation of the term 'interconnection' in the Interconnection Directive (97/33/EC).[48]

In 3G the scope for commercial agreements between mobile network operators and MVNOs appears to be greater because an effective use of the new capacity to provide data services at high bandwidth will require the creation of ventures that can successfully target particular groups of users. In this context, providing network access to MVNOs may be an attractive proposition for operators seeking to optimize the use of their networks. Moreover, given the scope for differentiation in data services, the risk of MVNOs undercutting network operators on price may be less serious than in 2G. However, where existing voice telephony markets are concerned, mobile network operators are likely to be less willing to grant MVNOs access for fear of price competition and cannibalizing their customer base.

Should effective competition fail to develop in 3G, regulators may consider a wider range of remedies, including mandating MVNOs access to mobile operators' networks. As a highly interventionist measure it would require careful consideration.

### *Mandating MVNO access*

Under EU competition law mandated third-party access is generally restricted to so-called 'essential facilities', and requires a high standard of proof – essentially that it would be uneconomical to reproduce the network concerned even if the volume of services provided over the network were equivalent to that provided by its owner.[49] Moreover, it would be difficult to imagine an essential facility where several comparable mobile infrastructures existed in parallel. On the other hand, licensing requirements severely restrict the possibilities for entry in mobile markets.

Under the new sector-specific regulatory framework, the significant market position threshold for intervention has been raised to dominance and the dynamics of the market as well as the scope for using competition instruments must be taken into consideration. However, the new framework specifies that mandatory access to specific facilities and/or associated services can be imposed 'inter alia in situations where the national regulatory authority considers that denial of access would hinder the emergence of a sustainable competitive market at the retail level, or would not be in the end-user's interest'.[50] Hence if there is a market failure in 3G, mandating MVNO access, on appropriate terms, may be possible and an effective remedy in particular cases by generating dynamic competitive pressure on the market.

## 5. CONCLUSION

It is remarkable that even when 3G licences were issued throughout the EU consistent with a common timetable there was little to no co-ordination on the methods used to assign the relevant frequencies, and issues that were subsequently found to be key such as spectrum trading and network sharing had not been considered a priority, or apparently at all.

Since then, the EU has provided the initial building blocks of a common approach, by means of the Spectrum Policy Decision, and the possibility of a harmonized licence awards procedure in the new Authorisation Directive. It has also become clear that in the EU, as worldwide, public policy on the assignment of frequencies is developing toward promoting the emergence of secondary markets based on spectrum trading. Whether this will coincide with the trend towards harmonization in the area of spectrum policy and lead to the creation of EU-wide secondary markets in frequency rights is not yet clear. It appears likely, however, that on this issue, as in telecommunications more generally, liberalization and harmonization will continue to go hand in hand.

Regulatory and competition problems in 3G are likely to be equally prevalent as in 2G and probably more difficult given the greater complexity of the technology. As we are at the time of writing at the roll-out stage, the most important competition issues affecting 3G mobile communications relate to network sharing. The preliminary analysis of planned network-sharing arrangements suggests that, provided competition between core networks is maintained, the competition problems involved are not insurmountable.

At a later stage, access issues and industry consolidation are expected to become more relevant. Question marks surround the role MVNOs will play in 3G. MVNOs can compensate for the entry barriers based on frequency scarcity by stimulating competition based on new types of service provision, and can ensure efficient use of frequency resources. Whether MVNO access will be mandated depends to a large extent on the development of effective competition in 3G markets and therefore on whether they can obtain entry on commercial terms or whether network operators maintain a united front to protect the *status quo* of vertical integration and oligopolistic returns downstream.

## NOTES

* The views expressed here are those of the authors alone and as such do not reflect or predict European Commission policy.

1. In ITU (International Telecommunications Union) terminology, (1) *allocation* is a term of art for the process carried out at global and regional level under ITU auspices that reserves frequency bands for certain services; next, (2) *allotment* of radio frequency channels to particular countries is co-ordinated between countries by international agreement based on a plan established by intergovernmental conference; and (3) *assignment* is the final stage of the process, by which a national government awards an entity the right to use a radio frequency (that is, licensing). Cf. C. Koenig, A. Bartosch and J.-D. Braun (eds), *EC Competition and Telecommunications Law* (Kluwer, 2002) ch. 11. Accordingly 'allocation' is also used for reserving frequencies for a particular purpose at European or national level. This usage clashes with the economic notion of 'allocation' of scarce resources. We have sought to use these terms consistent with their technical meaning, although we occasionally use 'allocation' in the economic sense of distribution.
2. Universal mobile telecommunications system (UMTS) is one of the third generation (3G) mobile systems being developed within the ITU's IMT-2000 framework. It is being developed by European Telecommunications Standards Institute (ETSI), and will be based on Wideband Code Division Multiple Access (WCDMA) technology. For further information, see www.umts-forum.org, www.umtsworld.com and www.imt-2000.org.
3. The other recognized scarce 'natural' resources in the telecommunications sector are numbering, Internet

addresses, and domain names. Unlike spectrum, however, their capacity can be increased, in principle infinitely, by common agreement. This does not of course account for the attractiveness or uniqueness of particular numbers, names or addresses. Radio spectrum can be extended by bringing new bandwidth into use, and by improving the capacity of frequencies already in use by technical means such as compression and multiplexing.

4. The ITU, headquartered in Geneva, Switzerland, is an international organization within the United Nations system where governments and the private sector co-ordinate global telecoms networks and services. Radiocommunications matters are dealt with by the permanent Radiocommunications Sector, supervised by a World Radiocommunications Conference. Within the ITU framework regional matters for Europe are co-ordinated by the CEPT. CEPT was established in 1959 by 19 countries and currently has 44 members. See www.cept.org. Both radiocommunications and non-radiocommunications tasks of the CEPT are co-ordinated by the Electronic Communications Committee (ECC) and its various working groups and project teams, including, for UMTS, 'Project Team IMT-2000 and Systems Beyond'.
5. Council Directive of 25 June 1987 on the frequency bands to be reserved for the co-ordinated introduction of public pan-European cellular digital land-based mobile communications in the Community (87/372/EEC, OJ L 196, 17.7.1987, p. 85); Council Recommendation on the co-ordinated introduction of public pan-European cellular digital land-based mobile communications in the Community (87/371/EEC, OJ L 196, 17.7.87, p. 81) (GSM). A similar approach to EU-wide frequency allocation to a specific purpose by means of Council Directive and co-ordinated introduction of the relevant service based on a Council Recommendation was adopted to the European Radio Messaging System (ERMES) in 1990 and the system for Digital European Cordless Telecommunications (DECT) in 1991.
6. Services Directive, Commission Directive 90/388/EC (OJ L 192/10, 24.7.1990), as amended by Satellite Directive, Commission Directive 94/46/EC (OJ L 268/15; 19.10.1994); Cable Directive, Commission Directive 95/51/EG (OJ L 256/49, 26.10.1995); Mobile Communications Directive, Commission Directive 62/2/EC (OJ L 20/59, 26.1.1996); Full Competition Directive, Commission Directive 96/19/EC (OJ L 74/13, 22.3.1996).
7. Commission Decision 95/489/EC concerning the conditions imposed on the second operator of GSM radiotelephony services in Italy (OJ L 280/49, 23.11.1995); Commission Decision 97/181/EC concerning the conditions imposed on the second operator of GSM radiotelephony services in Spain (OJ L 133/19, 24.5.1997).
8. European Parliament and Council Directive 97/13/EC (OJ L 117/15, 7.5.1997). This directive sets out the procedures associated with the granting of authorizations for the purpose of providing telecommunications services and the conditions attached to such authorizations.
9. These non-economic reasons of public interest include network security and integrity, the inter-operability of services, data protection, environmental protection and town and country planning, the effective use of frequency spectrum and the avoidance of radiomagnetic interference.
10. Decision No. 710/97/EC (24 March 1997) of the European Parliament and the Council on a co-ordinated authorization approach in the field of satellite personal-communication services (S-PCS) in the Community (OJ L 105/4, 23.4.1997).
11. Decision No. 128/1999/EC of the European Parliament and the Council of 14 December 1998 on the co-ordinated introduction of a third generation mobile and wireless communications system (UMTS) in the Community (OJ L 17/1, 22.1.1999).
12. The Committee set up by Article 14 of Directive 97/13/EC (Licensing Directive). This same approach was applied in the S-PCS Decision.
13. European Parliament and Council Directive 2002/20/EC of 7 March 2002 on the authorization of electronic communications networks and services (Authorisation Directive) (OJ L 108/21, 24.4.2002).
14. Article 5, Authorisation Directive and Article 9, 'management of radio frequencies for electronic communications services', European Parliament and Council Directive 2002/21/EC of 7 March 2002 on a common regulatory framework for electronic communications networks and services (Framework Directive) (OJ L 108/33, 24.4.2002).
15. European Parliament and Council Decision 676/2002/EC of 7 March 2002 on a regulatory framework for radio spectrum policy in the European Community (Radio Spectrum Decision) (OJ L 108/1, 24.4.2002).
16. The original Commission proposal had been more radical, aiming at establishing EU common positions outside the CEPT framework. Proposal for a Decision of the European Parliament and of the Council on a Regulatory Framework for Radio Spectrum Policy in the European Community, COM(2000) 407 final.
17. Commission Decision of 26 July 2002 establishing a Radio Spectrum Policy Group (OJ L 198/49, 27.7.2002). 'Commission creates Radio Spectrum Policy Group and European Regulators Group', Brussels 29 July 2002, IP/02/1171.
18. Radiocommunications Agency, 'Implementing Spectrum Trading', A Consultation Document, July 2002. See http://www.radio.gov.uk/topics/spectrum-strat/consult/implementingspectrumtrading.pdf
19. Notice of Proposed Rulemaking and Policy Statement on 'Principles for Promoting the Efficient Use of

Spectrum by Encouraging the Development of Secondary Markets', FCC, Adopted 9 November. See http://www.fcc.gov.

20. 'Review of Radio Spectrum Management', published 6 March 2002, see www.spectrumreview.radio.gov.uk
21. Ibid. See paras 70–79.
22. 'Implementing Spectrum Trading', Executive Summary, para. viii.
23. Communications Bill as introduced in the House of Commons on 19 November 2002. http://www.parliament.the-stationery-office.co.uk/pa/cm200203/cmbills/006/2003006.htm
24. 'Eighth report from the Commission on the Implementation of the telecommunications regulatory package', COM(2002) 695 final (3.12.2002), Annex II (Commission Services working document).
25. France reduced the fees for a 3G licence from €4.95bn to €619m, with an annual charge of 1 per cent of 3G revenues.
26. This prisoner's dilemma would lead incumbents to include the value of their existing operations (deemed to be under threat from the new technology) in their valuation of new spectrum, and lead to their bidding of licence fees out of all proportion to the business case for the new spectrum. See 'Comparative Assessment of the Licensing Regimes for 3G Mobile Communications in the European Union and their Impact on the Mobile Communications Sector'. European Commission, Directorate-General Information Society, Final Report – 25 June 2002. At: http://europa.eu.int/information_society/topics/telecoms/radiospec/mobile/studies/index_en.htm
27. Ibid.
28. Depending on the assumptions used (whether 3G income would eventually amount to 30 per cent or 100 per cent of current 2G income) this will affect between 12 and two Member States.
29. Commission Recommendation on relevant product and services markets within the electronic communications sector susceptible to *ex ante* regulation in accordance with Directive 2002/21/EC of the European Parliament and of the Council on a common regulatory framework for electronic communications networks and services, 11 February 2003, OJ L 114/45, 8.5.2003.
30. Commission Guidelines on market analysis and the assessment of significant market power under the community regulatory framework for electronic communications networks and services, OJ C 165/6, 11.7.2002.
31. In the past, the Commission would give precedence to sectoral regulation in the event that this could resolve the issues concerned within a reasonable period of time of generally six months. Commission Notice on the application of the competition rules to access agreements in the telecommunications sector: framework, relevant markets and principles (Access Notice), OJ 1998, C265/2, para. 30. Whether this is still the case is an open question.
32. Cf. Case No. IV/M.1430, *Vodafone/Airtouch*, Commission Decision of 21 May 1999; Case No. COMP/JV.17, *Mannesmann/Bell Atlantic/Omnitel*, Commission Decision of 21 May 1999; Case No. COMP/M.1439, *Telia/Telenor*, Commission Decision of 13 November 1999; Case No. COMP/M.1769, *Mannesmann/Orange*, Commission Decision of 20 December 1999; Case No. COMP/M.1795, *Vodafone Airtouch/Mannesmann*, Commission Decision of 12 April 2000; Case No. COMP/M.2053, *Telenor/BellSouth/Sonofon*, Commission Decision of 4 August 2000; Case No. COMP/M.2016, *France Telecom/Orange*, Commission Decision of 11 August 2000. For 3G services in particular see COMP/M.2130, *Belgacom/Tele Danmark/T-Mobile International/Ben Nederland Holding*, Commission Decision of 25 September 2000.
33. On the wholesale call termination and roaming markets, as well as on the retail distinction between business and residential customers see COMP/M.2803, *Telia/Sonera*, Commission Decision of 10 July 2002.
34. Cf. Case No. COMP/M.1795, *Vodafone Airtouch/Mannesmann*, Commission Decision of 12 April 2000; Case No. COMP/M.2016, *France Telecom/Orange*, Commission Decision of 11 August 2000.
35. Cf. also Guidelines on market analysis and the assessment of significant market power, para. 32.
36. Cf. Article 8, Directive 2002/21/EC (Framework Directive) and Article 8, Directive 2002/19/EC of the European Parliament and of the Council of 7 March 2002 on access to, an interconnection of, electronic communications networks and associated facilities (Access Directive), (OJ L 108/7, 24.2.2002).
37. See mmO2 press releases (i) 'Operator partnerships' (ii) 'mmO2 details 3G strategy as Isle of Man Network goes live' (05/12/01); available at www.mmo2.com
38. Cf. D. Morgan and D. Wilkins, 'Evaluating the Business Case for 3G Network Sharing', in R. Beale (ed.), *Analysys Telecommunication Consulting*, 2001.
39. 'The introduction of third generation mobile communications in the European Union: state of play and the way forward', COM (2001) 141 final (20.03.2001); 'Towards the Full Roll-Out of Third Generation Mobile Communication', COM (2002) 301 final (11.6.2002).
40. Commission's notices pursuant to Article 19(3) of Council Regulation No. 17, in Case COMP/C1/N.38.369, *Rahmenvertrag T-Mobile Deutschland/VIAG Interkom* (Notification) (OJ C 189/2, 9.8.2002); and in Case COMP/C1/N.38.370, *BT Cellnet & BT3G/One2One Personal Communications* (United Kingdom Agreement) (OJ C 214/17, 10.9.2002).
41. Commission Decision of 30 April 2003 in Case No. 38.370, O2 UK Limited/T-Mobile Limited (UK

Network Sharing Agreement OJ L 200/59, 7.8.2003). (Commission Decision of 16 July 2003 Case No. 38.369, T-Mobile Deutschland/O2 Germany): Network Sharing Rahmenvertrag, not yet published.

42. A number of national regulatory authorities (NRAs) in the EU issued guidance on the conditions on which infrastructure sharing would be consistent with national licensing and regulatory requirements. In the UK Oftel published a position in May 2001, available at http://www.oftel.gov.uk/publications/mobile/infrashare0501.htm. In Germany, RegTP issued its Interpretation of the UMTS Award Conditions in light of more recent technological advance, RegTP (6 June 2001), available at www.regtp.de. The Dutch and French NRAs have published similar documents, available at http://www.opta.nl/download/concept_notitie_nma_vw_opta_umts_netwerken_190701.pdf and at http://www.art-telecom.fr/dossiers/umts/partage-infras.htm. Moreover, information on national rules applicable to network sharing is available at http://europa/eu.int/information_society/topics/telecoms/radiospec/mobile/indes_en.htm.
43. Source: Commission Article 19(3), Notices on UK and German network sharing agreements; see above.
44. A pure MVNO would typically have a home location register (HLR), mobile switching centre (MSC), authentication centre (AUC), equipment identity register (EIR) and signalling.
45. Mobile service providers are defined as entities authorized to offer mobile services under their own brand name (dealing with marketing, billing, and so on), using a third party's mobile network.
46. For a mobile network operator, giving MVNO access may be profitable provided that the wholesale revenues they provide outstrip any related revenue losses at retail level, and consequently largely to the extent that the MVNO is taking (potential and actual) customers away for other network operators in greater proportion than from its host. This is likely to be an attractive proposition mainly for mobile operators with smaller market shares and significant excess capacity on their networks. However, incumbent mobile network operators with significant market share that are using their networks close to capacity are more likely to be concerned with the depressing effect that market entry by MVNOs can have on their retail price levels (which factors into the retail losses that a wholesale offer to MVNOs would have offset) than with any wholesale benefits.
47. 'Statement on Mobile Virtual Network Operators', October 1999, see Oftel website: www.oftel.gov.uk.
48. Decision 02-1192 of 17 December 2002: *Tele2 France SA/Orange France*, http://www.art-telecom.fr/textes/avis/index-02-1192.htm.
49. Case C-7/97, *Oscar Bronner* v. *Mediaprint Zeitungs- und Zeitschriftenverlag* [1998] ECR I-7791.
50. Article 12 of Directive 2002/19/EC (Access Directive).

# Roundtable Discussion

# 12. Introduction to the Roundtable on the economics of antitrust and regulation in the telecoms sector

**Philip Lowe**

The telecommunications sector has gone through a fast process of change in recent years. Both the liberalization process and technological progress have driven the sector through roads which were unimaginable a decade ago. In Europe, we are now in the process of implementing a new regulatory framework. This new regulatory framework combines 'traditional' sector-specific regulation with elements of antitrust methodology. I think this is an excellent moment to reflect on the role that economics should play in antitrust and regulatory policy to ensure that the new regulatory tools are applied consistently with competition policy instruments. On several issues I would like to listen to the opinion of the panellists on this roundtable:

The first issue that comes to my mind is the future role of antitrust in the telecommunications sector. What should be the respective roles of sector-specific regulation and competition law? The new European regulatory framework for electronic communications networks and services introduces competition tools for the determination and analysis of markets to be regulated and for the designation of those agents directly affected by regulatory obligations. For example, the regulatory concept of 'significant market power', which designates those operators that will be subject to regulatory obligations (mainly related to price regulation and to the granting of access to their network) is now made equivalent to the European competition-law concept of dominance.[1] An important issue to analyse is to what extent competition and regulatory policy converge; that is, under which circumstances are competition and regulatory instruments substitutes to deal with a problem. It is clear to me that some of the objectives of regulatory policy (such as universal service obligations and the facilitating of widespread consumption) are not common to those of competition policy and therefore that some degree of regulation may always be needed. At the same time, many provisions of the new framework aim to promote competition by granting access and interconnection rights under regulated conditions to the networks of dominant undertakings and duplicate the prohibition under competition law of refusal to deal or excessive pricing. In this regard, one has also to take into consideration that EU competition law does not prohibit the acquisition of a dominant position – except in the case of mergers – but the abuse of such dominant position. In this sense EU competition law and sector-specific regulation are much nearer than US competition law and regulatory intervention.

In such circumstances competition authorities will reflect on the appropriateness of applying in parallel competition-law procedures to issues which can be solved under sector-specific regulation. Do competition authorities have to tackle anticompetitive behaviour which national regulatory authorities monitor? Are the latter in most instances not better equipped to deal with complex telecommunications cases? At the same time, under the new framework there will only be an evolution from sector-specific regulation to the 'normal' application of competition-law remedies, when the latter are sufficient to deal with the problems concerned. If competition authorities refrain from using their powers in areas that can be dealt with under sector-specific rules, there will hardly be competition-law case-law in the sector and it will further be argued that competition-law remedies are not suited to address competition problems in the sector!

The complexity of cases in the telecommunications sector raises other questions and, in particular, whether specific economic reasoning is required in such sectors. Market assessments obviously require a very sophisticated analysis in the case of a technically complex industry such as the electronic communications industry. A sound competitive analysis of electronic communications markets is not the same as, for example, the assessment of the soft-drinks or packaging industry. The starting-point of market assessments in such industries is obviously the same as in other industries, but their application may not be as straightforward.

The electronic communications industry has complex market structures, such as the need for co-operation between players to ensure end-to-end inter-operability and compatibility, network effects, the key role of innovation, the presence of essential facilities and so on). Given these characteristics, we are obviously confronted with a number of conceptual challenges absent from more traditional sectors.

Three examples of economic questions to solve when addressing cases in the electronic communications sector are: (i) how broad should markets be defined to assess dominance? Should one consider networks and platforms as a separate market or take the view that, ultimately, there will be some substitution between these? (ii) How should access to networks be priced? Is the refusal of incumbents to share their economies of scale with entrants abusive? (iii) How far is co-operation economically justified? Above which threshold will co-operation foster concerted practices and harm consumer interest?

The question of how broad a market definition should be used under the new telecoms regulatory framework was central, in particular, for the drafting of the Recommendation of markets to be regulated *ex ante*, adopted by the Commission on 11 February 2003. This Recommendation defines the markets which will be subject to *ex ante* regulation as long as one or more operators are dominant on these markets. Under the relevant directive, the market definitions had to be listed according to competition-law criteria. Obviously, markets defined for the purpose of regulation do not necessarily coincide with those defined in specific cases under competition law. The starting-point for defining the markets of the Recommendation was indeed not the existence of an agreement or concerted practice, nor a concentration, nor an alleged abuse of dominance, but is based on an overall forward-looking assessment of the structure and the functioning of the market under examination for the purpose of its future regulation. Nevertheless, given that the methodology is the same, the outcome should generally be the same. One of the characteristics of competition-law market analysis is that it always starts from the smallest possible market definition and then examines whether there is possible substitution with other services. The 'smallest market principle' analysis ends when no further substitutable services can be identified, as Professor Katz explains in Chapter 13. This is why the Commission defined the markets for mobile call termination per individual network. However, as regards wholesale broadband access, incumbent operators stressed the possibility of platform competition where cable TV networks were available. Under a competition-law analysis it is none the less not sufficient that substitution is likely. It is required that substitution is probable in case of a specific increase in price. In this regard, contrary to the USA, cable networks – or at least upgraded cable networks – are available only in a small minority of the EU Member States (Belgium, the Netherlands and the UK), and are often not yet technically upgraded so as to allow them to provide a wholesale broadband access equivalent to 'bit-stream' services on the traditional telephony networks. For this reason, we were very reluctant to define the market broader than bit-stream access to the telephony networks. This example also illustrates that market definitions are not a theoretical issue. A broader market definition reduces the likelihood of the finding of dominance on the relevant market. As a consequence, unjustified broad market definitions prevent tackling possible abuses of dominant position and – since the equivalent access obligation under sector-specific regulation will only be imposed on dominant players – exempt the operators concerned from having to provide access at regulated conditions. For this reason, too broad an application of the economic concepts of 'bundled' and 'complementary' services, as set out by Professor Jordi Gual (see Chapter 4), could have far-reaching consequences for both antitrust and sector-specific enforcement.

This leads us to the second economic issue: how should access to telecoms networks be priced? This question is at the core of the slow take-up of local loop unbundling in Europe. Two years after the entry into force of the local loop unbundling Regulation in the EU, the figures are rather disappointing: by 1 October 2002, there were just over 1 million unbundled lines in the EU (out of a total of nearly 187 million subscriber lines). Addressing this problem is crucial, since in the EU unbundling is the only means to have competition in the local access networks, given that, as mentioned, cable networks are not rolled out or upgraded in most of the Member States in a way that allows for genuine competition with the incumbent telecommunications operators.

A probable explanation of the slow take-up of unbundling is that in certain Member States the tariffs applied do not allow for a real business case for entrants. This is reflected in the considerable variations in tariffs in the EU. When the average cost for a fully unbundled line is calculated (by amortizing these charges over a year[2]), the EU average is €22.6 per month, with prices varying from €12 in Denmark to €32.7 in Finland.

Incumbents plead for high prices, referring to the conflict of static versus dynamic efficiency. They argue that the promotion of short-term service competition would 'kill' long-term infrastructure-based competition. But, on the other hand, in the absence of service competition there is a risk that new entrants will have difficulties in putting a foot into the market and reaching the necessary economies of scale to make investments profitable. It is important to set the appropriate conditions to ensure that investment incentives remain while guaranteeing competition in services. Economists have a key role to play in identifying the pricing methodologies which will lead to prices that do not discourage market entry while compensating the incumbents for their genuine costs.

Finally, I will address the issue of competition and co-operation in the telecoms sector. The third economic issue which cannot be avoided in the electronic communications sector is that of co-operation among competitors. Should competition authorities and national regulatory authorities be concerned about the many co-operation agreements and structures in the sector? Competing operators co-operate with each other in different ways: by interconnecting their networks in order to guarantee service inter-operability (often to fulfil regulatory obligations) or by combining their efforts in order to make large investments (which might not be feasible in the absence of co-operation). The latter is the case of network-sharing agreements between 3G broadband mobile communications operators. Some 3G operators have proposed sharing part of their networks as a means to reduce investment costs. Although on the one hand such agreements can improve productive efficiency, on the other hand they can be conducive to anticompetitive behaviour. Such co-operation may be lawful, but where do we draw the line?

These questions are not only interesting in the context of the electronic communications industry. They also relate to issues of importance for other sectors and in particular for technology-based industries and other network industries. For this reason, the options taken in this sector will spill over to these other industries. It is therefore crucial that our approach is based on sound economic reasoning and that is why I highly appreciate the contributions of Professors Michael Katz and Jean Tirole to this book.

## NOTES

1. 'Guidelines for market analysis and the assessment of market power'.
2. This amortization period corresponds to the most common length of contract beyond which the operator has no guarantee that it will retain a customer.

# 13. Antitrust or regulation? US public policy in telecommunications markets

**Michael L. Katz**

## 1. INTRODUCTION

Telecommunications industries in the USA are subject to a complex set of laws administered by a variety of government organizations. In many instances there are overlapping jurisdictions at the federal, State, and local levels. Moreover, the actions of suppliers in these industries are often subject to both economy-wide antitrust laws and sector-specific regulations.

Historically, there was a tension between antitrust policy and telecommunications regulation. For the past century, antitrust policy has generally been based on the view that competition is desirable, and enforcement practices typically were aimed at protecting competition.[1] American telecommunications regulation has a rather different history. For most of the last century, State and federal regulators limited entry into telephone markets in the name of promoting universal service. And local and federal regulators limited entry into the provision of cable television service. In recent years, however, telecommunications policy-makers have embraced competition to a much greater degree, with the Telecommunications Act of 1996 being the most visible example.[2, 3]

As telecommunications regulation has joined antitrust policy in viewing competition as a tool rather than a threat, a number of commentators have called for an 'antitrust approach' toward public policy concerning telecommunications industries. There are several components that an antitrust approach to telecommunications public policy might include:

- One could rely solely on general antitrust *statutes* rather than sector-specific ones. Doing so would raise two broad issues. One is whether sector-specific statutes are needed to address a unique set of structural features possessed by telecommunications industries that affect the application of antitrust policy. The other is whether existing regulatory statutes serve objectives beyond the goals of traditional antitrust policy and, if so, how to meet these objectives going forward.
- One could apply antitrust principles of *analysis*, such as those enunciated in the *Horizontal Merger Guidelines*, to issues of telecommunications policy.[4] Similarly to the first point above, there are issues of whether telecommunications industries exhibit a unique set of structural features that necessitates a different analytical approach or if current regulations raise issues for which traditional antitrust analysis is ill suited.
- One could rely solely on the types of *remedies* utilized in antitrust enforcement. In comparison with regulation, antitrust typically makes more limited interventions re-

quiring less day-to-day oversight and adjustment. Antitrust also tends to be more reactive in comparison with the proactive nature of much sector-specific regulation.

- One could rely solely on government *organizations* charged with enforcing antitrust laws that apply to a broad range of sectors to conduct analyses, bring enforcement actions, and implement remedies in the telecommunications sector.

The first, third and fourth components all concern various institutional features. The second component concerns modes of economic analysis, and is thus of a rather different character. For that reason, I will consider these components in two groups. First, I characterize the current institutional structures and identify significant differences between broad antitrust policy and sector-specific telecommunications regulation in terms of mandates, processes and policy approaches. These institutional differences raise the question: are there significant differences in the underlying structures of telecommunications markets in comparison with other industries that make the institutional differences desirable? To answer this question, I first examine several important characteristics of telecommunications markets and discuss whether these differences create difficulties in the application of antitrust analytical principles to telecommunications markets. I argue that antitrust analytical principles are broad enough to apply to telecommunications markets as well as any other, but that – as always – care is needed to apply these principles correctly. These findings narrow the broader question down to the following: are sector-specific institutions needed to apply general principles to telecommunications markets? After discussing arguments both for and against sector-specific institutions, I close the chapter with a brief conclusion.

## 2. INSTITUTIONAL DIFFERENCES BETWEEN ANTITRUST AND REGULATION: A VIEW FROM THE USA

At the federal level, there are three principal agencies in the USA that either enforce the antitrust laws as applied to telecommunications industries or regulate telecommunications industries: the Antitrust Division of the US Department of Justice ('Division'), the Federal Communications Commission ('Commission'), and the Federal Trade Commission. I will focus on differences between the Division and the Commission. I do this in part because at present I am the chief economist of the Division and in the mid-1990s I was the Chief Economist of the Commission. On a less personal plane, the Division has responsibilities for telecommunications that the Federal Trade Commission does not.[5]

Simply put, are the Commission and the Division substitutes or complements? Is the Commission a duplicate antitrust authority for telecommunications markets? Or does the Commission supplement economy-wide antitrust policy by pursuing a different mandate and/or making use of special tools and procedures? An examination of the institutional features reveals elements of both substitutability and complementarity. While there are large overlaps in objectives and actions, the Commission and Division operate under significantly different statutes and procedures.

## 2.1 Statutes

Consider first the statutes underlying merger policy. In many cases, both the Commission and the Division have jurisdiction over mergers and review them to predict their competitive effects. The Division primarily does so under Section 7 of the Clayton Act.[6] Under Section 7, the standard is whether the effect of a proposed transaction 'may be substantially to lessen competition, or tend to create a monopoly.'[7] The Commission typically reviews a merger because it involves the transfer of a radio or facilities license and does so under a 'public interest standard.'[8]

An immediate question is whether the public interest standard coincides with protecting competition. In past decades, many of the Commission's policies were predicated on the belief that competition was not in the public interest. Now, however, competition appears to fall squarely within the public interest standard. Today, the two biggest differences between antitrust and the public interest standard are probably that the Commission interprets the latter to include 'universal service' and 'diversity' concerns that do not arise under antitrust enforcement.

For example, in announcing its approval of SBC's merger with Ameritech, the Commission summarized its 'statutory objectives to open all telecommunications markets to competition, to promote rapid deployment of advanced services, and to ensure that the public has access to efficient, high-quality telecommunications services.'[9] The latter two objectives appear to be more about facilitating widespread consumption than promoting competition.

More recently, in announcing its opposition to the proposed EchoStar–DirecTV merger, the Commission stated that the 'public interest standard includes an evaluation of the effect of the proposed transaction on competition, consistency with its spectrum policies, and in the case of [multichannel video programming distribution] services, a consideration of the impact on program and viewpoint diversity.'[10] While promoting competition under antitrust policy may indirectly promote diverse ownership and the expression of diverse media viewpoints, the protection of these types of diversity is not a direct objective of antitrust policy.

Now, consider non-merger policy. There are at least three broad classes of differences between the standards embodied in antitrust and telecommunications policy legislation. First, as in merger policy, sector-specific statutes embody objectives simply not present in the broad antitrust statutes. Perhaps the most significant examples for non-merger policy are common carriage and universal service policy. Under common carriage, a service provider is not permitted to engage in unreasonable discrimination based on the content of the messages that it transmits or the identities of the senders and receivers.[11] These restrictions go beyond those of antitrust law.[12] Turning to the promotion of universal service, antitrust enforcement can increase subscription to telecommunications services to the extent that it fosters competition that drives prices closer to costs. However, antitrust enforcement will not create the cross-subsidies and below-cost pricing that are sometimes used in telecommunications policy to promote greater consumption, ostensibly to internalize positive consumption externalities. As a general matter, antitrust does not attempt to internalize externalities, except to the extent that it can be viewed as internalizing changes in consumer surplus that are due to certain actions that create or sustain market power. There also are other differences between the objectives of antitrust and telecommunications policy, such as the regulation of content to limit indecency and to promote educational programming aimed at children, and requirements that telephone services be accessible to persons with disabilities.

A second category of differences between antitrust and telecommunications policy lies in the treatment of the exercise of market power. The possession and exercise of market power is itself not objectionable under American antitrust law.[13] What is objectionable is creating, enhancing, or extending market power in certain ways that have been deemed to be harmful to consumer interests. If a firm legitimately obtains a monopoly, it is generally free to charge consumers prices as high as it wishes.[14] In contrast, many regulatory policies are explicitly designed to limit the exercise of market power, rather than to attack the underlying source.[15] For example, price cap regulation limits the ability of local exchange carriers to price some of their telephone services, and benchmark regulation was applied to restrict the pricing of cable television services through much of the 1990s.

A third difference between general antitrust and sector-specific regulation is that general antitrust aims to *protect* competition, while today federal public policy toward local exchange telephone service markets is intended actively to *promote* competition. Specifically, consider the grounds for imposing mandatory interconnection, facilities sharing, or access. Some federal courts have found violations of antitrust law (that is, Section 2 of the Sherman Act) under an essential facilities doctrine. This doctrine has been applied sparingly to a limited set of circumstances in which control of a 'bottleneck facility' could otherwise extend monopoly power from one market to another.[16] In contrast, the Telecommunications Act of 1996 imposes a long list of obligations on incumbent local exchange carriers that arguably goes well beyond the application of an essential facilities doctrine.[17] Moreover, in passing the Telecommunications Act of 1996, Congress made sweeping decisions about the proper public policy toward incumbent local exchange carriers. In contrast, under the antitrust laws the Division would have to establish market power and illegal conduct on a case-by-case basis subject to high standards of proof.[18]

### 2.2 Organizations and Processes

There are other institutional differences between general-purpose antitrust and sector-specific telecommunications regulation. Here, I will briefly discuss two of them.

One difference lies in the procedures and standards of judicial review. The Commission's decisions are treated like those of a trial court: the Commission reaches an initial decision, which takes effect absent an appeal. Parties objecting to a Commission decision may apply to a federal court of appeals to have the decision reviewed. In theory, the Commission is given substantial deference by the courts because the Commission is an expert agency. Under the Chevron Standard, courts limit themselves to ensuring only that the Commission acts in a 'reasonable' manner and does not engage in 'arbitrary and capricious' behavior.[19] In contrast, when the Division chooses to bring an action, it must do so as the plaintiff in a federal district court. The Division typically bears the initial burden of proof in court and is not entitled to substantial deference. As with Commission decisions, the opinion of the district court judge is subject to appeal.

There also are important differences in the procedural approaches taken by the Commission and the Division. The Commission undertakes a significant number of industry-wide, prospective rule-makings, while the Division generally undertakes case-by-case enforcement of the antitrust laws. This broad difference leads to two specific differences. One, the

Commission historically often issued bright-line rules, such as the mobile telephony spectrum cap and the broadcast television station national ownership cap.[20] The Division occasionally issues guidelines (for example the *Merger Guidelines*), and these guidelines sometimes contain bright lines delineating safe harbors. However, the emphasis is much more on case-specific analysis without reliance on set numerical standards.[21] A second difference is sometimes characterized as the difference between *ex ante* and *ex post* intervention. The Commission often imposes rules designed to stop problems before they arise while, with the exception of merger policy, antitrust policy tends to attack conduct after the harm has taken place.

## 2.3 Remedies

The Commission and the Division also differ in the nature of the remedies they impose. To some extent, these differences stem from differences in the policy objectives embodied in the underlying legislation. For example, the Commission has imposed price caps to limit the exercise of market power – supplier behavior that would not be a stand-alone antitrust offense in the USA and thus not subject to an antitrust remedy. Other differences in remedy follow from different institutional competencies and resources. Antitrust remedies tend to be addressed to specific conduct or actions, and there is a strong preference among antitrust enforcers for imposing remedies that do not require extensive, ongoing monitoring or constant updating. For this reason, there is a preference for structural remedies over conduct remedies. While the Commission may impose structural remedies (for example block a license transfer), it is willing to impose detailed conduct remedies. For example, when it allowed SBC to acquire Ameritech, the Commission imposed 30 conditions on SBC, including conditions mandating the provision of certain wholesale services, wholesale and resale pricing levels and structures, deployment patterns for basic and advanced retail services, and extensive monitoring.[22]

While the distinction between structural and conduct remedies is a meaningful one, it should be recognized that the distinction is not sharp, particularly if one views ownership as one point on a contractual continuum. For example, as a structural condition for approving the SBC–Ameritech merger, the Commission required the creation of one or more separate affiliates to provide advanced services.[23] However, to ensure that the firm operated its affiliate at arm's length as intended by the Commission, a number of restrictions had to be placed on the conduct of the parent and affiliate, specifically that these affiliates 'will use the same processes as competitors and pay an equivalent price for facilities and services.'[24]

Another example is provided by intellectual property. Intellectual property can be an important asset, and compulsory licensing (a conduct remedy) may be as effective as actually transferring ownership (a structural remedy). Indeed, in contrast to transferring exclusive ownership, a non-exclusive license can allow more suppliers to make use of the technology and potentially lead to greater competition. With regard to the burdens of enforcement, one question is the extent to which private parties can reach licensing agreements that are then enforced under standard contract law, rather than as part of an antitrust or regulatory proceeding.

## 3. WHAT MAKES TELECOMMUNICATIONS SPECIAL?

Telecommunications regulation is based on sector-specific legislation that supplements general antitrust laws. The Commission also operates under different standards of review than does the Division, and the organizations have different procedures. These differences raise a fundamental question: are there significant differences in the underlying structures of telecommunications markets in comparison with other industries that make the institutional differences desirable? A first step toward answering this question is to identify characteristics of telecommunications industries that distinguish them from other sectors of the economy. A second step is to understand how various institutional features affect the accuracy, efficacy and timeliness of the policy process. Ideally, the process of institutional design would then match institutional structure to industry characteristics.

### 3.1 Role in Public Speech

This subsection and the next two briefly discuss several features of telecommunications that must be taken into account in designing and administering public policy toward telecommunications markets. One often-noted feature is the importance of telecommunications to the functioning of a democratic society. Telephony, television and the Internet all provide platforms for speech.

In discussing the policy implications of this feature, it is useful to distinguish one-way platforms over which the owners provide content, either directly or through contract, from two-way platforms that serve as conduits for user-generated content. Consider first two-way platforms, such as telephone networks and the Internet. It is not evident that market power should give rise to restrictions on user speech over these platforms. However, if there is a problem, a quick analysis suggests that it would be sufficient to rely on mandated common carriage.

One-way platforms include radio and terrestrial over-the-air, cable and direct-to-the-home satellite television. Arguably, market forces alone will not lead to desirable outcomes in terms of ensuring that a diverse set of viewpoints is aired. But past and current regulatory policies ostensibly aimed at promoting diversity have long been poorly analyzed and justified. There often is little linkage between the policy and its stated goal and, not surprisingly, there often is little indication that the policy serves it stated purpose. The broadcast television national multiple ownership rule is a prime example.[25] Under the rule, a single entity cannot control stations whose combined reach exceeds 35 percent of US television households.[26] Proponents of the rule claim that it prevents a single owner from acting as a media gatekeeper that limits viewer options.[27] However, 'relevant markets' for assessing diversity concerns are almost surely local, and the extent of diversity at the local level is unaffected by the national cap. One might reasonably ask whether it is inevitable that speech considerations will be handled badly because so much of the policy debate turns on concepts for which no measure is commonly agreed upon.

### 3.2 Economic Characteristics

A number of economic characteristics of telecommunications markets raise policy concerns:

- *Economies of density.* Sparse networks have higher costs, and local distribution networks exhibit natural monopoly cost conditions.[28]
- *Network effects.* Telecommunications services often exhibit network effects, whereby the value of subscribing to a service rises with the number of other subscribers. These demand-side increasing returns to scale can create a tendency toward monopoly, although inter-operability and cross-network compatibility can allow multiple providers to offer service without sacrificing the realization of network effects.
- *Economic importance.* Telecommunications services support vital commercial communications throughout the economy, as well as provide platforms for other services, such as e-commerce.

These features raise concerns with respect to both co-ordination and competition. With respect to co-ordination, the degree of inter-operability and interconnection is an important determinant of market performance. As is well known, the public and private incentives to interconnect may diverge. With respect to competition, a central question is whether there will be meaningful competition in the provision of some important local access services given the presence of various forms of increasing returns to scale and density.

These concerns have led many to call for compulsory access to local distribution facilities and to separating various vertical layers of service provision from one another. In evaluating the merits of compulsory access, a first question is: what is the policy trying to do? Is the aim to prevent an incumbent from exercising market power by setting interconnection and access prices that discourage even efficient suppliers from entering, or is the aim to subsidize entry (by efficient or inefficient suppliers) in order to further some other policy objective?[29] If the goal is to subsidize entry, there is unlikely to be an efficiency-based argument for making the incumbent bear the costs of funding the subsidies rather than relying on the general tax system.

If the goal is to limit the inefficient exercise of market power, then mandatory access can be viewed as overcoming transactions costs that prevent consumers from bargaining with suppliers to attain mutually beneficial outcomes. In the absence of transactions costs, the Coase theorem states that consumers, incumbent service providers and potential entrants will bargain their way to efficient outcomes, regardless of the initial allocations of property rights.[30] However, because of the large number of consumers affected, transactions costs are likely to be significant, and consumers are unlikely to be able to bargain their way to efficiency. When entry would raise consumer surplus, the joint incentives of an incumbent service provider and potential entrant to reach an agreement to interconnect or share facilities will fall short of the social incentives, as measured by the change in total surplus. A rule mandating access may imperfectly overcome this distortion.[31]

Some commentators argue the distortion will be particularly strong in markets with significant network effects.[32] The argument builds on the following observations. One, absent access, one network may become dominant and consumers may become locked in to that network. Moreover, even if multiple networks survive, absent interconnection there will be a loss of demand-side economies of scale due to network fragmentation. The argument for compulsory access is that a network owner might otherwise refuse access and thus stifle competition and/or the realization of network benefits.

Another argument builds on the fact that telecommunications services often are conceptualized as being provided in layers. One such taxonomy is that there is a physical transport layer, a logical layer, an applications layer and a content layer. Under vertical bundling, services at different layers are offered for purchase (at either the wholesale or retail levels) only as a package. Many service providers, particularly incumbent local exchange carriers and cable companies, would like to be able to choose the extent of vertical bundling in which they engage. Other participants in the public policy debate argue that there are at least three significant social benefits of vertical unbundling that might merit a mandatory unbundling policy. First, unbundling allows the realization of mix-and-match benefits. A consumer can take the best offering at one layer and combine it with the best offering at a second layer, even if the offerings are provided by different firms. A second benefit of vertical unbundling is that it facilitates innovation by allowing firms to engage in single-layer innovation. Unbundling stimulates and spreads innovation when it makes it feasible for a firm that is not vertically integrated to innovate at a single layer or allows an integrated firm to combine its innovative service at one layer with the services provided by different firms at other layers. Third, vertical unbundling allows competition at some layers, even if it is infeasible at others. The degree of economies of scale can differ across parts of the network or layers, and unbundling prevents the most concentrated layer from driving overall concentration. Moreover, vertical unbundling facilitates entry by allowing single-layer entry strategies, which reduce the risks associated with entry and lessen the need to acquire multiple skill sets.

Those who oppose compulsory access offer several objections. One is that, because such access almost certainly reduces the financial returns to network facility ownership, it reduces the incentives of an investor whose facilities would be subject to mandatory access.[33] For example, both cable companies and local exchange telephone companies argue that mandatory access to their broadband distribution facilities will inefficiently reduce their investment incentives.[34] Another reason is that other potential investors' incentives are lowered – rather than develop facilities of their own, rival suppliers can rely on compulsory access.[35] Finally, there can be tremendous practical difficulties of determining appropriate prices and terms of exchange.[36] Indeed, the first two incentives problems stem in large part from difficulties in setting appropriate prices.

Investment concerns extend beyond physical infrastructure. In markets subject to network effects, one must consider investments in building subscriber bases. While competition between incompatible networks may ultimately lead to tipping and monopoly, firms may compete extremely vigorously to become the monopolist and may be willing to make large investments in promoting their networks (for example, engaging in consumer education or penetration pricing). Moreover, firms may go through this cycle of competing to be the dominant supplier for each new generation of technology. In contrast, if various firms' products are compatible at the outset, they may compete less vigorously and be unwilling to subsidize network development.[37] Thus compulsory access may again lower the investment incentives of both incumbents and entrants.

In theory, settings with large amounts of potential complementary (rather than substitute) investment would be candidates for compulsory sharing. But one might argue that investments that create complementary opportunities for other investors are especially in need of protection in order to provide sufficient investment incentives.

One might also argue that property rights should be weaker where the rights holder would otherwise garner significant market power. But one would immediately confront the fact that investments conferring market power are in many circumstances those of the greatest social value and thus are investments for which private investors should be provided the greatest protection in order to encourage investment. This last point raises the issue of how the market got to its current state. I will turn to this issue in the next subsection.

### 3.3 Where We Are and How We Got Here

In determining how to move forward, optimal policy-making takes into account: (a) the current state of private and public institutions and, in some instances, (b) the process by which the industry reached its current state.

One difference between telecommunications and other economic sectors might irreverently be summarized by saying that the costs of red tape have already been sunk in telecommunications. That is, the start-up costs of creating sector-specific regulatory institutions have already been borne. The decision whether to keep regulation should be based on forward-looking costs. In doing so, it should be recognized that the largest and most important costs are likely to be the costs of unintended consequences (for example, the stifling of investment or innovation incentives) rather than administrative costs.

A second consideration might be labeled the *regulatory compact*, by which I mean a largely implicit agreement between regulators and service providers under which the former protect the latter's profits in return for service providers' undertaking certain socially desirable actions. Some participants in the policy debate have argued that incumbents have been promised particular returns on their investments and thus should not be forced to share their facilities at prices below those that would maximize the incumbents' profits. Other participants in the debate argue that low-priced sharing obligations should be imposed on local exchange carriers on the grounds that these carriers do not have the same claim to their investments as would other suppliers because local exchange carriers were explicitly shielded from competition by regulation. This argument is sometimes summarized as 'the rate-payers financed the investment, not the shareholders.'

In terms of efficiency considerations, the validity of this argument turns on the expectations of the regulated entities. If the investment costs of the facilities covered by the access or sharing policy already have been sunk, then those investments cannot be distorted by current policies, which suggests that facilities sharing at low prices would not adversely affect investment. But one must recognize that future investment incentives could be adversely affected. This concern arises with particular force when future investments are risky and the regulatory treatment of past investments leads potential investors to fear that the distribution of returns going forward will be truncated on the upside, with no corresponding downside adjustment. In this case, investment incentives can be inefficiently attenuated even if there is a meaningful sense in which rate-payers financed past investments.

## 4. APPLYING ANTITRUST ANALYSIS TO TELECOMS MARKETS

There are significant differences in the institutions underlying broad-based antitrust policy and sector-specific telecommunications regulation. Moreover, telecommunications industries possess a number of characteristics that distinguish them from many other industries and which can raise difficult policy issues. These differences raise two questions regarding the analytical tools used to make enforcement and policy decisions.

First, do the structural features necessitate a different analytical approach? Here, I think the answer is short and clear: no, because the general economics principles on which antitrust enforcement is based are just that, *general* principles. None of the special features of telecommunications markets makes them impervious to antitrust analysis. The real issue is applying the principles and tools correctly.

The second question is whether the differences in underlying statutes and current policies mean that telecommunications regulation raises issues for which traditional antitrust analysis is ill suited. For example, how should policy analysts deal with universal service and diversity concerns? Here, too, I think the answer is clear. Once objectives are well defined, basic tools of economic analysis are well suited to guiding enforcement and policy decisions. The problem with diversity has been that there is no consensus on how to measure it. And there has been little discipline in checking whether policies have any relationship to their stated objective. Universal service has often suffered from a similar lack of discipline. While these issues can be addressed using the standard tools of microeconomic analysis, the political will to do so is needed.

Having argued that the general principles of antitrust analysis are applicable to telecommunications policy issues, in the remainder of this section I will consider two examples.

### 4.1 Market Delineation

Market delineation provides a useful example of an antitrust analytical approach that can be – and indeed has been – applied to telecommunications markets. In delineating markets, it is critical to keep in mind that market definition is a tool of analysis, not an independent end in itself. Hence, from the outset, it is important to be cognizant of the use to which market definition will be put. There are at least three possibilities: (a) to guide the analysis used to project the competitive effects of a proposed merger, particularly when the underlying theory of harm is based on the possibility of co ordinated effects;[38] (b) to assess market power as part of analyzing the effects of non-merger conduct;[39] and (c) to identify arbitrage possibilities and distortions that can arise when competing suppliers are subject to different regulatory regimes.[40]

A fundamental principle by which economists define the product scope of a market is to include two goods or services in the same relevant market if consumers view them as sufficiently close substitutes and not to include them in the same relevant market if consumers do not view them as sufficiently close substitutes. The Commission, the Division and the Federal Trade Commission all use this approach.[41] A similar approach is also used to define the geographic scope of a relevant market.[42]

Delineation of geographic markets illustrates how general principles can be applied to the specific features of telecommunications markets. Although not unique to telecommunications markets, one important distinguishing feature of most such markets is that the service has to be delivered to a customer's location. For instance, wireline telephone service delivered to your next-door neighbor's house is generally a very poor substitute for telephone service to your house. And telephone termination to a clothing boutique is a very poor substitute indeed for telephone service to a doctor's office located down the street. Based on the principle of consumer substitution, there are literally trillions of telephone markets.

Clearly, analysis of each market on its own is impractical. Fortunately, it is also unnecessary. For purposes of assessing market power and the nature of competition, these markets can be aggregated for policy-making and judicial decision-making.

To my mind, the critical task is not to identify differences between market delineation in telecommunications industries and others. Rather it is to avoid pitfalls that are common to market delineation in a wide range of industries. As stated above, market definition is a tool of analysis, not an independent end in itself. To conduct sound policy analysis, it is essential not to get so caught up in the mechanics of various market-delineation algorithms that one loses sight of the ultimate purpose of the exercise.

Application of the so-called 'smallest-market principle' provides a good illustration of what can go wrong when one gets too caught up in process at the expense of substance. Under the hypothetical-monopolist test of the *Merger Guidelines*, one starts with one of the merging parties' products, combines that product with its best substitute, and then asks if a supplier with a hypothetical monopoly over those products could profitably raise price significantly for a sustained period of time.[43] If the answer is yes, then a relevant market has been found. If not, the product that is the next-best substitute for the merging party's product is added and the test repeated. This process continues until the hypothetical monopolist would find it profitable to increase the prices of some or all of its products, including the price of a product of one of the merging parties. In a differentiated-products market, it is possible that the merger of two single-product firms would allow them substantially to raise price above pre-merger levels, yet the firms would be found *not* to be in the same relevant market as one another if one mechanically applied the smallest-market principle.[44] One should not let mechanics obscure competitive realities.[45]

Rigidity with respect to the size of the price increase used in the hypothetical-monopoly test can also lead to unsound conclusions. The *Merger Guidelines* indicate that a 5 percent price increase often is a useful *threshold* for market delineation, although not the only one.[46] The choice of benchmark matters in part because, under some demand conditions, a larger percentage price increase may be profitable even when a 5 percent price increase is not. This could happen, for example, if there were one group of consumers who were near the margin of substituting to another product at the pre-merger price, while another group of consumers would not switch to substitutes even in response to a large price increase. On the practical level, it may be difficult to establish in court that a small price increase would be profitable because it is difficult to show that there would not be at least some lost sales in response. It may be easier to present evidence that a large price increase would not give rise to sufficiently large unit sales losses to render the price increase unprofitable.

A final piece of advice is that one should not place too heavy a burden on market delineation. Defining markets is not a substitute for conducting sound competitive analysis. In this regard, I think it is a mistake to bring supply-side considerations into market delineation because it introduces competitive analysis into the process and complicates what is an initial stage of analysis.[47] More generally, it is well to remember that the hypothetical-monopolist test articulated in the *Merger Guidelines* is an artificial exercise designed for a specific purpose. For example, under this approach, the prices of all other products are held constant, even if in reality the prices of some or all substitute products would be expected to change in response to changes in the price of the hypothetical monopolist's products.[48]

### 4.2 Two-sided consumption benefits and efficient service pricing

The public policy treatment of calling party pays has been a contentious issue in a number of countries. A central concern is whether a party's choosing to subscribe to a particular carrier creates a 'call-termination monopoly' (that is, anyone wanting to call that subscriber has only one choice of carrier for completing the call).

Antitrust teaches that one should think carefully about the sources and consequences of market power. There is a sense in which a carrier may have a termination monopoly, but it is important not to overstate this finding. Carriers may compete vigorously to become the monopolist. Thus the called party may be the one with market power, not the called party's carrier. For example, suppose that, from a carrier's perspective, a particular household is a highly desirable wireless subscriber because it will generate a lot of revenues from incoming calls. If there are several wireless carriers competing for that household's patronage, then they can be expected to offer attractive terms to the household, so that it – not the carriers – enjoys the rents extracted from callers. Note, too, that some called parties might enjoy low subscription prices, while others might rationally prefer to pay more themselves in order to have callers pay less to reach them.[49]

Antitrust economics also teaches that it is important to analyze economic problems fully. Consumption of communications services generally involves two (or more) parties who take actions, receive benefits and bear costs. When both parties benefit from a message exchange, it is feasible to charge both the sender and receiver positive prices. Moreover, both commercially and socially optimal pricing entails trading off the effects on the origination and acceptance of calls as the sending and receiving prices are varied. Under some conditions, the costs of a call should be shared equally. In other circumstances, it can be socially optimal to have one party pay more than marginal cost in order to subsidize the other party. Thus, in theory at least, it might be efficient for calling parties to subsidize called parties.[50]

An important lesson of this analysis is that, in general, socially optimal pricing depends on demand conditions, what each party knows about the value the other party places on communicating, and the costs of calling.[51] These are difficult issues. Given our current understanding of the forces at work, the most important conclusion is that one should not jump to conclusions. One should not automatically assume that any deviation from a cost-based standard is evidence of inefficiency or the exercise of market power. At the same time, one should recognize that there might be an efficiency-enhancing role for public restrictions on private pricing decisions in this area.

## 5. THE COMMISSION OR THE DIVISION?

Telecommunications industries have several distinctive features that may warrant a set of industry-specific policies, if not modes of analysis. In any event, the USA has a set of specialized statutes. Even taking the existence of specialized policies as given, however, it does not follow that these policies have to be administered by an industry-specific governmental agency.

In thinking about whether a generalist or specialist agency is more appropriate, it is worth noting at the outset the whole exercise is characterized by a degree of vagueness. After all, decisions are made by people, not by organizations. One could have specialist teams within a generalist agency. Indeed, both the Commission and Division are parts of a very large generalist organization, the US federal government. However, it is clear to me at least, that structure at the agency level has behavioral and performance consequences.

The tradeoffs can be broadly summarized as focus and expertise on the one hand, and broad perspective and independence on the other. A specialized institution can offer expertise built on experience. It has a toolkit of formal databases and models. It has at least some institutional memory about similar cases and issues handled in the past. It has trained, established teams of specialized personnel. The team aspect is important because a well-functioning organization is more than the sum of its parts. Specialized institutions can have staffs that collect and analyze detailed information on costs and demand conditions on an ongoing basis. They can also engage in regulatory intervention on an ongoing basis. Consequently, a specialized regulator may have ability to make enforcement judgments and apply remedies one would not want to leave to generalist organizations, such as the courts.[52] A specialized agency may also be able to rely on experience and expertise to reach decisions more quickly.[53]

A generalist agency can have advantages of a broad perspective, which may both improve its decision-making skills and enhance its independence. One reason is that the staff in a generalist agency may hear a wider range of arguments and perspectives. A second reason is that the political lines may not be as clearly drawn. It may be much harder for a member of Congress to determine how leading employers in his or her district or State will be affected by a change in general antitrust enforcement than by a change in a specific telecommunications policy. Third, a generalist institution may be less subject to capture.[54] This could be due to the exposure to a broader range of viewpoints, or because a generalist agency may build up fewer repeat relationships, which might have the effect of forcing parties to rely on sound arguments rather than personal contacts. Related to this concern is the popular sentiment that specialist agencies are more subject to a 'revolving door,' whereby former agency employees are hired by the companies whose behavior the agency oversees and these employees then trade on personal relationships built up while inside the agency.[55] However, the difference between the Division and the Commission is probably smaller than one might think. While the Division is responsible for a much broader set of industries than is the Commission, each deals repeatedly with a fairly small number of attorneys who serve as outside counsel for private entities. A final advantage of generalist agencies may be in a willingness to reduce oversight where appropriate. Experience suggests that a specialist agency may sometimes have trouble 'letting go' by relaxing detailed regulations and putting itself out of a job.[56]

This dichotomy suggests that it is desirable to move to a generalist agency once detailed regulation is unnecessary, but that a specialist organization can be useful when detailed regulation is needed.

## 6. CONCLUSION

If the USA were to make a wholesale switch to an antitrust approach today, there would be two broad consequences. First, in several cases policy-makers would have to accept less competition (or at least more concentration) than they apparently hoped for when passing the Telecommunications Act of 1996. There are several telecommunications markets –

including some of those supporting or building on Internet services – for which it is far from obvious that they are or will become competitive under the application of standard antitrust principles. Second, policy-makers would either have to abandon some goals or rely on supplements to antitrust policy to attain them, because even competitive markets can fail to internalize related externalities, disseminate diverse viewpoints, or redistribute income as policy-makers might like.[57]

Whatever institutions are the basis for creating and applying public policy in telecommunications markets going forward, policy-makers and enforcers will be successful only if they make careful use of sophisticated economic reasoning that builds on careful empirical analysis.

## NOTES

1. There have been exceptions over the years, notably the Robinson–Patman Act (15 U.S.C. §§13–13b, 21a), which arguably was aimed more at limiting competition than protecting it. Moreover, private firms may attempt to use antitrust litigation strategically to limit competition.
2. Telecommunications Act of 1996, Pub. L. No. 104–104, 110 Stat. 56. The 1996 Act amends the Communications Act of 1934, 47 U.S.C. §§151 *et seq.*
3. There remain areas in which telecommunications policy is explicitly anticompetitive, notably spectrum policy, where licensees are limited in the uses to which they can put spectrum.
4. US Department of Justice & Federal Trade Commission, *Horizontal Merger Guidelines* (rev. edn 1997) (*Merger Guidelines*).
5. For example, the Telecommunications Act of 1996 gives the Department of Justice a special status in proceedings before the Commission to consider whether certain incumbent local exchange carriers should be allowed to offer long-distance telephone services in their local exchange territories. In reaching its decision, the Commission must give 'substantial weight' to the Attorney-General's assessment of an application: 47 U.S.C. §271(d)(2)(A). Moreover, the Department of Justice and the Commission frequently file joint briefs in appeals stemming from rule-making orders, and the Solicitor-General represents the Commission in proceedings before the Supreme Court. Further, under the *ad hoc* division of labor between the Division and the Federal Trade Commission, the Division tends to undertake the major telecommunications investigations.
6. 15 U.S.C. §18. Mergers can also be challenged pursuant to the Sherman Act, 15 U.S.C. §§1, 2.
7. 15 U.S.C. §18.
8. 47 U.S.C. §§214, 310(d). The Commission recently noted that, '[i]n addition, while the Commission has never chosen to exercise it, the Commission has independent authority under sections 7 and 11 of Clayton Act to disapprove the acquisition of common carriers engaged in wire or radio communications or radio transmissions of energy in any line of commerce in any section of the country where the effects of such an acquisition may substantially lessen competition, or tend to create a monopoly.' (internal footnote omitted) (Report and Order, *In the Matter of 2000 Biennial Review, Spectrum Aggregation Limits for Commercial Mobile Radio Service*, WT Docket No. 01-14, rel. 18 December 2001, ¶ 63).
9. Press Release, Federal Communications Commission, FCC Approves SBC–Ameritech Merger Subject to Competition-Enhancing Conditions (6 October 1999), available at http://ftp.fcc.gov/Bureaus/Common_Carrier/News_Releases/1999/nrcc9077.html.
10. Press Release, Federal Communications Commission, FCC Declines to Approve EchoStar–DirecTV Merger (10 October 2002), available at http://hraunfoss.fcc.gov/edocs_public/attachmatch/DOC-227263A1.txt?date=021010.
11. 47 U.S.C. §202.
12. For example, the Robinson–Patman Act concerns only price discrimination that harms competition. And the federal courts have said that a firm generally is free to refuse to deal with its competitors. See *Monsanto Co.* v. *Spray-Rite Serv. Corp.*, 465 U.S. 752, 761 (1984). However, as noted below in the discussion of the essential facilities doctrine, there are limited circumstances in which courts have imposed duties on monopolists to deal.
13. See *United States* v. *Grinnell Corp.*, 384 U.S. 563, 570–71 (1996) (noting 'the willful acquisition or maintenance of that power as distinguished from growth or development as a consequence of superior product, business acumen, or historic accident' is an element of monopoly under the Sherman Act).

14. See *Blue Cross & Blue Shield United of Wisconsin* v. *Marshfield Clinic*, 65 F.3d 1406, 1413 (7th Cir. 1995) ('the antitrust laws are not a price-control statute or a public-utility or common-carrier rate-regulation statute').
15. Indeed, one can conceptualize the implicit agreement between regulators and service providers as follows: for years, service providers were protected from competition through the grant of monopoly franchises, in return for which they accepted restrictions on their pricing.
16. The Seventh Circuit identified the following critical elements, known as the 'MCI Factors': '(1) control of the essential facility by a monopolist; (2) a competitor's inability practically or reasonably to duplicate the essential facility; (3) the denial of the use of the facility to a competitor; and (4) the feasibility of providing the facility.' *MCI Communications Corp.* v. *AT&T*, 708 F.2d 1081, 1132–33 (7th Cir. 1983) (citations omitted), *cert. denied*, 464 U.S. 891 (1983).

    For a review of the essential facilities doctrine from economic and legal perspectives, see Gregory J. Werden, *The Law and Economics of the Essential Facility Doctrine*, *Saint Louis University Law Journal*, **32**(433) (1987), pp. 433–79 ('Werden').
17. According to the Division, '[a]lleged failures to meet the 1996 Act's affirmative obligations ... do not in themselves give rise to Sherman Act liability.' Brief of Amici Curiae United States and Federal Communications Commission, *Covad Communications Co.* v. *BellSouth Corp.*, 299 F.3d 1272 (11th Cir. 2002) (No. 01-16064-C) at 11–12. Justice Breyer asserted that, in implementing the local exchange facilities-sharing provisions of Telecommunications Act of 1996, the Commission went too far in extending 'the definition of what must be shared beyond that which is essential.' *AT&T* v. *Iowa Utilities Board*, 525 U.S. 366, 430 (1999) (Breyer, J., concurring in part and dissenting in part). No other Justice joined his opinion.
18. Based on *Goldwasser* v. *Ameritech Corp.*, 222 F.3d 390 (7th Cir. 2000), some local exchange carriers have argued that the Telecommunications Act of 1996 creates a form of implied antitrust immunity for an incumbent local exchange carrier's refusal to grant access to its facilities. The Division rejects this view. See, for example Brief of Amici Curiae United States and Federal Communications Commission, *Covad Communications Co.* v. *BellSouth Corp.*, 299 F.3d 1272 (11th Cir. 2002) (No. 01-16064-C). Generally '[i]mplied antitrust immunity is not favored, and can be justified only by a convincing showing of clear repugnancy between the antitrust laws and the regulatory system.' *Nat'l Gerimedical Hospital and Gerontology Center* v. *Blue Cross of Kansas City*, 452 U.S. 378, 388 (1981) (quoting *United States* v. *Nat'l Ass'n of Sec. Dealers*, 422 U.S. 694, 719–20 (1975)). Moreover, the Telecommunications Act of 1996 has both a general savings clause (Section 601(c)(1)) and an antitrust-specific one (Section 601(b)(1)). 47 U.S.C. §152 Historical and Statutory Notes. It is important to retain application of the antitrust laws because antitrust allows private enforcement to supplement public enforcement and because different remedies may be available to the courts than to regulators, among other reasons.
19. *Chevron, U.S.A., Inc.* v. *Natural Res. Def. Council, Inc.*, 467 U.S. 837, 843–45 (1984). (Where Congress explicitly delegates authority to an agency to elucidate a specific provision of a statute by regulation, such regulations are given controlling weight unless they are arbitrary, capricious or manifestly contrary to the statute, and where such legislative delegation is implicit, a court may not substitute its own construction of a statutory provision for a reasonable interpretation made by the agency administrator.)
20. 47 C.F.R. §20.6 (CMRS spectrum aggregation limit); 47 C.F.R. §73.3555(e) (national television ownership rule). The spectrum cap will be lifted as of 1 January 2003. Rep't and Order, *In re: 2000 Biennial Regulatory Review of Spectrum Aggregation Limits for Commercial Mobile Radio Services*, 16 FCC Rcd 22,668 (rel. 18 December 2001), ¶¶ 1–2. The national television ownership rule is currently under review. Notice of Proposed Rulemaking, *In re: 2002 Biennial Regulatory Review – Review of the Commission's Broadcast Ownership Rules and Other Rules Adopted Pursuant to Section 202 of the Telecommunications Act of 1996*, FCC MB Docket No. 02-277, et al. (rel. 23 September 2002).
21. See, for example, Letter from Charles James, Assistant Attorney-General, US Department of Justice Antitrust Division, to Michael Powell, Chairman, Federal Communications Commission, Re: In the Matter of 2000 Biennial Review, Spectrum Aggregation Limits for Commercial Mobile Radio Service, WT Docket No. 01-14 (7 November 2001).
22. Federal Communications Commission, Summary of SBC/Ameritech Conditions (6 October 1999), available at http://ftp.fcc.gov/Bureaus/Common_Carrier/News_Releases/1999/nrc9077a.html.
23. Ibid. (Condition I(1)).
24. Ibid.
25. In 1999, I was retained by several major television networks to analyze the economic effects of the national ownership cap. Michael L. Katz, 'Old Rules and New Rivals: An Examination of Broadcast Television Regulation and Competition' (September 1999) ('*Katz White Paper*'), attached to Supplemental Comments of Fox Television Stations, Inc. (18 November 1999), in Notice of Inquiry, *In re: 1998 Biennial Regulatory Review – Review of the Commission's Broadcast Ownership Rules and Other Rules Adopted Pursuant to Section 202 of the Telecommunications Act of 1996*, 13 FCC Rcd 11,276 (rel. 13 March 1998).
26. 47 C.F.R. §73.3555(e). When a group owner holds two licenses within a single Designated Market Area, that

audience is counted only once for purposes of the national reach cap. See Rep't and Order, *In re: Broadcast Television National Ownership Rules, et al.*, 15 FCC Rcd 20,743 (rel. 6 August 1999), ¶ 1. The cap is currently under review. See above, note 20.

27. Proponents of the broadcast television national cap also argue that it promotes competition, minority ownership and localism. I will not address those arguments here.
28. Recall that a natural monopoly exists when the costs of providing service are lower when the service is provided by a single provider rather than several.
29. Entry by a higher-cost supplier can, in theory, raise total surplus by driving an incumbent's price closer to costs and improving allocative efficiency even though productive efficiency is worsened.
30. See Ronald H. Coase, 'The Problem of Social Cost', *Journal of Law and Economics*, **3**(1) (1960), pp. 1–44.
31. One might wonder how giving an entrant the right to demand access would solve the problem if the collective incentives of the entrant and incumbent remain too low. Specifically, why couldn't the incumbent pay the entrant not to demand access? One answer is that such a policy would create a queue of would-be entrants that existed largely to collect the payments, and it might be too costly to pay off all of them. Second, antitrust enforcers could be expected to frown on payments made to competitors in return for their limiting their abilities to compete. For additional discussion of this point, see Michael L. Katz, 'Economic Efficiency, Public Policy, and the Pricing of Network Interconnection Under the Telecommunications Act of 1996,' in *Interconnection and the Internet: Selected Papers from the 1996 Telecommunications Policy Research Conference* (G. Rosston and D. Waterman, eds), Mawah, NJ: Lawrence Erlbaum Assocs. (1997).
32. Mark Lemley and David McGowan provide an insightful summary of the literature examining whether intellectual property rights should be weaker in industries with strong network effects. Mark A. Lemley and David McGowan, 'Legal Implications of Network Economic Effects', *Californian Law Review*, **86**, (May 1998), pp. 479–611 [pp. 523–41]; see also Joseph Farrell, 'Arguments for Weaker Intellectual Property Protection in Network Industries', *Standard View*, **3**(2) (June 1995), pp. 46–9.
33. In what follows, I restrict attention to compulsory access as a duty to deal rather than as a remedy to a separate competition problem. Compulsory access as a remedy to a separate antitrust injury is less likely to have the adverse investment incentives effects discussed below. To see why, suppose antitrust enforcers conclude that, absent the imposition of certain conditions, a proposed merger would reduce competition and harm consumers. To the extent that access is a means of preserving competition that is less costly to the merging parties than is blocking the merger entirely, the parties might benefit from being forced to grant access.
34. See, for example Comments of Cox Communications, Inc. at 16–17 (1 December 2000) (quoting Justice Breyer on the perils of mandatory access, *AT&T* v. *Iowa Utilities Bd.*, 525 U.S. 366, 427–30 (1999), and arguing that these concerns apply to broadband services), in Notice of Inquiry, *In re: Inquiry Concerning High-Speed Access to the Internet Over Cable and Other Facilities*, 15 FCC Rcd 19, 287 (rel. 28 September 2000), available at http://gullfoss2.fcc.gov/prod/ecfs/retrieve.cgi?native_or_pdf=pdf&id_document=6512159427; see also Comments of Verizon Communications Inc. at 5–7 (1 March 2002) (explaining the negative effects of mandatory unbundling and collocation, as well as other forms of government intervention, on broadband facilities investment), in Notice of Proposed Rulemaking, *In re: Review of Regulatory Requirements for Incumbent LEC Broadband Telecommunications Services*, 16 FCC Rcd 22,745 (rel. 20 December 2001), available at http://gullfoss2.fcc.gov/prod/ecfs/retrieve.cgi?native_or_pdf=pdf&id_document=6513079788.
35. For a general discussion, see Joseph Farrell and Michael L. Katz, 'Public Policy and Private Investment in Advanced Telecommunications Infrastructure', *IEEE Communications Magazine*, **36**(7) (July 1998), pp. 87–92 [89–90].
36. For a brief discussion of these problems, see Werden at 472–5.
37. See Michael L. Katz and Carl Shapiro, 'Technology Adoption in the Presence of Network Externalities', *Journal of Political Economy*, **94**(4) (1986), pp. 822–41.
38. While market delineation is a useful discipline for thinking about co-ordinated effects and systematically thinking about the nature of competition, it plays little role in a unilateral effects analysis that projects competitive effects based on a simulation model using econometric estimates of demand elasticities, margins and diversion ratios. Generally, boundaries of the relevant market have no influence on the resulting projections of post-merger equilibrium prices.
39. It is worth noting that market delineation is not legally required for assessing market power. On the lack of a legal requirement, see *Rebel Oil Co.* v. *Atlantic Richfield Co.*, 51 F.3d 1421, 1434 (9th Cir. 1995) ('Market power may be demonstrated through either of two types of proof. One type of proof is direct evidence of the injurious exercise of market power. If the plaintiff puts forth evidence of restricted output and supracompetitive prices, that is direct proof of injury to competition which a competitor with market power may inflict, and thus, of the actual exercise of market power. The more common type of proof is circumstantial evidence pertaining to the structure of the market.' (Citations omitted.)).
40. For a greater discussion of this role, see Michael L. Katz, 'Regulation: The Next 1000 Years,' in *Six Degrees*

*of Competition: Correlating Regulation with the Telecommunications Marketplace*, The Aspen Institute (2000).

41. See, for example, *Merger Guidelines*, §1.1; see also Memorandum Opinion and Order, *In re: Applications of NYNEX Corporation Transferor, and Bell Atlantic Corporation Transferee, For Consent to Transfer Control of NYNEX Corporation and Its Subsidiaries* (rel. 14 August 1997), ¶ 50.
42. *Merger Guidelines*, §1.2.
43. *Merger Guidelines*, §1.11.
44. Consider a differentiated products market comprising three suppliers, *A*, *B*, and *C*. Suppose that *A* is the largest supplier in the market and that *B* and *C* propose to merge. Also suppose that *A* is the 'closest competitor' to each of the other two firms, and that any two firms together would find it profitable to raise prices significantly above current levels for a sustained period of time. Then under the *Merger Guidelines* algorithm, *B* and *A* constitute a relevant market and *C* and *A* constitute a second relevant market. Thus, even though the merger would lead to significantly higher prices, *B* and *C* would not be found to have products in the same relevant market!
45. For a suggestion of how to deal with this issue, see Gregory J. Werden, 'Market Delineation Algorithms Based on the Hypothetical Monopolist Paradigm,' *Economic Analysis Group Discussion Paper*, EAG 02-8, US Department of Justice (August 2002), Section 4.
46. As the *Guidelines* explain, the question asked in determining whether a set of products constitutes a relevant market is whether 'a hypothetical monopolist over that group of products would profitably impose *at least* a "small but significant and nontransitory" increase, including the price of a product of one of the merging firms. ... [W]hat constitutes a "small but significant and nontransitory" increase in price will depend on the nature of the industry, and the Agency at times may use a price increase that is larger or smaller than five percent'. (Emphasis added.) (*Merger Guidelines*, §1.11). It also is important to recognize that it may be appropriate to use a different threshold for determining whether competitive effects are significant than for delineating markets. For instance, in some markets a 2 percent price increase would result in a significant loss in consumer welfare, but a 10 percent price increase might be a useful hypothetical for market delineation.
47. That said, some consideration of supply-side factors is inevitable. As noted above, the geographic scopes of telecommunications markets are often point to point, and a central task of market delineation is to find an appropriate level of aggregation. Finding this level typically involves an assessment of business strategies (for example the geographic scopes of suppliers) and the incentives and abilities of suppliers to vary their service prices, features, and quality levels across consumer locations.
48. *Merger Guidelines*, §1.11.
49. Freephone and what are known in the USA as '800 numbers' are examples in which receivers are willing to take on all of the marginal charges associated with a call.
50. The following somewhat exaggerated example illustrates some of the forces at work. Suppose that mobile telephony subscribers want to be called on their mobile telephones only in the event of an emergency and would rather wait to receive time-insensitive calls on their fixed lines. High mobile termination charges levied on calling parties might induce that calling pattern.
51. For recent work on the pricing in the presence of two-sided benefits, see Benjamin E. Hermalin and Michael L. Katz, 'Sender or Receiver: Who Should Pay to Exchange an Electronic Message?' (2002) (unpublished manuscript on file with the author), and references therein.
52. The US Supreme Court recently described the current division of labor along the following lines in reviewing an objection to the manner in which the Commission implemented certain elements of the Telecommunications Act of 1996: 'Whether the [Commission] picked the best way to set these rates is the stuff of debate for economists and regulators versed in the technology of telecommunications and microeconomic pricing theory. The job of judges is to ask whether the Commission made choices reasonably within the pale of statutory possibility ...' *Verizon Communications Inc.*, v. *Federal Communications Commission*, 535 U.S. 467, 122 S.Ct. 1646, 1687 (2002).
53. In the USA, regulators are subject to judicial review, although they conceivably could move more quickly in reaching an initial decision.
54. This point was made by Patrick Rey during the European Commission workshop.
55. Another concern voiced by some is that the employees are rewarded for having made favorable decisions while in the government. Casual empiricism, however, suggests that companies look for competence (which may have been manifest by skillfully arguing against a company), rather than someone who has taken their side in the past.
56. This effect should not be overstated, however. Over the years, for example, the Commission has on its own initiative dramatically reduced its regulation of long-distance markets.
57. Of course, one could argue that the regulatory policies flowing from the Telecommunications Act of 1996 have fallen far short of creating the competition for which many in Congress had hoped and that regulatory policies ostensibly promoting universal service and diversity have long been both costly and ineffectual.

# 14. Telecommunications and competition

**Jean Tirole**

This chapter is in two sections. The first and shorter one will go through a series of institutional remarks on the role of antitrust in telecoms.

The second and more substantial section will discuss the economic challenges that competition authorities face in the industry, with an emphasis on innovation, IP (intellectual property) and co-operation among competitors.

## 1. INSTITUTIONAL ISSUES

### 1.1 Does Antitrust Have a Role in Telecoms?

The answer, I believe, is a qualified yes. For one thing, regulation may not exist, at least not in its modern understanding (Germany, in a number of industries, is a case in point). In addition, regulators may lack expertise in competition issues or may be captured by the industry, in which case competition authorities bring some checks and balances into the system. Another reason why I feel that the role of competition policy in telecoms will continue to expand is that technology is moving fast. Rapid technological progress makes it difficult for regulators to issue detailed *ex ante* regulations. To take an example, regulators routinely control charges for access to old-fashioned local loops; regardless of the controversies that the setting of such charges may raise, all of us will agree that things will not get easier as more intelligence is brought into the networks and as technological alternatives grow in number and quickly become obsolete. Large but hard-to-measure economic depreciation associated with technological progress and substantial demand uncertainty will by no means facilitate the regulator's task in setting access charges.

We should, however, not forget that competition authorities lack the detailed knowledge and quantitative evidence associated with a continuous oversight of an industry by a large staff dedicated to the task. It is then perhaps not surprising that competition authorities often feel more at ease asserting the existence of anticompetitive behaviours than setting termination charges or measuring predatory prices. Second, competition authorities, except in merger analyses, operate *ex post*, and often not in a very expedient way. The delays and uncertainties as to the outcome of the legal process regarding, say, the appropriate level of access charges would make it more difficult (in the absence of regulation) for telecoms and Internet operators to build solid business plans and to obtain financing for investments.

### 1.2 Do we Need a *sui generis* Treatment for Telecoms and Other High-tech Industries?

My opinion is 'no'. Economic reasoning is not industry-specific, and there is no strong argument in favour of a *sui generis* treatment.

To be certain, telecoms and high-tech industries confront us with a number of challenges that I will discuss shortly: how should essential facilities be treated? Is competition an unmitigated blessing? What attitude should authorities adopt *vis-à-vis* the many instances of co-operation among competitors?

Those are extremely complex questions for which the 'obvious answers' are unfortunately often misleading. What we need is a sophisticated analysis, with the corollary that the role of economic analysis should be considerably reinforced within the European Commission (the appointment of a chief economist at DG Competition is a step in this direction).

Let me now turn to some of the more substantive points related to the *conceptual challenges* that oversight authorities face in regulating telecoms and high-tech industries. I have selected three topics that, I feel, are particularly interesting both for practitioners, who must adapt their views to these new issues, and for academics, who still have much work to do on specific aspects.

## 2. CONCEPTUAL CHALLENGES

### 2.1 Essential Facility Doctrine

The essential facility doctrine plays, explicitly or, at least as often, implicitly, a key role in a number of policy interventions in the industry: structural remedies (such as break-ups or mandatory joint ownership), access pricing regulations, compulsory licensing of IP and so forth.

In antitrust practice, an essential facility (or a bottleneck) is an essential input (that is, very costly to duplicate) controlled by a dominant firm that has no objective reason to deny access to competitors. Essential facility cases first identify whether a facility is indeed essential in this sense and whether access to this facility is foreclosed by either vertical integration *cum* refusal to deal, incompatibility or high wholesale prices or, in the absence of vertical integration, by sweet deals/discriminatory licensing. Such a finding usually leads to the conclusion that the facility should be shared with competitors (at presumably a low price – here, courts or authorities are not very precise: they only mandate a reasonable and non-discriminatory access charge).

It is easy to see that this doctrine clashes with another legal doctrine – that related to IP protection. Many patents – or more generally pieces of IP – technically are essential facilities according to the previous definition, especially if they are particularly innovative. Their duplication involves large fixed costs. The patent owner often has a monopoly on the technology. And diffusion entails negligible marginal costs. Yet (and despite some controversial recent IP cases in Europe), authorities have always taken a lenient attitude toward intellectual property.

Reconciling these two conflicting approaches isn't that difficult. The natural presumption is that market power creates economic distortions, and so opened access, by enhancing competition, increases welfare. This *ex post* reasoning must, however, be complemented by the *ex ante* or 'Schumpeterian' view: it may well be the case that market power is the price to pay for the very existence of the essential facility. Large investments require large rewards. Hence the key question is whether the essentiality is associated with a legal monopoly (port, airport and so on), scale economies (Gamco, Hecht and the like), 'pure network externalities' (WorldCom–MCI), or genuine investment or innovation. (I say 'pure network externalities' because network externalities may themselves result from investments – penetration pricing, subsidies to complementary products, and so on – that make the analysis more complex.)

The latter category (genuine investment or innovation) clearly should differ in treatment from the former three sources of essentiality. The key issue is thus, whether we want it to or not, one of regulating the return on investment (by the way, this is exactly what patent policy tries in theory – and often fails in practice – to do, through the requirements of novelty, non-obviousness and usefulness, or copyright law through sweat of the brow or creative act requirements).

To sum up: (a) simple economics unifies the *a priori* divergent doctrines concerning IP protection and essential facilities; (b) whether access should be given to essential facilities requires a fine economic analysis rather than rigid rules. I do not mind compulsory licensing, but, for the same reason that Ramsey pricing principles are sometimes forgotten when computing access charges, I'm concerned about too systematic an application of the essential facility doctrine in the IP realm (I'd much rather start fixing our dysfunctional IP system first); (c) we should face the issue that the courts cannot always pass the buck by 'mandating access against proper compensation' (it's hard to disagree with this!). The regulatory debates that we have witnessed in telecoms access in the last ten years will surface with a vengeance in antitrust policy!

## 2.2 Economics of Networks

Telecoms, software, hardware and a number of other industries under antitrust scrutiny exhibit strong network externalities. Network externalities alter our economic thinking in several ways. In particular, social welfare requires consumers to somehow be able to interconnect.

There are basically three ways of reaping the social benefits of network externalities. The first is to create a seamless interconnection among platforms, with competitors possibly charging each other for termination. This is the paradigm adopted by telecoms, the Internet or common listing in real estate. The second way of benefiting from network externalities is to take stock of the existence of multiple, non-interconnected platforms and to let users themselves connect to multiple platforms – 'multi-home' – in order to connect to a maximum number of other users. Merchants often accept Amex, Visa and Mastercard, and some card-holders carry multiple credit cards. Game developers may port their games to multiple game platforms, advertisers advertise in multiple medias, and so forth.

Sometimes, though, multi-homing may be too costly to users, while interconnection proves difficult to set up either because it would infringe on the platforms' IP, or more simply because there is no regulator, and antitrust authorities have trouble defining the nature of interconnection as the technology evolves rapidly. This third situation creates scope for platform incompatibility, no multi-homing and a winner-takes-all market. As distrustful as we (rightly) are of monopoly, we should not rush into quick conclusions. As Mike Katz and Carl Shapiro (1985, 1986) observed long ago, actual competition also involves a socially costly loss of network externalities; similarly, a successful entry leaves installed-based users stranded with a technology with few new complements and a dwindling number of users to communicate with. In a paper with Drew Fudenberg (*JIE*, 2000), we observed that, in such situations, maintaining the threat of entry is much more important than creating actual entry, as entrants don't internalize the cost to stranded users and may further enter primarily to grab the incumbent's rent (on the other hand, innovative entry creates spillovers that are not internalized by the entrant, and so there may be too little or too much actual entry). As I said, the most important thing is to keep the incumbent on its toes, fostering network externalities through innovation, low prices and the subsidization of complementary products. By contrast, a monopolist that creates incompatibilities that are artificial (understand, not based on the protection of significant innovations), preys or engages in other anticompetitive behaviours isn't that great. Distinguishing between the two situations is a key challenge for antitrust authorities.

Let me now come back to the first two forms of platform competition, starting with non-interconnected platforms. There has been a large amount of theoretical research in the last two years emphasizing the two-sidedness of such markets. Namely, network externalities arise from getting two sides of a market together: users and software or game developers, web sites and dial-up customers, sellers and buyers, merchants and card-holders, authors and readers, speakers and audiences, advertisers and eyeballs, and so on. Theoretical research and actual business models converge on the view that a (privately and socially) good policy often involves substantial cross-subsidization between the two sides, perhaps with below-cost pricing on one side (think of free newspapers and TV or below-cost game consoles). The key objective of this price discrimination is to 'bring both sides on board'. Prices are therefore unlikely to – and actually should not – reflect any fair cost allocation, in which each side would pay to platforms the cost of servicing it. This wide academic consensus will, I hope, find its way into antitrust thinking, which does not yet fully reflect the specificities of two-sided markets.

Last, let me turn to interconnection and termination charges. Again, pre-packaged thinking will do us no good. For example, consider the familiar idea that small players have no market power. Small players in telecoms are still monopolies with regard to call terminating on their networks, and the literature actually shows that the smaller the network, the stronger its monopoly power (simply because a small network can tax the other networks for terminating calls without triggering a calling price response of these other networks). Quite generally, letting networks (whatever their size) select their termination charges would be foolish, at least as long as the other networks cannot pass the corresponding termination charges through to their customers by using termination-based price discrimination. Such termination-based price discrimination helps solve this problem, by limiting networks' incentive to jack up their access charges, but it creates other hazards. As J.J. Laffont et al. (1998a, 1998b) and Jeon et al. (2003) showed in a sequence of papers, charging higher prices for calling off net recreates the big-network strategic benefits that had been eliminated by mutual interconnection. It can further be shown that interconnection breakdowns, through high off-net prices, is not unlikely even among equal-sized networks.

In this respect, I would like to submit to your consideration the case of mobile-to-mobile telecoms in Germany. Termination charges are negotiated rather than regulated and termination-based-price discrimination is allowed – and practised. For example, D1 (a big network) in 2000 charged 68 pfennigs per minute for peak-time calls to D1 and 189 pfennigs per minute to calls to E-Plus or VIAG Intercom (a one-to-three ratio!). And, guess what? People chose the network of their relatives; and French mobile phone users (who were not subject to such practices) called twice as much as their German counterparts.

## 2.3 Co-operation among Competitors

Telecoms and high-tech industries require competitors to co-operate on a number of issues: they must interconnect so as to allow their users to enjoy network externalities (as we have seen), set up joint ventures so as to avoid the duplication of large fixed costs, and set standards and create patent pools (or cross-license IP). Needless to say, authorities have good reason to feel nervous about such agreements, which have the potential to cartelize the industry. But in many ways these agreements, if well targeted, are key to a good functioning of the industries.

Regarding *patent pools*, I feel that US antitrust authorities have taken important steps in their reviews of several pools (for example the MPEG2 pool) in the late 1990s. Patent pools that had proved so crucial to many industries until the Second World War had pretty much disappeared afterwards in a hostile antitrust environment. Patent pools, though, will be crucial to address the patent thicket (understand: repeated-tax) problem, in particular in the high-tech industries. There are of course concerns. For example, it is difficult to know whether inventions are complements or substitutes (and, if they are current complements, whether they will not later become substitutes). Josh Lerner and I have shown in a recent paper (Lerner and Tirole, 2003) that the policy of allowing independent licensing by the pool members acts as a clever screen in this respect, since this provision is irrelevant for welfare-enhancing pools and annihilates the perverse effects of a welfare-reducing pool. Pools raise many other fascinating issues, such as the automatic assignment of future related patents, and discriminatory licensing, in which the European and US antitrust authorities will have to reflect in the years to come.

The economics of *joint ventures* and not-for-profit associations should also figure high on our agenda. Much progress has been made in the last ten years toward understanding their governance and their competitive impact, but this knowledge is still quite incomplete, and what has been learned has only imperfectly filtered through into policy. In Chapter 10, Jörn Kruse argued that UMTS licensees should be allowed to share radio access networks (75 per cent of their cost) on the ground that competition will really take place in the core networks and content, and that fixed costs duplication in RANs is wasteful. A recent McKinsey Global Institute report (McKinsey, 2002) on French and German competitiveness argues that in mobile telephony France has achieved labour productivity levels in the 1990s twice as great as the US levels, in part due to the fragmentation of the US mobile industry and the resulting duplication of fixed costs. Such numbers suggest reducing the number of players (as long as vigorous competition survives, as it did in the French mobile triopoly) or else, if some congruence/convergence of views exists as to the technological design of these shared facilities, allowing joint ventures, perhaps in the form of not-for-profit associations in order to prevent collusion in reducing joint investments.[1] Again, much is to be learned both from an economic analysis and from the many concrete examples of co-operative undertakings.

To conclude: as we have seen, telecoms and high-tech industries are complex (but extremely interesting!); what we need is sophisticated analysis, not a *sui generis* treatment. I'm optimistic that policy-makers and academics will work together to produce a better understanding of the corresponding issues.

## NOTE

1. Of course, we must recognize the possibility of perverse ancillary restraints or of the threat of terminating co-operation with mavericks. As many authors have emphasized, joint ventures should not be turned into naked collusive devices.

## REFERENCES

Fudenberg, D. and J. Tirole (2000), 'Under the Threat of Entry by a Sole Supplier of a Network Good', *Journal of Industrial Economics*, **48** (4), pp. 373–90.

Jeon, D.S., J.J. Laffont and J. Tirole (2003), 'On the Receiver Pays Principle', *Rand Journal of Economics*, Summer.

Katz, M. and C. Shapiro (1985), 'Network Externalities, Competition, and Compatibility', *American Economic Review*, **75**, pp. 424–40.

Katz, M. and C. Shapiro (1986), 'Technology Adoption in the Presence of Network Externalities', *Journal of Political Economy*, **94**, pp. 822–41.

Laffont, J.J., P. Rey and J. Tirole (1998a), 'Network Competition: I. Overview and Nondiscriminatory Pricing', *Rand Journal of Economics*, **29** (1), pp. 1–37.

Laffont, J.J., P. Rey and J. Tirole (1998b), 'Network Competition: II. Price Discrimination', *Rand Journal of Economics*, **29** (1), pp. 38–56.

Lerner, J. and J. Tirole (2003), 'Efficient Patent Pools', mimeo.

McKinsey (2002), *Reaching Higher Productivity Growth in France and Germany*, McKinsey Global Institute report, October.

# PART II

# Monitoring Competition in the Telecommunications Sector

# 15. Monitoring competition in the telecommunications sector: European Commission Sector Inquiries

**Dessislava Choumelova and Juan Delgado***

## 1. INTRODUCTION

Thorough knowledge of the market is fundamental for the analysis of any potential anticompetitive practice and for the application of competition-law remedies. For this aim, the European Commission (hereafter, the Commission) can make use of a sector inquiry, which is a legal instrument to perform sector-wide investigations into commercial practices relating to an entire sector of the EU economy.

There are circumstances where markets do not seem to work perfectly but the causes for the potential market failure are not easily identifiable. There are also situations, such as liberalization processes or major changes in legislation, where competition problems can be envisaged but not easily singled out in advance. It is in those circumstances where the use of a sector inquiry might be convenient.

A sector inquiry analyses the situation and evolution of a specific market or sector across the EU and allows for the comparison of the functioning of a product/service market in different Member States. As a result of this analysis and the comparison between different countries, it may be possible to identify and investigate breaches of competition law.

A sector inquiry is a preliminary step in the application of competition law itself although it does not necessarily have to result in the opening of an infringement procedure. A sector inquiry is no more than a powerful investigation tool and does not pre-empt the existence of infringements of competition law. If as a result of the investigation it is found that there exist agreements that restrict the development of competition in the sector at hand or if an abuse of dominant position is identified, then infringement procedures under Article 81 (restrictive agreements) or Article 82 (abuse of dominant position) of the EC Treaty can be initiated.

The purpose of this chapter is to describe the legal basis and the circumstances under which a sector inquiry is launched and to analyse the aims and conclusions of the sector inquiries launched by the European Commission into the telecommunications sector.

## 2. LEGAL BASIS, SCOPE AND OBJECTIVES OF A SECTOR INQUIRY

The possibility of conducting sector inquiries is provided for in Article 12 of Council Regulation 17/62[1] (hereafter, Regulation 17/62), which allows the Commission to carry out a general investigation into an economic sector 'if in any sector of the economy the trend of trade between Member States, price movements, inflexibility of prices or other circumstances suggest that in the economic sector concerned competition is being restricted or distorted within the common market'.

This legal instrument has rarely been used by the Commission since the adoption of Regulation 17/62 in 1962. Apart from two long-lasting and complex sector inquiries into the margarine and beer sectors in the 1960s, the sector inquiry launched in the telecommunications sector constitutes only the third sector inquiry ever conducted by the Commission. One of the aims of the Commission's inquiries is to establish whether existing commercial practices in a whole sector of the economy infringe EU competition rules, and in particular the prohibition of restrictive practices and abuses of dominant position (Articles 81, 82 and/or 86 of the EC Treaty), but sector inquiries do not necessarily have to lead to the opening of *ex officio* individual antitrust cases.

The Commission's means and powers in conducting a sector inquiry are those under Regulation 17/62 and in particular Article 11 empowering the Commission to collect information by requesting it officially, as well as to conduct on-the-spot investigations. More precisely, the Commission, even in the absence of a complaint or notification, may formally request information from market parties under the threat of fines, may request and receive information and assistance from national authorities, and may carry out on-site inspections of company premises and business records.

There are no specific procedural rules for the running of a sector inquiry apart from the general legal basis set by Regulation 17/62. A sector inquiry usually implies the gathering and analysis of information provided by market players, consumers and national authorities. Partial and final findings may be discussed in closed workshops where participation is restricted or in public hearings which any interested party can attend. Those findings can also be published periodically in the form of a report.

## 3. REASONS TO LAUNCH A SECTOR INQUIRY IN THE TELECOMMUNICATIONS SECTOR

The Commission decided to launch a sector inquiry in the telecommunications sector on 27 July 1999, following a number of formal and informal complaints from users, service providers and new entrants. The sector inquiry covers the pricing and availability of three different telecommunications services, namely (1) leased lines, (2) mobile roaming and (3) unbundled local loop access. In 1999, all those were identified as services where, in spite of formal liberalization on 1 January 1998, competition was slow to develop. Both fixed and mobile incumbent operators were suspected of employing anticompetitive practices, and, as a result, competition was not developing fully at European level, to the detriment of both competitors and consumers.

Leased lines means the provision of wholesale domestic and/or international transmission capacity between two network points. New entrants normally do not have the same level of ubiquity as incumbent operators and may have to rely on the transmission services provided by incumbents in order to link two points of their network. European legislation requires leased lines prices to be cost-oriented.[2] The Commission had detected that there were large price differences between Member States and that that could be an indication of infringement of the cost-orientation obligation and, in some cases, of charging excessive prices with the objective of deterring entry.

International roaming services provide access to a foreign mobile network operator for the purposes of enabling its subscribers to make and receive calls abroad (where the operator's network does not have coverage and therefore the operator cannot serve its own customers directly). International wholesale roaming services are thus provided by a domestic mobile network operator (visited network) to a mobile network operator in another country (home network). The Commission had received a number of complaints regarding the high prices charged for roaming services in comparison with other technically similar services provided by mobile operators.

The local loop is the 'last mile' of a telecommunications network, that is, the access network connection between the customer's premises and first exchange of the telecommunications operator. Replication of the local loop is costly and often uneconomic. Therefore access to the incumbent's local loop is crucial for the new entrants in order to have full control over the services provided to the end-user. At the time the local loop sector inquiry was launched, the provision of access to the local loop was recommended but not yet mandated at European level. The objective of the sector inquiry was to monitor the evolution of competition at local level and, once local loop unbundling was made mandatory, to ensure that the process was developing in respect of the competition rules.

The sector inquiry could be thought as serving two major goals: (1) to examine the reasons why the services under scrutiny appeared not to be subject to effective competition, and (2) to generate broader sectoral knowledge and understanding of the markets.

The first question boiled down to examining whether or not the observed market failures (that is, high and rigid prices unrelated to costs, lack of competitive offers, lack of pan-European services, allegations of discrimination, slow development of competition) were resulting from anticompetitive practices or could be explained otherwise. Before considering any action based on the EU competition law, the Commission had to ascertain that the problems displayed by the leased line, mobile roaming and local loop access services were indeed due to anticompetitive conduct and not caused by other factors. Factors such as the technical nature of services, market structure, stage of liberalization of the different national markets, country-specific problems and role of the national telecoms regulators have also been taken into account in approaching the market analysis.

Moreover, even if the sector inquiry was to show that there were certain anticompetitive practices in the marketplace, the Commission had to decide on the right remedies. Action could be taken either at national level or at Community level and either by resorting to the EU competition law or to the telecom sector-specific rules. Thus, even though it comes under the procedural rules implementing the antitrust provisions of the EC Treaty (Articles 81 and 82), a sector inquiry does not necessarily have to end in a classical antitrust investigation.

The second, but not less important, goal of the sector inquiry was to generate sectoral knowledge. In a highly technological field as telecommunications, a profound understanding of the market requires substantial technical knowledge. In addition, the proliferation of different services, tariffs and charging mechanisms underpinned by different accounting arrangements makes the economics of telecoms services complex. Further, the sector inquiry allows for the gathering of comparative market data necessary in order to study and analyse the market dynamics at hand.

In order to collect comparative market data for the entire European Economic Area (EEA), the Commission decided to act in close co-operation with the EFTA Surveillance Authority (ESA),[3] which has carried out a parallel sector inquiry into the telecommunications sector in the EFTA states (Iceland, Liechtenstein and Norway) covering the same services: leased lines, mobile roaming and access to the local loop. This is the first sector inquiry ever launched by the ESA.[4]

For logistical reasons the practical fact-finding work into the three categories of telecommunications services were launched successively in 1999 and 2000. In addition, although formally parts or 'phases' of one and the same sector inquiry into three areas of the EEA telecommunications sector, in practice the three investigations have been handled as three separate 'inquiries' into three different categories of telecommunications services. The first fact-finding exercise concerned the leased lines[5] (October 1999), followed by mobile roaming[6] (formal launch in January 2000), and finally, local loop access[7] (formally started in July 2000).

The Annex contains more specific information on the reasons to open the three sector inquiries, their respective procedures, the investigations, the Commission's reports and working documents, public hearings and the follow-up measures.

## 4. MAIN RESULTS AND CONCLUSIONS OF THE INQUIRIES

In summary, the leased lines, the mobile roaming and the local loop access inquiries led to a better understanding of the sectors covered and provided indications of some important issues that needed further consideration and/or investigation.

However, unlike in Article 81 and 82 cases, in a sector inquiry the Commission cannot make formal findings of infringements of the competition rules by undertakings or groups of undertakings, or order such infringements to be brought to an end by means of a decision. Therefore, individual procedures have to be opened. However, the sector inquiries have provided sufficient *prima facie* evidence to warrant opening of individual cases with full due process guarantees, in which such findings of infringements can be made.

The sector inquiries have permitted the Commission to fully analyse the market characteristics and dynamics, the behaviour and strategies of the different agents within the market and to reflect on the role of the institutional factors in the development of competition in each market.

### 4.1 Overall Market Analysis

The sector inquiries provided the Commission with extensive information on market structure and characteristics, in particular on the number and identity of the main players, their market shares and the relationships between wholesale and retail markets. For example, the mobile roaming sector inquiry showed that the wholesale roaming markets were strongly oligopolistic and characterized by high concentration ratios, even higher than those for retail mobile services. This revealed a need to investigate further the reasons for this fact.

Detailed information on the functioning of markets and on market dynamics, trends and developments was also provided. For example, the leased lines sector inquiry showed that competition was starting to develop at wholesale level in some market segments (mainly long-distance). At retail level, there appeared to be competitive pressure, demonstrated by substantial discounts offered by incumbent operators. However, the fairness of certain discount schemes was questionable and left scope for enforcement of competition rules, where the aim was to pre-empt competition.

In the case of roaming, the market evolution was very stationary in terms of distribution of market shares and evolution of prices. In the case of local access, in most countries, a very high proportion of network accesses were in the hands of incumbent operators and big regulatory efforts were needed to make them available to competitors. The competitive pressure of alternative technologies for the provision of broadband services (for example cable, fibre-optics, wireless local loop, satellite, and so on) was residual in most Member States.

Technology plays a major role in the telecommunications sector. The sector inquiries helped the Commission to understand the technical nature of the services, the ways in which they are provided and the existence of viable alternatives to provide such services. The development of competition is sometimes constrained by technological factors. In some other cases, uncertainty on the future evolution of technology conditions the behaviour and investment decisions of the market players.

The sector inquiries also provided background information useful in the context of other competition procedures. They proved that market definitions used in past procedures (that is, mergers and antitrust procedures) needed to be updated or amended to reflect the changing market reality. For example, domestic leased lines markets could be thought of as consisting of separate economic markets for short-distance leased lines and long-distance leased lines additionally segmented according to the bandwidth. The geographical markets for national leased lines could probably be defined as consisting of big metropolitan/rest-of-the-country segments, while the question of whether the international leased lines markets are global, EU-wide or narrower required a more detailed examination of relevant conditions of competition.

## 4.2 Market Players' Behaviour and Strategies

The Commission also gathered comprehensive information relating to the overall level of prices and the market behaviour of certain operators which gave rise to a number of competition concerns.

The comparison of prices of national leased lines confirmed very divergent ways of pricing leased lines by the incumbents in different Member States. Possibly excessive prices have been identified in the bandwidths of 2 Mbs, 34 Mbs and 155 Mbs. Non-price-related problems in the competitive provision of leased lines were also highlighted in the replies to the inquiry questionnaires. The two most important seemed to be potentially abusive strategic discounting and discriminative delays in the provision of leased lines.

The local loop sector inquiry revealed the existence of considerable variations in the prices charged for local-loop-related services across the EEA. In some cases, those differences seemed not to be fully justified by diverging cost structures.

The market monitoring conducted within the sector inquiries had sometimes immediate effects on the market: leased line prices in a number of EEA countries went down even in the absence of formal infringement procedures, while the mobile operators across the EEA took measures to improve retail tariff information for the customers. The discussion with national authorities at the different stages of the sector inquiry helped them to tackle some of the obstacles identified at national level.

### 4.3 Institutional Aspects

EEA-wide sector inquiries are particularly useful to address cross-border issues and problems that cannot be dealt with by one single national authority. A clear example of this is provided by the international roaming sector inquiry. The provision of international roaming implies the intervention of national and foreign operators. However, national authorities do not have powers to request information or to intervene outside their national territory.

In a sector involved in a liberalization process, it is important to analyse the interplay between sectoral regulation and application of competition law in order to use the relevant remedies for the problems identified. The telecommunications sector is subject to sector-specific legislation to promote the development of competition. Once a market failure has been identified it should be analysed whether it is better addressed through the application of competition law or sector-specific regulation.

Finally, the sector inquiries highlighted the importance of co-operation with the national authorities on the issues at stake and division of labour between the Commission and the national authorities in tackling the market failures identified by the sector inquiries.

## 5. FOLLOW-UP ACTIONS OF THE INQUIRIES

### 5.1 Leased Lines Inquiry

Five *ex officio* cases were opened at the end of 2000 to further investigate international leased lines prices in close co-operation with the respective national authorities in five Member States, Belgium, Greece, Italy, Portugal and Spain. International leased line prices appeared unjustifiably high in these Member States.

The Commission did not limit itself to examining international leased line prices, but extended the scope of the investigation to short-distance leased line prices and issues of delivery times and conditions, in order to address all competitive concerns voiced by the industry. In order to decide whether to pursue its investigation the Commission used the following four criteria: (1) price reductions since the results of the sector inquiry of September 2000; (2) competitiveness of the markets: new entries and market dynamics; (3) introduction/application of service-level agreements (SLA) and other conditions fostering competition; (4) both national telecoms regulator's and competition authority's stance and likelihood of problems to be solved at national level.

Following the opening of the investigations, the incumbent operators in the five concerned Member States significantly reduced their leased lines tariffs, although the pattern of price decreases varied between these countries.

In addition, the national regulators in the five Member States have taken steps to address competition problems in the provision of leased lines by the incumbent operators. Measures that have contributed to remedying existing problems are:

- more control over the service-level agreement (SLA) and its enforcement;

- imposing wholesale leased lines mandatory offers on the incumbent;
- improved monitoring of cost orientation through imposition of more detailed cost-orientation models and accounting systems;
- inclusion of an interconnect leased circuit offer in the reference interconnection offer and review of the conditions applied;
- in-depth sector investigations launched by the regulators.

In December 2002 the Commission decided to close its *ex officio* investigations regarding Belgium and Italy given the evidence of significant improvements in the competitive situation in these Member States. Similarly, subject to a further decrease of near-EU international leased line prices or further justification of their level, the Spanish case could be closed. The Commission will continue to follow up closely the situation in Portugal and Greece.[8]

At the same time, the Commission decided to close its leased lines inquiry,[9] since the concerns related to high prices and issues of possible discrimination were being adequately addressed, both at national level through the enforcement of the EU sector-specific regulation by the national regulatory authorities, and through *ex officio* procedures by the European Commission relating to specific EU Member States. Two years following its launch, one of the principal outcomes of the sector inquiry is a considerable decrease in leased lines prices across the EEA. For example, since the launch of the sector inquiry on 27 July 1999, 2 Mbps international leased line prices have gone down by 30 per cent to 40 per cent on average. A second important outcome is a proactive stance of the national authorities regarding both pricing and provisioning of leased lines.

### 5.2 Mobile Roaming Inquiry

As a follow-up to the mobile roaming inquiry, in 2001 the Commission opened, among others, *ex officio* infringement cases regarding roaming prices charged by the mobile network operators in Germany and the UK, and in parallel carried out the examination of two notifications by the GSM Association relating to standard roaming agreements and wholesale pricing arrangements.

On 11 July 2001 Commission officials conducted inspections at the premises of the roaming operators in the UK and Germany.[10] These inspections were aimed at discovering information about possible collusion on mark-ups from retail to wholesale prices under Article 81 ECT, but also about excessive pricing by dominant operators under Article 82 ECT in both Germany and the UK. The Commission has also requested additional information from all operators concerned about the evolution of revenues and profits from roaming since 1998.

### 5.3 Local Loop Unbundling Inquiry

The local loop sector inquiry has constituted a useful tool not only for monitoring the market but also for detecting anticompetitive practices and the subsequent opening of antitrust procedures. As a result of the findings of the sector inquiry a number of competition cases related to access to the local loop and broadband Internet access were initiated.

The Commission opened a procedure against Wanadoo (the Internet subsidiary of France Telecom) in France regarding the pricing of broadband Internet access services. The Commission suspected that those services were being charged below cost, which could represent abuse of a dominant position under Article 82 of the EC Treaty.

The Commission also initiated a case against Deutsche Telekom in Germany regarding local loop pricing. The Commission was concerned about Deutsche Telekom's practice of charging new entrants higher fees for wholesale access to the local loop than the fees that Deutsche Telekom's subscribers pay for retail access. This practice discouraged new companies from entering the market.

On 21 May 2003, the Commission adopted a decision against Deutsche Telekom for abusing its dominant position through unfair prices for the provision of local access to its fixed telecommunications network.

The local loop sector inquiry continues, since the effective implementation of local loop unbundling is still an ongoing process and competition problems may still arise.

## 6. CONCLUSIONS

The sector inquiry in the field of telecommunications helped to achieve a number of goals. It highlighted remaining differences in the level of competitiveness in the provision of leased lines, roaming and local loop access in the EEA. It provided the Commission and the national authorities in the EEA Member States with detailed knowledge about the way the markets for these services function, the underlying market structures, the competition problems at stake and the trends for the future. It provided comparative data that served as a basis for opening of individual investigations against companies in several Member States.

The growing importance of this legal instrument was recognized by a broadened scope of a sector inquiry in the new procedural regulation applying Articles 81 and 82 ECT.[11] Article 17 of the new regulation provides not only for an investigation into a sector of the economy, but also for a new type of investigation – into types of agreements across various sectors of the economy.

In a changed competition-law reality, with the current modernization of the European competition-law procedures, the role of sector inquiries could be even more important. There will not be notifications of potentially competition-restrictive agreements, as was the case under the past regime, and therefore it will be useful to gather market data and information in order to identify areas where competition-law-based action is needed. On the other hand, a greater knowledge of certain markets and services will be helpful to share with the functioning of the network of competition authorities.

## NOTES

* The views expressed by the authors are purely personal and do not necessarily reflect the official position of the institutions the authors are affiliated to.

1. OJ P 013, 21.02.1962, pp. 0204–0211.
2. See Council Directive 92/44/EEC of 5 June 1992 on the application of open network provision to leased lines. OJ L165, 19/06/1992 P. 0027–0036.
3. Co-operation under the competition rules of the EEA Agreement. These rules are broadly equivalent to those of the EC Treaty, which are enforced across the EEA by the EFTA Surveillance Authority (ESA) and the European Commission. In the fields of competition, state aid and public procurement, the powers of the ESA mirror the competencies of the European Commission.

4. http://www.eftasurv.int/information/pressreleases/1999pr/dbaFile1250.html; see also information on the preliminary results in the ESA sectoral inquiry at http://www.eftasurv.int/fieldsofwork/fieldcompetition/activities/dbaFile1132.html
5. Commission launches first phase of sectoral inquiry into telecommunications: leased line tariffs. *Press release* IP/99/786 – 22.10.1999, http://www.europa.eu.int/rapid/start/cgi/guesten.ksh?p_action.gettxt=gt&doc=IP/99/786|0|RAPID&lg=EN
6. Commission launches second phase of telecommunications sector inquiry under the competition rules: mobile roaming. *Press release IP/00/111 – 04.02.2000*, http://europa.eu.int/rapid/start/cgi/guesten.ksh?p_action.gettxt=gt&doc=IP/00/111|0|RAPID&lg=EN.
7. Commission asks incumbent operators to provide information on local loop access. Press release IP/00/765 – 12.07.2000, Questionnaire addressed to incumbent operators http://europa.eu.int/comm/competition/antitrust/others/sector_inquiries/local_loop/
8. For more details, see Explanatory Memorandum of 10 December 2002, http://europa.eu.int/comm/competition/antitrust/others/sector_inquiries/leased_lines/exmemonovember2002.pdf
9. See Commission press release IP/02/1852 of 11 December 2002.
10. See Statement on inquiry regarding mobile roaming, MEMO/01/262 – 11.07.2001, http://europa.eu.int/comm/competition/antitrust/others/sector_inquiries/
11. Council Regulation (EC) No. 1/2003 of 16 December 2002 on the implementation of the rules on competition laid down in Articles 81 and 82 of the Treaty (Text with EEA relevance), OJ L1 04/01/2003, pp. 1–25.

# ANNEX

## A.1 The Leased Lines Sector Inquiry

On 27 July 1999 the Commission decided to open an inquiry relating, among other things, to the provision and pricing of leased lines in the 15 Member States. This inquiry was in line with the general consensus that had emerged from international organizations, national regulatory and competition authorities, and user organizations in Europe, that prices for leased lines were in certain cases unreasonably high and should be reduced. However, the Commission was especially aware of the fact that special care should be used with any regulatory intervention since mandating price reductions in potentially competitive areas could dissuade investment. The new framework directives adopted on 7 March 2002 (which will apply in all Member States from 25 July 2003) will reduce regulatory intervention as regards leased lines pricing, taking into account the growing competition in the area.

In 1999 more than 200 formal requests for information pursuant to Article 11 of Regulation 17/62 were sent to all incumbent fixed network operators, a number of newly authorized fixed telecommunications network operators (new entrants), as well as large business users and Member States' national competition and regulatory authorities in the EEA.

The first phase of the leased lines inquiry concluded with a public hearing in Brussels on 22 September 2000.[1]

The main findings of the inquiry, although provisional, can be summarized as follows:

- Competition was growing in certain markets although some price reductions could respond to entry deterrence strategies rather than to increasing competitive pressure.
- Demand was dramatically increasing, with the biggest driver being the Internet. The two major categories of users were telecoms companies (alternative carriers, new fixed entrants and mobile networks), ISPs and big business users.
- Supply was growing and was expected to further increase once wireless local loop was commercially provided and alternative infrastructure (carrier's carriers) was further deployed.
- There was a correlation between the number of suppliers in a given market and the level of prices.
- Domestic leased lines revenue accounted for from below 1 per cent to above 17 per cent of the total revenue for selected EU incumbent operators. For international leased lines, the proportion of revenue in total turnover ranged between 3 per cent and 27 per cent. In the Member States where the liberalization had been early and decisive, large volumes of sales and higher revenue from leased lines could be due to the maturity of those markets and the expansion of the ISPs, a major demand-driver.
- The market definitions that had been used so far in antitrust or merger procedures should be narrowed. Domestic leased lines markets could be thought of as consisting of separate economic markets for short-distance leased lines and long-distance leased lines additionally segmented according to the bandwidth. The geographical markets for national leased lines could probably be defined as consisting of big metropolitan/rest-of-the-country segments, while the question of whether the international leased lines markets were global, EU-wide or narrower was thought to require a more detailed examination of relevant conditions of competition.

- Using a benchmark for international leased lines confirmed that prices among different Member States diverged widely, and this divergence could not be explained by variation in distance only. Regarding the pricing of leased lines, it appeared that distance was relatively unimportant in comparison to density.
- The comparison of prices of national leased lines confirmed very divergent ways of pricing leased lines by the incumbents in different Member States. Possibly excessive prices were identified in the bandwidths of 2 Mbs, 34 Mbs and 155 Mbs.
- Non-price-related problems in the competitive provision of leased lines were also highlighted in the replies to the various Commission questionnaires sent to market players. The two most important seemed to be potentially abusive strategic discounting and discriminative delays in the provision of leased lines.

## A.2 The Mobile Roaming Sector Inquiry

At the beginning of 2000 around 130 formal requests for information were sent to all licensed mobile network operators, as well as independent and tied service providers and Member States' national competition and regulatory authorities covering the entire EEA. In a later stage of the inquiry, questionnaires were also addressed to mobile roaming brokers and independent providers of new roaming solutions, as well as to Swisscom, which appeared to play a significant role in the provision of roaming services EEA-wide.

Data collection on roaming started in January 2000, and continued in the form of additional information requests in April–June 2000 and September 2000. A Working Document summarizing the preliminary findings reached was prepared in November, followed by a meeting with the national competition and regulatory authorities to discuss those preliminary findings. The meeting took place on 24 November 2000 in Brussels. In addition, in response to the very intense public interest in the results of the both inquiries, non-confidential versions of the Commission's Working Documents were made available on the Commission web site.[2]

The initial findings of the roaming inquiry can be summarized as follows:

- There were at least two clearly distinct relevant markets: the wholesale roaming market and the retail roaming market. Both markets were national. In both of these markets the findings confirmed the concerns that led the Commission to open the sector inquiry.
- The national wholesale markets were characterized by high concentration ratios and by a general lack of price competition.
- During the period 1997–2000, national wholesale roaming prices differed considerably between the EEA states, more than was likely to be justified by cost differences. In addition, there had been substantial absolute as well as relative increases in wholesale roaming prices throughout this period. Because roaming retail prices were calculated by adding a mark-up on the relevant wholesale rate, roaming retail prices had gone up as well, in contrast with domestic mobile retail prices, which had generally decreased.
- Most of the operators that had the lowest wholesale tariffs in 1997 had raised their tariffs gradually over the period 1997–2000, while most of the operators with relatively high tariffs had lowered their tariffs. Hence, over the period under review,

wholesale tariffs had clearly converged towards a higher overall level that did not appear to bear any relation to cost.

Those findings gave rise to a number of competition concerns both on the behaviour of the agents and on structural characteristics of the market.

### A.2.1 Behavioural aspects

- Excessive pricing and price collusion were likely concerning both the level of wholesale rates and the mark-ups applied in retail markets.
- In a number of national markets (including those where the highest inter-operator tariffs – IOTs – were charged) the different mobile operators charged almost identical wholesale rates which appeared to reflect co-ordinated pricing behaviour or tacit collusion. In some of these markets, mobile operators applied similar retail tariffs as well.

### A.2.2 Structural aspects

- The high level of transparency of wholesale roaming tariffs favoured collective dominance in national wholesale markets. Other relevant structural factors that might contribute to possible collective dominance included high concentration ratios, a homogeneous product, similar cost structures, high barriers to entry, inelastic market demand, and the absence of a competitive fringe.
- The cross-border nature of roaming resulted in non-typical vertical relationships between wholesale and retail markets in different countries, because excessive prices in the national wholesale market in one EEA state led to increasing retail prices in the national retail markets in other EEA states.
- The GSM Association's standard international roaming agreement (STIRA) and the Infocentre operated in the framework of that agreement fostered price transparency and reinforced the existing oligopolistic market structure. Use of the STIRA also induced operators not to conclude unilateral roaming agreements with any market parties other than licensed mobile network operators.
- The standard non-discrimination obligations introduced by the GSM Association appeared to remove incentives to introduce procompetitive roaming services, for example based on preferential wholesale roaming agreements and transparent one-rate retail tariffs.

The report offered a number of recommendations and follow-up actions. Given the broad scope of the sector inquiry, which covered the entire EEA, co-ordination on its follow-up with the national competition authorities and regulatory authorities was perceived as necessary.

Moreover, in view of the structural problems found, it could not be excluded that in addition to competition-law enforcement, action might be required to remove any regulatory constraints on mobile operators that impeded their ability to introduce more competition into the roaming markets. In addition, the inquiry clearly revealed cross-border issues that could not be effectively tackled by any national competition or regulatory authority acting alone. Because unilateral action might be ineffective and might in some cases even result in hampering competition, not only co-ordination but also bilateral or EEA-wide action might be necessary.

The Commission expressed its intentions to adopt specific steps such as:

- to provide legal certainty to the industry and the national competition authorities concerning differentiated roaming tariffs that would foster competition in the roaming markets. Market players had expressed a need for Commission guidance on the application of competition law to roaming agreements, which could take the form of guidelines. Given the complexity of the issues at stake and the cross-border nature of roaming, the national competition authorities could also benefit from such guidance;
- to revisit the terms and conditions of the standard international roaming agreement and the inter-operator tariff[3] (IOT) arrangement to ensure that they did not reduce competitive pressure. In order to encourage the emergence of competitive offers, it should be ensured that even operators with market power would be able to apply different tariffs to roaming counterparts licensed in the same country while at the same time not discriminating against small and new market entrants;
- to ensure that users are informed about roaming tariffs, including via consumer associations;
- to continue the monitoring of the evolution of prices in close co-operation with the national authorities.

## A.3 The Local Loop Access Sector Inquiry

The inquiry on local loop unbundling (LLU) followed on from the conclusions of the Lisbon European Council of March 2000[4] and the initiatives already taken by the Commission, in particular its Recommendation[5] to the Member States calling for unbundled access to incumbent operators' local loops by 31 December 2000.

On 23 September 2000, the Commission adopted a Communication[6] which identified potential anticompetitive practices related to the provision of access to the local loop which could constitute an abuse of dominant position in breach of Article 82 of the EC Treaty. The Communication made explicit that competition rules applied to local loop unbundling irrespective of the relevant sector-specific legislation.

The purpose of the inquiry was to detect possible anticompetitive practices and to monitor the evolution of the local loop unbundling process across the EU. The technical and economic complexity of the LLU process and the potentiality for competition problems to arise made necessary the use of this specific tool that provides a close picture of the evolution of the process and enables quick reaction in case of any breach of competition law. In the context of the inquiry, input from both incumbents and new entrants has been requested in order to have the views of all interested parties.

The provision of unbundled access to the local loop by operators with significant market power was declared mandatory by EC Regulation 2887/2000 of 18 December 2000 on unbundled access to the local loop which came into force on 2 January 2001.[7] The Regulation was of immediate application to all Member States. Under this Regulation incumbents had to publish a reference unbundling offer (RUO) setting the technical and financial conditions of access to their local loop.

The LLU sector inquiry was initiated before unbundling was mandated by European sector-specific legislation.[8] Unbundling was already mandated by national law in some Member States (Denmark, Finland, Germany, the Netherlands and Sweden) but in most Member States the incumbents' local networks remained inaccessible to new entrants. In consequence, the competitive situation across the EU was very heterogeneous and, obviously, the set of problems confronted by operators and regulators was of a different nature.

On the one hand, in those countries where LLU was already mandated by national law, the conditions of access proved to be not entirely satisfactory (mainly the conditions related to prices, the obstacles to equipment collocation,[9] the quality of the service and the delays in the provision of information and in the delivery of lines). On the other hand, and as a result of the publication of the Recommendation and the Communication, there had been requests for access to the local loop in countries where unbundling was not yet mandated. In all cases, incumbents refused to provide unbundled access to their local network and regulators lacked the appropriate legal tools to respond to those requests.

At this preliminary stage, pricing problems were envisaged in several Member States since the process of rebalancing of tariffs of the different services offered by incumbents was not fully completed.[10] The unbundling of the local loop without having completed the process of tariff rebalancing could result in price squeezes since the retail fees could happen to be below the wholesale fees. This could have a distorting effect on the new entrants' decisions to attack certain market segments. The rebalancing of tariffs was then a necessary preliminary step to the opening to competition of the incumbents' access network.[11]

Another issue of concern at this initial stage was the launch by several incumbent operators of broadband Internet access services through the copper pair. The fact that the local loop was not available to competitors implied that only the incumbent operators could offer such services, giving them a first-mover advantage with the consequent potential danger of monopolization of an emerging new market. This meant that the necessity to impose the obligation of LLU became even more pressing and necessary in order to promote innovation and lower prices in the provision of those services.

The adoption of the LLU Regulation changed the scenario across Europe. The nature of the problems faced by new entrants when requesting access to incumbents' local loop converged. However, there were still wide differences due to the different degree of effective implementation of the Regulation and the different dates of entry into force of the LLU obligation.[12]

A new survey carried out by the Commission after the adoption of the Regulation reflected in a detailed manner the problems faced by new entrants in this new regulatory phase.[13]

New entrants seeking to rent copper lines from incumbents to deploy their own services (in particular broadband services) to end-users claimed to be confronted mainly with two sets of problems:[14] tariff- and cost-related problems inherent in the unsatisfactory economic conditions of unbundling throughout Europe (for example discriminatory prices, excessive prices, predatory prices, price squeeze, and so on); and behavioural problems, reflecting the reluctance of telecoms incumbents to open their premises and facilities to their competitors (for example refusal to supply, unjustifiable delays, discriminatory terms of provision, tying of services, and so on).

The conditions for collocation[15] and other related facilities appeared to be a major obstacle for new entrants (for example minimum space requirements, high collocation space rental fees, necessity to build separate rooms to accommodate their equipment, high fees for escorted access to the incumbents' premises, and so on). Many new entrants also alleged that they were confronted with price squeezes between wholesale and retail fees.

The survey revealed several crucial issues for the analysis of anticompetitive practices in the context of local loop unbundling.

First, special attention should be paid to the pricing of facilities associated with local loop unbundling. Apart from the connection fee and the monthly rental fee, there are other recurrent and non-recurrent costs incurred by the new entrants, such as the collocation costs and other associated facilities. If those additional costs (which are often not subject to regulation) are high enough, new entrants may be driven out of the market even though there is apparently a 'sufficient' margin between the wholesale and retail incumbent fees for the copper pair. High prices for collocation and associated facilities may cause a price squeeze between the relevant wholesale prices and the retail price.

For example, the price for collocation differed broadly from country to country (for example the price for 5 $m^2$ ranged from €178.36 per month in Germany in some of the main cities to €515.06 per month, plus a one-off fee of €4468.48 in Portugal. Most incumbents did not publish their price list but determined their prices on a case-by-case basis). Excessive prices might deter new entrants from installing their equipment in certain exchanges since they would never be able to recover the investment. The combination of high prices and the fact that the incumbents (and in some countries their subsidiaries) were not subject to those high costs make it impossible for new entrants to compete.

Second, access to collocation space on the same terms and conditions as it is provided internally by the incumbents to themselves (or to their subsidiaries) is fundamental for the development of competition between incumbents and new entrants on an equal basis. In a number of countries new entrants alleged that incumbents had denied collocation in some of their premises. The refusal to provide collocation space made it impossible to unbundle the loops of the users connected to those exchanges and therefore deprived those consumers of the benefits of competition in the provision of services. New entrants also claimed that the retail arms of the incumbent might be enjoying preferential treatment in many respects *vis-à-vis* its competitors.

Third, strict deadlines must be set to ensure that access and other related services are provided to new entrants in due time. The speed at which new entrants are able to roll out their services is of great importance in such a dynamic market as the telecoms market. Furthermore, the delays in the provision of collocation and the delivery of lines give a first-mover advantage to the incumbent in the provision of broadband services and other advanced services.

The partial conclusions of the LLU sector inquiry were discussed in the context of a public hearing held in Brussels on 8 July 2002[16] in which market agents presented and discussed their views on the evolution of LLU and the remaining problems.[17]

## NOTES

1. See Sector Inquiry on the Competitive Provision of Leased Lines in EU Telecoms Markets – Public Hearing, 22 September 2000, Brussels, http://europa.eu.int/comm/competition/antitrust/others/sector_inquiries/leased_lines/hearing_2000_09_22/.
2. See Commission Working Document on the Initial Results of the Leased Lines Inquiry, 8.09.00, http://

europa.eu.int/comm/competition/antitrust/others/sector_inquiries/leased_lines/working_document_on_initial_results.pdf, and Commission Working Document on the Initial Findings of the Sector Inquiry into Mobile Roaming Charges, http://europa.eu.int/comm/competition/antitrust/others/sector_inquiries/roaming/working_document_on_initial_results.pdf.

3. System of roaming calls wholesale rates that network operators charge each other.
4. In Lisbon in March 2000, the European Council set out a ten-year strategy intended to transform the Union into the most competitive and dynamic knowledge-based economy in the world, capable of sustainable economic growth with more and better jobs and greater social cohesion. Among other things, the Lisbon European Council requested the adoption of measures to achieve greater competition in local access networks and the adoption of the unbundling of the local loop. See http://ue.eu.int/en/Info/eurocouncil/index.htm
5. Recommendation 2000/417/EC (OJ L 156, 28.6.2000, p. 44).
6. Communication (2000/C 272/10) from the Commission – Unbundled access to the local loop: enabling the competitive provision of a full range of electronic communication services, including broadband multimedia and high-speed Internet (OJ C 272/55 of 23.09.2000, p. 55).
7. OJ L 336/4, 31.12.2000.
8. In July 2000, the Commission sent requests for information to the incumbent operators of all 15 Member States. See Commission press release IP/00/765 of 12/07/2000.
9. Collocation basically consists of the rental of space in the local exchanges of the incumbent and the right for the staff of new entrants to access the incumbent's buildings. The Unbundling Regulation gives alternative telecoms operators the right to install their own switching and other technical equipment in these premises. Collocation is indispensable to implementing local loop unbundling.
10. Tariff rebalancing was required by Directive 90/388/EEC. The tariff structure of incumbent operators was traditionally out of line with cost. Certain categories of services (generally rental fee and local calls) were provided at a loss and were cross-subsidized out of the profits from other categories (mainly long-distance calls). Artificially low prices for some services impeded competition since potential competitors had no incentive to enter into the relevant segment of the voice telephony market.
11. The Commission opened infringement procedures against Spain and Italy for not having allowed their respective incumbent operators to rebalance their tariffs. The procedure against Italy was suspended once the appropriate measures had been adopted.
12. Belgium and the UK added themselves to the list of Member States where LLU was mandated by national law shortly before the adoption of the regulation.
13. In July 2001, the Commission sent requests for information to about 200 new entrants in all the European Economic Area to obtain information on the difficulties they were facing in obtaining access to the local loop of the dominant operators. The request for information was also posted on the web page to allow any interested party to express its opinion. More than one hundred replies were received. A number of entrants had in the meantime abandoned their plans to enter into the market. The country origin of the replies was rather unbalanced, depending mainly, apart from on the size of the market, on the state of negotiations to implement local loop unbundling (those countries in which the process was more advanced registered a higher number of replies).
14. The conclusions of this second phase were summarized in a report which is available at http://europa.eu.int/comm/dgs/competition/index_en.htm. The Study consists of an overall report summarizing the main findings, and of country reports dealing with the situation as perceived by alternative telecoms operators in each individual EEA state.
15. See note 6.
16. Commission press release IP/02/849 of 12/06/2002.
17. Presentations at the LLU hearing of 8 July 2002 can be consulted at http://europa.eu.int/comm/competition/liberalization/telecom/local_loop/hearing.html

# PART III

# The New Regulatory Framework: Main Regulatory Texts

# DIRECTIVE 2002/21/EC OF THE EUROPEAN PARLIAMENT AND OF THE COUNCIL

## of 7 March 2002

## on a common regulatory framework for electronic communications networks and services (Framework Directive)

THE EUROPEAN PARLIAMENT AND THE COUNCIL OF THE EUROPEAN UNION,

Having regard to the Treaty establishing the European Community, and in particular Article 95 thereof,

Having regard to the proposal from the Commission [1],

Having regard to the opinion of the Economic and Social Committee [2],

Acting in accordance with the procedure laid down in Article 251 of the Treaty [3],

Whereas:

(1) The current regulatory framework for telecommunications has been successful in creating the conditions for effective competition in the telecommunications sector during the transition from monopoly to full competition.

(2) On 10 November 1999, the Commission presented a communication to the European Parliament, the Council, the Economic and Social Committee and the Committee of the Regions entitled 'Towards a new framework for electronic communications infrastructure and associated services – the 1999 communications review'. In that communication, the Commission reviewed the existing regulatory framework for telecommunications, in accordance with its obligation under Article 8 of Council Directive 90/387/EEC of 28 June 1990 on the establishment of the internal market for telecommunications services through the implementation of open network provision [4]. It also presented a series of policy proposals for a new regulatory framework for electronic communications infrastructure and associated services for public consultation.

(3) On 26 April 2000 the Commission presented a communication to the European Parliament, the Council, the Economic and Social Committee and the Committee of the Regions on the results of the public consultation on the 1999 communications review and orientations for the new regulatory framework. The communication summarised the public consultation and set out certain key orientations for the preparation of a new framework for electronic communications infrastructure and associated services.

(4) The Lisbon European Council of 23 and 24 March 2000 highlighted the potential for growth, competitiveness and job creation of the shift to a digital, knowledge-based economy. In particular, it emphasised the importance for Europe's businesses and citizens of access to an inexpensive, world-class communications infrastructure and a wide range of services.

(5) The convergence of the telecommunications, media and information technology sectors means all transmission networks and services should be covered by a single regulatory framework. That regulatory framework consists of this Directive and four specific Directives: Directive 2002/20/EC of the European Parliament and of the Council of 7 March 2002 on the authorisation of electronic communications networks and services (Authorisation Directive) [5], Directive 2002/19/EC of the European Parliament and of the Council of 7 March 2002 on access to, and interconnection of, electronic communications networks and associated facilities (Access Directive) [6], Directive 2002/22/EC of the European Parliament and of the Council of 7 March 2002 on universal service and users' rights relating to electronic communications networks and services (Universal Service Directive) [7], Directive 97/66/EC of the European Parliament and of the Council of 15 December 1997 concerning the processing of personal data and the protection of privacy in the telecommunications sector [8], (hereinafter referred to as 'the Specific Directives'). It is necessary to separate the regulation of transmission from the regulation of content. This framework does not therefore cover the content of services delivered over electronic communications networks using electronic communications services, such as broadcasting content, financial services and certain information society services, and is therefore without prejudice to measures taken at Community or national level in respect of such services, in compliance with Community law, in order to promote cultural and linguistic diversity and to ensure the defence of media pluralism. The content of

[1] OJ C 365 E, 19.12.2000, p. 198 and OJ C 270 E, 25.9.2001, p. 199.
[2] OJ C 123, 25.4.2001, p. 56.
[3] Opinion of the European Parliament of 1 March 2001 (OJ C 277, 1.10.2001, p. 91), Council Common Position of 17 September 2001 (OJ C 337, 30.11.2001, p. 34) and Decision of the European Parliament of 12 December 2001 (not yet published in the Official Journal). Council Decision of 14 February 2002.
[4] OJ L 192, 24.7.1990, p. 1. Directive as amended by Directive 97/51/EC of the European Parliament and of the Council (OJ L 295, 29.10.1997, p. 23).

[5] See page 21 of this Official Journal.
[6] See page 7 of this Official Journal.
[7] See page 51 of this Official Journal.
[8] OJ L 24, 30.1.1998, p. 1.

television programmes is covered by Council Directive 89/552/EEC of 3 October 1989 on the coordination of certain provisions laid down by law, regulation or administrative action in Member States concerning the pursuit of television broadcasting activities (1). The separation between the regulation of transmission and the regulation of content does not prejudice the taking into account of the links existing between them, in particular in order to guarantee media pluralism, cultural diversity and consumer protection.

(6) Audiovisual policy and content regulation are undertaken in pursuit of general interest objectives, such as freedom of expression, media pluralism, impartiality, cultural and linguistic diversity, social inclusion, consumer protection and the protection of minors. The Commission communication 'Principles and guidelines for the Community's audio-visual policy in the digital age', and the Council conclusions of 6 June 2000 welcoming this communication, set out the key actions to be taken by the Community to implement its audio-visual policy.

(7) The provisions of this Directive and the Specific Directives are without prejudice to the possibility for each Member State to take the necessary measures to ensure the protection of its essential security interests, to safeguard public policy and public security, and to permit the investigation, detection and prosecution of criminal offences, including the establishment by national regulatory authorities of specific and proportional obligations applicable to providers of electronic communications services.

(8) This Directive does not cover equipment within the scope of Directive 1999/5/EC of the European Parliament and of the Council of 9 March 1999 on radio equipment and telecommunications terminal equipment and the mutual recognition of their conformity (2), but does cover consumer equipment used for digital television. It is important for regulators to encourage network operators and terminal equipment manufacturers to cooperate in order to facilitate access by disabled users to electronic communications services.

(9) Information society services are covered by Directive 2000/31/EC of the European Parliament and of the Council of 8 June 2000 on certain legal aspects of information society services, in particular electronic commerce, in the internal market (Directive on electronic commerce) (3).

(10) The definition of 'information society service' in Article 1 of Directive 98/34/EC of the European Parliament and of the Council of 22 June 1998 laying down a procedure for the provision of information in the field of technical standards and regulations and of rules of information society services (4) spans a wide range of economic activities which take place on-line. Most of these activities are not covered by the scope of this Directive because they do not consist wholly or mainly in the conveyance of signals on electronic communications networks. Voice telephony and electronic mail conveyance services are covered by this Directive. The same undertaking, for example an Internet service provider, can offer both an electronic communications service, such as access to the Internet, and services not covered under this Directive, such as the provision of web-based content.

(11) In accordance with the principle of the separation of regulatory and operational functions, Member States should guarantee the independence of the national regulatory authority or authorities with a view to ensuring the impartiality of their decisions. This requirement of independence is without prejudice to the institutional autonomy and constitutional obligations of the Member States or to the principle of neutrality with regard to the rules in Member States governing the system of property ownership laid down in Article 295 of the Treaty. National regulatory authorities should be in possession of all the necessary resources, in terms of staffing, expertise, and financial means, for the performance of their tasks.

(12) Any party who is the subject of a decision by a national regulatory authority should have the right to appeal to a body that is independent of the parties involved. This body may be a court. Furthermore, any undertaking which considers that its applications for the granting of rights to install facilities have not been dealt with in accordance with the principles set out in this Directive should be entitled to appeal against such decisions. This appeal procedure is without prejudice to the division of competences within national judicial systems and to the rights of legal entities or natural persons under national law.

(13) National regulatory authorities need to gather information from market players in order to carry out their tasks effectively. Such information may also need to be gathered on behalf of the Commission, to allow it to fulfil its obligations under Community law. Requests for information should be proportionate and not impose an undue burden on undertakings. Information

(1) OJ L 298, 17.10.1989, p. 23. Directive as amended by Directive 97/36/EC of the European Parliament and of the Council (OJ L 202, 30.7.1997, p. 60).
(2) OJ L 91, 7.4.1999, p. 10.
(3) OJ L 178, 17.7.2000, p. 1.
(4) OJ L 204, 21.7.1998, p. 37. Directive as amended by Directive 98/48/EC (OJ L 217, 5.8.1998, p. 18).

gathered by national regulatory authorities should be publicly available, except in so far as it is confidential in accordance with national rules on public access to information and subject to Community and national law on business confidentiality.

(14) Information that is considered confidential by a national regulatory authority, in accordance with Community and national rules on business confidentiality, may only be exchanged with the Commission and other national regulatory authorities where such exchange is strictly necessary for the application of the provisions of this Directive or the Specific Directives. The information exchanged should be limited to that which is relevant and proportionate to the purpose of such an exchange.

(15) It is important that national regulatory authorities consult all interested parties on proposed decisions and take account of their comments before adopting a final decision. In order to ensure that decisions at national level do not have an adverse effect on the single market or other Treaty objectives, national regulatory authorities should also notify certain draft decisions to the Commission and other national regulatory authorities to give them the opportunity to comment. It is appropriate for national regulatory authorities to consult interested parties on all draft measures which have an effect on trade between Member States. The cases where the procedures referred to in Articles 6 and 7 apply are defined in this Directive and in the Specific Directives. The Commission should be able, after consulting the Communications Committee, to require a national regulatory authority to withdraw a draft measure where it concerns definition of relevant markets or the designation or not of undertakings with significant market power, and where such decisions would create a barrier to the single market or would be incompatible with Community law and in particular the policy objectives that national regulatory authorities should follow. This procedure is without prejudice to the notification procedure provided for in Directive 98/34/EC and the Commission's prerogatives under the Treaty in respect of infringements of Community law.

(16) National regulatory authorities should have a harmonised set of objectives and principles to underpin, and should, where necessary, coordinate their actions with the regulatory authorities of other Member States in carrying out their tasks under this regulatory framework.

(17) The activities of national regulatory authorities established under this Directive and the Specific Directives contribute to the fulfilment of broader policies in the areas of culture, employment, the environment, social cohesion and town and country planning.

(18) The requirement for Member States to ensure that national regulatory authorities take the utmost account of the desirability of making regulation technologically neutral, that is to say that it neither imposes nor discriminates in favour of the use of a particular type of technology, does not preclude the taking of proportionate steps to promote certain specific services where this is justified, for example digital television as a means for increasing spectrum efficiency.

(19) Radio frequencies are an essential input for radio-based electronic communications services and, in so far as they relate to such services, should therefore be allocated and assigned by national regulatory authorities according to a set of harmonised objectives and principles governing their action as well as to objective, transparent and non-discriminatory criteria, taking into account the democratic, social, linguistic and cultural interests related to the use of frequency. It is important that the allocation and assignment of radio frequencies is managed as efficiently as possible. Transfer of radio frequencies can be an effective means of increasing efficient use of spectrum, as long as there are sufficient safeguards in place to protect the public interest, in particular the need to ensure transparency and regulatory supervision of such transfers. Decision No 676/2002/EC of the European Parliament and of the Council of 7 March 2002 on a regulatory framework for radio spectrum policy in the European Community (Radio Spectrum Decision) ([1]) establishes a framework for harmonisation of radio frequencies, and action taken under this Directive should seek to facilitate the work under that Decision.

(20) Access to numbering resources on the basis of transparent, objective and non-discriminatory criteria is essential for undertakings to compete in the electronic communications sector. All elements of national numbering plans should be managed by national regulatory authorities, including point codes used in network addressing. Where there is a need for harmonisation of numbering resources in the Community to support the development of pan-European services, the Commission may take technical implementing measures using its executive powers. Where this is appropriate to ensure full global interoperability of services, Member States should coordinate their national positions in accordance with the Treaty in international organisations and fora where numbering decisions are taken. The provisions of this Directive do not establish any new areas of responsibility for the national regulatory authorities in the field of Internet naming and addressing.

([1]) See page 1 of this Official Journal.

(21) Member States may use, *inter alia*, competitive or comparative selection procedures for the assignment of radio frequencies as well as numbers with exceptional economic value. In administering such schemes, national regulatory authorities should take into account the provisions of Article 8.

(22) It should be ensured that procedures exist for the granting of rights to install facilities that are timely, non-discriminatory and transparent, in order to guarantee the conditions for fair and effective competition. This Directive is without prejudice to national provisions governing the expropriation or use of property, the normal exercise of property rights, the normal use of the public domain, or to the principle of neutrality with regard to the rules in Member States governing the system of property ownership.

(23) Facility sharing can be of benefit for town planning, public health or environmental reasons, and should be encouraged by national regulatory authorities on the basis of voluntary agreements. In cases where undertakings are deprived of access to viable alternatives, compulsory facility or property sharing may be appropriate. It covers *inter alia*: physical co-location and duct, building, mast, antenna or antenna system sharing. Compulsory facility or property sharing should be imposed on undertakings only after full public consultation.

(24) Where mobile operators are required to share towers or masts for environmental reasons, such mandated sharing may lead to a reduction in the maximum transmitted power levels allowed for each operator for reasons of public health, and this in turn may require operators to install more transmission sites to ensure national coverage.

(25) There is a need for *ex ante* obligations in certain circumstances in order to ensure the development of a competitive market. The definition of significant market power in the Directive 97/33/EC of the European Parliament and of the Council of 30 June 1997 on interconnection in telecommunications with regard to ensuring universal service and interoperability through application of the principles of open network provision (ONP) [1] has proved effective in the initial stages of market opening as the threshold for *ex ante* obligations, but now needs to be adapted to suit more complex and dynamic markets. For this reason, the definition used in this Directive is equivalent to the concept of dominance as defined in the case law of the Court of Justice and the Court of First Instance of the European Communities.

(26) Two or more undertakings can be found to enjoy a joint dominant position not only where there exist structural or other links between them but also where the structure of the relevant market is conducive to coordinated effects, that is, it encourages parallel or aligned anti-competitive behaviour on the market.

(27) It is essential that *ex ante* regulatory obligations should only be imposed where there is not effective competition, i.e. in markets where there are one or more undertakings with significant market power, and where national and Community competition law remedies are not sufficient to address the problem. It is necessary therefore for the Commission to draw up guidelines at Community level in accordance with the principles of competition law for national regulatory authorities to follow in assessing whether competition is effective in a given market and in assessing significant market power. National regulatory authorities should analyse whether a given product or service market is effectively competitive in a given geographical area, which could be the whole or a part of the territory of the Member State concerned or neighbouring parts of territories of Member States considered together. An analysis of effective competition should include an analysis as to whether the market is prospectively competitive, and thus whether any lack of effective competition is durable. Those guidelines will also address the issue of newly emerging markets, where de facto the market leader is likely to have a substantial market share but should not be subjected to inappropriate obligations. The Commission should review the guidelines regularly to ensure that they remain appropriate in a rapidly developing market. National regulatory authorities will need to cooperate with each other where the relevant market is found to be transnational.

(28) In determining whether an undertaking has significant market power in a specific market, national regulatory authorities should act in accordance with Community law and take into the utmost account the Commission guidelines.

(29) The Community and the Member States have entered into commitments in relation to standards and the regulatory framework of telecommunications networks and services in the World Trade Organisation.

[1] OJ L 199, 26.7.1997, p. 32. Directive as amended by Directive 98/61/EC (OJ L 268, 3.10.1998, p. 37).

(30) Standardisation should remain primarily a market-driven process. However there may still be situations where it is appropriate to require compliance with specified standards at Community level to ensure interoperability in the single market. At national level, Member States are subject to the provisions of Directive 98/34/EC. Directive 95/47/EC of the European Parliament and of the Council of 24 October 1995 on the use of standards for the transmission of television signals [1] did not mandate any specific digital television transmission system or service requirement. Through the Digital Video Broadcasting Group, European market players have developed a family of television transmission systems that have been standardised by the European Telecommunications Standards Institute (ETSI) and have become International Telecommunication Union recommendations. Any decision to make the implementation of such standards mandatory should follow a full public consultation. Standardisation procedures under this Directive are without prejudice to the provisions of Directive 1999/5/EC, Council Directive 73/23/EEC of 19 February 1973 on the harmonisation of the laws of Member States relating to electrical equipment designed for use within certain voltage limits [2] and Council Directive 89/336/EEC of 3 May 1989 on the approximation of the laws of the Member States relating to electromagnetic compatibility [3].

(31) Interoperability of digital interactive television services and enhanced digital television equipment, at the level of the consumer, should be encouraged in order to ensure the free flow of information, media pluralism and cultural diversity. It is desirable for consumers to have the capability of receiving, regardless of the transmission mode, all digital interactive television services, having regard to technological neutrality, future technological progress, the need to promote the take-up of digital television, and the state of competition in the markets for digital television services. Digital interactive television platform operators should strive to implement an open application program interface (API) which conforms to standards or specifications adopted by a European standards organisation. Migration from existing APIs to new open APIs should be encouraged and organised, for example by Memoranda of Understanding between all relevant market players. Open APIs facilitate interoperability, i.e. the portability of interactive content between delivery mechanisms, and full functionality of this content on enhanced digital television equipment. However, the need not to hinder the functioning of the receiving equipment and to protect it from malicious attacks, for example from viruses, should be taken into account.

(32) In the event of a dispute between undertakings in the same Member State in an area covered by this Directive or the Specific Directives, for example relating to obligations for access and interconnection or to the means of transferring subscriber lists, an aggrieved party that has negotiated in good faith but failed to reach agreement should be able to call on the national regulatory authority to resolve the dispute. National regulatory authorities should be able to impose a solution on the parties. The intervention of a national regulatory authority in the resolution of a dispute between undertakings providing electronic communications networks or services in a Member State should seek to ensure compliance with the obligations arising under this Directive or the Specific Directives.

(33) In addition to the rights of recourse granted under national or Community law, there is a need for a simple procedure to be initiated at the request of either party in a dispute, to resolve cross-border disputes which lie outside the competence of a single national regulatory authority.

(34) A single Committee should replace the 'ONP Committee' instituted by Article 9 of Directive 90/387/EEC and the Licensing Committee instituted by Article 14 of Directive 97/13/EC of the European Parliament and of the Council of 10 April 1997 on a common framework for general authorisations and individual licences in the field of telecommunications services [4].

(35) National regulatory authorities and national competition authorities should provide each other with the information necessary to apply the provisions of this Directive and the Specific Directives, in order to allow them to cooperate fully together. In respect of the information exchanged, the receiving authority should ensure the same level of confidentiality as the originating authority.

(36) The Commission has indicated its intention to set up a European regulators group for electronic communications networks and services which would constitute a suitable mechanism for encouraging cooperation and coordination of national regulatory authorities, in order to promote the development of the internal market for electronic communications networks and services, and to seek to achieve consistent application, in all Member States, of the provisions set out in this Directive and the Specific Directives, in particular in areas where national law implementing Community law gives national regulatory authorities considerable discretionary powers in application of the relevant rules.

(1) OJ L 281, 23.11.1995, p. 51.
(2) OJ L 77, 26.3.1973, p. 29.
(3) OJ L 139, 23.5.1989, p. 19.
(4) OJ L 117, 7.5.1997, p. 15.

L 108/38 EN Official Journal of the European Communities 24.4.2002

(37) National regulatory authorities should be required to cooperate with each other and with the Commission in a transparent manner to ensure consistent application, in all Member States, of the provisions of this Directive and the Specific Directives. This cooperation could take place, *inter alia*, in the Communications Committee or in a group comprising European regulators. Member States should decide which bodies are national regulatory authorities for the purposes of this Directive and the Specific Directives.

(38) Measures that could affect trade between Member States are measures that may have an influence, direct or indirect, actual or potential, on the pattern of trade between Member States in a manner which might create a barrier to the single market. They comprise measures that have a significant impact on operators or users in other Member States, which include, *inter alia*: measures which affect prices for users in other Member States; measures which affect the ability of an undertaking established in another Member State to provide an electronic communications service, and in particular measures which affect the ability to offer services on a transnational basis; and measures which affect market structure or access, leading to repercussions for undertakings in other Member States.

(39) The provisions of this Directive should be reviewed periodically, in particular with a view to determining the need for modification in the light of changing technological or market conditions.

(40) The measures necessary for the implementation of this Directive should be adopted in accordance with Council Decision 1999/468/EC of 28 June 1999 laying down the procedures for the exercise of implementing powers conferred on the Commission [1].

(41) Since the objectives of the proposed action, namely achieving a harmonised framework for the regulation of electronic communications services, electronic communications networks, associated facilities and associated services cannot be sufficiently achieved by the Member States and can therefore, by reason of the scale and effects of the action, be better achieved at Community level, the Community may adopt measures in accordance with the principle of subsidiarity as set out in Article 5 of the Treaty. In accordance with the principle of proportionality, as set out in that Article, this Directive does not go beyond what is necessary for those objectives.

(42) Certain directives and decisions in this field should be repealed.

(43) The Commission should monitor the transition from the existing framework to the new framework, and may in particular, at an appropriate time, bring forward a proposal to repeal Regulation (EC) No 2887/2000 of the European Parliament and of the Council of 18 December 2000 on unbundled access to the local loop [2],

HAVE ADOPTED THIS DIRECTIVE:

CHAPTER I

## SCOPE, AIM AND DEFINITIONS

*Article 1*

### Scope and aim

1. This Directive establishes a harmonised framework for the regulation of electronic communications services, electronic communications networks, associated facilities and associated services. It lays down tasks of national regulatory authorities and establishes a set of procedures to ensure the harmonised application of the regulatory framework throughout the Community.

2. This Directive as well as the Specific Directives are without prejudice to obligations imposed by national law in accordance with Community law or by Community law in respect of services provided using electronic communications networks and services.

3. This Directive as well as the Specific Directives are without prejudice to measures taken at Community or national level, in compliance with Community law, to pursue general interest objectives, in particular relating to content regulation and audio-visual policy.

4. This Directive and the Specific Directives are without prejudice to the provisions of Directive 1999/5/EC.

*Article 2*

### Definitions

For the purposes of this Directive:

(a) 'electronic communications network' means transmission systems and, where applicable, switching or routing equipment and other resources which permit the conveyance of signals by wire, by radio, by optical or by other electromagnetic means, including satellite networks, fixed (circuit- and packet-switched, including Internet) and mobile terrestrial networks, electricity cable systems, to the extent that they are used for the purpose of transmitting

(1) OJ L 184, 17.7.1999, p. 23.

(2) OJ L 336, 30.12.2000, p. 4.

signals, networks used for radio and television broadcasting, and cable television networks, irrespective of the type of information conveyed;

(b) 'transnational markets' means markets identified in accordance with Article 15(4) covering the Community or a substantial part thereof;

(c) 'electronic communications service' means a service normally provided for remuneration which consists wholly or mainly in the conveyance of signals on electronic communications networks, including telecommunications services and transmission services in networks used for broadcasting, but exclude services providing, or exercising editorial control over, content transmitted using electronic communications networks and services; it does not include information society services, as defined in Article 1 of Directive 98/34/EC, which do not consist wholly or mainly in the conveyance of signals on electronic communications networks;

(d) 'public communications network' means an electronic communications network used wholly or mainly for the provision of publicly available electronic communications services;

(e) 'associated facilities' means those facilities associated with an electronic communications network and/or an electronic communications service which enable and/or support the provision of services via that network and/or service. It includes conditional access systems and electronic programme guides;

(f) 'conditional access system' means any technical measure and/or arrangement whereby access to a protected radio or television broadcasting service in intelligible form is made conditional upon subscription or other form of prior individual authorisation;

(g) 'national regulatory authority' means the body or bodies charged by a Member State with any of the regulatory tasks assigned in this Directive and the Specific Directives;

(h) 'user' means a legal entity or natural person using or requesting a publicly available electronic communications service;

(i) 'consumer' means any natural person who uses or requests a publicly available electronic communications service for purposes which are outside his or her trade, business or profession;

(j) 'universal service' means the minimum set of services, defined in Directive 2002/22/EC (Universal Service Directive), of specified quality which is available to all users regardless of their geographical location and, in the light of specific national conditions, at an affordable price;

(k) 'subscriber' means any natural person or legal entity who or which is party to a contract with the provider of publicly available electronic communications services for the supply of such services;

(l) 'Specific Directives' means Directive 2002/20/EC (Authorisation Directive), Directive 2002/19/EC (Access Directive), Directive 2002/22/EC (Universal Service Directive) and Directive 97/66/EC;

(m) 'provision of an electronic communications network' means the establishment, operation, control or making available of such a network;

(n) 'end-user' means a user not providing public communications networks or publicly available electronic communications services.

(o) 'enhanced digital television equipment' means set-top boxes intended for connection to television sets or integrated digital television sets, able to receive digital interactive television services;

(p) 'application program interface (API)' means the software interfaces between applications, made available by broadcasters or service providers, and the resources in the enhanced digital television equipment for digital television and radio services.

## CHAPTER II

## NATIONAL REGULATORY AUTHORITIES

### *Article 3*

### National regulatory authorities

1. Member States shall ensure that each of the tasks assigned to national regulatory authorities in this Directive and the Specific Directives is undertaken by a competent body.

2. Member States shall guarantee the independence of national regulatory authorities by ensuring that they are legally distinct from and functionally independent of all organisations providing electronic communications networks, equipment or services. Member States that retain ownership or control of undertakings providing electronic communications networks and/or services shall ensure effective structural separation of the regulatory function from activities associated with ownership or control.

3. Member States shall ensure that national regulatory authorities exercise their powers impartially and transparently.

4. Member States shall publish the tasks to be undertaken by national regulatory authorities in an easily accessible form, in particular where those tasks are assigned to more than one

body. Member States shall ensure, where appropriate, consultation and cooperation between those authorities, and between those authorities and national authorities entrusted with the implementation of competition law and national authorities entrusted with the implementation of consumer law, on matters of common interest. Where more than one authority has competence to address such matters, Member States shall ensure that the respective tasks of each authority are published in an easily accessible form.

5. National regulatory authorities and national competition authorities shall provide each other with the information necessary for the application of the provisions of this Directive and the Specific Directives. In respect of the information exchanged, the receiving authority shall ensure the same level of confidentiality as the originating authority.

6. Member States shall notify to the Commission all national regulatory authorities assigned tasks under this Directive and the Specific Directives, and their respective responsibilities.

*Article 4*

**Right of appeal**

1. Member States shall ensure that effective mechanisms exist at national level under which any user or undertaking providing electronic communications networks and/or services who is affected by a decision of a national regulatory authority has the right of appeal against the decision to an appeal body that is independent of the parties involved. This body, which may be a court, shall have the appropriate expertise available to it to enable it to carry out its functions. Member States shall ensure that the merits of the case are duly taken into account and that there is an effective appeal mechanism. Pending the outcome of any such appeal, the decision of the national regulatory authority shall stand, unless the appeal body decides otherwise.

2. Where the appeal body referred to in paragraph 1 is not judicial in character, written reasons for its decision shall always be given. Furthermore, in such a case, its decision shall be subject to review by a court or tribunal within the meaning of Article 234 of the Treaty.

*Article 5*

**Provision of information**

1. Member States shall ensure that undertakings providing electronic communications networks and services provide all the information, including financial information, necessary for national regulatory authorities to ensure conformity with the provisions of, or decisions made in accordance with, this Directive and the Specific Directives. These undertakings shall provide such information promptly on request and to the timescales and level of detail required by the national regulatory authority. The information requested by the national regulatory authority shall be proportionate to the performance of that task. The national regulatory authority shall give the reasons justifying its request for information.

2. Member States shall ensure that national regulatory authorities provide the Commission, after a reasoned request, with the information necessary for it to carry out its tasks under the Treaty. The information requested by the Commission shall be proportionate to the performance of those tasks. Where the information provided refers to information previously provided by undertakings at the request of the national regulatory authority, such undertakings shall be informed thereof. To the extent necessary, and unless the authority that provides the information has made an explicit and reasoned request to the contrary, the Commission shall make the information provided available to another such authority in another Member State.

Subject to the requirements of paragraph 3, Member States shall ensure that the information submitted to one national regulatory authority can be made available to another such authority in the same or different Member State, after a substantiated request, where necessary to allow either authority to fulfil its responsibilities under Community law.

3. Where information is considered confidential by a national regulatory authority in accordance with Community and national rules on business confidentiality, the Commission and the national regulatory authorities concerned shall ensure such confidentiality.

4. Member States shall ensure that, acting in accordance with national rules on public access to information and subject to Community and national rules on business confidentiality, national regulatory authorities publish such information as would contribute to an open and competitive market.

5. National regulatory authorities shall publish the terms of public access to information as referred to in paragraph 4, including procedures for obtaining such access.

*Article 6*

**Consultation and transparency mechanism**

Except in cases falling within Articles 7(6), 20 or 21 Member States shall ensure that where national regulatory authorities intend to take measures in accordance with this Directive or the Specific Directives which have a significant impact on the relevant market, they give interested parties the opportunity to comment on the draft measure within a reasonable period. National regulatory authorities shall publish their national consultation procedures. Member States shall ensure the establishment of a single information point through which all current consultations can be accessed. The results of the consultation procedure shall be made publicly available by the national regulatory authority, except in the case of confidential information in accordance with Community and national law on business confidentiality.

*Article 7*

**Consolidating the internal market for electronic communications**

1. In carrying out their tasks under this Directive and the Specific Directives, national regulatory authorities shall take the utmost account of the objectives set out in Article 8, including in so far as they relate to the functioning of the internal market.

2. National regulatory authorities shall contribute to the development of the internal market by cooperating with each other and with the Commission in a transparent manner to ensure the consistent application, in all Member States, of the provisions of this Directive and the Specific Directives. To this end, they shall, in particular, seek to agree on the types of instruments and remedies best suited to address particular types of situations in the market place.

3. In addition to the consultation referred to in Article 6, where a national regulatory authority intends to take a measure which:

(a) falls within the scope of Articles 15 or 16 of this Directive, Articles 5 or 8 of Directive 2002/19/EC (Access Directive) or Article 16 of Directive 2002/22/EC (Universal Service Directive), and

(b) would affect trade between Member States,

it shall at the same time make the draft measure accessible to the Commission and the national regulatory authorities in other Member States, together with the reasoning on which the measure is based, in accordance with Article 5(3), and inform the Commission and other national regulatory authorities thereof. National regulatory authorities and the Commission may make comments to the national regulatory authority concerned only within one month or within the period referred to in Article 6 if that period is longer. The one-month period may not be extended.

4. Where an intended measure covered by paragraph 3 aims at:

(a) defining a relevant market which differs from those defined in the recommendation in accordance with Article 15(1), or

(b) deciding whether or not to designate an undertaking as having, either individually or jointly with others, significant market power, under Article 16(3), (4) or (5),

and would affect trade between Member States and the Commission has indicated to the national regulatory authority that it considers that the draft measure would create a barrier to the single market or if it has serious doubts as to its compatibility with Community law and in particular the objectives referred to in Article 8, then the draft measure shall not be adopted for a further two months. This period may not be extended. Within this period the Commission may, in accordance with the procedure referred to in Article 22(2), take a decision requiring the national regulatory authority concerned to withdraw the draft measure. This decision shall be accompanied by a detailed and objective analysis of why the Commission considers that the draft measure should not be adopted together with specific proposals for amending the draft measure.

5. The national regulatory authority concerned shall take the utmost account of comments of other national regulatory authorities and the Commission and may, except in cases covered by paragraph 4, adopt the resulting draft measure and, where it does so, shall communicate it to the Commission.

6. In exceptional circumstances, where a national regulatory authority considers that there is an urgent need to act, by way of derogation from the procedure set out in paragraphs 3 and 4, in order to safeguard competition and protect the interests of users, it may immediately adopt proportionate and provisional measures. It shall, without delay, communicate those measures, with full reasons, to the Commission and the other national regulatory authorities. A decision by the national regulatory authority to render such measures permanent or extend the time for which they are applicable shall be subject to the provisions of paragraphs 3 and 4.

CHAPTER III

**TASKS OF NATIONAL REGULATORY AUTHORITIES**

*Article 8*

**Policy objectives and regulatory principles**

1. Member States shall ensure that in carrying out the regulatory tasks specified in this Directive and the Specific Directives, the national regulatory authorities take all reasonable measures which are aimed at achieving the objectives set out in paragraphs 2, 3 and 4. Such measures shall be proportionate to those objectives.

Member States shall ensure that in carrying out the regulatory tasks specified in this Directive and the Specific Directives, in particular those designed to ensure effective competition, national regulatory authorities take the utmost account of the desirability of making regulations technologically neutral.

National regulatory authorities may contribute within their competencies to ensuring the implementation of policies aimed at the promotion of cultural and linguistic diversity, as well as media pluralism.

2. The national regulatory authorities shall promote competition in the provision of electronic communications networks, electronic communications services and associated facilities and services by *inter alia*:

(a) ensuring that users, including disabled users, derive maximum benefit in terms of choice, price, and quality;

(b) ensuring that there is no distortion or restriction of competition in the electronic communications sector;

(c) encouraging efficient investment in infrastructure, and promoting innovation; and

(d) encouraging efficient use and ensuring the effective management of radio frequencies and numbering resources.

3. The national regulatory authorities shall contribute to the development of the internal market by *inter alia*:

(a) removing remaining obstacles to the provision of electronic communications networks, associated facilities and services and electronic communications services at European level;

(b) encouraging the establishment and development of trans-European networks and the interoperability of pan-European services, and end-to-end connectivity;

(c) ensuring that, in similar circumstances, there is no discrimination in the treatment of undertakings providing electronic communications networks and services;

(d) cooperating with each other and with the Commission in a transparent manner to ensure the development of consistent regulatory practice and the consistent application of this Directive and the Specific Directives.

4. The national regulatory authorities shall promote the interests of the citizens of the European Union by *inter alia*:

(a) ensuring all citizens have access to a universal service specified in Directive 2002/22/EC (Universal Service Directive);

(b) ensuring a high level of protection for consumers in their dealings with suppliers, in particular by ensuring the availability of simple and inexpensive dispute resolution procedures carried out by a body that is independent of the parties involved;

(c) contributing to ensuring a high level of protection of personal data and privacy;

(d) promoting the provision of clear information, in particular requiring transparency of tariffs and conditions for using publicly available electronic communications services;

(e) addressing the needs of specific social groups, in particular disabled users; and

(f) ensuring that the integrity and security of public communications networks are maintained.

*Article 9*

**Management of radio frequencies for electronic communications services**

1. Member States shall ensure the effective management of radio frequencies for electronic communication services in their territory in accordance with Article 8. They shall ensure that the allocation and assignment of such radio frequencies by national regulatory authorities are based on objective, transparent, non-discriminatory and proportionate criteria.

2. Member States shall promote the harmonisation of use of radio frequencies across the Community, consistent with the need to ensure effective and efficient use thereof and in accordance with the Decision No 676/2002/EC (Radio Spectrum Decision).

3. Member States may make provision for undertakings to transfer rights to use radio frequencies with other undertakings.

4. Member States shall ensure that an undertaking's intention to transfer rights to use radio frequencies is notified to the national regulatory authority responsible for spectrum assignment and that any transfer takes place in accordance with procedures laid down by the national regulatory authority and is made public. National regulatory authorities shall ensure that competition is not distorted as a result of any such transaction. Where radio frequency use has been harmonised through the application of Decision No 676/2002/EC (Radio Spectrum Decision) or other Community measures, any such transfer shall not result in change of use of that radio frequency.

*Article 10*

**Numbering, naming and addressing**

1. Member States shall ensure that national regulatory authorities control the assignment of all national numbering

resources and the management of the national numbering plans. Member States shall ensure that adequate numbers and numbering ranges are provided for all publicly available electronic communications services. National regulatory authorities shall establish objective, transparent and non-discriminatory assigning procedures for national numbering resources.

2. National regulatory authorities shall ensure that numbering plans and procedures are applied in a manner that gives equal treatment to all providers of publicly available electronic communications services. In particular, Member States shall ensure that an undertaking allocated a range of numbers does not discriminate against other providers of electronic communications services as regards the number sequences used to give access to their services.

3. Member States shall ensure that the national numbering plans, and all subsequent additions or amendments thereto, are published, subject only to limitations imposed on the grounds of national security.

4. Member States shall support the harmonisation of numbering resources within the Community where that is necessary to support the development of pan European services. The Commission may, in accordance with the procedure referred to in Article 22(3), take the appropriate technical implementing measures on this matter.

5. Where this is appropriate in order to ensure full global interoperability of services, Member States shall coordinate their positions in international organisations and forums in which decisions are taken on issues relating to the numbering, naming and addressing of electronic communications networks and services.

*Article 11*

**Rights of way**

1. Member States shall ensure that when a competent authority considers:

— an application for the granting of rights to install facilities on, over or under public or private property to an undertaking authorised to provide public communications networks, or

— an application for the granting of rights to install facilities on, over or under public property to an undertaking authorised to provide electronic communications networks other than to the public,

the competent authority:

— acts on the basis of transparent and publicly available procedures, applied without discrimination and without delay, and

— follows the principles of transparency and non-discrimination in attaching conditions to any such rights.

The abovementioned procedures can differ depending on whether the applicant is providing public communications networks or not.

2. Member States shall ensure that where public or local authorities retain ownership or control of undertakings operating electronic communications networks and/or services, there is effective structural separation of the function responsible for granting the rights referred to in paragraph 1 from activities associated with ownership or control.

3. Member States shall ensure that effective mechanisms exist to allow undertakings to appeal against decisions on the granting of rights to install facilities to a body that is independent of the parties involved.

*Article 12*

**Co-location and facility sharing**

1. Where an undertaking providing electronic communications networks has the right under national legislation to install facilities on, over or under public or private property, or may take advantage of a procedure for the expropriation or use of property, national regulatory authorities shall encourage the sharing of such facilities or property.

2. In particular where undertakings are deprived of access to viable alternatives because of the need to protect the environment, public health, public security or to meet town and country planning objectives, Member States may impose the sharing of facilities or property (including physical co-location) on an undertaking operating an electronic communications network or take measures to facilitate the coordination of public works only after an appropriate period of public consultation during which all interested parties must be given an opportunity to express their views. Such sharing or coordination arrangements may include rules for apportioning the costs of facility or property sharing.

*Article 13*

**Accounting separation and financial reports**

1. Member States shall require undertakings providing public communications networks or publicly available electronic communications services which have special or exclusive rights for the provision of services in other sectors in the same or another Member State to:

(a) keep separate accounts for the activities associated with the provision of electronic communications networks or services, to the extent that would be required if these activities were carried out by legally independent companies, so as to identify all elements of cost and revenue, with the basis of their calculation and the detailed attribution methods used, related to their activities associated with the provision of electronic communications networks or services including an itemised breakdown of fixed asset and structural costs, or

(b) have structural separation for the activities associated with the provision of electronic communications networks or services.

Member States may choose not to apply the requirements referred to in the first subparagraph to undertakings the annual turnover of which in activities associated with electronic communications networks or services in the Member States is less than EUR 50 million.

2. Where undertakings providing public communications networks or publicly available electronic communications services are not subject to the requirements of company law and do not satisfy the small and medium-sized enterprise criteria of Community law accounting rules, their financial reports shall be drawn up and submitted to independent audit and published. The audit shall be carried out in accordance with the relevant Community and national rules.

This requirement shall also apply to the separate accounts required under paragraph 1(a).

CHAPTER IV

**GENERAL PROVISIONS**

*Article 14*

**Undertakings with significant market power**

1. Where the Specific Directives require national regulatory authorities to determine whether operators have significant market power in accordance with the procedure referred to in Article 16, paragraphs 2 and 3 of this Article shall apply.

2. An undertaking shall be deemed to have significant market power if, either individually or jointly with others, it enjoys a position equivalent to dominance, that is to say a position of economic strength affording it the power to behave to an appreciable extent independently of competitors, customers and ultimately consumers.

In particular, national regulatory authorities shall, when assessing whether two or more undertakings are in a joint dominant position in a market, act in accordance with Community law and take into the utmost account the guidelines on market analysis and the assessment of significant market power published by the Commission pursuant to Article 15. Criteria to be used in making such an assessment are set out in Annex II.

3. Where an undertaking has significant market power on a specific market, it may also be deemed to have significant market power on a closely related market, where the links between the two markets are such as to allow the market power held in one market to be leveraged into the other market, thereby strengthening the market power of the undertaking.

*Article 15*

**Market definition procedure**

1. After public consultation and consultation with national regulatory authorities the Commission shall adopt a recommendation on relevant product and service markets (hereinafter 'the recommendation'). The recommendation shall identify in accordance with Annex I hereto those product and service markets within the electronic communications sector, the characteristics of which may be such as to justify the imposition of regulatory obligations set out in the Specific Directives, without prejudice to markets that may be defined in specific cases under competition law. The Commission shall define markets in accordance with the principles of competition law.

The Commission shall regularly review the recommendation.

2. The Commission shall publish, at the latest on the date of entry into force of this Directive, guidelines for market analysis and the assessment of significant market power (hereinafter 'the guidelines') which shall be in accordance with the principles of competition law.

3. National regulatory authorities shall, taking the utmost account of the recommendation and the guidelines, define relevant markets appropriate to national circumstances, in particular relevant geographic markets within their territory, in accordance with the principles of competition law. National regulatory authorities shall follow the procedures referred to in Articles 6 and 7 before defining the markets that differ from those defined in the recommendation.

4. After consultation with national regulatory authorities the Commission may, acting in accordance with the procedure referred to in Article 22(3), adopt a Decision identifying transnational markets.

*Article 16*

**Market analysis procedure**

1. As soon as possible after the adoption of the recommendation or any updating thereof, national regulatory authorities shall carry out an analysis of the relevant markets,

taking the utmost account of the guidelines. Member States shall ensure that this analysis is carried out, where appropriate, in collaboration with the national competition authorities.

2. Where a national regulatory authority is required under Articles 16, 17, 18 or 19 of Directive 2002/22/EC (Universal Service Directive), or Articles 7 or 8 of Directive 2002/19/EC (Access Directive) to determine whether to impose, maintain, amend or withdraw obligations on undertakings, it shall determine on the basis of its market analysis referred to in paragraph 1 of this Article whether a relevant market is effectively competitive.

3. Where a national regulatory authority concludes that the market is effectively competitive, it shall not impose or maintain any of the specific regulatory obligations referred to in paragraph 2 of this Article. In cases where sector specific regulatory obligations already exist, it shall withdraw such obligations placed on undertakings in that relevant market. An appropriate period of notice shall be given to parties affected by such a withdrawal of obligations.

4. Where a national regulatory authority determines that a relevant market is not effectively competitive, it shall identify undertakings with significant market power on that market in accordance with Article 14 and the national regulatory authority shall on such undertakings impose appropriate specific regulatory obligations referred to in paragraph 2 of this Article or maintain or amend such obligations where they already exist.

5. In the case of transnational markets identified in the Decision referred to in Article 15(4), the national regulatory authorities concerned shall jointly conduct the market analysis taking the utmost account of the guidelines and decide on any imposition, maintenance, amendment or withdrawal of regulatory obligations referred to in paragraph 2 of this Article in a concerted fashion.

6. Measures taken according to the provisions of paragraphs 3, 4 and 5 of this Article shall be subject to the procedures referred to in Articles 6 and 7.

*Article 17*

**Standardisation**

1. The Commission, acting in accordance with the procedure referred to in Article 22(2), shall draw up and publish in the *Official Journal of the European Communities* a list of standards and/or specifications to serve as a basis for encouraging the harmonised provision of electronic communications networks, electronic communications services and associated facilities and services. Where necessary, the Commission may, acting in accordance with the procedure referred to in Article 22(2) and following consultation of the Committee established by Directive 98/34/EC, request that standards be drawn up by the European standards organisations (European Committee for Standardisation (CEN), European Committee for Electrotechnical Standardisation (CENELEC), and European Telecommunications Standards Institute (ETSI)).

2. Member States shall encourage the use of the standards and/or specifications referred to in paragraph 1, for the provision of services, technical interfaces and/or network functions, to the extent strictly necessary to ensure interoperability of services and to improve freedom of choice for users.

As long as standards and/or specifications have not been published in accordance with paragraph 1, Member States shall encourage the implementation of standards and/or specifications adopted by the European standards organisations.

In the absence of such standards and/or specifications, Member States shall encourage the implementation of international standards or recommendations adopted by the International Telecommunication Union (ITU), the International Organisation for Standardisation (ISO) or the International Electrotechnical Commission (IEC).

Where international standards exist, Member States shall encourage the European standards organisations to use them, or the relevant parts of them, as a basis for the standards they develop, except where such international standards or relevant parts would be ineffective.

3. If the standards and/or specifications referred to in paragraph 1 have not been adequately implemented so that interoperability of services in one or more Member States cannot be ensured, the implementation of such standards and/or specifications may be made compulsory under the procedure laid down in paragraph 4, to the extent strictly necessary to ensure such interoperability and to improve freedom of choice for users.

4. Where the Commission intends to make the implementation of certain standards and/or specifications compulsory, it shall publish a notice in the *Official Journal of the European Communities* and invite public comment by all parties concerned. The Commission, acting in accordance with the procedure referred to in Article 22(3), shall make implementation of the relevant standards compulsory by making reference to them as compulsory standards in the list of standards and/or specifications published in the *Official Journal of the European Communities.*

5. Where the Commission considers that standards and/or specifications referred to in paragraph 1 no longer contribute to the provision of harmonised electronic communications services, or that they no longer meet consumers' needs or are hampering technological development, it shall, acting in accordance with the procedure referred to in Article 22(2), remove them from the list of standards and/or specifications referred to in paragraph 1.

6. Where the Commission considers that standards and/or specifications referred to in paragraph 4 no longer contribute to the provision of harmonised electronic communications services, or that they no longer meet consumers' needs or are hampering technological development, it shall, acting in

accordance with the procedure referred to in Article 22(3), remove them from this list of standards and/or specifications referred to in paragraph 1.

7. This Article does not apply in respect of any of the essential requirements, interface specifications or harmonised standards to which the provisions of Directive 1999/5/EC apply.

*Article 18*

**Interoperability of digital interactive television services**

1. In order to promote the free flow of information, media pluralism and cultural diversity, Member States shall encourage, in accordance with the provisions of Article 17(2):

(a) providers of digital interactive television services for distribution to the public in the Community on digital interactive television platforms, regardless of the transmission mode, to use an open API;

(b) providers of all enhanced digital television equipment deployed for the reception of digital interactive television services on interactive digital television platforms to comply with an open API in accordance with the minimum requirements of the relevant standards or specifications.

2. Without prejudice to Article 5(1)(b) of Directive 2002/19/ EC (Access Directive), Member States shall encourage proprietors of APIs to make available on fair, reasonable and non-discriminatory terms, and against appropriate remuneration, all such information as is necessary to enable providers of digital interactive television services to provide all services supported by the API in a fully functional form.

3. Within one year after the date of application referred to in Article 28(1), second subparagraph, the Commission shall examine the effects of this Article. If interoperability and freedom of choice for users have not been adequately achieved in one or more Member States, the Commission may take action in accordance with the procedure laid down in Article 17(3) and (4).

*Article 19*

**Harmonisation procedures**

1. Where the Commission, acting in accordance with the procedure referred to in Article 22(2), issues recommendations to Member States on the harmonised application of the provisions in this Directive and the Specific Directives in order to further the achievement of the objectives set out in Article 8, Member States shall ensure that national regulatory authorities take the utmost account of those recommendations in carrying out their tasks. Where a national regulatory authority chooses not to follow a recommendation, it shall inform the Commission giving the reasoning for its position.

2. Where the Commission finds that divergence at national level in regulations aimed at implementing Article 10(4) creates a barrier to the single market, the Commission may, acting in accordance with the procedure referred to in Article 22(3), take the appropriate technical implementing measures.

*Article 20*

**Dispute resolution between undertakings**

1. In the event of a dispute arising in connection with obligations arising under this Directive or the Specific Directives between undertakings providing electronic communications networks or services in a Member State, the national regulatory authority concerned shall, at the request of either party, and without prejudice to the provisions of paragraph 2, issue a binding decision to resolve the dispute in the shortest possible time frame and in any case within four months except in exceptional circumstances. The Member State concerned shall require that all parties cooperate fully with the national regulatory authority.

2. Member States may make provision for national regulatory authorities to decline to resolve a dispute through a binding decision where other mechanisms, including mediation, exist and would better contribute to resolution of the dispute in a timely manner in accordance with the provisions of Article 8. The national regulatory authority shall inform the parties without delay. If after four months the dispute is not resolved, and if the dispute has not been brought before the courts by the party seeking redress, the national regulatory authority shall issue, at the request of either party, a binding decision to resolve the dispute in the shortest possible time frame and in any case within four months.

3. In resolving a dispute, the national regulatory authority shall take decisions aimed at achieving the objectives set out in Article 8. Any obligations imposed on an undertaking by the national regulatory authority in resolving a dispute shall respect the provisions of this Directive or the Specific Directives.

4. The decision of the national regulatory authority shall be made available to the public, having regard to the requirements of business confidentiality. The parties concerned shall be given a full statement of the reasons on which it is based.

5. The procedure referred to in paragraphs 1, 3 and 4 shall not preclude either party from bringing an action before the courts.

*Article 21*

**Resolution of cross-border disputes**

1. In the event of a cross-border dispute arising under this Directive or the Specific Directives between parties in different

Member States, where the dispute lies within the competence of national regulatory authorities from more than one Member State, the procedure set out in paragraphs 2, 3 and 4 shall be applicable.

2. Any party may refer the dispute to the national regulatory authorities concerned. The national regulatory authorities shall coordinate their efforts in order to bring about a resolution of the dispute, in accordance with the objectives set out in Article 8. Any obligations imposed on an undertaking by the national regulatory authority in resolving a dispute shall respect the provisions of this Directive or the Specific Directives.

3. Member States may make provision for national regulatory authorities jointly to decline to resolve a dispute where other mechanisms, including mediation, exist and would better contribute to resolution of the dispute in a timely manner in accordance with the provisions of Article 8. They shall inform the parties without delay. If after four months the dispute is not resolved, if the dispute has not been brought before the courts by the party seeking redress, and if either party requests it, the national regulatory authorities shall coordinate their efforts in order to bring about a resolution of the dispute, in accordance with the provisions set out in Article 8.

4. The procedure referred to in paragraph 2 shall not preclude either party from bringing an action before the courts.

*Article 22*

**Committee**

1. The Commission shall be assisted by a Committee ('the Communications Committee').

2. Where reference is made to this paragraph, Articles 3 and 7 of Decision 1999/468/EC shall apply, having regard to the provisions of Article 8 thereof.

3. Where reference is made to this paragraph, Articles 5 and 7 of Decision 1999/468/EC shall apply, having regard to the provisions of Article 8 thereof.

The period laid down in Article 5(6) of Decision 1999/468/EC shall be three months.

4. The Committee shall adopt its rules of procedure.

*Article 23*

**Exchange of information**

1. The Commission shall provide all relevant information to the Communications Committee on the outcome of regular consultations with the representatives of network operators, service providers, users, consumers, manufacturers and trade unions, as well as third countries and international organisations.

2. The Communications Committee shall, taking account of the Community's electronic communications policy, foster the exchange of information between the Member States and between the Member States and the Commission on the situation and the development of regulatory activities regarding electronic communications networks and services.

*Article 24*

**Publication of information**

1. Member States shall ensure that up-to-date information pertaining to the application of this Directive and the Specific Directives is made publicly available in a manner that guarantees all interested parties easy access to that information. They shall publish a notice in their national official gazette describing how and where the information is published. The first such notice shall be published before the date of application referred to in Article 28(1), second subparagraph, and thereafter a notice shall be published whenever there is any change in the information contained therein.

2. Member States shall send to the Commission a copy of all such notices at the time of publication. The Commission shall distribute the information to the Communications Committee as appropriate.

*Article 25*

**Review procedures**

1. The Commission shall periodically review the functioning of this Directive and report to the European Parliament and to the Council, on the first occasion not later than three years after the date of application referred to in Article 28(1), second subparagraph. For this purpose, the Commission may request information from the Member States, which shall be supplied without undue delay.

CHAPTER V

**FINAL PROVISIONS**

*Article 26*

**Repeal**

The following Directives and Decisions are hereby repealed with effect from the date of application referred to in Article 28(1), second subparagraph:

— Directive 90/387/EEC,

— Council Decision 91/396/EEC of 29 July 1991 on the introduction of a single European emergency call number ([1]),

— Council Directive 92/44/EEC of 5 June 1992 on the application of open network provision to leased lines ([2]),

— Council Decision 92/264/EEC of 11 May 1992 on the introduction of a standard international telephone access code in the Community ([3]),

— Directive 95/47/EC,

— Directive 97/13/EC,

— Directive 97/33/EC,

— Directive 98/10/EC of the European Parliament and of the Council of 26 February 1998 on the application of open network provision (ONP) to voice telephony and on universal service for telecommunications in a competitive environment ([4]).

*Article 27*

**Transitional measures**

Member States shall maintain all obligations under national law referred to in Article 7 of Directive 2002/19/EC (Access Directive) and Article 16 of Directive 2002/22/EC (Universal Service Directive) until such time as a determination is made in respect of those obligations by a national regulatory authority in accordance with Article 16 of this Directive.

Operators of fixed public telephone networks that were designated by their national regulatory authority as having significant market power in the provision of fixed public telephone networks and services under Annex I, Part 1 of Directive 97/33/EC or Directive 98/10/EC shall continue to be considered 'notified operators' for the purposes of Regulation (EC) No 2887/2000 until such a time as the market analysis procedure referred to in Article 16 has been completed. Thereafter they shall cease to be considered 'notified operators' for the purposes of the Regulation.

*Article 28*

**Transposition**

1. Member States shall adopt and publish the laws, regulations and administrative provisions necessary to comply with this Directive not later than 24 July 2003. They shall forthwith inform the Commission thereof.

They shall apply those measures from 25 July 2003.

2. When Member States adopt these measures, they shall contain a reference to this Directive or be accompanied by such a reference on the occasion of their official publication. The methods of making such a reference shall be laid down by the Member States.

3. Member States shall communicate to the Commission the text of the provisions of national law which they adopt in the field governed by this Directive and of any subsequent amendments to those provisions.

*Article 29*

**Entry into force**

This Directive shall enter into force on the day of its publication in the *Official Journal of the European Communities*.

*Article 30*

**Addressees**

This Directive is addressed to the Member States.

Done at Brussels, 7 March 2002.

| *For the European Parliament* | *For the Council* |
|---|---|
| *The President* | *The President* |
| P. COX | J. C. APARICIO |

---

([1]) OJ L 217, 6.8.1991, p. 31.
([2]) OJ L 165, 19.6.1992, p. 27. Directive as last amended by Commission Decision 98/80/EC (OJ L 14, 20.1.1998, p. 27).
([3]) OJ L 137, 20.5.1992, p. 21.
([4]) OJ L 101, 1.4.1998, p. 24.

*ANNEX I*

**List of markets to be included in the initial Commission recommendation on relevant product and service markets referred to in Article 15**

1. *Markets referred to in Directive 2002/22/EC (Universal Service Directive)*

Article 16 — Markets defined under the former regulatory framework, where obligations should be reviewed.

The provision of connection to and use of the public telephone network at fixed locations.

The provision of leased lines to end users.

2. *Markets referred to in Directive 2002/19/EC (Access Directive)*

Article 7 — Markets defined under the former regulatory framework, where obligations should be reviewed.

Interconnection (Directive 97/33/EC)

call origination in the fixed public telephone network

call termination in the fixed public telephone network

transit services in the fixed public telephone network

call origination on public mobile telephone networks

call termination on public mobile telephone networks

leased line interconnection (interconnection of part circuits)

Network access and special network access (Directive 97/33/EC, Directive 98/10/EC)

access to the fixed public telephone network, including unbundled access to the local loop

access to public mobile telephone networks, including carrier selection

Wholesale leased line capacity (Directive 92/44/EEC)

wholesale provision of leased line capacity to other suppliers of electronic communications networks or services

3. *Markets referred to in Regulation (EC) No 2887/2000*

Services provided over unbundled (twisted metallic pair) loops.

4. *Additional markets*

The national market for international roaming services on public mobile telephone networks.

L 108/50 EN Official Journal of the European Communities 24.4.2002

*ANNEX II*

**Criteria to be used by national regulatory authorities in making an assessment of joint dominance in accordance with Article 14(2), second subparagraph**

Two or more undertakings can be found to be in a joint dominant position within the meaning of Article 14 if, even in the absence of structural or other links between them, they operate in a market the structure of which is considered to be conducive to coordinated effects. Without prejudice to the case law of the Court of Justice on joint dominance, this is likely to be the case where the market satisfies a number of appropriate characteristics, in particular in terms of market concentration, transparency and other characteristics mentioned below:

— mature market,

— stagnant or moderate growth on the demand side,

— low elasticity of demand,

— homogeneous product,

— similar cost structures,

— similar market shares,

— lack of technical innovation, mature technology,

— absence of excess capacity,

— high barriers to entry,

— lack of countervailing buying power,

— lack of potential competition,

— various kinds of informal or other links between the undertakings concerned,

— retaliatory mechanisms,

— lack or reduced scope for price competition.

The above is not an exhaustive list, nor are the criteria cumulative. Rather, the list is intended to illustrate only the sorts of evidence that could be used to support assertions concerning the existence of joint dominance.

# DIRECTIVE 2002/20/EC OF THE EUROPEAN PARLIAMENT AND OF THE COUNCIL

## of 7 March 2002

## on the authorisation of electronic communications networks and services (Authorisation Directive)

THE EUROPEAN PARLIAMENT AND THE COUNCIL OF THE EUROPEAN UNION,

Having regard to the Treaty establishing the European Community, and in particular Article 95 thereof,

Having regard to the proposal from the Commission [1],

Having regard to the opinion of the Economic and Social Committee [2],

Acting in accordance with the procedure laid down in Article 251 of the Treaty [3],

Whereas:

(1) The outcome of the public consultation on the 1999 review of the regulatory framework for electronic communications, as reflected in the Commission communication of 26 April 2000, and the findings reported by the Commission in its communications on the fifth and sixth reports on the implementation of the telecommunications regulatory package, has confirmed the need for a more harmonised and less onerous market access regulation for electronic communications networks and services throughout the Community.

(2) Convergence between different electronic communications networks and services and their technologies requires the establishment of an authorisation system covering all comparable services in a similar way regardless of the technologies used.

(3) The objective of this Directive is to create a legal framework to ensure the freedom to provide electronic communications networks and services, subject only to the conditions laid down in this Directive and to any restrictions in conformity with Article 46(1) of the Treaty, in particular measures regarding public policy, public security and public health.

(4) This Directive covers authorisation of all electronic communications networks and services whether they are provided to the public or not. This is important to ensure that both categories of providers may benefit from objective, transparent, non-discriminatory and proportionate rights, conditions and procedures.

(5) This Directive only applies to the granting of rights to use radio frequencies where such use involves the provision of an electronic communications network or service, normally for remuneration. The self-use of radio terminal equipment, based on the non-exclusive use of specific radio frequencies by a user and not related to an economic activity, such as use of a citizen's band by radio amateurs, does not consist of the provision of an electronic communications network or service and is therefore not covered by this Directive. Such use is covered by the Directive 1999/5/EC of the European Parliament and of the Council of 9 March 1999 on radio equipment and telecommunications terminal equipment and the mutual recognition of their conformity [4].

(6) Provisions regarding the free movement of conditional access systems and the free provision of protected services based on such systems are laid down in Directive 98/84/EC of the European Parliament and of the Council of 20 November 1998 on the legal protection of services based on, or consisting of, conditional access [5]. The authorisation of such systems and services therefore does not need to be covered by this Directive.

(7) The least onerous authorisation system possible should be used to allow the provision of electronic communications networks and services in order to stimulate the development of new electronic communications services and pan-European communications networks and services and to allow service providers and consumers to benefit from the economies of scale of the single market.

(8) Those aims can be best achieved by general authorisation of all electronic communications networks and services without requiring any explicit decision or administrative act by the national regulatory authority and by limiting any procedural requirements to notification only. Where Member States require notification by providers of electronic communication networks or services when they start their activities,

---

[1] OJ C 365 E, 19.12.2000, p. 230 and OJ C 270 E, 25.9.2001, p. 182.

[2] OJ C 123, 25.4.2001, p. 55.

[3] Opinion of the European Parliament of 1 March 2001 (OJ C 277, 1.10.2001, p. 116), Council Common Position of 17 September 2001 (OJ C 337, 30.11.2001, p. 18) and Decision of the European Parliament of 12 December 2001(not yet published in the Official Journal). Council Decision of 14 February 2002.

[4] OJ L 91, 7.4.1999, p. 10.

[5] OJ L 320, 28.11.1998, p. 54.

they may also require proof of such notification having been made by means of any legally recognised postal or electronic acknowledgement of receipt of the notification. Such acknowledgement should in any case not consist of or require an administrative act by the national regulatory authority to which the notification must be made.

(9) It is necessary to include the rights and obligations of undertakings under general authorisations explicitly in such authorisations in order to ensure a level playing field throughout the Community and to facilitate cross-border negotiation of interconnection between public communications networks.

(10) The general authorisation entitles undertakings providing electronic communications networks and services to the public to negotiate interconnection under the conditions of Directive 2002/19/EC of the European Parliament and of the Council of 7 March 2002 on access to, and interconnection of, electronic communication networks and associated facilities (Access Directive) ([1]). Undertakings providing electronic communications networks and services other than to the public can negotiate interconnection on commercial terms.

(11) The granting of specific rights may continue to be necessary for the use of radio frequencies and numbers, including short codes, from the national numbering plan. Rights to numbers may also be allocated from a European numbering plan, including for example the virtual country code '3883' which has been attributed to member countries of the European Conference of Post and Telecommunications (CEPT). Those rights of use should not be restricted except where this is unavoidable in view of the scarcity of radio frequencies and the need to ensure the efficient use thereof.

(12) This Directive does not prejudice whether radio frequencies are assigned directly to providers of electronic communication networks or services or to entities that use these networks or services. Such entities may be radio or television broadcast content providers. Without prejudice to specific criteria and procedures adopted by Member States to grant rights of use for radio frequencies to providers of radio or television broadcast content services, to pursue general interest objectives in conformity with Community law, the procedure for assignment of radio frequencies should in any event be objective, transparent, non-discriminatory and proportionate. In accordance with case law of the Court of Justice, any national restrictions on the rights guaranteed by Article 49 of the Treaty should be objectively justified, proportionate and not exceed what is necessary to achieve general interest objectives as defined by Member States in conformity with Community law. The responsibility for compliance with the conditions attached to the right to use a radio frequency and the relevant conditions attached to the general authorisation should in any case lie with the undertaking to whom the right of use for the radio frequency has been granted.

(13) As part of the application procedure for granting rights to use a radio frequency, Member States may verify whether the applicant will be able to comply with the conditions attached to such rights. For this purpose the applicant may be requested to submit the necessary information to prove his ability to comply with these conditions. Where such information is not provided, the application for the right to use a radio frequency may be rejected.

(14) Member States are neither obliged to grant nor prevented from granting rights to use numbers from the national numbering plan or rights to install facilities to undertakings other than providers of electronic communications networks or services.

(15) The conditions, which may be attached to the general authorisation and to the specific rights of use, should be limited to what is strictly necessary to ensure compliance with requirements and obligations under Community law and national law in accordance with Community law.

(16) In the case of electronic communications networks and services not provided to the public it is appropriate to impose fewer and lighter conditions than are justified for electronic communications networks and services provided to the public.

(17) Specific obligations which may be imposed on providers of electronic communications networks and services in accordance with Community law by virtue of their significant market power as defined in Directive 2002/21/EC of the European Parliament and of the Council of 7 March 2002 on a common regulatory framework for electronic communications networks and services (Framework Directive) ([2]) should be imposed separately from the general rights and obligations under the general authorisation.

(18) The general authorisation should only contain conditions which are specific to the electronic communications sector. It should not be made subject

([1]) See page 7 of this Official Journal.

([2]) See page 33 of this Official Journal.

to conditions which are already applicable by virtue of other existing national law which is not specific to the electronic communications sector. Nevertheless, the national regulatory authorities may inform network operators and service providers about other legislation concerning their business, for instance through references on their websites.

(19) The requirement to publish decisions on the granting of rights to use frequencies or numbers may be fulfilled by making these decisions publicly accessible via a website.

(20) The same undertaking, for example a cable operator, can offer both an electronic communications service, such as the conveyance of television signals, and services not covered under this Directive, such as the commercialisation of an offer of sound or television broadcasting content services, and therefore additional obligations can be imposed on this undertaking in relation to its activity as a content provider or distributor, according to provisions other than those of this Directive, without prejudice to the list of conditions laid in the Annex to this Directive.

(21) When granting rights of use for radio frequencies, numbers or rights to install facilities, the relevant authorities may inform the undertakings to whom they grant such rights of the relevant conditions in the general authorisation.

(22) Where the demand for radio frequencies in a specific range exceeds their availability, appropriate and transparent procedures should be followed for the assignment of such frequencies in order to avoid any discrimination and optimise use of those scarce resources.

(23) National regulatory authorities should ensure, in establishing criteria for competitive or comparative selection procedures, that the objectives in Article 8 of Directive 2002/21/EC (Framework Directive) are met. It would therefore not be contrary to this Directive if the application of objective, non-discriminatory and proportionate selection criteria to promote the development of competition would have the effect of excluding certain undertakings from a competitive or comparative selection procedure for a particular radio frequency.

(24) Where the harmonised assignment of radio frequencies to particular undertakings has been agreed at European level, Member States should strictly implement such agreements in the granting of rights of use of radio frequencies from the national frequency usage plan.

(25) Providers of electronic communications networks and services may need a confirmation of their rights under the general authorisation with respect to interconnection and rights of way, in particular to facilitate negotiations with other, regional or local, levels of government or with service providers in other Member States. For this purpose the national regulatory authorities should provide declarations to undertakings either upon request or alternatively as an automatic response to a notification under the general authorisation. Such declarations should not by themselves constitute entitlements to rights nor should any rights under the general authorisation or rights of use or the exercise of such rights depend upon a declaration.

(26) Where undertakings find that their applications for rights to install facilities have not been dealt with in accordance with the principles set out in Directive 2002/21/EC (Framework Directive) or where such decisions are unduly delayed, they should have the right to appeal against decisions or delays in such decisions in accordance with that Directive.

(27) The penalties for non-compliance with conditions under the general authorisation should be commensurate with the infringement. Save in exceptional circumstances, it would not be proportionate to suspend or withdraw the right to provide electronic communications services or the right to use radio frequencies or numbers where an undertaking did not comply with one or more of the conditions under the general authorisation. This is without prejudice to urgent measures which the relevant authorities of the Member States may need to take in case of serious threats to public safety, security or health or to economic and operational interests of other undertakings. This Directive should also be without prejudice to any claims between undertakings for compensation for damages under national law.

(28) Subjecting service providers to reporting and information obligations can be cumbersome, both for the undertaking and for the national regulatory authority concerned. Such obligations should therefore be proportionate, objectively justified and limited to what is strictly necessary. It is not necessary to require systematic and regular proof of compliance with all conditions under the general authorisation or attached to rights of use. Undertakings have a right to know the purposes for which the information they should provide will be used. The provision of information should not be a condition for market access. For statistical purposes a notification may be required from providers of electronic communication networks or services when they cease activities.

(29) This Directive should be without prejudice to Member States' obligations to provide any information necessary for the defence of Community interests within the context of international agreements. This Directive should also be without prejudice to any reporting obligations under legislation which is not specific to the electronic communications sector such as competition law.

(30) Administrative charges may be imposed on providers of electronic communications services in order to finance the activities of the national regulatory authority in managing the authorisation system and for the granting of rights of use. Such charges should be limited to cover the actual administrative costs for those activities. For this purpose transparency should be created in the income and expenditure of national regulatory authorities by means of annual reporting about the total sum of charges collected and the administrative costs incurred. This will allow undertakings to verify that administrative costs and charges are in balance.

(31) Systems for administrative charges should not distort competition or create barriers for entry into the market. With a general authorisation system it will no longer be possible to attribute administrative costs and hence charges to individual undertakings except for the granting of rights to use numbers, radio frequencies and for rights to install facilities. Any applicable administrative charges should be in line with the principles of a general authorisation system. An example of a fair, simple and transparent alternative for these charge attribution criteria could be a turnover related distribution key. Where administrative charges are very low, flat rate charges, or charges combining a flat rate basis with a turnover related element could also be appropriate.

(32) In addition to administrative charges, usage fees may be levied for the use of radio frequencies and numbers as an instrument to ensure the optimal use of such resources. Such fees should not hinder the development of innovative services and competition in the market. This Directive is without prejudice to the purpose for which fees for rights of use are employed. Such fees may for instance be used to finance activities of national regulatory authorities that cannot be covered by administrative charges. Where, in the case of competitive or comparative selection procedures, fees for rights of use for radio frequencies consist entirely or partly of a one-off amount, payment arrangements should ensure that such fees do not in practice lead to selection on the basis of criteria unrelated to the objective of ensuring optimal use of radio frequencies. The Commission may publish on a regular basis benchmark studies with regard to best practices for the assignment of radio frequencies, the assignment of numbers or the granting of rights of way.

(33) Member States may need to amend rights, conditions, procedures, charges and fees relating to general authorisations and rights of use where this is objectively justified. Such changes should be duly notified to all interested parties in good time, giving them adequate opportunity to express their views on any such amendments.

(34) The objective of transparency requires that service providers, consumers and other interested parties have easy access to any information regarding rights, conditions, procedures, charges, fees and decisions concerning the provision of electronic communications services, rights of use of radio frequencies and numbers, rights to install facilities, national frequency usage plans and national numbering plans. The national regulatory authorities have an important task in providing such information and keeping it up to date. Where such rights are administered by other levels of government the national regulatory authorities should endeavour to create a user-friendly instrument for access to information regarding such rights.

(35) The proper functioning of the single market on the basis of the national authorisation regimes under this Directive should be monitored by the Commission.

(36) In order to arrive at a single date of application of all elements of the new regulatory framework for the electronic communications sector, it is important that the process of national transposition of this Directive and of alignment of the existing licences with the new rules take place in parallel. However, in specific cases where the replacement of authorisations existing on the date of entry into force of this Directive by the general authorisation and the individual rights of use in accordance with this Directive would lead to an increase in the obligations for service providers operating under an existing authorisation or to a reduction of their rights, Member States may avail themselves of an additional nine months after the date of application of this Directive for alignment of such licences, unless this would have a negative effect on the rights and obligations of other undertakings.

(37) There may be circumstances under which the abolition of an authorisation condition regarding access to electronic communications networks would create serious hardship for one or more undertakings that have benefited from the condition. In such cases further transitional arrangements may be granted by the Commission, upon request by a Member State.

(38) Since the objectives of the proposed action, namely the harmonisation and simplification of electronic communications rules and conditions for the authorisation of networks and services cannot be sufficiently achieved by the Member States and can therefore, by reason of the scale and effects of the action, be better achieved at Community level, the Community may adopt measures in accordance with the principle of subsidiarity as set out in Article 5 of the Treaty. In accordance with the principle of proportionality, as set out in that Article, this Directive does not go beyond what is necessary for those objectives,

HAVE ADOPTED THIS DIRECTIVE:

*Article 1*

**Objective and scope**

1. The aim of this Directive is to implement an internal market in electronic communications networks and services through the harmonisation and simplification of authorisation rules and conditions in order to facilitate their provision throughout the Community.

2. This Directive shall apply to authorisations for the provision of electronic communications networks and services.

*Article 2*

**Definitions**

1. For the purposes of this Directive, the definitions set out in Article 2 of Directive 2002/21/EC (Framework Directive) shall apply.

2. The following definitions shall also apply:

(a) 'general authorisation' means a legal framework established by the Member State ensuring rights for the provision of electronic communications networks or services and laying down sector specific obligations that may apply to all or to specific types of electronic communications networks and services, in accordance with this Directive;

(b) 'harmful interference' means interference which endangers the functioning of a radionavigation service or of other safety services or which otherwise seriously degrades, obstructs or repeatedly interrupts a radiocommunications service operating in accordance with the applicable Community or national regulations.

*Article 3*

**General authorisation of electronic communications networks and services**

1. Member States shall ensure the freedom to provide electronic communications networks and services, subject to the conditions set out in this Directive. To this end, Member States shall not prevent an undertaking from providing electronic communications networks or services, except where this is necessary for the reasons set out in Article 46(1) of the Treaty.

2. The provision of electronic communications networks or the provision of electronic communications services may, without prejudice to the specific obligations referred to in Article 6(2) or rights of use referred to in Article 5, only be subject to a general authorisation. The undertaking concerned may be required to submit a notification but may not be required to obtain an explicit decision or any other administrative act by the national regulatory authority before exercising the rights stemming from the authorisation. Upon notification, when required, an undertaking may begin activity, where necessary subject to the provisions on rights of use in Articles 5, 6 and 7.

3. The notification referred to in paragraph 2 shall not entail more than a declaration by a legal or natural person to the national regulatory authority of the intention to commence the provision of electronic communication networks or services and the submission of the minimal information which is required to allow the national regulatory authority to keep a register or list of providers of electronic communications networks and services. This information must be limited to what is necessary for the identification of the provider, such as company registration numbers, and the provider's contact persons, the provider's address, a short description of the network or service, and an estimated date for starting the activity.

*Article 4*

**Minimum list of rights derived from the general authorisation**

1. Undertakings authorised pursuant to Article 3, shall have the right to:

(a) provide electronic communications networks and services;

(b) have their application for the necessary rights to install facilities considered in accordance with Article 11 of Directive 2002/21/EC (Framework Directive).

2. When such undertakings provide electronic communications networks or services to the public the general authorisation shall also give them the right to:

(a) negotiate interconnection with and where applicable obtain access to or interconnection from other providers of publicly available communications networks and services covered by a general authorisation anywhere in the Community under the conditions of and in accordance with Directive 2002/19/EC (Access Directive);

(b) be given an opportunity to be designated to provide different elements of a universal service and/or to cover different parts of the national territory in accordance with Directive 2002/22/EC of the European Parliament and of the Council of 7 March 2002 on universal service and users' rights relating to electronic communications networks and services (Universal Service Directive) ([1]).

*Article 5*

**Rights of use for radio frequencies and numbers**

1. Member States shall, where possible, in particular where the risk of harmful interference is negligible, not make the use of radio frequencies subject to the grant of individual rights of use but shall include the conditions for usage of such radio frequencies in the general authorisation.

2. Where it is necessary to grant individual rights of use for radio frequencies and numbers, Member States shall grant such rights, upon request, to any undertaking providing or using networks or services under the general authorisation, subject to the provisions of Articles 6, 7 and 11(1)(c) of this Directive and any other rules ensuring the efficient use of those resources in accordance with Directive 2002/21/EC (Framework Directive).

Without prejudice to specific criteria and procedures adopted by Member States to grant rights of use of radio frequencies to providers of radio or television broadcast content services with a view to pursuing general interest objectives in conformity with Community law, such rights of use shall be granted through open, transparent and non-discriminatory procedures. When granting rights of use, Member States shall specify whether those rights can be transferred at the initiative of the right holder, and under which conditions, in the case of radio frequencies, in accordance with Article 9 of Directive 2002/21/EC (Framework Directive). Where Member States grant rights of use for a limited period of time, the duration shall be appropriate for the service concerned.

3. Decisions on rights of use shall be taken, communicated and made public as soon as possible after receipt of the complete application by the national regulatory authority, within three weeks in the case of numbers that have been allocated for specific purposes within the national numbering plan and within six weeks in the case of radio frequencies that have been allocated for specific purposes within the national frequency plan. The latter time limit shall be without prejudice to any applicable international agreements relating to the use of radio frequencies or of orbital positions.

4. Where it has been decided, after consultation with interested parties in accordance with Article 6 of Directive 2002/21/EC (Framework Directive), that rights for use of numbers of exceptional economic value are to be granted through competitive or comparative selection procedures, Member States may extend the maximum period of three weeks by up to three weeks.

With regard to competitive or comparative selection procedures for radio frequencies Article 7 shall apply.

5. Member States shall not limit the number of rights of use to be granted except where this is necessary to ensure the efficient use of radio frequencies in accordance with Article 7.

*Article 6*

**Conditions attached to the general authorisation and to the rights of use for radio frequencies and for numbers, and specific obligations**

1. The general authorisation for the provision of electronic communications networks or services and the rights of use for radio frequencies and rights of use for numbers may be subject only to the conditions listed respectively in parts A, B and C of the Annex. Such conditions shall be objectively justified in relation to the network or service concerned, non-discriminatory, proportionate and transparent.

2. Specific obligations which may be imposed on providers of electronic communications networks and services under Articles 5(1), 5(2), 6 and 8 of Directive 2002/19/EC (Access Directive) and Articles 16, 17, 18 and 19 of Directive 2002/22/EC (Universal Service Directive) or on those designated to provide universal service under the said Directive shall be legally separate from the rights and obligations under the general authorisation. In order to achieve transparency for undertakings, the criteria and procedures for imposing such specific obligations on individual undertakings shall be referred to in the general authorisation.

3. The general authorisation shall only contain conditions which are specific for that sector and are set out in Part A of the Annex and shall not duplicate conditions which are applicable to undertakings by virtue of other national legislation.

4. Member States shall not duplicate the conditions of the general authorisation where they grant the right of use for radio frequencies or numbers.

([1]) See page 51 of this Official Journal.

*Article 7*

**Procedure for limiting the number of rights of use to be granted for radio frequencies**

1. Where a Member State is considering whether to limit the number of rights of use to be granted for radio frequencies, it shall *inter alia*:

(a) give due weight to the need to maximise benefits for users and to facilitate the development of competition;

(b) give all interested parties, including users and consumers, the opportunity to express their views on any limitation in accordance with Article 6 of Directive 2002/21/EC (Framework Directive);

(c) publish any decision to limit the granting of rights of use, stating the reasons therefor;

(d) after having determined the procedure, invite applications for rights of use; and

(e) review the limitation at reasonable intervals or at the reasonable request of affected undertakings.

2. Where a Member State concludes that further rights of use for radio frequencies can be granted, it shall publish that conclusion and invite applications for such rights.

3. Where the granting of rights of use for radio frequencies needs to be limited, Member States shall grant such rights on the basis of selection criteria which must be objective, transparent, non-discriminatory and proportionate. Any such selection criteria must give due weight to the achievement of the objectives of Article 8 of Directive 2002/21/EC (Framework Directive).

4. Where competitive or comparative selection procedures are to be used, Member States may extend the maximum period of six weeks referred to in Article 5(3) for as long as necessary to ensure that such procedures are fair, reasonable, open and transparent to all interested parties, but by no longer than eight months.

These time limits shall be without prejudice to any applicable international agreements relating to the use of radio frequencies and satellite coordination.

5. This Article is without prejudice to the transfer of rights of use for radio frequencies in accordance with Article 9 of Directive 2002/21/EC (Framework Directive).

*Article 8*

**Harmonised assignment of radio frequencies**

Where the usage of radio frequencies has been harmonised, access conditions and procedures have been agreed, and undertakings to which the radio frequencies shall be assigned have been selected in accordance with international agreements and Community rules, Member States shall grant the right of use for such radio frequencies in accordance therewith. Provided that all national conditions attached to the right to use the radio frequencies concerned have been satisfied in the case of a common selection procedure, Member States shall not impose any further conditions, additional criteria or procedures which would restrict, alter or delay the correct implementation of the common assignment of such radio frequencies.

*Article 9*

**Declarations to facilitate the exercise of rights to install facilities and rights of interconnection**

At the request of an undertaking, national regulatory authorities shall, within one week, issue standardised declarations, confirming, where applicable, that the undertaking has submitted a notification under Article 3(2) and detailing under what circumstances any undertaking providing electronic communications networks or services under the general authorisation has the right to apply for rights to install facilities, negotiate interconnection, and/or obtain access or interconnection in order to facilitate the exercise of those rights for instance at other levels of government or in relation to other undertakings. Where appropriate such declarations may also be issued as an automatic reply following the notification referred to in Article 3(2).

*Article 10*

**Compliance with the conditions of the general authorisation or of rights of use and with specific obligations**

1. National regulatory authorities may require undertakings providing electronic communications networks or services covered by the general authorisation or enjoying rights of use for radio frequencies or numbers to provide information necessary to verify compliance with the conditions of the general authorisation or of rights of use or with the specific obligations referred to in Article 6(2), in accordance with Article 11.

2. Where a national regulatory authority finds that an undertaking does not comply with one or more of the conditions of the general authorisation, or of rights of use or with the specific obligations referred to in Article 6(2), it shall notify the undertaking of those findings and give the undertaking a reasonable opportunity to state its views or remedy any breaches within:

— one month after notification, or

— a shorter period agreed by the undertaking or stipulated by the national regulatory authority in case of repeated breaches, or

— a longer period decided by the national regulatory authority.

3. If the undertaking concerned does not remedy the breaches within the period as referred to in paragraph 2, the relevant authority shall take appropriate and proportionate measures aimed at ensuring compliance. In this regard, Member States may empower the relevant authorities to impose financial penalties where appropriate. The measures and the reasons on which they are based shall be communicated to the undertaking concerned within one week of their adoption and shall stipulate a reasonable period for the undertaking to comply with the measure.

4. Notwithstanding the provisions of paragraphs 2 and 3, Member States may empower the relevant authority to impose financial penalties where appropriate on undertakings for failure to provide information in accordance with obligations imposed under Article 11(1)(a) or (b) of this Directive or Article 9 of Directive 2002/19/EC (Access Directive) within a reasonable period stipulated by the national regulatory authority.

5. In cases of serious and repeated breaches of the conditions of the general authorisation, the rights of use or specific obligations referred to in Article 6(2), where measures aimed at ensuring compliance as referred to in paragraph 3 of this Article have failed, national regulatory authorities may prevent an undertaking from continuing to provide electronic communications networks or services or suspend or withdraw rights of use.

6. Irrespective of the provisions of paragraphs 2, 3 and 5, where the relevant authority has evidence of a breach of the conditions of the general authorisation, rights of use or specific obligations referred to in Article 6(2) that represents an immediate and serious threat to public safety, public security or public health or will create serious economic or operational problems for other providers or users of electronic communications networks or services, it may take urgent interim measures to remedy the situation in advance of reaching a final decision. The undertaking concerned shall thereafter be given a reasonable opportunity to state its view and propose any remedies. Where appropriate, the relevant authority may confirm the interim measures.

7. Undertakings shall have the right to appeal against measures taken under this Article in accordance with the procedure referred to in Article 4 of Directive 2002/21/EC (Framework Directive).

*Article 11*

**Information required under the general authorisation, for rights of use and for the specific obligations**

1. Without prejudice to information and reporting obligations under national legislation other than the general authorisation, national regulatory authorities may only require undertakings to provide information under the general authorisation, for rights of use or the specific obligations referred to in Article 6(2) that is proportionate and objectively justified for:

(a) systematic or case-by-case verification of compliance with conditions 1 and 2 of Part A, condition 6 of Part B and condition 7 of Part C of the Annex and of compliance with obligations as referred to in Article 6(2);

(b) case-by-case verification of compliance with conditions as set out in the Annex where a complaint has been received or where the national regulatory authority has other reasons to believe that a condition is not complied with or in case of an investigation by the national regulatory authority on its own initiative;

(c) procedures for and assessment of requests for granting rights of use;

(d) publication of comparative overviews of quality and price of services for the benefit of consumers;

(e) clearly defined statistical purposes;

(f) market analysis for the purposes of Directive 2002/19/EC (Access Directive) or Directive 2002/22/EC (Universal Service Directive).

The information referred to in points (a), (b), (d), (e) and (f) of the first subparagraph may not be required prior to or as a condition for market access.

2. Where national regulatory authorities require undertakings to provide information as referred to in paragraph 1, they shall inform them of the specific purpose for which this information is to be used.

*Article 12*

**Administrative charges**

1. Any administrative charges imposed on undertakings providing a service or a network under the general authorisation or to whom a right of use has been granted shall:

(a) in total, cover only the administrative costs which will be incurred in the management, control and enforcement of the general authorisation scheme and of rights of use and of specific obligations as referred to in Article 6(2), which may include costs for international cooperation, harmonisation and standardisation, market analysis, monitoring compliance and other market control, as well as regulatory work involving preparation and enforcement of secondary legislation and administrative decisions, such as decisions on access and interconnection; and

(b) be imposed upon the individual undertakings in an objective, transparent and proportionate manner which minimises additional administrative costs and attendant charges.

2. Where national regulatory authorities impose administrative charges, they shall publish a yearly overview of their administrative costs and of the total sum of the charges collected. In the light of the difference between the total sum of the charges and the administrative costs, appropriate adjustments shall be made.

*Article 13*

**Fees for rights of use and rights to install facilities**

Member States may allow the relevant authority to impose fees for the rights of use for radio frequencies or numbers or rights to install facilities on, over or under public or private property which reflect the need to ensure the optimal use of these resources. Member States shall ensure that such fees shall be objectively justified, transparent, non-discriminatory and proportionate in relation to their intended purpose and shall take into account the objectives in Article 8 of Directive 2002/21/EC (Framework Directive).

*Article 14*

**Amendment of rights and obligations**

1. Member States shall ensure that the rights, conditions and procedures concerning general authorisations and rights of use or rights to install facilities may only be amended in objectively justified cases and in a proportionate manner. Notice shall be given in an appropriate manner of the intention to make such amendments and interested parties, including users and consumers, shall be allowed a sufficient period of time to express their views on the proposed amendments, which shall be no less than four weeks except in exceptional circumstances.

2. Member States shall not restrict or withdraw rights to install facilities before expiry of the period for which they were granted except where justified and where applicable in conformity with relevant national provisions regarding compensation for withdrawal of rights.

*Article 15*

**Publication of information**

1. Member States shall ensure that all relevant information on rights, conditions, procedures, charges, fees and decisions concerning general authorisations and rights of use is published and kept up to date in an appropriate manner so as to provide easy access to that information for all interested parties.

2. Where information as referred to in paragraph 1 is held at different levels of government, in particular information regarding procedures and conditions on rights to install facilities, the national regulatory authority shall make all reasonable efforts, bearing in mind the costs involved, to create a user-friendly overview of all such information, including information on the relevant levels of government and the responsible authorities, in order to facilitate applications for rights to install facilities.

*Article 16*

**Review procedures**

The Commission shall periodically review the functioning of the national authorisation systems and the development of cross-border service provision within the Community and report to the European Parliament and to the Council on the first occasion not later than three years after the date of application of this Directive referred to in Article 18(1), second subparagraph. For this purpose, the Commission may request from the Member States information, which shall be supplied without undue delay.

*Article 17*

**Existing authorisations**

1. Member States shall bring authorisations already in existence on the date of entry into force of this Directive into line with the provisions of this Directive by at the latest the date of application referred to in Article 18(1), second subparagraph.

2. Where application of paragraph 1 results in a reduction of the rights or an extension of the obligations under authorisations already in existence, Member States may extend the validity of those rights and obligations until at the latest nine months after the date of application referred to in Article 18(1), second subparagraph, provided that the rights of other undertakings under Community law are not affected thereby. Member States shall notify such extensions to the Commission and state the reasons therefor.

3. Where the Member State concerned can prove that the abolition of an authorisation condition regarding access to electronic communications networks, which was in force before the date of entry into force of this Directive, creates excessive difficulties for undertakings that have benefited from mandated access to another network, and where it is not possible for these undertakings to negotiate new agreements on reasonable commercial terms before the date of application referred to in Article 18(1), second subparagraph, Member States may request a temporary prolongation of the relevant condition(s). Such requests shall be submitted by the date of application referred to in Article 18(1), second subparagraph,

L 108/30 EN Official Journal of the European Communities 24.4.2002

at the latest, and shall specify the condition(s) and period for which the temporary prolongation is requested.

The Member State shall inform the Commission of the reasons for requesting a prolongation. The Commission shall consider such a request, taking into account the particular situation in that Member State and of the undertaking(s) concerned, and the need to ensure a coherent regulatory environment at a Community level. It shall take a decision on whether to grant or reject the request, and where it decides to grant the request, on the scope and duration of the prolongation to be granted. The Commission shall communicate its decision to the Member State concerned within six months after receipt of the application for a prolongation. Such decisions shall be published in the *Official Journal of the European Communities*.

*Article 18*

**Transposition**

1. Member States shall adopt and publish the laws, regulations and administrative provisions necessary to comply with this Directive by 24 July 2003 at the latest. They shall forthwith inform the Commission thereof.

They shall apply those measures from 25 July 2003.

When Member States adopt these measures, they shall contain a reference to this Directive or be accompanied by such reference on the occasion of their official publication. The methods of making such reference shall be laid down by Member States.

2. Member States shall communicate to the Commission the text of the provisions of national law which they adopt in the field governed by this Directive and of any subsequent amendments to those provisions.

*Article 19*

**Entry into force**

This Directive shall enter into force on the day of its publication in the *Official Journal of the European Communities*.

*Article 20*

**Addressees**

This Directive is addressed to the Member States.

Done at Brussels, 7 March 2002.

| *For the European Parliament* | *For the Council* |
| --- | --- |
| *The President* | *The President* |
| P. COX | J. C. APARICIO |

*ANNEX*

The conditions listed in this Annex provide the maximum list of conditions which may be attached to general authorisations (Part A), rights to use radio frequencies (Part B) and rights to use numbers (Part C) as referred to in Article 6(1) and Article 11(1)(a).

A. **Conditions which may be attached to a general authorisation**

1. Financial contributions to the funding of universal service in conformity with Directive 2002/22/EC (Universal Service Directive).

2. Administrative charges in accordance with Article 12 of this Directive.

3. Interoperability of services and interconnection of networks in conformity with Directive 2002/19/EC (Access Directive).

4. Accessibility of numbers from the national numbering plan to end-users including conditions in conformity with Directive 2002/22/EC (Universal Service Directive).

5. Environmental and town and country planning requirements, as well as requirements and conditions linked to the granting of access to or use of public or private land and conditions linked to co-location and facility sharing in conformity with Directive 2002/22/EC (Framework Directive) and including, where applicable, any financial or technical guarantees necessary to ensure the proper execution of infrastructure works.

6. 'Must carry' obligations in conformity with Directive 2002/22/EC (Universal Service Directive).

7. Personal data and privacy protection specific to the electronic communications sector in conformity with Directive 97/66/EC of the European Parliament and of the Council of 15 December 1997 concerning the processing of personal data and the protection of privacy in the telecommunications sector ([1]).

8. Consumer protection rules specific to the electronic communications sector including conditions in conformity with Directive 2002/22/EC (Universal Service Directive).

9. Restrictions in relation to the transmission of illegal content, in accordance with Directive 2000/31/EC of the European Parliament and of the Council of 8 June 2000 on certain legal aspects of information society services, in particular electronic commerce, in the internal market ([2]) and restrictions in relation to the transmission of harmful content in accordance with Article 2a(2) of Council Directive 89/552/EEC of 3 October 1989 on the coordination of certain provisions laid down by law, regulation or administrative action in Member States concerning the pursuit of television broadcasting activities ([3]).

10. Information to be provided under a notification procedure in accordance with Article 3(3) of this Directive and for other purposes as included in Article 11 of this Directive.

11. Enabling of legal interception by competent national authorities in conformity with Directive 97/66/EC and Directive 95/46/EC of the European Parliament and of the Council of 24 October 1995 on the protection of individuals with regard to the processing of personal data and on the free movement of such data ([4]).

12. Terms of use during major disasters to ensure communications between emergency services and authorities and broadcasts to the general public.

13. Measures regarding the limitation of exposure of the general public to electromagnetic fields caused by electronic communications networks in accordance with Community law.

14. Access obligations other than those provided for in Article 6(2) of this Directive applying to undertakings providing electronic communications networks or services, in conformity with Directive 2002/19/EC (Access Directive).

---

([1]) OJ L 24, 30.1.1998, p. 1.
([2]) OJ L 178, 17.7.2000, p. 1.
([3]) OJ L 298, 17.10.1989, p. 23. Directive as amended by Directive 97/36/EC of the European Parliament and of the Council (OJ L 202, 30.7.1997, p. 60).
([4]) OJ L 281, 23.11.1995, p. 31.

15. Maintenance of the integrity of public communications networks in accordance with Directive 2002/19/EC (Access Directive) and Directive 2002/22/EC (Universal Service Directive) including by conditions to prevent electromagnetic interference between electronic communications networks and/or services in accordance with Council Directive 89/336/EEC of 3 May 1989 on the approximation of the laws of the Member States relating to electromagnetic compatibility ([1]).

16. Security of public networks against unauthorised access according to Directive 97/66/EC.

17. Conditions for the use of radio frequencies, in conformity with Article 7(2) of Directive 1999/5/EC, where such use is not made subject to the granting of individual rights of use in accordance with Article 5(1) of this Directive.

18. Measures designed to ensure compliance with the standards and/or specifications referred to in Article 17 of Directive 2002/21/EC (Framework Directive).

B. **Conditions which may be attached to rights of use for radio frequencies**

1. Designation of service or type of network or technology for which the rights of use for the frequency has been granted, including, where applicable, the exclusive use of a frequency for the transmission of specific content or specific audiovisual services.

2. Effective and efficient use of frequencies in conformity with Directive 2002/21/EC (Framework Directive), including, where appropriate, coverage requirements.

3. Technical and operational conditions necessary for the avoidance of harmful interference and for the limitation of exposure of the general public to electromagnetic fields, where such conditions are different from those included in the general authorisation.

4. Maximum duration in conformity with Article 5 of this Directive, subject to any changes in the national frequency plan.

5. Transfer of rights at the initiative of the right holder and conditions for such transfer in conformity with Directive 2002/21/EC (Framework Directive).

6. Usage fees in accordance with Article 13 of this Directive.

7. Any commitments which the undertaking obtaining the usage right has made in the course of a competitive or comparative selection procedure.

8. Obligations under relevant international agreements relating to the use of frequencies.

C. **Conditions which may be attached to rights of use for numbers**

1. Designation of service for which the number shall be used, including any requirements linked to the provision of that service.

2. Effective and efficient use of numbers in conformity with Directive 2002/21/EC (Framework Directive).

3. Number portability requirements in conformity with Directive 2002/22/EC (Universal Service Directive).

4. Obligation to provide public directory subscriber information for the purposes of Articles 5 and 25 of Directive 2002/22/EC (Universal Service Directive).

5. Maximum duration in conformity with Article 5 of this Directive, subject to any changes in the national numbering plan.

6. Transfer of rights at the initiative of the right holder and conditions for such transfer in conformity with Directive 2002/21/EC (Framework Directive).

7. Usage fees in accordance with Article 13 of this Directive.

8. Any commitments which the undertaking obtaining the usage right has made in the course of a competitive or comparative selection procedure.

9. Obligations under relevant international agreements relating to the use of numbers.

---

([1]) OJ L 139, 23.5.1989, p. 19. Directive as last amended by Directive 93/68/EEC (OJ L 220, 30.8.1993, p. 1).

# DIRECTIVE 2002/19/EC OF THE EUROPEAN PARLIAMENT AND OF THE COUNCIL

## of 7 March 2002

## on access to, and interconnection of, electronic communications networks and associated facilities (Access Directive)

THE EUROPEAN PARLIAMENT AND THE COUNCIL OF THE EUROPEAN UNION,

Having regard to the Treaty establishing the European Community, and in particular Article 95 thereof,

Having regard to the proposal from the Commission [1],

Having regard to the opinion of the Economic and Social Committee [2],

Acting in accordance with the procedure laid down in Article 251 of the Treaty [3],

Whereas:

(1) Directive 2002/21/EC of the European Parliament and of the Council of 7 March 2002 on a common regulatory framework for electronic communications networks and services (Framework Directive) [4] lays down the objectives of a regulatory framework to cover electronic communications networks and services in the Community, including fixed and mobile telecommunications networks, cable television networks, networks used for terrestrial broadcasting, satellite networks and Internet networks, whether used for voice, fax, data or images. Such networks may have been authorised by Member States under Directive 2002/20/EC of the European Parliament and of the Council of 7 March 2002 on the authorisation of electronic communications networks and services (Authorisation Directive) [5] or have been authorised under previous regulatory measures. The provisions of this Directive apply to those networks that are used for the provision of publicly available electronic communications services. This Directive covers access and interconnection arrangements between service suppliers. Non-public networks do not have obligations under this Directive except where, in benefiting from access to public networks, they may be subject to conditions laid down by Member States.

(2) Services providing content such as the offer for sale of a package of sound or television broadcasting content are not covered by the common regulatory framework for electronic communications networks and services.

(3) The term 'access' has a wide range of meanings, and it is therefore necessary to define precisely how that term is used in this Directive, without prejudice to how it may be used in other Community measures. An operator may own the underlying network or facilities or may rent some or all of them.

(4) Directive 95/47/EC of the European Parliament and of the Council of 24 October 1995 on the use of standards for the transmission of television signals [6] did not mandate any specific digital television transmission system or service requirement, and this opened up an opportunity for the market actors to take the initiative and develop suitable systems. Through the Digital Video Broadcasting Group, European market actors have developed a family of television transmission systems that have been adopted by broadcasters throughout the world. These transmissions systems have been standardised by the European Telecommunications Standards Institute (ETSI) and have become International Telecommunication Union recommendations. In relation to wide-screen digital television, the 16:9 aspect ratio is the reference format for wide-format television services and programmes, and is now established in Member States' markets as a result of Council Decision 93/424/EEC of 22 July 1993 on an action plan for the introduction of advanced television services in Europe [7].

(5) In an open and competitive market, there should be no restrictions that prevent undertakings from negotiating access and interconnection arrangements between themselves, in particular on cross-border agreements,

(1) OJ C 365 E, 19.12.2000, p. 215 and OJ C 270 E, 25.9.2001, p. 161.
(2) OJ C 123, 25.4.2001, p. 50.
(3) Opinion of the European Parliament of 1 March 2001 (OJ C 277, 1.10.2001, p. 72), Council Common Position of 17 September 2001 (OJ C 337, 30.11.2001, p. 1) and Decision of the European Parliament of 12 December 2001 (not yet published in the Official Journal). Council Decision of 14 February 2002.
(4) See page 33 of this Official Journal.
(5) See page 21 of this Official Journal.
(6) OJ L 281, 23.11.1995, p. 51.
(7) OJ L 196, 5.8.1993, p. 48.

subject to the competition rules of the Treaty. In the context of achieving a more efficient, truly pan-European market, with effective competition, more choice and competitive services to consumers, undertakings which receive requests for access or interconnection should in principle conclude such agreements on a commercial basis, and negotiate in good faith.

(6) In markets where there continue to be large differences in negotiating power between undertakings, and where some undertakings rely on infrastructure provided by others for delivery of their services, it is appropriate to establish a framework to ensure that the market functions effectively. National regulatory authorities should have the power to secure, where commercial negotiation fails, adequate access and interconnection and interoperability of services in the interest of end-users. In particular, they may ensure end-to-end connectivity by imposing proportionate obligations on undertakings that control access to end-users. Control of means of access may entail ownership or control of the physical link to the end-user (either fixed or mobile), and/or the ability to change or withdraw the national number or numbers needed to access an end-user's network termination point. This would be the case for example if network operators were to restrict unreasonably end-user choice for access to Internet portals and services.

(7) National legal or administrative measures that link the terms and conditions for access or interconnection to the activities of the party seeking interconnection, and specifically to the degree of its investment in network infrastructure, and not to the interconnection or access services provided, may cause market distortion and may therefore not be compatible with competition rules.

(8) Network operators who control access to their own customers do so on the basis of unique numbers or addresses from a published numbering or addressing range. Other network operators need to be able to deliver traffic to those customers, and so need to be able to interconnect directly or indirectly to each other. The existing rights and obligations to negotiate interconnection should therefore be maintained. It is also appropriate to maintain the obligations formerly laid down in Directive 95/47/EC requiring fully digital electronic communications networks used for the distribution of television services and open to the public to be capable of distributing wide-screen television services and programmes, so that users are able to receive such programmes in the format in which they were transmitted.

(9) Interoperability is of benefit to end-users and is an important aim of this regulatory framework. Encouraging interoperability is one of the objectives for national regulatory authorities as set out in this framework, which also provides for the Commission to publish a list of standards and/or specifications covering the provision of services, technical interfaces and/or network functions, as the basis for encouraging harmonisation in electronic communications. Member States should encourage the use of published standards and/or specifications to the extent strictly necessary to ensure interoperability of services and to improve freedom of choice for users.

(10) Competition rules alone may not be sufficient to ensure cultural diversity and media pluralism in the area of digital television. Directive 95/47/EC provided an initial regulatory framework for the nascent digital television industry which should be maintained, including in particular the obligation to provide conditional access on fair, reasonable and non-discriminatory terms, in order to make sure that a wide variety of programming and services is available. Technological and market developments make it necessary to review these obligations on a regular basis, either by a Member State for its national market or the Commission for the Community, in particular to determine whether there is justification for extending obligations to new gateways, such as electronic programme guides (EPGs) and application program interfaces (APIs), to the extent that is necessary to ensure accessibility for end-users to specified digital broadcasting services. Member States may specify the digital broadcasting services to which access by end-users must be ensured by any legislative, regulatory or administrative means that they deem necessary.

(11) Member States may also permit their national regulatory authority to review obligations in relation to conditional access to digital broadcasting services in order to assess through a market analysis whether to withdraw or amend conditions for operators that do not have significant market power on the relevant market. Such withdrawal or amendment should not adversely affect access for end-users to such services or the prospects for effective competition.

(12) In order to ensure continuity of existing agreements and to avoid a legal vacuum, it is necessary to ensure that obligations for access and interconnection imposed under Articles 4, 6, 7, 8, 11, 12, and 14 of Directive 97/33/EC of the European Parliament and of the Council of 30 June 1997 on interconnection in telecommunications with regard to ensuring universal service and interoperability through application of the

principles of open network provision (ONP) ([1]), obligations on special access imposed under Article 16 of Directive 98/10/EC of the European Parliament and of the Council of 26 February 1998 on the application of open network provision (ONP) to voice telephony and on universal service for telecommunications in a competitive environment ([2]), and obligations concerning the provision of leased line transmission capacity under Council Directive 92/44/EEC of 5 June 1992 on the application of open network provision to leased lines ([3]), are initially carried over into the new regulatory framework, but are subject to immediate review in the light of prevailing market conditions. Such a review should also extend to those organisations covered by Regulation (EC) No 2887/2000 of the European Parliament and of the Council of 18 December 2000 on unbundled access to the local loop ([4]).

(13) The review should be carried out using an economic market analysis based on competition law methodology. The aim is to reduce *ex ante* sector specific rules progressively as competition in the market develops. However the procedure also takes account of transitional problems in the market such as those related to international roaming and of the possibility of new bottlenecks arising as a result of technological development, which may require *ex ante* regulation, for example in the area of broadband access networks. It may well be the case that competition develops at different speeds in different market segments and in different Member States, and national regulatory authorities should be able to relax regulatory obligations in those markets where competition is delivering the desired results. In order to ensure that market players in similar circumstances are treated in similar ways in different Member States, the Commission should be able to ensure harmonised application of the provisions of this Directive. National regulatory authorities and national authorities entrusted with the implementation of competition law should, where appropriate, coordinate their actions to ensure that the most appropriate remedy is applied. The Community and its Member States have entered into commitments on interconnection of telecommunications networks in the context of the World Trade Organisation agreement on basic telecommunications and these commitments need to be respected.

(14) Directive 97/33/EC laid down a range of obligations to be imposed on undertakings with significant market power, namely transparency, non-discrimination, accounting separation, access, and price control including cost orientation. This range of possible obligations should be maintained but, in addition, they should be established as a set of maximum obligations that can be applied to undertakings, in order to avoid over-regulation. Exceptionally, in order to comply with international commitments or Community law, it may be appropriate to impose obligations for access or interconnection on all market players, as is currently the case for conditional access systems for digital television services.

(15) The imposition of a specific obligation on an undertaking with significant market power does not require an additional market analysis but a justification that the obligation in question is appropriate and proportionate in relation to the nature of the problem identified.

(16) Transparency of terms and conditions for access and interconnection, including prices, serve to speed-up negotiation, avoid disputes and give confidence to market players that a service is not being provided on discriminatory terms. Openness and transparency of technical interfaces can be particularly important in ensuring interoperability. Where a national regulatory authority imposes obligations to make information public, it may also specify the manner in which the information is to be made available, covering for example the type of publication (paper and/or electronic) and whether or not it is free of charge, taking into account the nature and purpose of the information concerned.

(17) The principle of non-discrimination ensures that undertakings with market power do not distort competition, in particular where they are vertically integrated undertakings that supply services to undertakings with whom they compete on downstream markets.

(18) Accounting separation allows internal price transfers to be rendered visible, and allows national regulatory authorities to check compliance with obligations for non-discrimination where applicable. In this regard the Commission published Recommendation 98/322/EC of 8 April 1998 on interconnection in a liberalised telecommunications market (Part 2 — accounting separation and cost accounting) ([5]).

(19) Mandating access to network infrastructure can be justified as a means of increasing competition, but national regulatory authorities need to balance the rights of an infrastructure owner to exploit its infrastructure forits own benefit, and the rights of other service

([1]) OJ L 199, 26.7.1997, p. 32. Directive as last amended by Directive 98/61/EC (OJ L 268, 3.10.1998, p. 37).

([2]) OJ L 101, 1.4.1998, p. 24.

([3]) OJ L 165, 19.6.1992, p. 27. Directive as last amended by Commission Decision No 98/80/EC (OJ L 14, 20.1.1998, p. 27).

([4]) OJ L 366, 30.12.2000, p. 4.

([5]) OJ L 141, 13.5.1998, p. 6.

providers to access facilities that are essential for the provision of competing services. Where obligations are imposed on operators that require them to meet reasonable requests for access to and use of networks elements and associated facilities, such requests should only be refused on the basis of objective criteria such as technical feasibility or the need to maintain network integrity. Where access is refused, the aggrieved party may submit the case to the dispute resolutions procedure referred to in Articles 20 and 21 of Directive 2002/21/EC (Framework Directive). An operator with mandated access obligations cannot be required to provide types of access which are not within its powers to provide. The imposition by national regulatory authorities of mandated access that increases competition in the short-term should not reduce incentives for competitors to invest in alternative facilities that will secure more competition in the long-term. The Commission has published a Notice on the application of the competition rules to access agreements in the telecommunications sector ([1]) which addresses these issues. National regulatory authorities may impose technical and operational conditions on the provider and/or beneficiaries of mandated access in accordance with Community law. In particular the imposition of technical standards should comply with Directive 98/34/EC of the European Parliament and of the Council of 22 June 1998 laying down a procedure for the provision of information in the field of technical standards and regulations and of rules of Information Society Services ([2]).

(20) Price control may be necessary when market analysis in a particular market reveals inefficient competition. The regulatory intervention may be relatively light, such as an obligation that prices for carrier selection are reasonable as laid down in Directive 97/33/EC, or much heavier such as an obligation that prices are cost oriented to provide full justification for those prices where competition is not sufficiently strong to prevent excessive pricing. In particular, operators with significant market power should avoid a price squeeze whereby the difference between their retail prices and the interconnection prices charged to competitors who provide similar retail services is not adequate to ensure sustainable competition. When a national regulatory authority calculates costs incurred in establishing a service mandated under this Directive, it is appropriate to allow a reasonable return on the capital employed including appropriate labour and building costs, with the value of capital adjusted where necessary to reflect the current valuation of assets and efficiency of operations. The method of cost recovery should be appropriate to the circumstances taking account of the need to promote efficiency and sustainable competition and maximise consumer benefits.

(21) Where a national regulatory authority imposes obligations to implement a cost accounting system in order to support price controls, it may itself undertake an annual audit to ensure compliance with that cost accounting system, provided that it has the necessary qualified staff, or it may require the audit to be carried out by another qualified body, independent of the operator concerned.

(22) Publication of information by Member States will ensure that market players and potential market entrants understand their rights and obligations, and know where to find the relevant detailed information. Publication in the national gazette helps interested parties in other Member States to find the relevant information.

(23) In order to ensure that the pan-European electronic communications market is effective and efficient, the Commission should monitor and publish information on charges which contribute to determining prices to end-users.

(24) The development of the electronic communications market, with its associated infrastructure, could have adverse effects on the environment and the landscape. Member States should therefore monitor this process and, if necessary, take action to minimise any such effects by means of appropriate agreements and other arrangements with the relevant authorities.

(25) In order to determine the correct application of Community law, the Commission needs to know which undertakings have been designated as having significant market power and what obligations have been placed upon market players by national regulatory authorities. In addition to national publication of this information, it is therefore necessary for Member States to send this information to the Commission. Where Member States are required to send information to the Commission, this may be in electronic form, subject to appropriate authentication procedures being agreed.

(26) Given the pace of technological and market developments, the implementation of this Directive

([1]) OJ C 265, 22.8.1998, p. 2.

([2]) OJ L 204, 21.7.1998, p. 37. Directive as amended by Directive 98/48/EC (OJ L 217, 5.8.1998, p. 18).

should be reviewed within three years of its date of application to determine if it is meeting its objectives.

(27) The measures necessary for the implementation of this Directive should be adopted in accordance with Council Decision 1999/468/EC of 28 June 1999 laying down the procedures for the exercise of implementing powers conferred on the Commission ([1]).

(28) Since the objectives of the proposed action, namely establishing a harmonised framework for the regulation of access to and interconnection of electronic communications networks and associated facilities, cannot be sufficiently achieved by the Member States and can therefore, by reason of the scale and effects of the action, be better achieved at Community level, the Community may adopt measures, in accordance with the principle of subsidiarity as set out in Article 5 of the Treaty. In accordance with the principle of proportionality, as set out in that Article, this Directive does not go beyond what is necessary in order to achieve those objectives,

HAVE ADOPTED THIS DIRECTIVE:

CHAPTER I

**SCOPE, AIM AND DEFINITIONS**

*Article 1*

**Scope and aim**

1. Within the framework set out in Directive 2002/21/EC (Framework Directive), this Directive harmonises the way in which Member States regulate access to, and interconnection of, electronic communications networks and associated facilities. The aim is to establish a regulatory framework, in accordance with internal market principles, for the relationships between suppliers of networks and services that will result in sustainable competition, interoperability of electronic communications services and consumer benefits.

2. This Directive establishes rights and obligations for operators and for undertakings seeking interconnection and/or access to their networks or associated facilities. It sets out objectives for national regulatory authorities with regard to access and interconnection, and lays down procedures to ensure that obligations imposed by national regulatory authorities are reviewed and, where appropriate, withdrawn once the desired objectives have been achieved. Access in this Directive does not refer to access by end-users.

*Article 2*

**Definitions**

For the purposes of this Directive the definitions set out in Article 2 of Directive 2002/21/EC (Framework Directive) shall apply.

The following definitions shall also apply:

(a) 'access' means the making available of facilities and/or services, to another undertaking, under defined conditions, on either an exclusive or non-exclusive basis, for the purpose of providing electronic communications services. It covers *inter alia*: access to network elements and associated facilities, which may involve the connection of equipment, by fixed or non-fixed means (in particular this includes access to the local loop and to facilities and services necessary to provide services over the local loop), access to physical infrastructure including buildings, ducts and masts; access to relevant software systems including operational support systems, access to number translation or systems offering equivalent functionality, access to fixed and mobile networks, in particular for roaming, access to conditional access systems for digital television services; access to virtual network services;

(b) 'interconnection' means the physical and logical linking of public communications networks used by the same or a different undertaking in order to allow the users of one undertaking to communicate with users of the same or another undertaking, or to access services provided by another undertaking. Services may be provided by the parties involved or other parties who have access to the network. Interconnection is a specific type of access implemented between public network operators;

(c) 'operator' means an undertaking providing or authorised to provide a public communications network or an associated facility;

(d) 'wide-screen television service' means a television service that consists wholly or partially of programmes produced and edited to be displayed in a full height wide-screen format. The 16:9 format is the reference format for wide-screen television services;

(e) 'local loop' means the physical circuit connecting the network termination point at the subscriber's premises to the main distribution frame or equivalent facility in the fixed public telephone network.

([1]) OJ L 184, 17.7.1999, p. 23.

L 108/12 EN Official Journal of the European Communities 24.4.2002

CHAPTER II

**GENERAL PROVISIONS**

*Article 3*

**General framework for access and interconnection**

1. Member States shall ensure that there are no restrictions which prevent undertakings in the same Member State or in different Member States from negotiating between themselves agreements on technical and commercial arrangements for access and/or interconnection, in accordance with Community law. The undertaking requesting access or interconnection does not need to be authorised to operate in the Member State where access or interconnection is requested, if it is not providing services and does not operate a network in that Member State.

2. Without prejudice to Article 31 of Directive 2002/22/EC of the European Parliament and of the Council of 7 March 2002 on universal service and users' rights relating to electronic communications networks and services (Universal Service Directive) ([1]), Member States shall not maintain legal or administrative measures which oblige operators, when granting access or interconnection, to offer different terms and conditions to different undertakings for equivalent services and/or imposing obligations that are not related to the actual access and interconnection services provided without prejudice to the conditions fixed in the Annex of Directive 2002/20/EC (Authorisation Directive).

*Article 4*

**Rights and obligations for undertakings**

1. Operators of public communications networks shall have a right and, when requested by other undertakings so authorised, an obligation to negotiate interconnection with each other for the purpose of providing publicly available electronic communications services, in order to ensure provision and interoperability of services throughout the Community. Operators shall offer access and interconnection to other undertakings on terms and conditions consistent with obligations imposed by the national regulatory authority pursuant to Articles 5, 6, 7 and 8.

2. Public electronic communications networks established for the distribution of digital television services shall be capable of distributing wide-screen television services and programmes. Network operators that receive and redistribute wide-screen television services or programmes shall maintain that wide-screen format.

3. Without prejudice to Article 11 of Directive 2002/20/EC (Authorisation Directive), Member States shall require that undertakings which acquire information from another undertaking before, during or after the process of negotiating access or interconnection arrangements use that information solely for the purpose for which it was supplied and respect at all times the confidentiality of information transmitted or stored. The received information shall not be passed on to any other party, in particular other departments, subsidiaries or partners, for whom such information could provide a competitive advantage.

([1]) See page 51 of this Official Journal.

*Article 5*

**Powers and responsibilities of the national regulatory authorities with regard to access and interconnection**

1. National regulatory authorities shall, acting in pursuit of the objectives set out in Article 8 of Directive 2002/21/EC (Framework Directive), encourage and where appropriate ensure, in accordance with the provisions of this Directive, adequate access and interconnection, and interoperability of services, exercising their responsibility in a way that promotes efficiency, sustainable competition, and gives the maximum benefit to end-users.

In particular, without prejudice to measures that may be taken regarding undertakings with significant market power in accordance with Article 8, national regulatory authorities shall be able to impose:

(a) to the extent that is necessary to ensure end-to-end connectivity, obligations on undertakings that control access to end-users, including in justified cases the obligation to interconnect their networks where this is not already the case;

(b) to the extent that is necessary to ensure accessibility for end-users to digital radio and television broadcasting services specified by the Member State, obligations on operators to provide access to the other facilities referred to in Annex I, Part II on fair, reasonable and non-discriminatory terms.

2. When imposing obligations on an operator to provide access in accordance with Article 12, national regulatory authorities may lay down technical or operational conditions to be met by the provider and/or beneficiaries of such access, in accordance with Community law, where necessary to ensure normal operation of the network. Conditions that refer to implementation of specific technical standards or specifications shall respect Article 17 of Directive 2002/21/EC (Framework Directive).

3. Obligations and conditions imposed in accordance with paragraphs 1 and 2 shall be objective, transparent, proportionate and non-discriminatory, and shall be implemented in accordance with the procedures referred to in Articles 6 and 7 of Directive 2002/21/EC (Framework Directive).

4. With regard to access and interconnection, Member States shall ensure that the national regulatory authority is empowered to intervene at its own initiative where justified or, in the absence of agreement between undertakings, at the request of either of the parties involved, in order to secure the policy objectives of Article 8 of Directive 2002/21/EC (Framework Directive), in accordance with the provisions of this Directive and the procedures referred to in Articles 6 and 7, 20 and 21 of Directive 2002/21/EC (Framework Directive).

CHAPTER III

**OBLIGATIONS ON OPERATORS AND MARKET REVIEW PROCEDURES**

*Article 6*

**Conditional access systems and other facilities**

1. Member States shall ensure that, in relation to conditional access to digital television and radio services broadcast to viewers and listeners in the Community, irrespective of the means of transmission, the conditions laid down in Annex I, Part I apply.

2. In the light of market and technological developments, Annex I may be amended in accordance with the procedure referred to in Article 14(3).

3. Notwithstanding the provisions of paragraph 1, Member States may permit their national regulatory authority, as soon as possible after the entry into force of this Directive and periodically thereafter, to review the conditions applied in accordance with this Article, by undertaking a market analysis in accordance with the first paragraph of Article 16 of Directive 2002/21/EC (Framework Directive) to determine whether to maintain, amend or withdraw the conditions applied.

Where, as a result of this market analysis, a national regulatory authority finds that one or more operators do not have significant market power on the relevant market, it may amend or withdraw the conditions with respect to those operators, in accordance with the procedures referred to in Articles 6 and 7 of Directive 2002/21/EC (Framework Directive), only to the extent that:

(a) accessibility for end-users to radio and television broadcasts and broadcasting channels and services specified in accordance with Article 31 of Directive 2002/22/EC (Universal Service Directive) would not be adversely affected by such amendment or withdrawal, and

(b) the prospects for effective competition in the markets for:

(i) retail digital television and radio broadcasting services, and

(ii) conditional access systems and other associated facilities,

would not be adversely affected by such amendment or withdrawal.

An appropriate period of notice shall be given to parties affected by such amendment or withdrawal of conditions.

4. Conditions applied in accordance with this Article are without prejudice to the ability of Member States to impose obligations in relation to the presentational aspect of electronic programme guides and similar listing and navigation facilities.

*Article 7*

**Review of former obligations for access and interconnection**

1. Member States shall maintain all obligations on undertakings providing public communications networks and/or services concerning access and interconnection that were in force prior to the date of entry into force of this Directive under Articles 4, 6, 7, 8, 11, 12, and 14 of Directive 97/33/EC, Article 16 of Directive 98/10/EC, and Articles 7 and 8 of Directive 92/44/EC, until such time as these obligations have been reviewed and a determination made in accordance with paragraph 3.

2. The Commission will indicate relevant markets for the obligations referred to in paragraph 1 in the initial recommendation on relevant product and service markets and the Decision identifying transnational markets to be adopted in accordance with Article 15 of Directive 2002/21/EC (Framework Directive).

3. Member States shall ensure that, as soon as possible after the entry into force of this Directive, and periodically thereafter, national regulatory authorities undertake a market analysis, in accordance with Article 16 of Directive 2002/21/EC (Framework Directive) to determine whether to maintain, amend or withdraw these obligations. An appropriate period of notice shall be given to parties affected by such amendment or withdrawal of obligations.

*Article 8*

**Imposition, amendment or withdrawal of obligations**

1. Member States shall ensure that national regulatory authorities are empowered to impose the obligations identified in Articles 9 to 13.

2. Where an operator is designated as having significant market power on a specific market as a result of a market analysis carried out in accordance with Article 16 of Directive 2002/21/EC (Framework Directive), national regulatory

authorities shall impose the obligations set out in Articles 9 to 13 of this Directive as appropriate.

3. Without prejudice to:

— the provisions of Articles 5(1), 5(2) and 6,

— the provisions of Articles 12 and 13 of Directive 2002/21/EC (Framework Directive), Condition 7 in Part B of the Annex to Directive 2002/20/EC (Authorisation Directive) as applied by virtue of Article 6(1) of that Directive, Articles 27, 28 and 30 of Directive 2002/22/EC (Universal Service Directive) and the relevant provisions of Directive 97/66/EC of the European Parliament and of the Council of 15 December 1997 concerning the processing of personal data and the protection of privacy in the telecommunications sector [1] containing obligations on undertakings other than those designated as having significant market power, or

— the need to comply with international commitments,

national regulatory authorities shall not impose the obligations set out in Articles 9 to 13 on operators that have not been designated in accordance with paragraph 2.

In exceptional circumstances, when a national regulatory authority intends to impose on operators with significant market power other obligations for access or interconnection than those set out in Articles 9 to 13 in this Directive it shall submit this request to the Commission. The Commission, acting in accordance with Article 14(2), shall take a decision authorising or preventing the national regulatory authority from taking such measures.

4. Obligations imposed in accordance with this Article shall be based on the nature of the problem identified, proportionate and justified in the light of the objectives laid down in Article 8 of Directive 2002/21/EC (Framework Directive). Such obligations shall only be imposed following consultation in accordance with Articles 6 and 7 of that Directive.

5. In relation to the third indent of the first subparagraph of paragraph 3, national regulatory authorities shall notify decisions to impose, amend or withdraw obligations on market players to the Commission, in accordance with the procedure referred to in Article 7 of Directive 2002/21/EC (Framework Directive).

*Article 9*

**Obligation of transparency**

1. National regulatory authorities may, in accordance with the provisions of Article 8, impose obligations for transparency in relation to interconnection and/or access, requiring operators to make public specified information, such as accounting information, technical specifications, network characteristics, terms and conditions for supply and use, and prices.

2. In particular where an operator has obligations of non-discrimination, national regulatory authorities may require that operator to publish a reference offer, which shall be sufficiently unbundled to ensure that undertakings are not required to pay for facilities which are not necessary for the service requested, giving a description of the relevant offerings broken down into components according to market needs, and the associated terms and conditions including prices. The national regulatory authority shall, *inter alia*, be able to impose changes to reference offers to give effect to obligations imposed under this Directive.

3. National regulatory authorities may specify the precise information to be made available, the level of detail required and the manner of publication.

4. Notwithstanding paragraph 3, where an operator has obligations under Article 12 concerning unbundled access to the twisted metallic pair local loop, national regulatory authorities shall ensure the publication of a reference offer containing at least the elements set out in Annex II.

5. In the light of market and technological developments, Annex II may be amended in accordance with the procedure referred to in Article 14(3).

*Article 10*

**Obligation of non-discrimination**

1. A national regulatory authority may, in accordance with the provisions of Article 8, impose obligations of non-discrimination, in relation to interconnection and/or access.

2. Obligations of non-discrimination shall ensure, in particular, that the operator applies equivalent conditions in equivalent circumstances to other undertakings providing equivalent services, and provides services and information to others under the same conditions and of the same quality as it provides for its own services, or those of it subsidiaries or partners.

*Article 11*

**Obligation of accounting separation**

1. A national regulatory authority may, in accordance with the provisions of Article 8, impose obligations for accounting separation in relation to specified activities related to interconnection and/or access.

[1] OJ L 24, 30.1.1998, p. 1.

In particular, a national regulatory authority may require a vertically integrated company to make transparent its wholesale prices and its internal transfer prices *inter alia* to ensure compliance where there is a requirement for non-discrimination under Article 10 or, where necessary, to prevent unfair cross-subsidy. National regulatory authorities may specify the format and accounting methodology to be used.

2. Without prejudice to Article 5 of Directive 2002/21/EC (Framework Directive), to facilitate the verification of compliance with obligations of transparency and non-discrimination, national regulatory authorities shall have the power to require that accounting records, including data on revenues received from third parties, are provided on request. National regulatory authorities may publish such information as would contribute to an open and competitive market, while respecting national and Community rules on commercial confidentiality.

*Article 12*

**Obligations of access to, and use of, specific network facilities**

1. A national regulatory authority may, in accordance with the provisions of Article 8, impose obligations on operators to meet reasonable requests for access to, and use of, specific network elements and associated facilities, *inter alia* in situations where the national regulatory authority considers that denial of access or unreasonable terms and conditions having a similar effect would hinder the emergence of a sustainable competitive market at the retail level, or would not be in the end-user's interest.

Operators may be required *inter alia*:

(a) to give third parties access to specified network elements and/or facilities, including unbundled access to the local loop;

(b) to negotiate in good faith with undertakings requesting access;

(c) not to withdraw access to facilities already granted;

(d) to provide specified services on a wholesale basis for resale by third parties;

(e) to grant open access to technical interfaces, protocols or other key technologies that are indispensable for the interoperability of services or virtual network services;

(f) to provide co-location or other forms of facility sharing, including duct, building or mast sharing;

(g) to provide specified services needed to ensure interoperability of end-to-end services to users, including facilities for intelligent network services or roaming on mobile networks;

(h) to provide access to operational support systems or similar software systems necessary to ensure fair competition in the provision of services;

(i) to interconnect networks or network facilities.

National regulatory authorities may attach to those obligations conditions covering fairness, reasonableness and timeliness.

2. When national regulatory authorities are considering whether to impose the obligations referred in paragraph 1, and in particular when assessing whether such obligations would be proportionate to the objectives set out in Article 8 of Directive 2002/21/EC (Framework Directive), they shall take account in particular of the following factors:

(a) the technical and economic viability of using or installing competing facilities, in the light of the rate of market development, taking into account the nature and type of interconnection and access involved;

(b) the feasibility of providing the access proposed, in relation to the capacity available;

(c) the initial investment by the facility owner, bearing in mind the risks involved in making the investment;

(d) the need to safeguard competition in the long term;

(e) where appropriate, any relevant intellectual property rights;

(f) the provision of pan-European services.

*Article 13*

**Price control and cost accounting obligations**

1. A national regulatory authority may, in accordance with the provisions of Article 8, impose obligations relating to cost recovery and price controls, including obligations for cost orientation of prices and obligations concerning cost accounting systems, for the provision of specific types of interconnection and/or access, in situations where a market analysis indicates that a lack of effective competition means

that the operator concerned might sustain prices at an excessively high level, or apply a price squeeze, to the detriment of end-users. National regulatory authorities shall take into account the investment made by the operator and allow him a reasonable rate of return on adequate capital employed, taking into account the risks involved.

2. National regulatory authorities shall ensure that any cost recovery mechanism or pricing methodology that is mandated serves to promote efficiency and sustainable competition and maximise consumer benefits. In this regard national regulatory authorities may also take account of prices available in comparable competitive markets.

3. Where an operator has an obligation regarding the cost orientation of its prices, the burden of proof that charges are derived from costs including a reasonable rate of return on investment shall lie with the operator concerned. For the purpose of calculating the cost of efficient provision of services, national regulatory authorities may use cost accounting methods independent of those used by the undertaking. National regulatory authorities may require an operator to provide full justification for its prices, and may, where appropriate, require prices to be adjusted.

4. National regulatory authorities shall ensure that, where implementation of a cost accounting system is mandated in order to support price controls, a description of the cost accounting system is made publicly available, showing at least the main categories under which costs are grouped and the rules used for the allocation of costs. Compliance with the cost accounting system shall be verified by a qualified independent body. A statement concerning compliance shall be published annually.

CHAPTER IV

## PROCEDURAL PROVISIONS

*Article 14*

### Committee

1. The Commission shall be assisted by the Communications Committee set up by Article 22 of Directive 2002/21/EC (Framework Directive).

2. Where reference is made to this paragraph, Articles 3 and 7 of Decision 1999/468/EC shall apply, having regard to the provisions of Article 8 thereof.

3. Where reference is made to this paragraph, Articles 5 and 7 of Decision 1999/468/EC shall apply, having regard to the provisions of Article 8 thereof.

The period laid down in Article 5(6) of Decision 1999/468/EC shall be set at three months.

4. The Committee shall adopt its rules of procedure.

*Article 15*

### Publication of, and access to, information

1. Member States shall ensure that the specific obligations imposed on undertakings under this Directive are published and that the specific product/service and geographical markets are identified. They shall ensure that up-to-date information, provided that the information is not confidential and, in particular, does not comprise business secrets, is made publicly available in a manner that guarantees all interested parties easy access to that information.

2. Member States shall send to the Commission a copy of all such information published. The Commission shall make this information available in a readily accessible form, and shall distribute the information to the Communications Committee as appropriate.

*Article 16*

### Notification

1. Member States shall notify to the Commission by at the latest the date of application referred to in Article 18(1) second subparagraph the national regulatory authorities responsible for the tasks set out in this Directive.

2. National regulatory authorities shall notify to the Commission the names of operators deemed to have significant market power for the purposes of this Directive, and the obligations imposed upon them under this Directive. Any changes affecting the obligations imposed upon undertakings or of the undertakings affected under the provisions of this Directive shall be notified to the Commission without delay.

*Article 17*

### Review procedures

The Commission shall periodically review the functioning of this Directive and report to the European Parliament and to the Council, on the first occasion not later than three years after the date of application referred to in Article 18(1), second subparagraph. For this purpose, the Commission may request from the Member States information, which shall be supplied without undue delay.

*Article 18*

### Transposition

1. Member States shall adopt and publish the laws, regulations and administrative provisions necessary to comply with this Directive by not later than 24 July 2003. They shall forthwith inform the Commission thereof.

They shall apply those measures from 25 July 2003.

When Member States adopt these measures, they shall contain a reference to this Directive or be accompanied by such a reference on the occasion of their official publication. The methods of making such reference shall be laid down by Member States.

2. Member States shall communicate to the Commission the text of the provisions of national law which they adopt in the field governed by this Directive and of any subsequent amendments to those provisions.

*Article 19*

**Entry into force**

This Directive shall enter into force on the day of its publication in the *Official Journal of the European Communities*.

*Article 20*

**Addressees**

This Directive is addressed to the Member States.

Done at Brussels, 7 March 2002.

| | |
|---|---|
| *For the European Parliament* | *For the Council* |
| *The President* | *The President* |
| P. COX | J. C. APARICIO |

L 108/18 EN Official Journal of the European Communities 24.4.2002

*ANNEX I*

## CONDITIONS FOR ACCESS TO DIGITAL TELEVISION AND RADIO SERVICES BROADCAST TO VIEWERS AND LISTENERS IN THE COMMUNITY

### Part I: Conditions for conditional access systems to be applied in accordance with Article 6(1)

In relation to conditional access to digital television and radio services broadcast to viewers and listeners in the Community, irrespective of the means of transmission, Member States must ensure in accordance with Article 6 that the following conditions apply:

(a) conditional access systems operated on the market in the Community are to have the necessary technical capability for cost-effective transcontrol allowing the possibility for full control by network operators at local or regional level of the services using such conditional access systems;

(b) all operators of conditional access services, irrespective of the means of transmission, who provide access services to digital television and radio services and whose access services broadcasters depend on to reach any group of potential viewers or listeners are to:

— offer to all broadcasters, on a fair, reasonable and non-discriminatory basis compatible with Community competition law, technical services enabling the broadcasters' digitally-transmitted services to be received by viewers or listeners authorised by means of decoders administered by the service operators, and comply with Community competition law,

— keep separate financial accounts regarding their activity as conditional access providers.

(c) when granting licences to manufacturers of consumer equipment, holders of industrial property rights to conditional access products and systems are to ensure that this is done on fair, reasonable and non-discriminatory terms. Taking into account technical and commercial factors, holders of rights are not to subject the granting of licences to conditions prohibiting, deterring or discouraging the inclusion in the same product of:

— a common interface allowing connection with several other access systems, or

— means specific to another access system, provided that the licensee complies with the relevant and reasonable conditions ensuring, as far as he is concerned, the security of transactions of conditional access system operators.

### Part II: Other facilities to which conditions may be applied under Article 5(1)(b)

(a) Access to application program interfaces (APIs);

(b) Access to electronic programme guides (EPGs).

*ANNEX II*

**MINIMUM LIST OF ITEMS TO BE INCLUDED IN A REFERENCE OFFER FOR UNBUNDLED ACCESS TO THE TWISTED METALLIC PAIR LOCAL LOOP TO BE PUBLISHED BY NOTIFIED OPERATORS**

For the purposes of this Annex the following definitions apply:

(a) 'local sub-loop' means a partial local loop connecting the network termination point at the subscriber's premises to a concentration point or a specified intermediate access point in the fixed public telephone network;

(b) 'unbundled access to the local loop' means full unbundled access to the local loop and shared access to the local loop; it does not entail a change in ownership of the local loop;

(c) 'full unbundled access to the local loop' means the provision to a beneficiary of access to the local loop or local sub-loop of the notified operator authorising the use of the full frequency spectrum of the twisted metallic pair;

(d) 'shared access to the local loop' means the provision to a beneficiary of access to the local loop or local sub-loop of the notified operator, authorising the use of the non-voice band frequency spectrum of the twisted metallic pair; the local loop continues to be used by the notified operator to provide the telephone service to the public;

A. **Conditions for unbundled access to the local loop**

1. Network elements to which access is offered covering in particular the following elements:

   (a) access to local loops;

   (b) access to non-voice band frequency spectrum of a local loop, in the case of shared access to the local loop;

2. Information concerning the locations of physical access sites (1), availability of local loops in specific parts of the access network;

3. Technical conditions related to access and use of local loops, including the technical characteristics of the twisted metallic pair in the local loop;

4. Ordering and provisioning procedures, usage restrictions.

B. **Co-location services**

1. Information on the notified operator's relevant sites (1).

2. Co-location options at the sites indicated under point 1 (including physical co-location and, as appropriate, distant co-location and virtual co-location).

3. Equipment characteristics: restrictions, if any, on equipment that can be co-located.

4. Security issues: measures put in place by notified operators to ensure the security of their locations.

5. Access conditions for staff of competitive operators.

6. Safety standards.

7. Rules for the allocation of space where co-location space is limited.

8. Conditions for beneficiaries to inspect the locations at which physical co-location is available, or sites where co-location has been refused on grounds of lack of capacity.

(1) Availability of this information may be restricted to interested parties only, in order to avoid public security concerns.

C. **Information systems**

Conditions for access to notified operator's operational support systems, information systems or databases for pre-ordering, provisioning, ordering, maintenance and repair requests and billing.

D. **Supply conditions**

1. Lead time for responding to requests for supply of services and facilities; service level agreements, fault resolution, procedures to return to a normal level of service and quality of service parameters.

2. Standard contract terms, including, where appropriate, compensation provided for failure to meet lead times.

3. Prices or pricing formulae for each feature, function and facility listed above.

# DIRECTIVE 2002/22/EC OF THE EUROPEAN PARLIAMENT AND OF THE COUNCIL

## of 7 March 2002

## on universal service and users' rights relating to electronic communications networks and services (Universal Service Directive)

THE EUROPEAN PARLIAMENT AND THE COUNCIL OF THE EUROPEAN UNION,

Having regard to the Treaty establishing the European Community, and in particular Article 95 thereof,

Having regard to the proposal from the Commission [1],

Having regard to the opinion of the Economic and Social Committee [2],

Having regard to the opinion of the Committee of the Regions [3],

Acting in accordance with the procedure laid down in Article 251 of the Treaty [4],

Whereas:

(1) The liberalisation of the telecommunications sector and increasing competition and choice for communications services go hand in hand with parallel action to create a harmonised regulatory framework which secures the delivery of universal service. The concept of universal service should evolve to reflect advances in technology, market developments and changes in user demand. The regulatory framework established for the full liberalisation of the telecommunications market in 1998 in the Community defined the minimum scope of universal service obligations and established rules for its costing and financing.

(2) Under Article 153 of the Treaty, the Community is to contribute to the protection of consumers.

(3) The Community and its Member States have undertaken commitments on the regulatory framework of telecommunications networks and services in the context of the World Trade Organisation (WTO) agreement on basic telecommunications. Any member of the WTO has the right to define the kind of universal service obligation it wishes to maintain. Such obligations will not be regarded as anti-competitive per se, provided they are administered in a transparent, non-discriminatory and competitively neutral manner and are not more burdensome than necessary for the kind of universal service defined by the member.

(4) Ensuring universal service (that is to say, the provision of a defined minimum set of services to all end-users at an affordable price) may involve the provision of some services to some end-users at prices that depart from those resulting from normal market conditions. However, compensating undertakings designated to provide such services in such circumstances need not result in any distortion of competition, provided that designated undertakings are compensated for the specific net cost involved and provided that the net cost burden is recovered in a competitively neutral way.

(5) In a competitive market, certain obligations should apply to all undertakings providing publicly available telephone services at fixed locations and others should apply only to undertakings enjoying significant market power or which have been designated as a universal service operator.

(6) The network termination point represents a boundary for regulatory purposes between the regulatory framework for electronic communication networks and services and the regulation of telecommunication terminal equipment. Defining the location of the network termination point is the responsibility of the national regulatory authority, where necessary on the basis of a proposal by the relevant undertakings.

(7) Member States should continue to ensure that the services set out in Chapter II are made available with the quality specified to all end-users in their territory, irrespective of their geographical location, and, in the light of specific national conditions, at an affordable price. Member States may, in the context of universal service obligations and in the light of national conditions, take specific measures for consumers in rural or geographically isolated areas to ensure their access to the services set out in the Chapter II and the affordability of those services, as well as ensure under

[1] OJ C 365 E, 19.12.2000, p. 238 and OJ C 332 E, 27.11.2001, p. 292.
[2] OJ C 139, 11.5.2001, p. 15.
[3] OJ C 144, 16.5.2001, p. 60.
[4] Opinion of the European Parliament of 13 June 2001 (not yet published in the Official Journal), Council Common Position of 17 September 2001 (OJ C 337, 30.11.2001, p. 55) and Decision of the European Parliament of 12 December 2001 (not yet published in the Official Journal). Council Decision of 14 February 2002.

the same conditions this access, in particular for the elderly, the disabled and for people with special social needs. Such measures may also include measures directly targeted at consumers with special social needs providing support to identified consumers, for example by means of specific measures, taken after the examination of individual requests, such as the paying off of debts.

(8) A fundamental requirement of universal service is to provide users on request with a connection to the public telephone network at a fixed location, at an affordable price. The requirement is limited to a single narrowband network connection, the provision of which may be restricted by Member States to the end-user's primary location/residence, and does not extend to the Integrated Services Digital Network (ISDN) which provides two or more connections capable of being used simultaneously. There should be no constraints on the technical means by which the connection is provided, allowing for wired or wireless technologies, nor any constraints on which operators provide part or all of universal service obligations. Connections to the public telephone network at a fixed location should be capable of supporting speech and data communications at rates sufficient for access to online services such as those provided via the public Internet. The speed of Internet access experienced by a given user may depend on a number of factors including the provider(s) of Internet connectivity as well as the given application for which a connection is being used. The data rate that can be supported by a single narrowband connection to the public telephone network depends on the capabilities of the subscriber's terminal equipment as well as the connection. For this reason it is not appropriate to mandate a specific data or bit rate at Community level. Currently available voice band modems typically offer a data rate of 56 kbit/s and employ automatic data rate adaptation to cater for variable line quality, with the result that the achieved data rate may be lower than 56 kbit/s. Flexibility is required on the one hand to allow Member States to take measures where necessary to ensure that connections are capable of supporting such a data rate, and on the other hand to allow Member States where relevant to permit data rates below this upper limit of 56 kbits/s in order, for example, to exploit the capabilities of wireless technologies (including cellular wireless networks) to deliver universal service to a higher proportion of the population. This may be of particular importance in some accession countries where household penetration of traditional telephone connections remains relatively low. In specific cases where the connection to the public telephony network at a fixed location is clearly insufficient to support satisfactory Internet access, Member States should be able to require the connection to be brought up to the level enjoyed by the majority of subscribers so that it supports data rates sufficient for access to the Internet. Where such specific measures produce a net cost burden for those consumers concerned, the net effect may be included in any net cost calculation of universal service obligations.

(9) The provisions of this Directive do not preclude Member States from designating different undertakings to provide the network and service elements of universal service. Designated undertakings providing network elements may be required to ensure such construction and maintenance as are necessary and proportionate to meet all reasonable requests for connection at a fixed location to the public telephone network and for access to publicly available telephone services at a fixed location.

(10) Affordable price means a price defined by Member States at national level in the light of specific national conditions, and may involve setting common tariffs irrespective of location or special tariff options to deal with the needs of low-income users. Affordability for individual consumers is related to their ability to monitor and control their expenditure.

(11) Directory information and a directory enquiry service constitute an essential access tool for publicly available telephone services and form part of the universal service obligation. Users and consumers desire comprehensive directories and a directory enquiry service covering all listed telephone subscribers and their numbers (including fixed and mobile numbers) and want this information to be presented in a non-preferential fashion. Directive 97/66/EC of the European Parliament and of the Council of 15 December 1997 concerning the processing of personal data and the protection of privacy in the telecommunications sector [1] ensures the subscribers' right to privacy with regard to the inclusion of their personal information in a public directory.

(12) For the citizen, it is important for there to be adequate provision of public pay telephones, and for users to be able to call emergency telephone numbers and, in particular, the single European emergency call number ('112') free of charge from any telephone, including public pay telephones, without the use of any means of

[1] OJ L 24, 30.1.1998, p. 1.

payment. Insufficient information about the existence of '112' deprives citizens of the additional safety ensured by the existence of this number at European level especially during their travel in other Member States.

(13) Member States should take suitable measures in order to guarantee access to and affordability of all publicly available telephone services at a fixed location for disabled users and users with special social needs. Specific measures for disabled users could include, as appropriate, making available accessible public telephones, public text telephones or equivalent measures for deaf or speech-impaired people, providing services such as directory enquiry services or equivalent measures free of charge for blind or partially sighted people, and providing itemised bills in alternative format on request for blind or partially sighted people. Specific measures may also need to be taken to enable disabled users and users with special social needs to access emergency services '112' and to give them a similar possibility to choose between different operators or service providers as other consumers. Quality of service standards have been developed for a range of parameters to assess the quality of services received by subscribers and how well undertakings designated with universal service obligations perform in achieving these standards. Quality of service standards do not yet exist in respect of disabled users. Performance standards and relevant parameters should be developed for disabled users and are provided for in Article 11 of this Directive. Moreover, national regulatory authorities should be enabled to require publication of quality of service performance data if and when such standards and parameters are developed. The provider of universal service should not take measures to prevent users from benefiting fully from services offered by different operators or service providers, in combination with its own services offered as part of universal service.

(14) The importance of access to and use of the public telephone network at a fixed location is such that it should be available to anyone reasonably requesting it. In accordance with the principle of subsidiarity, it is for Member States to decide on the basis of objective criteria which undertakings have universal service obligations for the purposes of this Directive, where appropriate taking into account the ability and the willingness of undertakings to accept all or part of the universal service obligations. It is important that universal service obligations are fulfilled in the most efficient fashion so that users generally pay prices that correspond to efficient cost provision. It is likewise important that universal service operators maintain the integrity of the network as well as service continuity and quality. The development of greater competition and choice provide more possibilities for all or part of the universal service obligations to be provided by undertakings other than those with significant market power. Therefore, universal service obligations could in some cases be allocated to operators demonstrating the most cost-effective means of delivering access and services, including by competitive or comparative selection procedures. Corresponding obligations could be included as conditions in authorisations to provide publicly available services.

(15) Member States should monitor the situation of consumers with respect to their use of publicly available telephone services and in particular with respect to affordability. The affordability of telephone service is related to the information which users receive regarding telephone usage expenses as well as the relative cost of telephone usage compared to other services, and is also related to their ability to control expenditure. Affordability therefore means giving power to consumers through obligations imposed on undertakings designated as having universal service obligations. These obligations include a specified level of itemised billing, the possibility for consumers selectively to block certain calls (such as high-priced calls to premium services), the possibility for consumers to control expenditure via pre-payment means and the possibility for consumers to offset up-front connection fees. Such measures may need to be reviewed and changed in the light of market developments. Current conditions do not warrant a requirement for operators with universal service obligations to alert subscribers where a predetermined limit of expenditure is exceeded or an abnormal calling pattern occurs. Review of the relevant legislative provisions in future should consider whether there is a possible need to alert subscribers for these reasons.

(16) Except in cases of persistent late payment or non-payment of bills, consumers should be protected from immediate disconnection from the network on the grounds of an unpaid bill and, particularly in the case of disputes over high bills for premium rate services, should continue to have access to essential telephone services pending resolution of the dispute. Member States may decide that such access may continue to be provided only if the subscriber continues to pay line rental charges.

(17) Quality and price are key factors in a competitive market and national regulatory authorities should be able to monitor achieved quality of service for undertakings which have been designated as having universal service obligations. In relation to the quality of service attained by such undertakings, national

regulatory authorities should be able to take appropriate measures where they deem it necessary. National regulatory authorities should also be able to monitor the achieved quality of services of other undertakings providing public telephone networks and/or publicly available telephone services to users at fixed locations.

(18) Member States should, where necessary, establish mechanisms for financing the net cost of universal service obligations in cases where it is demonstrated that the obligations can only be provided at a loss or at a net cost which falls outside normal commercial standards. It is important to ensure that the net cost of universal service obligations is properly calculated and that any financing is undertaken with minimum distortion to the market and to undertakings, and is compatible with the provisions of Articles 87 and 88 of the Treaty.

(19) Any calculation of the net cost of universal service should take due account of costs and revenues, as well as the intangible benefits resulting from providing universal service, but should not hinder the general aim of ensuring that pricing structures reflect costs. Any net costs of universal service obligations should be calculated on the basis of transparent procedures.

(20) Taking into account intangible benefits means that an estimate in monetary terms, of the indirect benefits that an undertaking derives by virtue of its position as provider of universal service, should be deducted from the direct net cost of universal service obligations in order to determine the overall cost burden.

(21) When a universal service obligation represents an unfair burden on an undertaking, it is appropriate to allow Member States to establish mechanisms for efficiently recovering net costs. Recovery via public funds constitutes one method of recovering the net costs of universal service obligations. It is also reasonable for established net costs to be recovered from all users in a transparent fashion by means of levies on undertakings. Member States should be able to finance the net costs of different elements of universal service through different mechanisms, and/or to finance the net costs of some or all elements from either of the mechanisms or a combination of both. In the case of cost recovery by means of levies on undertakings, Member States should ensure that that the method of allocation amongst them is based on objective and non-discriminatory criteria and is in accordance with the principle of proportionality. This principle does not prevent Member States from exempting new entrants which have not yet achieved any significant market presence. Any funding mechanism should ensure that market participants only contribute to the financing of universal service obligations and not to other activities which are not directly linked to the provision of the universal service obligations. Recovery mechanisms should in all cases respect the principles of Community law, and in particular in the case of sharing mechanisms those of non-discrimination and proportionality. Any funding mechanism should ensure that users in one Member State do not contribute to universal service costs in another Member State, for example when making calls from one Member State to another.

(22) Where Member States decide to finance the net cost of universal service obligations from public funds, this should be understood to comprise funding from general government budgets including other public financing sources such as state lotteries.

(23) The net cost of universal service obligations may be shared between all or certain specified classes of undertaking. Member States should ensure that the sharing mechanism respects the principles of transparency, least market distortion, non-discrimination and proportionality. Least market distortion means that contributions should be recovered in a way that as far as possible minimises the impact of the financial burden falling on end-users, for example by spreading contributions as widely as possible.

(24) National regulatory authorities should satisfy themselves that those undertakings benefiting from universal service funding provide a sufficient level of detail of the specific elements requiring such funding in order to justify their request. Member States' schemes for the costing and financing of universal service obligations should be communicated to the Commission for verification of compatibility with the Treaty. There are incentives for designated operators to raise the assessed net cost of universal service obligations. Therefore Member States should ensure effective transparency and control of amounts charged to finance universal service obligations.

(25) Communications markets continue to evolve in terms of the services used and the technical means used to deliver them to users. The universal service obligations, which are defined at a Community level, should be periodically reviewed with a view to proposing that the scope be changed or redefined. Such a review should take account of evolving social, commercial and technological conditions and the fact that any change of scope should be subject to the twin test of services that become available to a substantial majority of the

population, with a consequent risk of social exclusion for those who can not afford them. Care should be taken in any change of the scope of universal service obligations to ensure that certain technological choices are not artificially promoted above others, that a disproportionate financial burden is not imposed on sector undertakings (thereby endangering market developments and innovation) and that any financing burden does not fall unfairly on consumers with lower incomes. Any change of scope automatically means that any net cost can be financed via the methods permitted in this Directive. Member States are not permitted to impose on market players financial contributions which relate to measures which are not part of universal service obligations. Individual Member States remain free to impose special measures (outside the scope of universal service obligations) and finance them in conformity with Community law but not by means of contributions from market players.

(26) More effective competition across all access and service markets will give greater choice for users. The extent of effective competition and choice varies across the Community and varies within Member States between geographical areas and between access and service markets. Some users may be entirely dependent on the provision of access and services by an undertaking with significant market power. In general, for reasons of efficiency and to encourage effective competition, it is important that the services provided by an undertaking with significant market power reflect costs. For reasons of efficiency and social reasons, end-user tariffs should reflect demand conditions as well as cost conditions, provided that this does not result in distortions of competition. There is a risk that an undertaking with significant market power may act in various ways to inhibit entry or distort competition, for example by charging excessive prices, setting predatory prices, compulsory bundling of retail services or showing undue preference to certain customers. Therefore, national regulatory authorities should have powers to impose, as a last resort and after due consideration, retail regulation on an undertaking with significant market power. Price cap regulation, geographical averaging or similar instruments, as well as non-regulatory measures such as publicly available comparisons of retail tariffs, may be used to achieve the twin objectives of promoting effective competition whilst pursuing public interest needs, such as maintaining the affordability of publicly available telephone services for some consumers. Access to appropriate cost accounting information is necessary, in order for national regulatory authorities to fulfil their regulatory duties in this area, including the imposition of any tariff controls. However, regulatory controls on retail services should only be imposed where national regulatory authorities consider that relevant wholesale measures or measures regarding carrier selection or pre-selection would fail to achieve the objective of ensuring effective competition and public interest.

(27) Where a national regulatory authority imposes obligations to implement a cost accounting system in order to support price controls, it may itself undertake an annual audit to ensure compliance with that cost accounting system, provided that it has the necessary qualified staff, or it may require the audit to be carried out by another qualified body, independent of the operator concerned.

(28) It is considered necessary to ensure the continued application of the existing provisions relating to the minimum set of leased line services in Community telecommunications legislation, in particular in Council Directive 92/44/EEC of 5 June 1992 on the application of open network provision to leased lines ([1]), until such time as national regulatory authorities determine, in accordance with the market analysis procedures laid down in Directive 2002/21/EC of the European Parliament and of the Council of 7 March 2002 on a common regulatory framework for electronic communications networks and services (Framework Directive) ([2]), that such provisions are no longer needed because a sufficiently competitive market has developed in their territory. The degree of competition is likely to vary between different markets of leased lines in the minimum set, and in different parts of the territory. In undertaking the market analysis, national regulatory authorities should make separate assessments for each market of leased lines in the minimum set, taking into account their geographic dimension. Leased lines services constitute mandatory services to be provided without recourse to any compensation mechanisms. The provision of leased lines outside of the minimum set of leased lines should be covered by general retail regulatory provisions rather than specific requirements covering the supply of the minimum set.

(29) National regulatory authorities may also, in the light of an analysis of the relevant market, require mobile operators with significant market power to enable their subscribers to access the services of any interconnected provider of publicly available telephone services on a call-by-call basis or by means of pre-selection.

([1]) OJ L 165, 19.6.1992, p. 27. Directive as last amended by Commission Decision No 98/80/EC (OJ L 14, 20.1.1998, p. 27).

([2]) See page 33 of this Official Journal.

(30) Contracts are an important tool for users and consumers to ensure a minimum level of transparency of information and legal security. Most service providers in a competitive environment will conclude contracts with their customers for reasons of commercial desirability. In addition to the provisions of this Directive, the requirements of existing Community consumer protection legislation relating to contracts, in particular Council Directive 93/13/EEC of 5 April 1993 on unfair terms in consumer contracts ([1]) and Directive 97/7/EC of the European Parliament and of the Council of 20 May 1997 on the protection of consumers in respect of distance contracts ([2]), apply to consumer transactions relating to electronic networks and services. Specifically, consumers should enjoy a minimum level of legal certainty in respect of their contractual relations with their direct telephone service provider, such that the contractual terms, conditions, quality of service, condition for termination of the contract and the service, compensation measures and dispute resolution are specified in their contracts. Where service providers other than direct telephone service providers conclude contracts with consumers, the same information should be included in those contracts as well. The measures to ensure transparency on prices, tariffs, terms and conditions will increase the ability of consumers to optimise their choices and thus to benefit fully from competition.

(31) End-users should have access to publicly available information on communications services. Member States should be able to monitor the quality of services which are offered in their territories. National regulatory authorities should be able systematically to collect information on the quality of services offered in their territories on the basis of criteria which allow comparability between service providers and between Member States. Undertakings providing communications services, operating in a competitive environment, are likely to make adequate and up-to-date information on their services publicly available for reasons of commercial advantage. National regulatory authorities should nonetheless be able to require publication of such information where it is demonstrated that such information is not effectively available to the public.

(32) End-users should be able to enjoy a guarantee of interoperability in respect of all equipment sold in the Community for the reception of digital television. Member States should be able to require minimum harmonised standards in respect of such equipment. Such standards could be adapted from time to time in the light of technological and market developments.

([1]) OJ L 95, 21.4.1993, p. 29.
([2]) OJ L 144, 4.6.1997, p. 19.

(33) It is desirable to enable consumers to achieve the fullest connectivity possible to digital television sets. Interoperability is an evolving concept in dynamic markets. Standards bodies should do their utmost to ensure that appropriate standards evolve along with the technologies concerned. It is likewise important to ensure that connectors are available on television sets that are capable of passing all the necessary elements of a digital signal, including the audio and video streams, conditional access information, service information, application program interface (API) information and copy protection information. This Directive therefore ensures that the functionality of the open interface for digital television sets is not limited by network operators, service providers or equipment manufacturers and continues to evolve in line with technological developments. For display and presentation of digital interactive television services, the realisation of a common standard through a market-driven mechanism is recognised as a consumer benefit. Member States and the Commission may take policy initiatives, consistent with the Treaty, to encourage this development.

(34) All end-users should continue to enjoy access to operator assistance services whatever organisation provides access to the public telephone network.

(35) The provision of directory enquiry services and directories is already open to competition. The provisions of this Directive complement the provisions of Directive 97/66/EC by giving subscribers a right to have their personal data included in a printed or electronic directory. All service providers which assign telephone numbers to their subscribers are obliged to make relevant information available in a fair, cost-oriented and non-discriminatory manner.

(36) It is important that users should be able to call the single European emergency number '112', and any other national emergency telephone numbers, free of charge, from any telephone, including public pay telephones, without the use of any means of payment. Member States should have already made the necessary organisational arrangements best suited to the national organisation of the emergency systems, in order to ensure that calls to this number are adequately answered and handled. Caller location information, to be made available to the emergency services, will improve the level of protection and the security of users of '112' services and assist the emergency services, to the extent technically feasible, in the discharge of their duties, provided that the transfer of calls and associated data to the emergency services concerned is guaranteed. The reception and use of such information should comply with relevant Community law on the processing of personal data. Steady information technology

improvements will progressively support the simultaneous handling of several languages over the networks at a reasonable cost. This in turn will ensure additional safety for European citizens using the '112' emergency call number.

(37) Easy access to international telephone services is vital for European citizens and European businesses. '00' has already been established as the standard international telephone access code for the Community. Special arrangements for making calls between adjacent locations across borders between Member States may be established or continued. The ITU has assigned, in accordance with ITU Recommendation E.164, code '3883' to the European Telephony Numbering Space (ETNS). In order to ensure connection of calls to the ETNS, undertakings operating public telephone networks should ensure that calls using '3883' are directly or indirectly interconnected to ETNS serving networks specified in the relevant European Telecommunications Standards Institute (ETSI) standards. Such interconnection arrangements should be governed by the provisions of Directive 2002/19/EC of the European Parliament and of the Council of 7 March 2002 on access to, and interconnection of, electronic communications networks and associated facilities (Access Directive) ([1]).

(38) Access by end-users to all numbering resources in the Community is a vital pre-condition for a single market. It should include freephone, premium rate, and other non-geographic numbers, except where the called subscriber has chosen, for commercial reasons, to limit access from certain geographical areas. Tariffs charged to parties calling from outside the Member State concerned need not be the same as for those parties calling from inside that Member State.

(39) Tone dialling and calling line identification facilities are normally available on modern telephone exchanges and can therefore increasingly be provided at little or no expense. Tone dialling is increasingly being used for user interaction with special services and facilities, including value added services, and the absence of this facility can prevent the user from making use of these services. Member States are not required to impose obligations to provide these facilities when they are already available. Directive 97/66/EC safeguards the privacy of users with regard to itemised billing, by giving them the means to protect their right to privacy when calling line identification is implemented. The development of these services on a pan-European basis would benefit consumers and is encouraged by this Directive.

(40) Number portability is a key facilitator of consumer choice and effective competition in a competitive telecommunications environment such that end-users who so request should be able to retain their number(s) on the public telephone network independently of the organisation providing service. The provision of this facility between connections to the public telephone network at fixed and non-fixed locations is not covered by this Directive. However, Member States may apply provisions for porting numbers between networks providing services at a fixed location and mobile networks.

(41) The impact of number portability is considerably strengthened when there is transparent tariff information, both for end-users who port their numbers and also for end-users who call those who have ported their numbers. National regulatory authorities should, where feasible, facilitate appropriate tariff transparency as part of the implementation of number portability.

(42) When ensuring that pricing for interconnection related to the provision of number portability is cost-oriented, national regulatory authorities may also take account of prices available in comparable markets.

(43) Currently, Member States impose certain 'must carry' obligations on networks for the distribution of radio or television broadcasts to the public. Member States should be able to lay down proportionate obligations on undertakings under their jurisdiction, in the interest of legitimate public policy considerations, but such obligations should only be imposed where they are necessary to meet general interest objectives clearly defined by Member States in conformity with Community law and should be proportionate, transparent and subject to periodical review. 'Must carry' obligations imposed by Member States should be reasonable, that is they should be proportionate and transparent in the light of clearly defined general interest objectives, and could, where appropriate, entail a provision for proportionate remuneration. Such 'must carry' obligations may include the transmission of services specifically designed to enable appropriate access by disabled users.

(44) Networks used for the distribution of radio or television broadcasts to the public include cable, satellite and terrestrial broadcasting networks. They might also include other networks to the extent that a significant number of end-users use such networks as their principal means to receive radio and television broadcasts.

([1]) See page 7 of this Official Journal.

(45) Services providing content such as the offer for sale of a package of sound or television broadcasting content are not covered by the common regulatory framework for electronic communications networks and services. Providers of such services should not be subject to universal service obligations in respect of these activities. This Directive is without prejudice to measures taken at national level, in compliance with Community law, in respect of such services.

(46) Where a Member State seeks to ensure the provision of other specific services throughout its national territory, such obligations should be implemented on a cost efficient basis and outside the scope of universal service obligations. Accordingly, Member States may undertake additional measures (such as facilitating the development of infrastructure or services in circumstances where the market does not satisfactorily address the requirements of end-users or consumers), in conformity with Community law. As a reaction to the Commission's e-Europe initiative, the Lisbon European Council of 23 and 24 March 2000 called on Member States to ensure that all schools have access to the Internet and to multimedia resources.

(47) In the context of a competitive environment, the views of interested parties, including users and consumers, should be taken into account by national regulatory authorities when dealing with issues related to end-users' rights. Effective procedures should be available to deal with disputes between consumers, on the one hand, and undertakings providing publicly available communications services, on the other. Member States should take full account of Commission Recommendation 98/257/EC of 30 March 1998 on the principles applicable to the bodies responsible for out-of-court settlement of consumer disputes ([1]).

(48) Co-regulation could be an appropriate way of stimulating enhanced quality standards and improved service performance. Co-regulation should be guided by the same principles as formal regulation, i.e. it should be objective, justified, proportional, non-discriminatory and transparent.

(49) This Directive should provide for elements of consumer protection, including clear contract terms and dispute resolution, and tariff transparency for consumers. It should also encourage the extension of such benefits to other categories of end-users, in particular small and medium-sized enterprises.

(50) The provisions of this Directive do not prevent a Member State from taking measures justified on grounds set out in Articles 30 and 46 of the Treaty, and in particular on grounds of public security, public policy and public morality.

([1]) OJ L 115, 17.4.1998, p. 31.

(51) Since the objectives of the proposed action, namely setting a common level of universal service for telecommunications for all European users and of harmonising conditions for access to and use of public telephone networks at a fixed location and related publicly available telephone services and also achieving a harmonised framework for the regulation of electronic communications services, electronic communications networks and associated facilities, cannot be sufficiently achieved by the Member States and can therefore by reason of the scale or effects of the action be better achieved at Community level, the Community may adopt measures in accordance with the principles of subsidiarity as set out in Article 5 of the Treaty. In accordance with the principle of proportionality, as set out in that Article, this Directive does not go beyond what is necessary in order to achieve those objectives.

(52) The measures necessary for the implementation of this Directive should be adopted in accordance with Council Decision 1999/468/EC of 28 June 1999 laying down the procedures for the exercise of implementing powers conferred on the Commission ([2]),

HAVE ADOPTED THIS DIRECTIVE:

CHAPTER I

## SCOPE, AIMS AND DEFINITIONS

*Article 1*

### Scope and aims

1. Within the framework of Directive 2002/21/EC (Framework Directive), this Directive concerns the provision of electronic communications networks and services to end-users. The aim is to ensure the availability throughout the Community of good quality publicly available services through effective competition and choice and to deal with circumstances in which the needs of end-users are not satisfactorily met by the market.

2. This Directive establishes the rights of end-users and the corresponding obligations on undertakings providing publicly available electronic communications networks and services. With regard to ensuring provision of universal service within an environment of open and competitive markets, this Directive defines the minimum set of services of specified quality to which all end-users have access, at an affordable

([2]) OJ L 184, 17.7.1999, p. 23.

price in the light of specific national conditions, without distorting competition. This Directive also sets out obligations with regard to the provision of certain mandatory services such as the retail provision of leased lines.

*Article 2*

**Definitions**

For the purposes of this Directive, the definitions set out in Article 2 of Directive 2002/21/EC (Framework Directive) shall apply.

The following definitions shall also apply:

(a) 'public pay telephone' means a telephone available to the general public, for the use of which the means of payment may include coins and/or credit/debit cards and/or pre-payment cards, including cards for use with dialling codes;

(b) 'public telephone network' means an electronic communications network which is used to provide publicly available telephone services; it supports the transfer between network termination points of speech communications, and also other forms of communication, such as facsimile and data;

(c) 'publicly available telephone service' means a service available to the public for originating and receiving national and international calls and access to emergency services through a number or numbers in a national or international telephone numbering plan, and in addition may, where relevant, include one or more of the following services: the provision of operator assistance, directory enquiry services, directories, provision of public pay phones, provision of service under special terms, provision of special facilities for customers with disabilities or with special social needs and/or the provision of non-geographic services;

(d) 'geographic number' means a number from the national numbering plan where part of its digit structure contains geographic significance used for routing calls to the physical location of the network termination point (NTP);

(e) 'network termination point' (NTP) means the physical point at which a subscriber is provided with access to a public communications network; in the case of networks involving switching or routing, the NTP is identified by means of a specific network address, which may be linked to a subscriber number or name;

(f) 'non-geographic numbers' means a number from the national numbering plan that is not a geographic number. It includes *inter alia* mobile, freephone and premium rate numbers.

CHAPTER II

**UNIVERSAL SERVICE OBLIGATIONS INCLUDING SOCIAL OBLIGATIONS**

*Article 3*

**Availability of universal service**

1. Member States shall ensure that the services set out in this Chapter are made available at the quality specified to all end-users in their territory, independently of geographical location, and, in the light of specific national conditions, at an affordable price.

2. Member States shall determine the most efficient and appropriate approach for ensuring the implementation of universal service, whilst respecting the principles of objectivity, transparency, non-discrimination and proportionality. They shall seek to minimise market distortions, in particular the provision of services at prices or subject to other terms and conditions which depart from normal commercial conditions, whilst safeguarding the public interest.

*Article 4*

**Provision of access at a fixed location**

1. Member States shall ensure that all reasonable requests for connection at a fixed location to the public telephone network and for access to publicly available telephone services at a fixed location are met by at least one undertaking.

2. The connection provided shall be capable of allowing end-users to make and receive local, national and international telephone calls, facsimile communications and data communications, at data rates that are sufficient to permit functional Internet access, taking into account prevailing technologies used by the majority of subscribers and technological feasibility.

*Article 5*

**Directory enquiry services and directories**

1. Member States shall ensure that:

(a) at least one comprehensive directory is available to end-users in a form approved by the relevant authority, whether printed or electronic, or both, and is updated on a regular basis, and at least once a year;

(b) at least one comprehensive telephone directory enquiry service is available to all end-users, including users of public pay telephones.

2. The directories in paragraph 1 shall comprise, subject to the provisions of Article 11 of Directive 97/66/EC, all subscribers of publicly available telephone services.

3. Member States shall ensure that the undertaking(s) providing the services referred to in paragraph 1 apply the principle of non-discrimination to the treatment of information that has been provided to them by other undertakings.

*Article 6*

**Public pay telephones**

1. Member States shall ensure that national regulatory authorities can impose obligations on undertakings in order to ensure that public pay telephones are provided to meet the reasonable needs of end-users in terms of the geographical coverage, the number of telephones, the accessibility of such telephones to disabled users and the quality of services.

2. A Member State shall ensure that its national regulatory authority can decide not to impose obligations under paragraph 1 in all or part of its territory, if it is satisfied that these facilities or comparable services are widely available, on the basis of a consultation of interested parties as referred to in Article 33.

3. Member States shall ensure that it is possible to make emergency calls from public pay telephones using the single European emergency call number '112' and other national emergency numbers, all free of charge and without having to use any means of payment.

*Article 7*

**Special measures for disabled users**

1. Member States shall, where appropriate, take specific measures for disabled end-users in order to ensure access to and affordability of publicly available telephone services, including access to emergency services, directory enquiry services and directories, equivalent to that enjoyed by other end-users.

2. Member States may take specific measures, in the light of national conditions, to ensure that disabled end-users can also take advantage of the choice of undertakings and service providers available to the majority of end-users.

*Article 8*

**Designation of undertakings**

1. Member States may designate one or more undertakings to guarantee the provision of universal service as identified in Articles 4, 5, 6 and 7 and, where applicable, Article 9(2) so that the whole of the national territory can be covered. Member States may designate different undertakings or sets of undertakings to provide different elements of universal service and/or to cover different parts of the national territory.

2. When Member States designate undertakings in part or all of the national territory as having universal service obligations, they shall do so using an efficient, objective, transparent and non-discriminatory designation mechanism, whereby no undertaking is a priori excluded from being designated. Such designation methods shall ensure that universal service is provided in a cost-effective manner and may be used as a means of determining the net cost of the universal service obligation in accordance with Article 12.

*Article 9*

**Affordability of tariffs**

1. National regulatory authorities shall monitor the evolution and level of retail tariffs of the services identified in Articles 4, 5, 6 and 7 as falling under the universal service obligations and provided by designated undertakings, in particular in relation to national consumer prices and income.

2. Member States may, in the light of national conditions, require that designated undertakings provide tariff options or packages to consumers which depart from those provided under normal commercial conditions, in particular to ensure that those on low incomes or with special social needs are not prevented from accessing or using the publicly available telephone service.

3. Member States may, besides any provision for designated undertakings to provide special tariff options or to comply with price caps or geographical averaging or other similar schemes, ensure that support is provided to consumers identified as having low incomes or special social needs.

4. Member States may require undertakings with obligations under Articles 4, 5, 6 and 7 to apply common tariffs, including geographical averaging, throughout the territory, in the light of national conditions or to comply with price caps.

5. National regulatory authorities shall ensure that, where a designated undertaking has an obligation to provide special

tariff options, common tariffs, including geographical averaging, or to comply with price caps, the conditions are fully transparent and are published and applied in accordance with the principle of non-discrimination. National regulatory authorities may require that specific schemes be modified or withdrawn.

*Article 10*

**Control of expenditure**

1. Member States shall ensure that designated undertakings, in providing facilities and services additional to those referred to in Articles 4, 5, 6, 7 and 9(2), establish terms and conditions in such a way that the subscriber is not obliged to pay for facilities or services which are not necessary or not required for the service requested.

2. Member States shall ensure that designated undertakings with obligations under Articles 4, 5, 6, 7 and 9(2) provide the specific facilities and services set out in Annex I, Part A, in order that subscribers can monitor and control expenditure and avoid unwarranted disconnection of service.

3. Member States shall ensure that the relevant authority is able to waive the requirements of paragraph 2 in all or part of its national territory if it is satisfied that the facility is widely available.

*Article 11*

**Quality of service of designated undertakings**

1. National regulatory authorities shall ensure that all designated undertakings with obligations under Articles 4, 5, 6, 7 and 9(2) publish adequate and up-to-date information concerning their performance in the provision of universal service, based on the quality of service parameters, definitions and measurement methods set out in Annex III. The published information shall also be supplied to the national regulatory authority.

2. National regulatory authorities may specify, *inter alia*, additional quality of service standards, where relevant parameters have been developed, to assess the performance of undertakings in the provision of services to disabled end-users and disabled consumers. National regulatory authorities shall ensure that information concerning the performance of undertakings in relation to these parameters is also published and made available to the national regulatory authority.

3. National regulatory authorities may, in addition, specify the content, form and manner of information to be published, in order to ensure that end-users and consumers have access to comprehensive, comparable and user-friendly information.

4. National regulatory authorities shall be able to set performance targets for those undertakings with universal service obligations at least under Article 4. In so doing, national regulatory authorities shall take account of views of interested parties, in particular as referred to in Article 33.

5. Member States shall ensure that national regulatory authorities are able to monitor compliance with these performance targets by designated undertakings.

6. Persistent failure by an undertaking to meet performance targets may result in specific measures being taken in accordance with Directive 2002/20/EC of the European Parliament and of the Council of 7 March 2002 on the authorisation of electronic communications networks and services (Authorisation Directive) ([1]). National regulatory authorities shall be able to order independent audits or similar reviews of the performance data, paid for by the undertaking concerned, in order to ensure the accuracy and comparability of the data made available by undertakings with universal service obligations.

*Article 12*

**Costing of universal service obligations**

1. Where national regulatory authorities consider that the provision of universal service as set out in Articles 3 to 10 may represent an unfair burden on undertakings designated to provide universal service, they shall calculate the net costs of its provision.

For that purpose, national regulatory authorities shall:

(a) calculate the net cost of the universal service obligation, taking into account any market benefit which accrues to an undertaking designated to provide universal service, in accordance with Annex IV, Part A; or

(b) make use of the net costs of providing universal service identified by a designation mechanism in accordance with Article 8(2).

2. The accounts and/or other information serving as the basis for the calculation of the net cost of universal service obligations under paragraph 1(a) shall be audited or verified by

([1]) See page 21 of this Official Journal.

the national regulatory authority or a body independent of the relevant parties and approved by the national regulatory authority. The results of the cost calculation and the conclusions of the audit shall be publicly available.

*Article 13*

**Financing of universal service obligations**

1. Where, on the basis of the net cost calculation referred to in Article 12, national regulatory authorities find that an undertaking is subject to an unfair burden, Member States shall, upon request from a designated undertaking, decide:

(a) to introduce a mechanism to compensate that undertaking for the determined net costs under transparent conditions from public funds; and/or

(b) to share the net cost of universal service obligations between providers of electronic communications networks and services.

2. Where the net cost is shared under paragraph 1(b), Member States shall establish a sharing mechanism administered by the national regulatory authority or a body independent from the beneficiaries under the supervision of the national regulatory authority. Only the net cost, as determined in accordance with Article 12, of the obligations laid down in Articles 3 to 10 may be financed.

3. A sharing mechanism shall respect the principles of transparency, least market distortion, non-discrimination and proportionality, in accordance with the principles of Annex IV, Part B. Member States may choose not to require contributions from undertakings whose national turnover is less than a set limit.

4. Any charges related to the sharing of the cost of universal service obligations shall be unbundled and identified separately for each undertaking. Such charges shall not be imposed or collected from undertakings that are not providing services in the territory of the Member State that has established the sharing mechanism.

*Article 14*

**Transparency**

1. Where a mechanism for sharing the net cost of universal service obligations as referred to in Article 13 is established, national regulatory authorities shall ensure that the principles for cost sharing, and details of the mechanism used, are publicly available.

2. Subject to Community and national rules on business confidentiality, national regulatory authorities shall ensure that an annual report is published giving the calculated cost of universal service obligations, identifying the contributions made by all the undertakings involved, and identifying any market benefits, that may have accrued to the undertaking(s) designated to provide universal service, where a fund is actually in place and working.

*Article 15*

**Review of the scope of universal service**

1. The Commission shall periodically review the scope of universal service, in particular with a view to proposing to the European Parliament and the Council that the scope be changed or redefined. A review shall be carried out, on the first occasion within two years after the date of application referred to in Article 38(1), second subparagraph, and subsequently every three years.

2. This review shall be undertaken in the light of social, economic and technological developments, taking into account, *inter alia*, mobility and data rates in the light of the prevailing technologies used by the majority of subscribers. The review process shall be undertaken in accordance with Annex V. The Commission shall submit a report to the European Parliament and the Council regarding the outcome of the review.

CHAPTER III

**REGULATORY CONTROLS ON UNDERTAKINGS WITH SIGNIFICANT MARKET POWER IN SPECIFIC MARKETS**

*Article 16*

**Review of obligations**

1. Member States shall maintain all obligations relating to:

(a) retail tariffs for the provision of access to and use of the public telephone network, imposed under Article 17 of Directive 98/10/EC of the European Parliament and of the Council of 26 February 1998 on the application of open network provision (ONP) to voice telephony and on universal service for telecommunications in a competitive environment [1];

(b) carrier selection or pre-selection, imposed under Directive 97/33/EC of the European Parliament and of the Council of 30 June 1997 on interconnection in telecommunications with regard to ensuring universal

[1] OJ L 101, 1.4.1998, p. 24.

service and interoperability through application of the principles of open network provision (ONP) ([1]);

(c) leased lines, imposed under Articles 3, 4, 6, 7, 8 and 10 of Directive 92/44/EEC,

until a review has been carried out and a determination made in accordance with the procedure in paragraph 3 of this Article.

2. The Commission shall indicate relevant markets for the obligations relating to retail markets in the initial recommendation on relevant product and service markets and the Decision identifying transnational markets to be adopted in accordance with Article 15 of Directive 2002/21/EC (Framework Directive).

3. Member States shall ensure that, as soon as possible after the entry into force of this Directive, and periodically thereafter, national regulatory authorities undertake a market analysis, in accordance with the procedure set out in Article 16 of Directive 2002/21/EC (Framework Directive) to determine whether to maintain, amend or withdraw the obligations relating to retail markets. Measures taken shall be subject to the procedure referred to in Article 7 of Directive 2002/21/EC (Framework Directive).

*Article 17*

**Regulatory controls on retail services**

1. Member States shall ensure that, where:

(a) as a result of a market analysis carried out in accordance with Article 16(3) a national regulatory authority determines that a given retail market identified in accordance with Article 15 of Directive 2002/21/EC (Framework Directive) is not effectively competitive, and

(b) the national regulatory authority concludes that obligations imposed under Directive 2002/19/EC (Access Directive), or Article 19 of this Directive would not result in the achievement of the objectives set out in Article 8 of Directive 2002/21/EC (Framework Directive),

national regulatory authorities shall impose appropriate regulatory obligations on undertakings identified as having significant market power on a given retail market in accordance with Article 14 of Directive 2002/21/EC (Framework Directive).

2. Obligations imposed under paragraph 1 shall be based on the nature of the problem identified and be proportionate and justified in the light of the objectives laid down in Article 8 of Directive 2002/21/EC (Framework Directive). The obligations imposed may include requirements that the identified undertakings do not charge excessive prices, inhibit market entry or restrict competition by setting predatory prices, show undue preference to specific end-users or unreasonably bundle services. National regulatory authorities may apply to such undertakings appropriate retail price cap measures, measures to control individual tariffs, or measures to orient tariffs towards costs or prices on comparable markets, in order to protect end-user interests whilst promoting effective competition.

3. National regulatory authorities shall, on request, submit information to the Commission concerning the retail controls applied and, where appropriate, the cost accounting systems used by the undertakings concerned.

4. National regulatory authorities shall ensure that, where an undertaking is subject to retail tariff regulation or other relevant retail controls, the necessary and appropriate cost accounting systems are implemented. National regulatory authorities may specify the format and accounting methodology to be used. Compliance with the cost accounting system shall be verified by a qualified independent body. National regulatory authorities shall ensure that a statement concerning compliance is published annually.

5. Without prejudice to Article 9(2) and Article 10, national regulatory authorities shall not apply retail control mechanisms under paragraph 1 of this Article to geographical or user markets where they are satisfied that there is effective competition.

*Article 18*

**Regulatory controls on the minimum set of leased lines**

1. Where, as a result of the market analysis carried out in accordance with Article 16(3), a national regulatory authority determines that the market for the provision of part or all of the minimum set of leased lines is not effectively competitive, it shall identify undertakings with significant market power in the provision of those specific elements of the minimum set of leased lines services in all or part of its territory in accordance with Article 14 of Directive 2002/21/EC (Framework Directive). The national regulatory authority shall impose obligations regarding the provision of the minimum set of leased lines, as identified in the list of standards published in the *Official Journal of the European Communities* in accordance with Article 17 of Directive 2002/21/EC (Framework Directive), and the conditions for such provision set out in

([1]) OJ L 199, 26.7.1997, p. 32. Directive as amended by Directive 98/61/EC (OJ L 268, 3.10.1998, p. 37).

Annex VII to this Directive, on such undertakings in relation to those specific leased line markets.

2. Where as a result of the market analysis carried out in accordance with Article 16(3), a national regulatory authority determines that a relevant market for the provision of leased lines in the minimum set is effectively competitive, it shall withdraw the obligations referred to in paragraph 1 in relation to this specific leased line market.

3. The minimum set of leased lines with harmonised characteristics, and associated standards, shall be published in the *Official Journal of the European Communities* as part of the list of standards referred to in Article 17 of Directive 2002/21/EC (Framework Directive). The Commission may adopt amendments necessary to adapt the minimum set of leased lines to new technical developments and to changes in market demand, including the possible deletion of certain types of leased line from the minimum set, acting in accordance with the procedure referred to in Article 37(2) of this Directive.

*Article 19*

**Carrier selection and carrier pre-selection**

1. National regulatory authorities shall require undertakings notified as having significant market power for the provision of connection to and use of the public telephone network at a fixed location in accordance with Article 16(3) to enable their subscribers to access the services of any interconnected provider of publicly available telephone services:

(a) on a call-by-call basis by dialling a carrier selection code; and

(b) by means of pre-selection, with a facility to override any pre-selected choice on a call-by-call basis by dialling a carrier selection code.

2. User requirements for these facilities to be implemented on other networks or in other ways shall be assessed in accordance with the market analysis procedure laid down in Article 16 of Directive 2002/21/EC (Framework Directive) and implemented in accordance with Article 12 of Directive 2002/19/EC (Access Directive).

3. National regulatory authorities shall ensure that pricing for access and interconnection related to the provision of the facilities in paragraph 1 is cost oriented and that direct charges to subscribers, if any, do not act as a disincentive for the use of these facilities.

CHAPTER IV

**END-USER INTERESTS AND RIGHTS**

*Article 20*

**Contracts**

1. Paragraphs 2, 3 and 4 apply without prejudice to Community rules on consumer protection, in particular Directives 97/7/EC and 93/13/EC, and national rules in conformity with Community law.

2. Member States shall ensure that, where subscribing to services providing connection and/or access to the public telephone network, consumers have a right to a contract with an undertaking or undertakings providing such services. The contract shall specify at least:

(a) the identity and address of the supplier;

(b) services provided, the service quality levels offered, as well as the time for the initial connection;

(c) the types of maintenance service offered;

(d) particulars of prices and tariffs and the means by which up-to-date information on all applicable tariffs and maintenance charges may be obtained;

(e) the duration of the contract, the conditions for renewal and termination of services and of the contract;

(f) any compensation and the refund arrangements which apply if contracted service quality levels are not met; and

(g) the method of initiating procedures for settlement of disputes in accordance with Article 34.

Member States may extend these obligations to cover other end-users.

3. Where contracts are concluded between consumers and electronic communications services providers other than those providing connection and/or access to the public telephone network, the information in paragraph 2 shall also be included in such contracts. Member States may extend this obligation to cover other end-users.

4. Subscribers shall have a right to withdraw from their contracts without penalty upon notice of proposed modifications in the contractual conditions. Subscribers shall be given adequate notice, not shorter than one month, ahead

of any such modifications and shall be informed at the same time of their right to withdraw, without penalty, from such contracts, if they do not accept the new conditions.

*Article 21*

**Transparency and publication of information**

1. Member States shall ensure that transparent and up-to-date information on applicable prices and tariffs, and on standard terms and conditions, in respect of access to and use of publicly available telephone services is available to end-users and consumers, in accordance with the provisions of Annex II.

2. National regulatory authorities shall encourage the provision of information to enable end-users, as far as appropriate, and consumers to make an independent evaluation of the cost of alternative usage patterns, by means of, for instance, interactive guides.

*Article 22*

**Quality of service**

1. Member States shall ensure that national regulatory authorities are, after taking account of the views of interested parties, able to require undertakings that provide publicly available electronic communications services to publish comparable, adequate and up-to-date information for end-users on the quality of their services. The information shall, on request, also be supplied to the national regulatory authority in advance of its publication.

2. National regulatory authorities may specify, *inter alia*, the quality of service parameters to be measured, and the content, form and manner of information to be published, in order to ensure that end-users have access to comprehensive, comparable and user-friendly information. Where appropriate, the parameters, definitions and measurement methods given in Annex III could be used.

*Article 23*

**Integrity of the network**

Member States shall take all necessary steps to ensure the integrity of the public telephone network at fixed locations and, in the event of catastrophic network breakdown or in cases of *force majeure*, the availability of the public telephone network and publicly available telephone services at fixed locations. Member States shall ensure that undertakings providing publicly available telephone services at fixed locations take all reasonable steps to ensure uninterrupted access to emergency services.

*Article 24*

**Interoperability of consumer digital television equipment**

In accordance with the provisions of Annex VI, Member States shall ensure the interoperability of the consumer digital television equipment referred to therein.

*Article 25*

**Operator assistance and directory enquiry services**

1. Member States shall ensure that subscribers to publicly available telephone services have the right to have an entry in the publicly available directory referred to in Article 5(1)(a).

2. Member States shall ensure that all undertakings which assign telephone numbers to subscribers meet all reasonable requests to make available, for the purposes of the provision of publicly available directory enquiry services and directories, the relevant information in an agreed format on terms which are fair, objective, cost oriented and non-discriminatory.

3. Member States shall ensure that all end-users provided with a connection to the public telephone network can access operator assistance services and directory enquiry services in accordance with Article 5(1)(b).

4. Member States shall not maintain any regulatory restrictions which prevent end-users in one Member State from accessing directly the directory enquiry service in another Member State.

5. Paragraphs 1, 2, 3 and 4 apply subject to the requirements of Community legislation on the protection of personal data and privacy and, in particular, Article 11 of Directive 97/66/EC.

*Article 26*

**Single European emergency call number**

1. Member States shall ensure that, in addition to any other national emergency call numbers specified by the national regulatory authorities, all end-users of publicly available telephone services, including users of public pay telephones, are able to call the emergency services free of charge, by using the single European emergency call number '112'.

2. Member States shall ensure that calls to the single European emergency call number '112' are appropriately answered and handled in a manner best suited to the national organisation of emergency systems and within the technological possibilities of the networks.

3. Member States shall ensure that undertakings which operate public telephone networks make caller location information available to authorities handling emergencies, to the extent technically feasible, for all calls to the single European emergency call number '112'.

4. Member States shall ensure that citizens are adequately informed about the existence and use of the single European emergency call number '112'.

*Article 27*

**European telephone access codes**

1. Member States shall ensure that the '00' code is the standard international access code. Special arrangements for making calls between adjacent locations across borders between Member States may be established or continued. The end-users of publicly available telephone services in the locations concerned shall be fully informed of such arrangements.

2. Member States shall ensure that all undertakings that operate public telephone networks handle all calls to the European telephony numbering space, without prejudice to the need for an undertaking that operates a public telephone network to recover the cost of the conveyance of calls on its network.

*Article 28*

**Non-geographic numbers**

Member States shall ensure that end-users from other Member States are able to access non-geographic numbers within their territory where technically and economically feasible, except where a called subscriber has chosen for commercial reasons to limit access by calling parties located in specific geographical areas.

*Article 29*

**Provision of additional facilities**

1. Member States shall ensure that national regulatory authorities are able to require all undertakings that operate public telephone networks to make available to end-users the facilities listed in Annex I, Part B, subject to technical feasibility and economic viability.

2. A Member State may decide to waive paragraph 1 in all or part of its territory if it considers, after taking into account the views of interested parties, that there is sufficient access to these facilities.

3. Without prejudice to Article 10(2), Member States may impose the obligations in Annex I, Part A, point (e), concerning disconnection as a general requirement on all undertakings.

*Article 30*

**Number portability**

1. Member States shall ensure that all subscribers of publicly available telephone services, including mobile services, who so request can retain their number(s) independently of the undertaking providing the service:

(a) in the case of geographic numbers, at a specific location; and

(b) in the case of non-geographic numbers, at any location.

This paragraph does not apply to the porting of numbers between networks providing services at a fixed location and mobile networks.

2. National regulatory authorities shall ensure that pricing for interconnection related to the provision of number portability is cost oriented and that direct charges to subscribers, if any, do not act as a disincentive for the use of these facilities.

3. National regulatory authorities shall not impose retail tariffs for the porting of numbers in a manner that would distort competition, such as by setting specific or common retail tariffs.

*Article 31*

**'Must carry' obligations**

1. Member States may impose reasonable 'must carry' obligations, for the transmission of specified radio and television broadcast channels and services, on undertakings under their jurisdiction providing electronic communications networks used for the distribution of radio or television broadcasts to the public where a significant number of end-users of such networks use them as their principal means to receive radio and television broadcasts. Such obligations shall only be imposed where they are necessary to meet clearly defined general interest objectives and shall be proportionate and transparent. The obligations shall be subject to periodical review.

2. Neither paragraph 1 of this Article nor Article 3(2) of Directive 2002/19/EC (Access Directive) shall prejudice the ability of Member States to determine appropriate remuneration, if any, in respect of measures taken in accordance with this Article while ensuring that, in similar

circumstances, there is no discrimination in the treatment of undertakings providing electronic communications networks. Where remuneration is provided for, Member States shall ensure that it is applied in a proportionate and transparent manner.

CHAPTER V

## GENERAL AND FINAL PROVISIONS

*Article 32*

### Additional mandatory services

Member States may decide to make additional services, apart from services within the universal service obligations as defined in Chapter II, publicly available in its own territory but, in such circumstances, no compensation mechanism involving specific undertakings may be imposed.

*Article 33*

### Consultation with interested parties

1. Member States shall ensure as far as appropriate that national regulatory authorities take account of the views of end-users, and consumers (including, in particular, disabled users), manufacturers, undertakings that provide electronic communications networks and/or services on issues related to all end-user and consumer rights concerning publicly available electronic communications services, in particular where they have a significant impact on the market.

2. Where appropriate, interested parties may develop, with the guidance of national regulatory authorities, mechanisms, involving consumers, user groups and service providers, to improve the general quality of service provision by, *inter alia*, developing and monitoring codes of conduct and operating standards.

*Article 34*

### Out-of-court dispute resolution

1. Member States shall ensure that transparent, simple and inexpensive out-of-court procedures are available for dealing with unresolved disputes, involving consumers, relating to issues covered by this Directive. Member States shall adopt measures to ensure that such procedures enable disputes to be settled fairly and promptly and may, where warranted, adopt a system of reimbursement and/or compensation. Member States may extend these obligations to cover disputes involving other end-users.

2. Member States shall ensure that their legislation does not hamper the establishment of complaints offices and the provision of on-line services at the appropriate territorial level to facilitate access to dispute resolution by consumers and end-users.

3. Where such disputes involve parties in different Member States, Member States shall coordinate their efforts with a view to bringing about a resolution of the dispute.

4. This Article is without prejudice to national court procedures.

*Article 35*

### Technical adjustment

Amendments necessary to adapt Annexes I, II, III, VI and VII to technological developments or to changes in market demand shall be adopted by the Commission, acting in accordance with the procedure referred to in Article 37(2).

*Article 36*

### Notification, monitoring and review procedures

1. National regulatory authorities shall notify to the Commission by at the latest the date of application referred to in Article 38(1), second subparagraph, and immediately in the event of any change thereafter in the names of undertakings designated as having universal service obligations under Article 8(1).

The Commission shall make the information available in a readily accessible form, and shall distribute it to the Communications Committee referred to in Article 37.

2. National regulatory authorities shall notify to the Commission the names of operators deemed to have significant market power for the purposes of this Directive, and the obligations imposed upon them under this Directive. Any changes affecting the obligations imposed upon undertakings or of the undertakings affected under the provisions of this Directive shall be notified to the Commission without delay.

3. The Commission shall periodically review the functioning of this Directive and report to the European Parliament and to the Council, on the first occasion not later than three years

after the date of application referred to in Article 38(1), second subparagraph. The Member States and national regulatory authorities shall supply the necessary information to the Commission for this purpose.

*Article 37*

**Committee**

1. The Commission shall be assisted by the Communications Committee, set up by Article 22 of Directive 2002/21/EC (Framework Directive).

2. Where reference is made to this paragraph, Articles 5 and 7 of Decision 1999/468/EC shall apply, having regard to the provisions of Article 8 thereof.

The period laid down in Article 5(6) of Decision 1999/468/EC shall be three months.

3. The Committee shall adopt its rules of procedure.

*Article 38*

**Transposition**

1. Member States shall adopt and publish the laws, regulations and administrative provisions necessary to comply with this Directive by 24 July 2003 at the latest. They shall forthwith inform the Commission thereof.

They shall apply those measures from 25 July 2003.

2. When Member States adopt these measures, they shall contain a reference to this Directive or be accompanied by such a reference on the occasion of their official publication. The methods of making such a reference shall be laid down by the Member States.

3. Member States shall communicate to the Commission the text of the provisions of national law which they adopt in the field governed by this Directive and of any subsequent modifications to those provisions.

*Article 39*

**Entry into force**

This Directive shall enter into force on the day of its publication in the *Official Journal of the European Communities*.

*Article 40*

**Addressees**

This Directive is addressed to the Member States.

Done at Brussels, 7 March 2002.

| | |
|---|---|
| *For the European Parliament* | *For the Council* |
| *The President* | *The President* |
| P. COX | J. C. APARICIO |

*ANNEX I*

**DESCRIPTION OF FACILITIES AND SERVICES REFERRED TO IN ARTICLE 10 (CONTROL OF EXPENDITURE) AND ARTICLE 29 (ADDITIONAL FACILITIES)**

**Part A: Facilities and services referred to in Article 10**

(a) *Itemised billing*

Member States are to ensure that national regulatory authorities, subject to the requirements of relevant legislation on the protection of personal data and privacy, may lay down the basic level of itemised bills which are to be provided by designated undertakings (as established in Article 8) to consumers free of charge in order that they can:

(i) allow verification and control of the charges incurred in using the public telephone network at a fixed location and/or related publicly available telephone services, and

(ii) adequately monitor their usage and expenditure and thereby exercise a reasonable degree of control over their bills.

Where appropriate, additional levels of detail may be offered to subscribers at reasonable tariffs or at no charge.

Calls which are free of charge to the calling subscriber, including calls to helplines, are not to be identified in the calling subscriber's itemised bill.

(b) *Selective call barring for outgoing calls, free of charge*

I.e. the facility whereby the subscriber can, on request to the telephone service provider, bar outgoing calls of defined types or to defined types of numbers free of charge.

(c) *Pre-payment systems*

Member States are to ensure that national regulatory authorities may require designated undertakings to provide means for consumers to pay for access to the public telephone network and use of publicly available telephone services on pre-paid terms.

(d) *Phased payment of connection fees*

Member States are to ensure that national regulatory authorities may require designated undertakings to allow consumers to pay for connection to the public telephone network on the basis of payments phased over time.

(e) *Non-payment of bills*

Member States are to authorise specified measures, which are to be proportionate, non-discriminatory and published, to cover non-payment of telephone bills for use of the public telephone network at fixed locations. These measures are to ensure that due warning of any consequent service interruption or disconnection is given to the subscriber beforehand. Except in cases of fraud, persistent late payment or non-payment, these measures are to ensure, as far as is technically feasible, that any service interruption is confined to the service concerned. Disconnection for non-payment of bills should take place only after due warning is given to the subscriber. Member States may allow a period of limited service prior to complete disconnection, during which only calls that do not incur a charge to the subscriber (e.g. '112' calls) are permitted.

**Part B: List of facilities referred to in Article 29**

(a) *Tone dialling or DTMF (dual-tone multi-frequency operation)*

I.e. the public telephone network supports the use of DTMF tones as defined in ETSI ETR 207 for end-to-end signalling throughout the network both within a Member State and between Member States.

(b) *Calling-line identification*

I.e. the calling party's number is presented to the called party prior to the call being established.

This facility should be provided in accordance with relevant legislation on protection of personal data and privacy, in particular Directive 97/66/EC.

To the extent technically feasible, operators should provide data and signals to facilitate the offering of calling-line identity and tone dialling across Member State boundaries.

*ANNEX II*

**INFORMATION TO BE PUBLISHED IN ACCORDANCE WITH ARTICLE 21 (TRANSPARENCY AND PUBLICATION OF INFORMATION)**

The national regulatory authority has a responsibility to ensure that the information in this Annex is published, in accordance with Article 21. It is for the national regulatory authority to decide which information is to be published by the undertakings providing public telephone networks and/or publicly available telephone services and which information is to be published by the national regulatory authority itself, so as to ensure that consumers are able to make informed choices.

1. Name(s) and address(es) of undertaking(s)

   I.e. names and head office addresses of undertakings providing public telephone networks and/or publicly available telephone services.

2. Publicly available telephone services offered

2.1. Scope of the publicly available telephone service

   Description of the publicly available telephone services offered, indicating what is included in the subscription charge and the periodic rental charge (e.g. operator services, directories, directory enquiry services, selective call barring, itemised billing, maintenance, etc.).

2.2. Standard tariffs covering access, all types of usage charges, maintenance, and including details of standard discounts applied and special and targeted tariff schemes.

2.3. Compensation/refund policy, including specific details of any compensation/refund schemes offered.

2.4. Types of maintenance service offered.

2.5. Standard contract conditions, including any minimum contractual period, if relevant.

3. Dispute settlement mechanisms including those developed by the undertaking.

4. Information about rights as regards universal service, including the facilities and services mentioned in Annex I.

*ANNEX III*

## QUALITY OF SERVICE PARAMETERS

### Supply-time and quality-of-service parameters, definitions and measurement methods referred to Articles 11 and 22

| Parameter [1] | Definition | Measurement method |
|---|---|---|
| Supply time for initial connection | ETSI EG 201 769-1 | ETSI EG 201 769-1 |
| Fault rate per access line | ETSI EG 201 769-1 | ETSI EG 201 769-1 |
| Fault repair time | ETSI EG 201 769-1 | ETSI EG 201 769-1 |
| Unsuccessful call ratio [2] | ETSI EG 201 769-1 | ETSI EG 201 769-1 |
| Call set up time [2] | ETSI EG 201 769-1 | ETSI EG 201 769-1 |
| Response times for operator services | ETSI EG 201 769-1 | ETSI EG 201 769-1 |
| Response times for directory enquiry services | ETSI EG 201 769-1 | ETSI EG 201 769-1 |
| Proportion of coin and card operated public pay telephones in working order | ETSI EG 201 769-1 | ETSI EG 201 769-1 |
| Bill correctness complaints | ETSI EG 201 769-1 | ETSI EG 201 769-1 |

(1) Parameters should allow for performance to be analysed at a regional level (i.e. no less than Level 2 in the Nomenclature of Territorial Units for Statistics (NUTS) established by Eurostat).

(2) Member States may decide not to require that up-to-date information concerning the performance for these two parameters be kept, if evidence is available to show that performance in these two areas is satisfactory.

*Note:* Version number of ETSI EG 201 769-1 is 1.1.1 (April 2000).

*ANNEX IV*

## CALCULATING THE NET COST, IF ANY, OF UNIVERSAL SERVICE OBLIGATIONS AND ESTABLISHING ANY RECOVERY OR SHARING MECHANISM IN ACCORDANCE WITH ARTICLES 12 AND 13

### Part A: Calculation of net cost

Universal service obligations refer to those obligations placed upon an undertaking by a Member State which concern the provision of a network and service throughout a specified geographical area, including, where required, averaged prices in that geographical area for the provision of that service or provision of specific tariff options for consumers with low incomes or with special social needs.

National regulatory authorities are to consider all means to ensure appropriate incentives for undertakings (designated or not) to provide universal service obligations cost efficiently. In undertaking a calculation exercise, the net cost of universal service obligations is to be calculated as the difference between the net cost for a designated undertaking of operating with the universal service obligations and operating without the universal service obligations. This applies whether the network in a particular Member State is fully developed or is still undergoing development and expansion. Due attention is to be given to correctly assessing the costs that any designated undertaking would have chosen to avoid had there been no universal service obligation. The net cost calculation should assess the benefits, including intangible benefits, to the universal service operator.

The calculation is to be based upon the costs attributable to:

(i) elements of the identified services which can only be provided at a loss or provided under cost conditions falling outside normal commercial standards.

This category may include service elements such as access to emergency telephone services, provision of certain public pay telephones, provision of certain services or equipment for disabled people, etc;

(ii) specific end-users or groups of end-users who, taking into account the cost of providing the specified network and service, the revenue generated and any geographical averaging of prices imposed by the Member State, can only be served at a loss or under cost conditions falling outside normal commercial standards.

This category includes those end-users or groups of end-users which would not be served by a commercial operator which did not have an obligation to provide universal service.

The calculation of the net cost of specific aspects of universal service obligations is to be made separately and so as to avoid the double counting of any direct or indirect benefits and costs. The overall net cost of universal service obligations to any undertaking is to be calculated as the sum of the net costs arising from the specific components of universal service obligations, taking account of any intangible benefits. The responsibility for verifying the net cost lies with the national regulatory authority.

### Part B: Recovery of any net costs of universal service obligations

The recovery or financing of any net costs of universal service obligations requires designated undertakings with universal service obligations to be compensated for the services they provide under non-commercial conditions. Because such a compensation involves financial transfers, Member States are to ensure that these are undertaken in an objective, transparent, non-discriminatory and proportionate manner. This means that the transfers result in the least distortion to competition and to user demand.

In accordance with Article 13(3), a sharing mechanism based on a fund should use a transparent and neutral means for collecting contributions that avoids the danger of a double imposition of contributions falling on both outputs and inputs of undertakings.

The independent body administering the fund is to be responsible for collecting contributions from undertakings which are assessed as liable to contribute to the net cost of universal service obligations in the Member State and is to oversee the transfer of sums due and/or administrative payments to the undertakings entitled to receive payments from the fund.

*ANNEX V*

**PROCESS FOR REVIEWING THE SCOPE OF UNIVERSAL SERVICE IN ACCORDANCE WITH ARTICLE 15**

In considering whether a review of the scope of universal service obligations should be undertaken, the Commission is to take into consideration the following elements:

— social and market developments in terms of the services used by consumers,

— social and market developments in terms of the availability and choice of services to consumers,

— technological developments in terms of the way services are provided to consumers.

In considering whether the scope of universal service obligations be changed or redefined, the Commission is to take into consideration the following elements:

— are specific services available to and used by a majority of consumers and does the lack of availability or non-use by a minority of consumers result in social exclusion, and

— does the availability and use of specific services convey a general net benefit to all consumers such that public intervention is warranted in circumstances where the specific services are not provided to the public under normal commercial circumstances?

*ANNEX VI*

**INTEROPERABILITY OF DIGITAL CONSUMER EQUIPMENT REFERRED TO IN ARTICLE 24**

1. *The common scrambling algorithm and free-to-air reception*

All consumer equipment intended for the reception of digital television signals, for sale or rent or otherwise made available in the Community, capable of descrambling digital television signals, is to possess the capability to:

— allow the descrambling of such signals according to the common European scrambling algorithm as administered by a recognised European standards organisation, currently ETSI;

— display signals that have been transmitted in clear provided that, in the event that such equipment is rented, the rentee is in compliance with the relevant rental agreement.

2. *Interoperability for analogue and digital television sets*

Any analogue television set with an integral screen of visible diagonal greater than 42 cm which is put on the market for sale or rent in the Community is to be fitted with at least one open interface socket, as standardised by a recognised European standards organisation, e.g. as given in the CENELEC EN 50 049-1:1997 standard, permitting simple connection of peripherals, especially additional decoders and digital receivers.

Any digital television set with an integral screen of visible diagonal greater than 30 cm which is put on the market for sale or rent in the Community is to be fitted with at least one open interface socket (either standardised by, or conforming to a standard adopted by, a recognised European standards organisation, or conforming to an industry-wide specification) e.g. the DVB common interface connector, permitting simple connection of peripherals, and able to pass all the elements of a digital television signal, including information relating to interactive and conditionally accessed services.

*ANNEX VII*

**CONDITIONS FOR THE MINIMUM SET OF LEASED LINES REFERRED TO IN ARTICLE 18**

*Note:* In accordance with the procedure in Article 18, provision of the minimum set of leased lines under the conditions established by Directive 92/44/EC should continue until such time as the national regulatory authority determines that there is effective competition in the relevant leased lines market.

National regulatory authorities are to ensure that provision of the minimum set of leased lines referred to in Article 18 follows the basic principles of non-discrimination, cost orientation and transparency.

1. *Non discrimination*

National regulatory authorities are to ensure that the organisations identified as having significant market power pursuant to Article 18(1) adhere to the principle of non-discrimination when providing leased lines referred to in Article 18. Those organisations are to apply similar conditions in similar circumstances to organisations providing similar services, and are to provide leased lines to others under the same conditions and of the same quality as they provide for their own services, or those of their subsidiaries or partners, where applicable.

2. *Cost orientation*

National regulatory authorities are, where appropriate, to ensure that tariffs for leased lines referred to in Article 18 follow the basic principles of cost orientation.

To this end, national regulatory authorities are to ensure that undertakings identified as having significant market power pursuant to Article 18(1) formulate and put in practice a suitable cost accounting system.

National regulatory authorities are to keep available, with an adequate level of detail, information on the cost accounting systems applied by such undertakings. They are to submit this information to the Commission on request.

3. *Transparency*

National regulatory authorities are to ensure that the following information in respect of the minimum set of leased lines referred to in Article 18 is published in an easily accessible form.

3.1. Technical characteristics, including the physical and electrical characteristics as well as the detailed technical and performance specifications which apply at the network termination point.

3.2. Tariffs, including the initial connection charges, the periodic rental charges and other charges. Where tariffs are differentiated, this must be indicated.

Where, in response to a particular request, an organisation identified as having significant market power pursuant to Article 18(1) considers it unreasonable to provide a leased line in the minimum set under its published tariffs and supply conditions, it must seek the agreement of the national regulatory authority to vary those conditions in that case.

3.3. Supply conditions, including at least the following elements:

— information concerning the ordering procedure,

— the typical delivery period, which is the period, counted from the date when the user has made a firm request for a leased line, in which 95 % of all leased lines of the same type have been put through to the customers.

This period will be established on the basis of the actual delivery periods of leased lines during a recent time interval of reasonable duration. The calculation must not include cases where late delivery periods were requested by users,

— the contractual period, which includes the period which is in general laid down in the contract and the minimum contractual period which the user is obliged to accept,

— the typical repair time, which is the period, counted from the time when a failure message has been given to the responsible unit within the undertaking identified as having significant market power pursuant to Article 18(1) up to the moment in which 80 % of all leased lines of the same type have been re-established and in appropriate cases notified back in operation to the users. Where different classes of quality of repair are offered for the same type of leased lines, the different typical repair times shall be published,

— any refund procedure.

In addition where a Member State considers that the achieved performance for the provision of the minimum set of leased lines does not meet users' needs, it may define appropriate targets for the supply conditions listed above.

# COMMISSION DIRECTIVE 2002/77/EC

## of 16 September 2002

## on competition in the markets for electronic communications networks and services

**(Text with EEA relevance)**

THE COMMISSION OF THE EUROPEAN COMMUNITIES,

Having regard to the Treaty establishing the European Community, and in particular Article 86(3) thereof,

Whereas:

(1) Commission Directive 90/388/EEC of 28 June 1990 on competition in the markets for telecommunications services [1], as last amended by Directive 1999/64/EC [2], has been substantially amended several times. Since further amendments are to be made, it should be recast in the interest of clarity.

(2) Article 86 of the Treaty entrusts the Commission with the task of ensuring that, in the case of public undertakings and undertakings enjoying special or exclusive rights, Member States comply with their obligations under Community law. Pursuant to Article 86(3), the Commission can specify and clarify the obligations arising from that Article and, in that framework, set out the conditions which are necessary to allow the Commission to perform effectively the duty of surveillance imposed upon it by that paragraph.

(3) Directive 90/388/EEC required Member States to abolish special and exclusive rights for the provision of telecommunications services, initially for other services than voice telephony, satellite services and mobile radiocommunications, and then it gradually established full competition in the telecommunications market.

(4) A number of other Directives in this field have also been adopted under Article 95 of the Treaty by the European Parliament and the Council aiming, principally, at the establishment of an internal market for telecommunications services through the implementation of open network provision and the provision of a universal service in an environment of open and competitive markets. Those Directives should be repealed with effect from 25 July 2003 when the new regulatory framework for electronic communications networks and services is applied.

(5) The new electronic communications regulatory framework consists of one general Directive, Directive 2002/21/EC of the European Parliament and of the Council of 7 March 2002 on a common regulatory framework for electronic communications networks and services (Framework Directive) [3] and four specific Directives: Directive 2002/20/EC of the European Parliament and of the Council of 7 March 2002 on the authorisation of electronic communications networks and services (Authorisation Directive) [4], Directive 2002/19/EC of the European Parliament and of the Council of 7 March 2002 on access to, and interconnection of, electronic communications networks and associated facilities (Access Directive) [5], Directive 2002/22/EC of the European Parliament and of the Council of 7 March 2002 on universal service and users' rights relating to electronic communications networks and services (Universal Service Directive) [6], and Directive 2002/58/EC of the European Parliament and of the Council of 12 July 2002 concerning the processing of personal data and the protection of privacy in the electronic communications (Directive on privacy and electronic communications) sector [7].

(6) In the light of the developments which have marked the liberalisation process and the gradual opening of the telecommunications markets in Europe since 1990, certain definitions used in Directive 90/388/EEC and its amending acts should be adjusted in order to reflect the latest technological developments in the telecommunications field, or replaced in order to take account of the convergence phenomenon which has shaped the information technology, media and telecommunications industries over recent years. The wording of certain provisions should, where possible, be clarified in order to facilitate their application, taking into account, where appropriate, the relevant Directives adopted under Article 95 of the Treaty, and the experience acquired through the implementation of Directive 90/388/EEC as amended.

(7) This Directive makes reference to 'electronic communications services' and 'electronic communications networks' rather than the previously used terms 'telecommunications services' and 'telecommunications networks'. These new definitions are indispensable in order to take account of the convergence phenomenon by bringing together under one single definition all electronic communications services and/or networks which are concerned with the conveyance of signals by wire, radio, optical or other electromagnetic means (i.e. fixed, wireless, cable television, satellite networks). Thus, the transmission and broadcasting of radio and television

(1) OJ L 192, 24.7.1990, p. 10.
(2) OJ L 175, 10.7.1999, p. 39.
(3) OJ L 108, 24.4.2002, p. 33.
(4) OJ L 108, 24.4.2002, p. 21.
(5) OJ L 108, 24.4.2002, p. 7.
(6) OJ L 108, 24.4.2002, p. 51.
(7) OJ L 201, 31.7.2002, p. 37.

programmes should be recognised as an electronic communication service and networks used for such transmission and broadcasting should likewise be recognised as electronic communications networks. Furthermore, it should be made clear that the new definition of electronic communications networks also covers fibre networks which enable third parties, using their own switching or routing equipment, to convey signals.

(8) In this context, it should be made clear that Member States must remove (if they have not already done so) exclusive and special rights for the provision of all electronic communications networks, not just those for the provision of electronic communications services and should ensure that undertakings are entitled to provide such services without prejudice to the provisions of Directives 2002/19/EC, 2002/20/EC, 2002/21/EC and 2002/22/EC. The definition of electronic communications networks should also mean that Member States are not permitted to restrict the right of an operator to establish, extend and/or provide a cable network on the ground that such network could also be used for the transmission of radio and television programming. In particular, special or exclusive rights which amount to restricting the use of electronic communications networks for the transmission and distribution of television signals are contrary to Article 86(1), read in conjunction with Article 43 (right of establishment) and/or Article 82(b) of the EC Treaty insofar as they have the effect of permitting a dominant undertaking to limit 'production, markets or technical development to the prejudice of consumers'. This is, however, without prejudice to the specific rules adopted by the Member States in accordance with Community law, and, in particular, in accordance with Council Directive 89/552/EEC of 3 October 1989 [1], on the coordination of certain provisions laid down by law, regulation or administrative action in Member States concerning the pursuit of television broadcasting activities, as amended by Directive 97/36/EC of the European Parliament and of the Council [2], governing the distribution of audiovisual programmes intended for the general public.

(9) Pursuant to the principle of proportionality, Member States should no longer make the provision of electronic communications services and the establishment and provision of electronic communications networks subject to a licensing regime but to a general authorisation regime. This is also required by Directive 2002/20/EC, according to which electronic communications services or networks should be provided on the basis of a general authorisation and not on the basis of a license. An aggrieved party should have the right to challenge a decision preventing him from providing electronic communications services or networks before an independent body and, ultimately, before a court or a tribunal. It is a fundamental principle of Community law that an individual is entitled to effective judicial protection whenever a State measure violates rights conferred upon him by the provisions of a Directive.

(10) Public authorities may exercise a dominant influence on the behaviour of public undertakings, as a result either of the rules governing the undertaking or of the manner in which the shareholdings are distributed. Therefore, where Member States control vertically integrated network operators which operate networks which have been established under special or exclusive rights, those Member States should ensure that, in order to avoid potential breaches of the Treaty competition rules, such operators, when they enjoy a dominant position in the relevant market, do not discriminate in favour of their own activities. It follows that Member States should take all measures necessary to prevent any discrimination between such vertically integrated operators and their competitors.

(11) This Directive should also clarify the principle derived from Commission Directive 96/2/EC of 16 January 1996 amending Directive 90/388/EC with regard to mobile and personal communications [3], by providing that Member States should not grant exclusive or special rights of use of radio frequencies and that the rights of use of those frequencies should be assigned according to objective, non-discriminatory and transparent procedures. This should be without prejudice to specific criteria and procedures adopted by Member States to grant such rights to providers of radio or television broadcast content services with a view to pursuing general interest objectives in conformity with Community law.

(12) Any national scheme pursuant to Directive 2002/22/EC, serving to share the net cost of the provision of universal service obligations shall be based on objective, transparent and non-discriminatory criteria and shall be consistent with the principles of proportionality and of least market distortion. Least market distortion means that contributions should be recovered in a way that as far as possible minimises the impact of the financial burden falling on end-users, for example by spreading contributions as widely as possible.

(13) Where rights and obligations arising from international conventions setting up international satellite organisations are not compatible with the competition rules of the Treaty, Member States should take, in accordance with Article 307 of the EC Treaty, all appropriate steps to eliminate such incompatibilities. This Directive should clarify this obligation because Article 3 of Directive 94/46/EC [4], merely required Member States to 'communicate to the Commission' the information they possessed on such incompatibilities. Article 11 of this Directive should clarify the obligation on Member States to remove any restrictions which could still be in force because of those international conventions.

[1] OJ L 298, 17.10.1989, p. 23.
[2] OJ L 202, 30.7.1997, p. 60.
[3] OJ L 20, 26.1.1996, p. 59.
[4] OJ L 268, 19.10.1994, p. 15.

(14) This Directive should maintain the obligation imposed on Member States by Directive 1999/64/EC, so as to ensure that dominant providers of electronic communications networks and publicly available telephone services operate their public electronic communication network and cable television network as separate legal entities.

(15) This Directive should be without prejudice to obligations of the Member States concerning the time limits set out in Annex I, Part B, within which the Member States are to comply with the preceding Directives.

(16) Member States should supply to the Commission any information which is necessary to demonstrate that existing national implementing legislation reflects the clarifications provided for in this Directive as compared with Directives 90/388/EC, 94/46/EC, 95/51/EC (1), 96/2/EC, 96/19/EC (2) and 1999/64/EC.

(17) In the light of the above, Directive 90/388/EC should be repealed,

HAS ADOPTED THIS DIRECTIVE:

*Article 1*

**Definitions**

For the purposes of this Directive the following definitions shall apply:

1. 'electronic communications network' shall mean transmission systems and, where applicable, switching or routing equipment and other resources which permit the conveyance of signals by wire, by radio, by optical or by other electromagnetic means, including satellite networks, fixed (circuit — and packet — switched, including Internet) and mobile terrestrial networks, and electricity cable systems, to the extent that they are used for the purpose of transmitting signals, networks used for radio and television broadcasting, and cable television networks, irrespective of the type of information conveyed;

2. 'public communications network' shall mean an electronic communications network used wholly or mainly for the provision of public electronic communications services;

3. 'electronic communications services' shall mean a service normally provided for remuneration which consists wholly or mainly in the conveyance of signals on electronic communications networks, including telecommunications services and transmission services in networks used for broadcasting but exclude services providing or exercising editorial control over, content transmitted using electronic communications networks and services; it does not include information society services as defined in Article 1 of Directive 98/34/EC which do not consist wholly or mainly in the conveyance of signals on electronic communications networks;

4. 'publicly available electronic communications services' shall mean electronic communications services available to the public;

5. 'exclusive rights' shall mean the rights that are granted by a Member State to one undertaking through any legislative, regulatory or administrative instrument, reserving it the right to provide an electronic communications service or to undertake an electronic communications activity within a given geographical area;

6. 'special rights' shall mean the rights that are granted by a Member State to a limited number of undertakings through any legislative, regulatory or administrative instrument which, within a given geographical area:

   (a) designates or limits to two or more the number of such undertakings authorised to provide an electronic communications service or undertake an electronic communications activity, otherwise than according to objective, proportional and non-discriminatory criteria, or

   (b) confers on undertakings, otherwise than according to such criteria, legal or regulatory advantages which substantially affect the ability of any other undertaking to provide the same electronic communications service or to undertake the same electronic communications activity in the same geographical area under substantially equivalent conditions;

7. 'satellite earth station network' shall mean a configuration of two or more earth stations which interwork by means of a satellite;

8. 'cable television networks' shall mean any mainly wire-based infrastructure established primarily for the delivery or distribution of radio or television broadcast to the public.

*Article 2*

**Exclusive and special rights for electronic communications networks and electronic communications services**

1. Member States shall not grant or maintain in force exclusive or special rights for the establishment and/or the provision of electronic communications networks, or for the provision of publicly available electronic communications services.

2. Member States shall take all measures necessary to ensure that any undertaking is entitled to provide electronic communications services or to establish, extend or provide electronic communications networks.

3. Member States shall ensure that no restrictions are imposed or maintained on the provision of electronic communications services over electronic communications networks established by the providers of electronic communications services, over infrastructures provided by third parties, or by means of sharing networks, other facilities or sites without prejudice to the provisions of Directives 2002/19/EC, 2002/20/EC, 2002/21/EC and 2002/22/EC.

(1) OJ L 256, 26.10.1995, p. 49.
(2) OJ L 74, 22.3.1996, p. 13.

4. Member States shall ensure that a general authorisation granted to an undertaking to provide electronic communications services or to establish and/or provide electronic communications networks, as well as the conditions attached thereto, shall be based on objective, non-discriminatory, proportionate and transparent criteria.

5. Reasons shall be given for any decision taken on the grounds set out in Article 3(1) of Directive 2002/20/EC preventing an undertaking from providing electronic communications services or networks.

Any aggrieved party should have the possibility to challenge such a decision before a body that is independent of the parties involved and ultimately before a court or a tribunal.

*Article 3*

**Vertically integrated public undertakings**

In addition to the requirements set out in Article 2(2), and without prejudice to Article 14 of Directive 2002/21/EC, Member States, shall ensure that vertically integrated public undertakings which provide electronic communications networks and which are in a dominant position do not discriminate in favour of their own activities.

*Article 4*

**Rights of use of frequencies**

Without prejudice to specific criteria and procedures adopted by Member States to grant rights of use of radio frequencies to providers of radio or television broadcast content services with a view to pursuing general interest objectives in conformity with Community law:

1. Member States shall not grant exclusive or special rights of use of radio frequencies for the provision of electronic communications services.

2. The assignment of radio frequencies for electronic communication services shall be based on objective, transparent, non-discriminatory and proportionate criteria.

*Article 5*

**Directory services**

Member States shall ensure that all exclusive and/or special rights with regard to the establishment and provision of directory services on their territory, including both the publication of directories and directory enquiry services, are abolished.

*Article 6*

**Universal service obligations**

1. Any national scheme pursuant to Directive 2002/22/EC, serving to share the net cost of the provision of universal service obligations shall be based on objective, transparent and non-discriminatory criteria and shall be consistent with the principle of proportionality and of least market distortion. In particular, where universal service obligations are imposed in whole or in part on public undertakings providing electronic communications services, this shall be taken into consideration in calculating any contribution to the net cost of universal service obligations.

2. Member States shall communicate any scheme of the kind referred to in paragraph 1 to the Commission.

*Article 7*

**Satellites**

1. Member States shall ensure that any regulatory prohibition or restriction on the offer of space segment capacity to any authorised satellite earth station network operator are abolished, and shall authorise within their territory any space-segment supplier to verify that the satellite earth station network for use in connection with the space segment of the supplier in question is in conformity with the published conditions for access to such person's space segment capacity.

2. Member States which are party to international conventions setting up international satellite organisations shall, where such conventions are not compatible with the competition rules of the EC Treaty, take all appropriate steps to eliminate such incompatibilities.

*Article 8*

**Cable television networks**

1. Each Member State shall ensure that no undertaking providing public electronic communications networks operates its cable television network using the same legal entity as it uses for its other public electronic communications network, when such undertaking:

(a) is controlled by that Member State or benefits from special rights; and

(b) is dominant in a substantial part of the common market in the provision of public electronic communications networks and publicly available telephone services; and

(c) operates a cable television network which has been established under special or exclusive right in the same geographic area.

2. The term 'publicly available telephone services' shall be considered synonymous with the term 'public voice telephony services' referred to in Article 1 of Directive 1999/64/EC.

3. Member States which consider that there is sufficient competition in the provision of local loop infrastructure and services in their territory shall inform the Commission accordingly.

Such information shall include a detailed description of the market structure. The information provided shall be made available to any interested party on demand, regard being had to the legitimate interest of undertakings in the protection of their business secrets.

17.9.2002 EN Official Journal of the European Communities L 249/25

4. The Commission shall decide within a reasonable period, after having heard the comments of these parties, whether the obligation of legal separation may be ended in the Member State concerned.

5. The Commission shall review the application of this Article not later than 31 December 2004.

*Article 9*

Member States shall supply to the Commission not later than 24 July 2003 such information as will allow the Commission to confirm that the provisions of this Directive have been complied with.

*Article 10*

**Repeal**

Directive 90/388/EC, as amended by the Directives listed in Annex I, Part A, is repealed with effect from 25 July 2003, without prejudice to the obligations of the Member States in respect of the time limits for transposition laid down in Annex I, Part B.

References to the repealed Directives shall be construed as references to this Directive and shall be read in accordance with the correlation table in Annex II.

*Article 11*

This Directive shall enter into force on the 20th day following that of its publication in the *Official Journal of the European Communities*.

*Article 12*

This Directive is addressed to the Member States.

Done at Brussels, 16 September 2002.

*For the Commission*
Mario MONTI
*Member of the Commission*

*ANNEX I*

PART A

**List of Directives to be repealed**

Directive 90/388/EEC (OJ L 192, 24.7.1990, p. 10)

Articles 2 and 3 of Directive 94/46/EC (OJ L 268, 19.1.1994, p. 15)

Directive 95/51/EC (OJ L 256, 26.10.1995, p. 49)

Directive 96/2/EC (OJ L 20, 26.1.1996, p. 59)

Directive 96/19/EC (OJ L 74, 22.3.1996, p. 13)

Directive 1999/64/EC (OJ L 175, 10.7.1999, p. 39)

PART B

**Transposition dates for the above Directives**

| | |
|---|---|
| Directive 90/388/EEC: transposition date: | 31 December 1990 |
| Directive 94/46/EC: transposition date: | 8 August 1995 |
| Directive 95/51/EC: transposition date: | 1 October 1996 |
| Directive 96/2/EC: transposition date: | 15 November 1996 |
| Directive 96/19/EC: transposition date: | 11 January 1997 |
| Directive 1999/64/EC: transposition date: | 30 April 2000 |

*ANNEX II*

**Correlation table**

| This Directive | Directive 90/388/EEC |
|---|---|
| Article 1 (Definitions) | Article 1 |
| Article 2 (withdrawal of exclusive/special rights) | Article 2 |
| Article 3 (vertically integrated public undertakings) | Article 3(a)(ii) |
| Article 4 (rights of use of radio frequencies) | Article 3(b) |
| Article 5 (directory services) | Article 4(b) |
| Article 6 (universal service obligations) | Article 4(c) |
| Article 7 (satellites) | Article 3 of Directive 94/46/EC |
| Article 8 (cable networks) | Article 9 |

I

*(Acts whose publication is obligatory)*

# REGULATION (EC) No 2887/2000 OF THE EUROPEAN PARLIAMENT AND OF THE COUNCIL of 18 December 2000 on unbundled access to the local loop

**(Text with EEA relevance)**

THE EUROPEAN PARLIAMENT AND THE COUNCIL OF THE EUROPEAN UNION,

Having regard to the Treaty establishing the European Community, and in particular Article 95 thereof,

Having regard to the proposal from the Commission,

Having regard to the opinion of the Economic and Social Committee [1],

Acting in accordance with the procedure laid down in Article 251 of the Treaty [2],

Whereas:

(1) The conclusions of the European Council of Lisbon of 23 and 24 March 2000 note that, for Europe to fully seize the growth and job potential of the digital, knowledge-based economy, businesses and citizens must have access to an inexpensive, world-class communications infrastructure and a wide range of services. The Member States, together with the Commission, are called upon to work towards introducing greater competition in local access networks before the end of 2000 and unbundling the local loop, in order to help bring about a substantial reduction in the costs of using the Internet. The Feira European Council of 20 June 2000 endorsed the proposed 'e-Europe' Action Plan which identifies unbundled access to the local loop as a short-term priority.

(2) Local loop unbundling should complement the existing provisions in Community law guaranteeing universal service and affordable access for all citizens by enhancing competition, ensuring economic efficiency and bringing maximum benefit to users.

(3) The 'local loop' is the physical twisted metallic pair circuit in the fixed public telephone network connecting the network termination point at the subscriber's premises to the main distribution frame or equivalent facility. As noted in the Commission's Fifth Report on the implementation of the telecommunications regulatory package, the local access network remains one of the least competitive segments of the liberalised telecommunications market. New entrants do not have widespread alternative network infrastructures and are unable, with traditional technologies, to match the economies of scale and the coverage of operators designated as having significant market power in the fixed public telephone network market. This results from the fact that these operators rolled out their metallic local access infrastructures over significant periods of time protected by exclusive rights and were able to fund investment costs through monopoly rents.

(4) The European Parliament Resolution of 13 June 2000 on the Commission communication on the 1999 Communications review stresses the importance of enabling the sector to develop infrastructures which promote the growth of electronic communications and e-commerce and the importance of regulating in a way that supports this growth. It notes that the unbundling of the local loop currently concerns mainly the metallic infrastructure of a dominant entity and that investment in alternative infrastructures must have the possibility of ensuring a reasonable rate of return, since that might facilitate the expansion of these infrastructures in areas where their penetration is still low.

(5) The provision of new loops with high capacity optical fibre directly to major users is a specific market that is developing under competitive conditions with new investments. This Regulation therefore addresses access to metallic local loops, without prejudice to national obligations regarding other types of access to local infrastructures.

[1] Opinion delivered on 19 October 2000 (not yet published in the Official Journal).

[2] Opinion of the European Parliament of 26 October 2000 (not yet published in the Official Journal) and Decision of the Council of 5 December 2000.

(6) It would not be economically viable for new entrants to duplicate the incumbent's metallic local access infrastructure in its entirety within a reasonable time. Alternative infrastructures such as cable television, satellite, wireless local loops do not generally offer the same functionality or ubiquity for the time being, though situations in Member States may differ.

(7) Unbundled access to the local loop allows new entrants to compete with notified operators in offering high bit-rate data transmission services for continuous Internet access and for multimedia applications based on digital subscriber line (DSL) technology as well as voice telephony services. A reasonable request for unbundled access implies that the access is necessary for the provision of the services of the beneficiary, and that refusal of the request would prevent, restrict or distort competition in this sector.

(8) This Regulation mandates unbundled access to the metallic local loops only of notified operators that have been designated by their national regulatory authorities as having significant market power in the fixed public telephone network supply market under the relevant Community provisions (hereinafter referred to as 'notified operators'). Member States have already notified to the Commission the names of those fixed public network operators which have significant market power under Annex I, Part 1, of Directive 97/33/EC of the European Parliament and of the Council of 30 June 1997 on interconnection in telecommunications with regard to ensuring universal service and interoperability through application of the principles of open network provision (ONP) [1], and Directive 98/10/EC of the European Parliament and of the Council of 26 February 1998 on the application of open network provision to voice telephony and on universal service for telecommunications in a competitive environment [2].

(9) A notified operator cannot be required to provide types of access which are not within its powers to provide, for example where fulfilment of a request would cause a violation of the legal rights of an independent third party. The obligation to provide unbundled access to the local loop does not imply that notified operators have to install entirely new local network infrastructure specifically to meet beneficiaries' requests.

(10) Although commercial negotiation is the preferred method for reaching agreement on technical and pricing issues for local loop access, experience shows that in most cases regulatory intervention is necessary due to imbalance in negotiating power between the new entrant and the notified operator, and lack of other alternatives. In certain circumstances the national regulatory authority may, in accordance with Community law, intervene on its own initiative in order to ensure fair competition, economic efficiency and maximum benefit for end-users. Failure of the notified operator to meet lead times should entitle the beneficiary to receive compensation.

(11) Costing and pricing rules for local loops and related facilities should be transparent, non-discriminatory and objective to ensure fairness. Pricing rules should ensure that the local loop provider is able to cover its appropriate costs in this regard plus a reasonable return, in order to ensure the long term development and upgrade of local access infrastructure. Pricing rules for local loops should foster fair and sustainable competition, bearing in mind the need for investment in alternative infrastructures, and ensure that there is no distortion of competition, in particular no margin squeeze between prices of wholesale and retail services of the notified operator. In this regard, it is considered important that competition authorities be consulted.

(12) Notified operators should provide information and unbundled access to third parties under the same conditions and of the same quality as they provide for their own services or to their associated companies. To this end, the publication by the notified operator of an adequate reference offer for unbundled access to the local loop, within a short time-frame and ideally on the Internet, and under the supervisory control of the national regulatory authority, would contribute to the creation of transparent and non-discriminatory market conditions.

(13) In its Recommendation 2000/417/EC of 25 May 2000 on unbundled access to the local loop enabling the competitive provision of a full range of electronic communications services including broadband multimedia and high-speed Internet [3] and its Communication of 26 April 2000 [4], the Commission set out detailed guidance to assist national regulatory authorities on the fair regulation of different forms of unbundled access to the local loop.

(14) In accordance with the principle of subsidiarity as set out in Article 5 of the Treaty, the objective of achieving a harmonised framework for unbundled access to the local loop in order to enable the competitive provision of an inexpensive, world-class communications infrastructure and a wide range of services for all businesses and citizens in the Community cannot be achieved by the Member States in a secure, harmonised and timely manner and can therefore be better achieved by the Community. In accordance with the principle of proportionality as set out in that Article, the provisions of this

[1] OJ L 199, 26.7.1997, p. 32. Directive as amended by Directive 98/61/EC (OJ L 268, 3.10.1998, p. 37).
[2] OJ L 101, 1.4.1998, p. 24.
[3] OJ L 156, 29.6.2000, p. 44.
[4] OJ C 272, 23.9.2000, p. 55.

Regulation do not go beyond what is necessary in order to achieve this objective for that purpose. They are adopted without prejudice to national provisions complying with Community law which set out more detailed measures, for example dealing with virtual collocation.

(15) This Regulation complements the regulatory framework for telecommunications, in particular Directives 97/33/EC and 98/10/EC. The new regulatory framework for electronic communications should include appropriate provisions to replace this Regulation,

HAVE ADOPTED THIS REGULATION:

*Article 1*

**Aim and Scope**

1. This Regulation aims at intensifying competition and stimulating technological innovation on the local access market, through the setting of harmonised conditions for unbundled access to the local loop, to foster the competitive provision of a wide range of electronic communications services.

2. This Regulation shall apply to unbundled access to the local loops and related facilities of notified operators as defined in Article 2(a).

3. This Regulation shall apply without prejudice to the obligations for notified operators to comply with the principle of non-discrimination, when using the fixed public telephone network in order to provide high speed access and transmission services to third parties in the same manner as they provide for their own services or to their associated companies, in accordance with Community provisions.

4. This Regulation is without prejudice to the rights of Member States to maintain or introduce measures in conformity with Community law which contain more detailed provisions than those set out in this Regulation and/or are outside the scope of this Regulation *inter alia* with respect to other types of access to local infrastructures.

*Article 2*

**Definitions**

For the purposes of this Regulation the following definitions apply:

(a) 'notified operator' means operators of fixed public telephone networks that have been designated by their national regulatory authority as having significant market power in the provision of fixed public telephone networks and services under Annex I, Part 1, of Directive 97/33/EC or Directive 98/10/EC;

(b) 'beneficiary' means a third party duly authorised in accordance with Directive 97/13/EC (1) or entitled to provide communications services under national legislation, and which is eligible for unbundled access to a local loop;

(c) 'local loop' means the physical twisted metallic pair circuit connecting the network termination point at the subscriber's premises to the main distribution frame or equivalent facility in the fixed public telephone network;

(d) 'local sub-loop' means a partial local loop connecting the network termination point at the subscriber's premises to a concentration point or a specified intermediate access point in the fixed public telephone network;

(e) 'unbundled access to the local loop' means full unbundled access to the local loop and shared access to the local loop; it does not entail a change in ownership of the local loop;

(f) 'full unbundled access to the local loop' means the provision to a beneficiary of access to the local loop or local sub loop of the notified operator authorising the use of the full frequency spectrum of the twisted metallic pair;

(g) 'shared access to the local loop' means the provision to a beneficiary of access to the local loop or local sub loop of the notified operator, authorising the use of the non-voice band frequency spectrum of the twisted metallic pair; the local loop continues to be used by the notified operator to provide the telephone service to the public;

(h) 'collocation' means the provision of physical space and technical facilities necessary to reasonably accommodate and connect the relevant equipment of a beneficiary, as mentioned in Section B of the Annex;

(i) 'related facilities' means the facilities associated with the provision of unbundled access to the local loop, notably collocation, cable connections and relevant information technology systems, access to which is necessary for a beneficiary to provide services on a competitive and fair basis.

*Article 3*

**Provision of unbundled access**

1. Notified operators shall publish from 31 December 2000, and keep updated, a reference offer for unbundled access to their local loops and related facilities, which shall include at least the items listed in the Annex. The offer shall be sufficiently unbundled so that the beneficiary does not have to pay for network elements or facilities which are not necessary for the supply of its services, and shall contain a description of the components of the offer, associated terms and conditions, including charges.

(1) Directive 97/13/EC of the European Parliament and of the Council of 10 April 1997 on a common framework for general authorisations and individual licences in the field of telecommunications services (OJ L 117, 7.5.1997, p. 15).

2. Notified operators shall from 31 December 2000 meet reasonable requests from beneficiaries for unbundled access to their local loops and related facilities, under transparent, fair and non-discriminatory conditions. Requests shall only be refused on the basis of objective criteria, relating to technical feasibility or the need to maintain network integrity. Where access is refused, the aggrieved party may submit the case to the dispute resolution procedure referred to in Article 4(5). Notified operators shall provide beneficiaries with facilities equivalent to those provided for their own services or to their associated companies, and with the same conditions and time-scales.

3. Without prejudice to Article 4(4), notified operators shall charge prices for unbundled access to the local loop and related facilities set on the basis of cost-orientation.

*Article 4*

**Supervision by the national regulatory authority**

1. The national regulatory authority shall ensure that charging for unbundled access to the local loop fosters fair and sustainable competition.

2. The national regulatory authority shall have the power to:

(a) impose changes on the reference offer for unbundled access to the local loop and related facilities, including prices, where such changes are justified; and

(b) require notified operators to supply information relevant for the implementation of this Regulation.

3. The national regulatory authority may, where justified, intervene on its own initiative in order to ensure non-discrimination, fair competition, economic efficiency and maximum benefit for users.

4. When the national regulatory authority determines that the local access market is sufficiently competitive, it shall relieve the notified operators of the obligation laid down in Article 3(3) for prices to be set on the basis of cost-orientation.

5. Disputes between undertakings concerning issues included in this Regulation shall be subject to the national dispute resolution procedures established in conformity with Directive 97/33/EC and shall be handled promptly, fairly and transparently.

*Article 5*

**Entry into force**

This Regulation shall enter into force on the third day following that of its publication in the *Official Journal of the European Communities*.

This Regulation shall be binding in its entirety and directly applicable in all Member States.

Done at Brussels, 18 December 2000.

| | |
|---|---|
| *For the European Parliament* | *For the Council* |
| *The President* | *The President* |
| N. FONTAINE | D. VOYNET |

*ANNEX*

**MINIMUM LIST OF ITEMS TO BE INCLUDED IN A REFERENCE OFFER FOR UNBUNDLED ACCESS TO THE LOCAL LOOP TO BE PUBLISHED BY NOTIFIED OPERATORS**

A. Conditions for unbundled access to the local loop

1. Network elements to which access is offered covering in particular the following elements:

   (a) access to local loops;

   (b) access to non-voice band frequency spectrum of a local loop, in the case of shared access to the local loop;

2. Information concerning the locations of physical access sites ([1]), availability of local loops in specific parts of the access network;

3. Technical conditions related to access and use of local loops, including the technical characteristics of the twisted metallic pair in the local loop;

4. Ordering and provisioning procedures, usage restrictions.

B. Collocation services

1. Information on the notified operator's relevant sites ([1]);

2. Collocation options at the sites indicated under point 1 (including physical collocation and, as appropriate, distant collocation and virtual collocation);

3. Equipment characteristics: restrictions, if any, on equipment that can be collocated;

4. Security issues: measures put in place by notified operators to ensure the security of their locations;

5. Access conditions for staff of competitive operators;

6. Safety standards;

7. Rules for the allocation of space where collocation space is limited;

8. Conditions for beneficiaries to inspect the locations at which physical collocation is available, or sites where collocation has been refused on grounds of lack of capacity.

C. Information systems

Conditions for access to notified operator's operational support systems, information systems or databases for pre-ordering, provisioning, ordering, maintenance and repair requests and billing.

D. Supply conditions

1. Lead time for responding to requests for supply of services and facilities; service level agreements, fault resolution, procedures to return to a normal level of service and quality of service parameters;

2. Standard contract terms, including, where appropriate, compensation provided for failure to meet lead times;

3. Prices or pricing formulae for each feature, function and facility listed above.

---

([1]) Availability of this information may be restricted to interested parties only, in order to avoid public security concerns.

# COMMISSION DECISION

## of 29 July 2002

## establishing the European Regulators Group for Electronic Communications Networks and Services

**(Text with EEA relevance)**

(2002/627/EC)

THE COMMISSION OF THE EUROPEAN COMMUNITIES,

Having regard to the Treaty establishing the European Community,

Whereas:

(1) A new regulatory framework for electronic communications networks and services has been established in accordance with European Parliament and Council Directives 2002/21/EC of 7 March 2002 on a common regulatory framework for electronic communications networks and services (Framework Directive) [1], 2002/19/EC of 7 March 2002 on access to, and interconnection of, electronic communications networks and associated facilities (Access Directive) [2], 2002/20/EC of 7 March 2002 on the authorisation of electronic communications networks and services (Authorisation Directive) [3] and 2002/22/EC of 7 March 2002 on the universal services and users' rights related to electronic communications networks and services (Universal Service Directive) [4].

(2) National regulatory authorities have been set up in all Member States to carry out the regulatory tasks specified in these Directives and as to be notified to the Commission in accordance with Article 3(6) of the Framework Directive. In accordance with the Framework Directive, Member States must guarantee the independence of national regulatory authorities by ensuring that they are legally distinct from and functionally independent of all organisations providing electronic communications networks, equipment or services. Member States that retain ownership or control of undertakings providing electronic communications networks and/or services must also ensure effective structural separation of the regulatory function from activities associated with ownership or control.

(3) Detailed responsibilities and tasks of the national regulatory authorities differ among the various Member States, but all of them have at least one national regulatory authority who is charged with application of the rules once they have been transposed into national law, in particular the rules concerning day-to-day supervision of the market.

(4) The need for the relevant rules to be consistently applied in all Member States is essential for the successful development of an internal market for electronic communications networks and services. The new regulatory framework sets out objectives to be achieved and provides a framework for action by national regulatory authorities, whilst granting them flexibility in certain areas to apply the rules in the light of national conditions.

(5) A European Regulators Group for Electronic Communications Networks and Services (hereinafter referred to as the Group) should be established to provide an interface for advising and assisting the Commission in the electronic communications field.

(6) The Group should provide an interface between national regulatory authorities and the Commission in such a way as to contribute to the development of the internal market. It should also allow cooperation between national regulatory authorities and the Commission in a transparent manner so as to ensure the consistent application in all Member States of the regulatory framework for electronic communications networks and services.

(7) The Group should serve as a body for reflection, debate and advice for the Commission in the electronic communications field, including on matters related to the implementation and revision of Recommendation on Relevant Product and Service Markets and in drawing up the Decision on transnational markets.

(8) Close cooperation should be maintained between the Group and the Communications Committee established under the Framework Directive. The work of the Group should not interfere with the work of the Committee.

(9) Coordination should be ensured with the Radio Spectrum Committee established under a Decision No 676/2002/EC of the European Parliament and Council of 7 March 2002 on a regulatory framework for radio spectrum policy in the European Community (Radio Spectrum Decision) [5], the Radio Spectrum Policy Group established under the Commission Decision 2002/622/EC of 26 July 2002 establishing a Radio Spectrum Policy

[1] OJ L 108, 24.4.2002, p. 33.
[2] OJ L 108, 24.4.2002, p. 7.
[3] OJ L 108, 24.4.2002, p. 21.
[4] OJ L 108, 24.4.2002, p. 51.
[5] OJ L 108, 24.4.2002, p. 1.

30.7.2002 EN Official Journal of the European Communities L 200/39

Group ([1]) and the Television Without frontiers Contact Committee, created pursuant to Directive 97/36/EC of the European Parliament and of the Council ([2]) on the coordination of certain provisions laid down by law, regulation or administrative action in Member States concerning the pursuit of television broadcasting activities,

HAS DECIDED AS FOLLOWS:

*Article 1*

**Subject matter**

An advisory group of the independent national regulatory authorities on electronic communications networks and services, called the European Regulators Group for Electronic Communications Networks and Services (hereinafter referred to as the Group), is hereby established.

*Article 2*

**Definition**

For the purpose of this Decision: 'relevant national regulatory authority' means the public authority established in each Member State to oversee the day-to-day interpretation and application of the provisions of the Directives relating to electronic communications network and services as defined in 'Framework' Directive.

*Article 3*

**Aims**

The role of the Group shall be to advise and assist the Commission in consolidating the internal market for electronic communications networks and services.

The Group shall provide an interface between national regulatory authorities and the Commission in such a way as to contribute to the development of the internal market and to the consistent application in all Member States of the regulatory framework for electronic communications networks and services.

*Article 4*

**Membership**

The Group shall be composed of the heads of each relevant national regulatory authority in each Member State or their representatives.

The Commission shall be represented at an appropriate level and shall provide the secretariat to the Group.

*Article 5*

**Operational arrangements**

At its own initiative or at the Commission's request the Group shall advise and assist the Commission on any matter related to electronic communications networks and services.

The Group shall elect a chairperson from among its members. The work of the group may be organised into subgroups and expert working groups as appropriate.

The chairperson shall convene the meetings of the Group in agreement with the Commission.

The Group shall adopt its rules of procedure by consensus or, in the absence of consensus, by a two-thirds majority vote, one vote being expressed per Member State, subject to the approval of the Commission.

The Commission shall be represented at all meetings of the Group and be able to attend all meetings of its subgroups and expert working groups.

Experts from EEA States and those states that are candidates for accession to the European Union may participate as observers in the Group. The Group may invite other experts and observers to attend its meetings

*Article 6*

**Consultation**

The Group shall consult extensively and at an early stage with market participants, consumers and end-users in an open and transparent manner.

*Article 7*

**Confidentiality**

Without prejudice to the provisions of Article 287 of the Treaty, where the Commission informs them that the advice requested or the question raised is of a confidential nature, members of the Group as well as observers and any other person shall be under an obligation not to disclose information which has come to their knowledge through the work of the Group, its subgroups or expert groups. The Commission may decide in such cases that only members of the Group may be present at meetings.

*Article 8*

**Annual report**

The Group shall submit an annual report of its activities to the Commission. The Commission shall transmit the report to the European Parliament and to the Council, where appropriate with comments.

*Article 9*

**Entry into force**

This Decision shall enter into force the day of its publication in the *Official Journal of the European Communities*.

([1]) OJ L 198, 27.7.2002, p. 49.
([2]) OJ L 202, 30.7.1997, p. 60.

The Group shall take up its duties on the date of entry into force of this Decision.

Done at Brussels, 29 July 2002.

*For the Commission*

Erkki LIIKANEN

*Member of the Commission*

**Commission guidelines on market analysis and the assessment of significant market power under the Community regulatory framework for electronic communications networks and services**

(2002/C 165/03)

**(Text with EEA relevance)**

## 1. INTRODUCTION

### 1.1. Scope and purpose of the guidelines

1. These guidelines set out the principles for use by national regulatory authorities (NRAs) in the analysis of markets and effective competition under the new regulatory framework for electronic communications networks and services.

2. This new regulatory framework comprises five Directives: Directive 2002/21/EC of the European Parliament and of the Council of 7 March 2002 on a common regulatory framework for electronic communications networks and services [1], hereinafter the framework Directive; Directive 2002/20/EC of the European Parliament and of the Council of 7 March 2002 on the authorisation of electronic communications networks and services [2], hereinafter the authorisation Directive; Directive 2002/19/EC of the European Parliament and of the Council of 7 March 2002 on access to, and interconnection of, electronic communications networks and associated facilities [3], hereinafter the access Directive; Directive 2002/22/EC of the European Parliament and of the Council of 7 March 2002 on universal service and users' rights relating to electronic communications networks and services [4], hereinafter the universal service Directive; a Directive of the European Parliament and of the Council concerning the processing of personal data and the protection of privacy in the electronic communications sector [5]. However, until this last Directive is formally adopted, Directive 97/66/EC of the European Parliament and the Council concerning the processing of personal data and protection of privacy in the telecommunications sector [6], hereinafter the data protection Directive, remains the relevant Directive.

3. Under the 1998 regulatory framework, the market areas of the telecommunications sector that were subject to *ex-ante* regulation were laid down in the relevant directives, but were not markets defined in accordance with the principles of competition law. In these areas defined under the 1998 regulatory framework, NRAs had the power to designate undertakings as having significant market power when they possessed 25 % market share, with the possibility to deviate from this threshold taking into account the undertaking's ability to influence the market, its turnover relative to the size of the market, its control of the means of access to end-users, its access to financial resources and its experience in providing products and services in the market.

4. Under the new regulatory framework, the markets to be regulated are defined in accordance with the principles of European competition law. They are identified by the Commission in its recommendation on relevant product and service markets pursuant to Article 15(1) of the framework Directive (hereinafter 'the Recommendation'). When justified by national circumstances, other markets can also be identified by the NRAs, in accordance with the procedures set out in Articles 6 and 7 of the framework Directive. In case of transnational markets which are susceptible to *ex-ante* regulation, they will where appropriate be identified by the Commission in a decision on relevant transnational markets pursuant to Article 15(4) of the framework Directive (hereinafter 'the Decision on transnational markets').

5. On all of these markets, NRAs will intervene to impose obligations on undertakings only where the markets are considered not to be effectively competitive [7] as a result of such undertakings being in a position equivalent to dominance within the meaning of Article 82 of the EC Treaty [8]. The notion of dominance has been defined in the case-law of the Court of Justice as a position of economic strength affording an undertaking the power to behave to an appreciable extent independently of competitors, customers and ultimately consumers. Therefore, under the new regulatory framework, in contrast with the 1998 framework, the Commission and the NRAs will rely on competition law principles and methodologies to define the markets to be regulated *ex-ante* and to assess whether undertakings have significant market power ('SMP') on those markets.

6. These guidelines are intended to guide NRAs in the exercise of their new responsibilities for defining markets and assessing SMP. They have been adopted by the Commission in accordance with Article 15(2) of the framework Directive, after consultation of the relevant national authorities and following a public consultation, the results of which have been duly taken into account.

7. Under Article 15(3) of the framework Directive, NRAs should take the utmost account of these guidelines. This will be an important factor in any assessment by the Commission of the proportionality and legality of proposed decisions by NRAs, taking into account the policy objectives laid down in Article 8 of the framework Directive.

8. These guidelines specifically address the following subjects: (a) market definition; (b) assessment of SMP; (c) SMP designation; and (d) procedural issues related to all of these subjects.

9. The guidelines have been designed for NRAs to use as follows:

— to define the geographical dimension of those product and service markets identified in the Recommendation. NRAs will not define the geographic scope of any transnational markets, as any Decision on transnational markets will define their geographic dimension,

— to carry out, using the methodology set out in Section 3 of the guidelines, a market analysis of the conditions of competition prevailing in the markets identified in the Recommendation and Decision and by NRAs,

— to identify relevant national or sub-national product and service markets which are not listed in the Recommendation when this is justified by national circumstances and following the procedures set out in Articles 6 and 7 of the framework Directive,

— to designate, following the market analysis, undertakings with SMP in the relevant market and to impose proportionate *ex-ante* measures consistent with the terms of the regulatory framework as described in Sections 3 and 4 of the guidelines,

— to assist Member States and NRAs in applying Article 11(1f) of the authorisation Directive, and Article 5(1) of the framework Directive, and thus ensure that undertakings comply with the obligation to provide information necessary for NRAs to determine relevant markets and assess significant market power thereon,

— to guide NRAs when dealing with confidential information, which is likely to be provided by:

— undertakings under Article 11(1f) of the authorisation Directive and Article 5(1) of the framework Directive,

— national competition authorities (NCAs) as part of the cooperation foreseen in Article 3(5) of the framework Directive, and

— the Commission and a NRA in another Member State as part of the cooperation foreseen in Article 5(2) of the framework Directive.

10. The guidelines are structured in the following way:

**Section 1** provides an introduction and overview of the background, purpose, scope and content of the guidelines. **Section 2** describes the methodology to be used by NRAs to define the geographic scope of the markets identified in the market Recommendation as well as to define relevant markets outside this Recommendation. **Section 3** describes the criteria for assessing SMP in a relevant market. **Section 4** outlines the possible conclusions that NRAs may reach in their market analyses and describes the possible actions that may result. **Section 5** describes the powers of investigation of NRAs, suggests procedures for coordination between NRAs and between NRAs and NCAs, and describes coordination and cooperation procedures between NRAs and the Commission. Finally, **Section 6** describes procedures for public consultation and publication of NRAs' proposed decisions.

11. The major objective of these guidelines is to ensure that NRAs use a consistent approach in applying the new regulatory framework, and especially when designating undertakings with SMP in application of the provisions of the regulatory framework.

12. By issuing these guidelines, the Commission also intends to explain to interested parties and undertakings operating in the electronic communications sector how NRAs should undertake their assessments of SMP under the framework Directive, thereby maximising the transparency and legal certainty of the application of the sector specific legislation.

13. The Commission will amend these guidelines, whenever appropriate, taking into account experience with the application of the regulatory framework and future developments in the jurisprudence of the Court of First Instance and the European Court of Justice.

14. These guidelines do not in any way restrict the rights conferred by Community law on individuals or undertakings. They are entirely without prejudice to the application of Community law, and in particular of the competition rules, by the Commission and the relevant national authorities, and to its interpretation by the European Court of Justice and the Court of First Instance. These guidelines do not prejudice any action the Commission may take or any guidelines the Commission may issue in the future with regard to the application of European competition law.

### 1.2. Principles and policy objectives behind sector specific measures

15. NRAs must seek to achieve the policy objectives identified in Article 8(2), (3) and (4) of the framework Directive. These fall into three categories:

— promotion of an open and competitive market for electronic communications networks, services and associated facilities,

— development of the internal market, and

— promotion of the interests of European citizens.

C 165/8 EN Official Journal of the European Communities 11.7.2002

16. The purpose of imposing *ex-ante* obligations on undertakings designated as having SMP is to ensure that undertakings cannot use their market power either to restrict or distort competition on the relevant market, or to leverage such market power onto adjacent markets.

17. These regulatory obligations should only be imposed on those electronic communications markets whose characteristics may be such as to justify sector-specific regulation and in which the relevant NRA has determined that one or more operators have SMP.

18. The product and service markets whose characteristics may be such as to justify sector-specific regulation are identified by the Commission in its Recommendation and, when the definition of different relevant markets is justified by national circumstances, by the NRAs following the procedures set out in Articles 6 and 7 of the framework Directive (9). In addition, certain other markets are specifically identified in Article 6 of the access Directive and Articles 18 and 19 of the universal service Directive.

19. In respect of each of these relevant markets, NRAs will assess whether the competition is effective. A finding that effective competition exists on a relevant market is equivalent to a finding that no operator enjoys a single or joint dominant position on that market. Therefore, for the purposes of applying the new regulatory framework, effective competition means that there is no undertaking in the relevant market which holds alone or together with other undertakings a single or collective dominant position. When NRAs conclude that a relevant market is not effectively competitive, they will designate undertakings with SMP on that market, and will either impose appropriate specific obligations, or maintain or amend such obligations where they already exist, in accordance with Article 16(4) of the framework Directive.

20. In carrying out the market analysis under the terms of Article 16 of the framework Directive, NRAs will conduct a forward looking, structural evaluation of the relevant market, based on existing market conditions. NRAs should determine whether the market is prospectively competitive, and thus whether any lack of effective competition is durable (10), by taking into account expected or foreseeable market developments over the course of a reasonable period. The actual period used should reflect the specific characteristics of the market and the expected timing for the next review of the relevant market by the NRA. NRAs should take past data into account in their analysis when such data are relevant to the developments in that market in the foreseeable future.

21. If NRAs designate undertakings as having SMP, they must impose on them one or more regulatory obligations, in accordance with the relevant Directives and taking into account the principle of proportionality. Exceptionally, NRAs may impose obligations for access and interconnection that go beyond those specified in the access Directive, provided this is done with the prior agreement of the Commission, as provided by Article 8(3) of that Directive.

22. In the exercise of their regulatory tasks under Article 15 and 16 of the framework Directive, NRAs enjoy discretionary powers which reflect the complexity of all the relevant factors that must be assessed (economic, factual and legal) when identifying the relevant market and determining the existence of undertakings with SMP. These discretionary powers remain subject, however, to the procedures provided for in Article 6 and 7 of the framework Directive.

23. Regulatory decisions adopted by NRAs pursuant to the Directives will have an impact on the development of the internal market. In order to prevent any adverse effects on the functioning of the internal market, NRAs must ensure that they implement the provisions to which these guidelines apply in a consistent manner. Such consistency can only be achieved by close coordination and cooperation with other NRAs, with NCAs and with the Commission, as provided in the framework Directive and as recommended in Section 5.3 of these guidelines.

### 1.3. **Relationship with competition law**

24. Under the regulatory framework, markets will be defined and SMP will be assessed using the same methodologies as under competition law. Therefore the definition of the geographic scope of markets identified in the Recommendation, the definition where necessary of relevant product/services markets outside the Recommendation, and the assessment of effective competition by NRAs should be consistent with competition case-law and practice. To ensure such consistency, these guidelines are based on (1) existing case-law of the Court of First Instance and the European Court of Justice concerning market definition and the notion of dominant position within the meaning of Article 82 of the EC Treaty and Article 2 of the merger control Regulation (11); (2) the 'Guidelines on the application of EEC competition rules in the telecommunications sector' (12); (3) the 'Commission notice on the definition of relevant markets for the purposes of Community competition law' (13), hereinafter the 'Notice on market definition'; and (4) the 'Notice on the application of competition rules to access agreements in the telecommunications sector' (14), hereinafter the 'Access notice'.

25. The use of the same methodologies ensures that the relevant market defined for the purpose of sector-specific regulation will in most cases correspond to the market definitions that would apply under competition law. In some cases, and for the reasons set out in Section 2 of these guidelines, markets defined by the Commission and competition authorities in competition cases may differ from those identified in the Recommendation and Decision, and/or from markets defined by NRAs under Article 15(3) of the framework Directive. Article 15(1) of the framework Directive makes clear that the markets to be defined by NRAs for the purpose of *ex-ante* regulation are without prejudice to those defined by NCAs and by the Commission in the exercise of their respective powers under competition law in specific cases.

26. For the purposes of the application of Community competition law, the Commission's Notice on market definition explains that the concept of the relevant market is closely linked to the objectives pursued under Community policies. Markets defined under Articles 81 and 82 EC Treaty are generally defined on an ex-post basis. In these cases, the analysis will consider events that have already taken place in the market and will not be influenced by possible future developments. Conversely, under the merger control provisions of EC competition law, markets are generally defined on a forward-looking basis.

27. On the other hand, relevant markets defined for the purposes of sector-specific regulation will always be assessed on a forward looking basis, as the NRA will include in its assessment an appreciation of the future development of the market. However, NRAs' market analyses should not ignore, where relevant, past evidence when assessing the future prospects of the relevant market (see also Section 2, below). The starting point for carrying out a market analysis for the purpose of Article 15 of the framework Directive is not the existence of an agreement or concerted practice within the scope of Article 81 EC Treaty, nor a concentration within the scope of the Merger Regulation, nor an alleged abuse of dominance within the scope of Article 82 EC Treaty, but is based on an overall forward-looking assessment of the structure and the functioning of the market under examination. Although NRAs and competition authorities, when examining the same issues in the same circumstances and with the same objectives, should in principle reach the same conclusions, it cannot be excluded that, given the differences outlined above, and in particular the broader focus of the NRAs' assessment, markets defined for the purposes of competition law and markets defined for the purpose of sector-specific regulation may not always be identical.

28. Although merger analysis is also applied *ex ante*, it is not carried out periodically as is the case with the analysis of the NRAs under the new regulatory framework. A competition authority does not, in principle, have the opportunity to conduct a periodic review of its decision in the light of market developments, whereas NRAs are bound to review their decisions periodically under Article 16(1) of the framework Directive. This factor can influence the scope and breadth of the market analysis and the competitive assessment carried out by NRAs, and for this reason, market definitions under the new regulatory framework, even in similar areas, may in some cases, be different from those markets defined by competition authorities.

29. It is considered that markets which are not identified in the Recommendation will not warrant *ex-ante* sector specific regulation, except where the NRA is able to justify such regulation of an additional or different relevant market in accordance with the procedure in Article 7 of the framework Directive.

30. The designation of an undertaking as having SMP in a market identified for the purpose of *ex-ante* regulation does not automatically imply that this undertaking is also dominant for the purpose of Article 82 EC Treaty or similar national provisions. Moreover, the SMP designation has no bearing on whether that undertaking has committed an abuse of a dominant position within the meaning of Article 82 of the EC Treaty or national competition laws. It merely implies that, from a structural perspective, and in the short to medium term, the operator has and will have, on the relevant market identified, sufficient market power to behave to an appreciable extent independently of competitors, customers, and ultimately consumers, and this, solely for purposes of Article 14 of the framework Directive.

31. In practice, it cannot be excluded that parallel procedures under *ex-ante* regulation and competition law may arise with respect to different kinds of problems in relevant markets [15]. Competition authorities may therefore carry out their own market analysis and impose appropriate competition law remedies alongside any sector specific measures applied by NRAs. However, it must be noted that such simultaneous application of remedies by different regulators would address different problems in such markets. *Ex-ante* obligations imposed by NRAs on undertakings with SMP aim to fulfil the specific objectives set out in the relevant directives, whereas competition law remedies aim to sanction agreements or abusive behaviour which restrict or distort competition in the relevant market.

32. As far as emerging markets are concerned, recital 27 of the framework Directive notes that emerging markets, where *de facto* the market leader is likely to have a substantial market share, should not be subject to inappropriate *ex-ante* regulation. This is because premature imposition of *ex-ante* regulation may unduly influence the competitive conditions taking shape within a new and emerging market. At the same time, foreclosure of such emerging markets by the leading undertaking should be prevented. Without prejudice to the appropriateness of intervention by the competition authorities in individual cases, NRAs should ensure that they can fully justify any form of early, *ex-ante* intervention in an emerging market, in particular since they retain the ability to intervene at a later stage, in the context of the periodic re-assessment of the relevant markets.

## 2. MARKET DEFINITION

### 2.1. Introduction

33. In the Competition guidelines issued in 1991 [16], the Commission recognised the difficulties inherent in defining the relevant market in an area of rapid technological change, such as the telecommunications sector. Whilst this statement still holds true today as far as the electronic communications sector is concerned, the Commission since the publication of those guidelines has gained considerable experience in applying the competition rules in a dynamic sector shaped by constant technological changes and innovation, as a result of its role in managing the transition from monopoly to competition in this sector. It should however be recalled that the present guidelines do not purport to explain how the competition rules apply, generally, in the electronic communications sector, but focus only on issues related to (i) market definition; and (ii) the assessment of significant market power within the meaning of Article 14 of the framework Directive (hereafter SMP).

34. In assessing whether an undertaking has SMP, that is whether it 'enjoys a position of economic strength affording it the power to behave to an appreciable extent independently of its competitors, customers and ultimately consumers' [17], the definition of the relevant market is of fundamental importance since effective competition can only be assessed by reference to the market thus defined [18]. The use of the term 'relevant market' implies the description of the products or services that make up the market and the assessment of the geographical scope of that market (the terms 'products' and 'services' are used interchangeably throughout this text). In that regard, it should be recalled that relevant markets defined under the 1998 regulatory framework were distinct from those identified for competition-law purposes, since they were based on certain specific aspects of end-to-end communications rather than on the demand and supply criteria used in a competition law analysis [19].

35. Market definition is not a mechanical or abstract process but requires an analysis of any available evidence of past market behaviour and an overall understanding of the mechanics of a given sector. In particular, a dynamic rather than a static approach is required when carrying out a prospective, or forward-looking, market analysis [20]. In this respect, any experience gained by NRAs, NCAs and the Commission through the application of competition rules to the telecommunication sector clearly will be of particular relevance in applying Article 15 of the framework Directive. Thus, any information gathered, any findings made and any studies or reports commissioned or relied upon by NRAs (or NCAs) in the exercise of their tasks, in relation to the conditions of competition in the telecommunications markets (provided of course that market conditions have since remained unchanged), should serve as a starting point for the purposes of applying Article 15 of the framework Directive and carrying out a prospective market analysis [21].

36. The main product and service markets whose characteristics may be such as to justify the imposition of *ex-ante* regulatory obligations are identified in the Recommendation which the Commission is required to adopt pursuant to Article 15(1) of the framework Directive, as well as any Decision on transnational markets which the Commission decides to adopt pursuant to Article 15(4) of the framework Directive. Therefore, in practice the task of NRAs will normally be to define the geographical scope of the relevant market, although NRAs have the possibility under Article 15(3) of the framework Directive to define markets other than those listed in the Recommendation in accordance with Article 7 of the framework Directive (see below, Section 6).

37. Whilst a prospective analysis of market conditions may in some cases lead to a market definition different from that resulting from a market analysis based on past behaviour [22], NRAs should nonetheless seek to preserve, where possible, consistency in the methodology adopted between, on the one hand, market definitions developed for the purposes of *ex-ante* regulation, and on the other hand, market definitions developed for the purposes of the application of the competition rules. Nevertheless, as stated in Article 15(1) of the framework Directive and Section 1 of the guidelines, markets defined under sector-specific regulation are defined without prejudice to markets that may be defined in specific cases under competition law.

### 2.2. Main criteria for defining the relevant market

38. The extent to which the supply of a product or the provision of a service in a given geographical area constitutes the relevant market depends on the existence of competitive constraints on the price-setting behaviour of the producer(s) or service provider(s)

concerned. There are two main competitive constraints to consider in assessing the behaviour of undertakings on the market, (i) demand-side; and (ii) supply-side substitution. A third source of competitive constraint on an operator's behaviour exists, namely potential competition. The difference between potential competition and supply-substitution lies in the fact that supply-side substitution responds promptly to a price increase whereas potential entrants may need more time before starting to supply the market. Supply substitution involves no additional significant costs whereas potential entry occurs at significant sunk costs ([23]). The existence of potential competition should thus be examined for the purpose of assessing whether a market is effectively competitive within the meaning of the framework Directive, that is whether there exist undertakings with SMP ([24]).

39. Demand-side substitutability is used to measure the extent to which consumers are prepared to substitute other services or products for the service or product in question ([25]), whereas supply-side substitutability indicates whether suppliers other than those offering the product or services in question would switch in the immediate to short term their line of production or offer the relevant products or services without incurring significant additional costs.

40. One possible way of assessing the existence of any demand and supply-side substitution is to apply the so-called 'hypothetical monopolist test' ([26]). Under this test, an NRA should ask what would happen if there were a small but significant, lasting increase in the price of a given product or service, assuming that the prices of all other products or services remain constant (hereafter, 'relative price increase'). While the significance of a price increase will depend on each individual case, in practice, NRAs should normally consider customers' (consumers or undertakings) reactions to a permanent price increase of between 5 to 10 % ([27]). The responses by consumers or undertakings concerned will aid in determining whether substitutable products do exist and, if so, where the boundaries of the relevant product market should be delineated ([28]).

41. As a starting point, an NRA should apply this test firstly to an electronic communications service or product offered in a given geographical area, the characteristics of which may be such as to justify the imposition of regulatory obligations, and having done so, add additional products or areas depending on whether competition from those products or areas constrains the price of the main product or service in question. Since a relative price increase of a set of products ([29]) is likely to lead to some sales being lost, the key issue is to determine whether the loss of sales would be sufficient to offset the increased profits which would otherwise be made from sales made following the price increase. Assessing the demand-side and supply-side substitution provides a way of measuring the quantity of the sales likely to be lost and consequently of determining the scope of the relevant market.

42. In principle, the 'hypothetical monopolist test' is relevant only with regard to products or services, the price of which is freely determined and not subject to regulation. Thus, the working assumption will be that current prevailing prices are set at competitive levels. If, however, a service or product is offered at a regulated, cost-based price, then such price is presumed, in the absence of indications to the contrary, to be set at what would otherwise be a competitive level and should therefore be taken as the starting point for applying the 'hypothetical monopolist test' ([30]). In theory, if the demand elasticity of a given product or service is significant, even at relative competitive prices, the firm in question lacks market power. If, however, elasticity is high even at current prices, that may mean only that the firm in question has already exercised market power to the point that further price increases will not increase its profits. In this case, the application of the hypothetical monopoly test may lead to a different market definition from that which would be produced if the prices were set at a competitive level ([31]). Any assessment of market definition must therefore take into account this potential difficulty. However, NRAs should proceed on the basis that the prevailing price levels provide a reasonable basis from which to start the relevant analysis unless there is evidence that this is not in fact the case.

43. If an NRA chooses to have recourse to the hypothetical monopolist test, it should then apply this test up to the point where it can be established that a relative price increase within the geographic and product markets defined will not lead consumers to switch to readily available substitutes or to suppliers located in other areas.

2.2.1. *The relevant product/service market*

44. According to settled case-law, the relevant product/service market comprises all those products or services that are sufficiently interchangeable or substitutable, not only in terms of their objective characteristics, by virtue of which they are particularly suitable for satisfying the constant needs of consumers, their prices or their intended use, but also in terms of the conditions of competition and/or the structure of supply and demand on the market in question ([32]). Products or services which are only to a small, or relative degree interchangeable with each other do not form part of the same market ([33]). NRAs should thus commence the exercise of defining the relevant product or service market by grouping together products or services that are used by consumers for the same purposes (end use).

45. Although the aspect of the end use of a product or service is closely related to its physical characteristics, different kind of products or services may be used for the same end. For instance, consumers may use dissimilar services such as cable and satellite connections for the same purpose, namely to access the Internet. In such a case, both services (cable and satellite access services) may be included in the same product market. Conversely, paging services and mobile telephony services, which may appear to be capable of offering the same service, that is, dispatching of two-way short messages, may be found to belong to distinct product markets in view of their different perceptions by consumers as regards their functionality and end use.

46. Differences in pricing models and offerings for a given product or service may also imply different groups of consumers. Thus, by looking into prices, NRAs may define separate markets for business and residential customers for essentially the same service. For instance, the ability of operators engaged in providing international retail electronic communications services to discriminate between residential and business customers, by applying different sets of prices and discounts, has led the Commission to decide that these two groups form separate markets as far as such services are concerned (see below). However, in order for products to be viewed as demand-side substitutes it is not necessary that they are offered at the same price. A low quality product or service sold at a low price could well be an effective substitute to a higher quality product sold at higher prices. What matters in this case is the likely responses of consumers following a relative price increase ([34]).

47. Furthermore, product substitutability between different electronic communications services will arise increasingly through the convergence of various technologies. Use of digital systems leads to an increasing similarity in the performance and characteristics of network services using distinct technologies. A packet-switched network, for instance, such as Internet, may be used to transmit digitised voice signals in competition with traditional voice telephony services ([35]).

48. In order, therefore, to complete the market-definition analysis, an NRA, in addition to considering products or services whose objective characteristics, prices and intended use make them sufficiently interchangeable, should also examine, where necessary, the prevailing conditions of demand and supply substitution by applying the hypothetical monopolist test.

2.2.1.1. Demand-side substitution

49. Demand-side substitution enables NRAs to determine the substitutable products or range of products to which consumers could easily switch in case of a relative price increase. In determining the existence of demand substitutability, NRAs should make use of any previous evidence of consumers' behaviour. Where available, an NRA should examine historical price fluctuations in potentially competing products, any records of price movements, and relevant tariff information. In such circumstances evidence showing that consumers have in the past promptly shifted to other products or services, in response to past price changes, should be given appropriate consideration. In the absence of such records, and where necessary, NRAs will have to seek and assess the likely response of consumers and suppliers to a relative price increase of the service in question.

50. The possibility for consumers to substitute a product or a service for another because of a small, but significant lasting price increase may, however, be hindered by considerable switching costs. Consumers who have invested in technology or made any other necessary investments in order to receive a service or use a product may be unwilling to incur any additional costs involved in switching to an otherwise substitutable service or product. In the same vein, customers of existing providers may also be 'locked in' by long-term contracts or by the prohibitively high cost of switching terminals. Accordingly, in a situation where end users face significant switching costs in order to substitute product A for product B, these two products should not be included in the same relevant market ([36]).

51. Demand substitutability focuses on the interchangeable character of products or services from the buyer's point of view. Proper delineation of the product market may, however, require further consideration of potential substitutability from the supply side.

2.2.1.2. Supply-side substitution

52. In assessing the scope for supply substitution, NRAs may also take into account the likelihood that undertakings not currently active on the relevant product market may decide to enter the market, within a reasonable time frame ([37]), following a relative price increase, that is, a small but significant, lasting price increase. In circumstances where the overall costs of switching production to the product in question are relatively negligible, then that product may be incorporated into the product market definition. The fact that a rival firm possesses some of the assets required to provide a given service is immaterial if significant additional investment is needed to market and offer profitably the services in question ([38]). Furthermore, NRAs will need to ascertain whether a given supplier would actually use or switch its productive assets to produce the relevant product or offer the relevant service (for instance, whether their capacity is committed under long-term supply agreements, etc.). Mere hypothetical supply-side substitution is not sufficient for the purposes of market definition.

53. Account should also be taken of any existing legal, statutory or other regulatory requirements which could defeat a time-efficient entry into the relevant market and as a result discourage supply-side substitution. For instance, delays and obstacles in concluding interconnection or co-location agreements, negotiating any other form of network access, or obtaining rights of ways for network expansion (39), may render unlikely in the short term the provision of new services and the deployment of new networks by potential competitors.

54. As can been seen from the above considerations, supply substitution may serve not only for defining the relevant market but also for identifying the number of market participants.

2.2.2. *Geographic market*

55. Once the relevant product market is identified, the next step to be undertaken is the definition of the geographical dimension of the market. It is only when the geographical dimension of the product or service market has been defined that a NRA may properly assess the conditions of effective competition therein.

56. According to established case-law, the relevant geographic market comprises an area in which the undertakings concerned are involved in the supply and demand of the relevant products or services, in which area the conditions of competition are similar or sufficiently homogeneous and which can be distinguished from neighbouring areas in which the prevailing conditions of competition are appreciably different (40). The definition of the geographic market does not require the conditions of competition between traders or providers of services to be perfectly homogeneous. It is sufficient that they are similar or sufficiently homogeneous, and accordingly, only those areas in which the conditions of competition are 'heterogeneous' may not be considered to constitute a uniform market (41).

57. The process of defining the limits of the geographic market proceeds along the same lines as those discussed above in relation to the assessment of the demand and supply-side substitution in response to a relative price increase.

58. Accordingly, with regard to demand-side substitution, NRAs should assess mainly consumers' preferences as well as their current geographic patterns of purchase. In particular, linguistic reasons may explain why certain services are not available or marketed in different language areas. As far as supply-side substitution is concerned, where it can be established that operators which are not currently engaged or present on the relevant market, will, however, decide to enter that market in the short term in the event of a relative price increase, then the market definition should be expanded to incorporate those 'outside' operators.

59. In the electronic communications sector, the geographical scope of the relevant market has traditionally been determined by reference to two main criteria (42):

(a) the area covered by a network (43); and

(b) the existence of legal and other regulatory instruments (44).

60. On the basis of these two main criteria (45), geographic markets can be considered to be local, regional, national or covering territories of two or more countries (for instance, pan-European, EEA-wide or global markets).

2.2.3. *Other issues of market definition*

61. For the purposes of *ex-ante* regulation, in certain exceptional cases, the relevant market may be defined on a route-by-route basis. In particular, when considering the dimension of markets for international retail or wholesale electronic communications services, it may be appropriate to treat paired countries or paired cities as separate markets (46). Clearly, from the demand side, the delivery of a call to one country is not a substitute for the delivery of the same to another country. On the other hand, the question of whether indirect transmission services, that is, re-routing or transit of the same call via a third country, represent effective supply-side substitutes depends on the specificities of the market and should be decided on a case-by-case basis (47). However, a market for the provision of services on a bilateral route would be national in scope since supply and demand patterns in both ends of the route would most likely correspond to different market structures (48).

62. In its Notice on market definition, the Commission drew attention to certain cases where the boundaries of the relevant market may be expanded to take into consideration products or geographical areas which, although not directly substitutable, should be included in the market definition because of so-called 'chain substitutability' (49). In essence, chain substitutability occurs where it can be demonstrated that although products A and C are not directly substitutable, product B is a substitute for both product A and product C and therefore products A and C may be in the same product market since their pricing might be constrained by the substitutability of product B. The same reasoning also applies for defining the geographic market. Given the inherent risk of unduly widening the scope of the relevant market, findings of chain substitutability should be adequately substantiated (50).

C 165/14 EN Official Journal of the European Communities 11.7.2002

### 2.3. **The Commission's own practice**

63. The Commission has adopted a number of decisions under Regulation No 17 and the merger control Regulation relating to the electronic communications sector. These decisions may be of particular relevance for NRAs with regard to the methodology applied by the Commission in defining the relevant market (51). As stated above, however, in a sector characterised by constant innovation and rapid technological convergence, it is clear that any current market definition runs the risk of becoming inaccurate or irrelevant in the near future (52). Furthermore, markets defined under competition law are without prejudice to markets defined under the new regulatory framework as the context and the timeframe within which a market analysis is conducted may be different (53).

64. As stated in the Access notice, there are in the electronic communications sector at least two main types of relevant markets to consider, that of services provided to end users (services market) and that of access to facilities necessary to provide such services (access market) (54). Within these two broad market definitions further market distinctions may be made depending on demand and supply side patterns.

65. In particular, in its decision-making practice, the Commission will normally make a distinction between the provision of services and the provision of underlying network infrastructure. For instance, as regards the provision of infrastructure, the Commission has identified separate markets for the provision of local loop, long distance and international infrastructure (55). As regards fixed services, the Commission has distinguished between subscriber (retail) access to switched voice telephony services (local, long distance and international), operator (wholesale) access to networks (local, long distance and international) and business data communications services (56). In the market for fixed telephony retail services, the Commission has also distinguished between the initial connection and the monthly rental (57). Retail services are offered to two distinct classes of consumers, namely, residential and business users, the latter possibly being broken down further into a market for professional, small and medium sized business customers and another for large businesses (58). With regard to fixed telephony retail services offered to residential users, demand and supply patterns seem to indicate that two main types of services are currently being offered, traditional fixed telephony services (voice and narrowband data transmissions) on the one hand, and high speed communications services (currently in the form of xDSL services) on the other hand (59).

66. As regards the provision of mobile communications services, the Commission has found that, from a demand-side point of view, mobile telephony services and fixed telephony services constitute separate markets (60). Within the mobile market, evidence gathered from the Commission has indicated that the market for mobile communications services encompasses both GSM 900 and GSM 1800 and possibly analogue platforms (61).

67. The Commission has found that with regard to the 'access' market, the latter comprises all types of infrastructure that can be used for the provision of a given service (62). Whether the market for network infrastructures should be divided into as many separate submarkets as there are existing categories of network infrastructure, depends clearly on the degree of substitutability among such (alternative) networks (63). This exercise should be carried out in relation to the class of users to which access to the network is provided. A distinction should, therefore, be made between provision of infrastructure to other operators (wholesale level) and provision to end users (retail level) (64). At the retail level, a further segmentation may take place between business and residential customers (65).

68. When the service to be provided concerns only end users subscribed to a particular network, access to the termination points of that network may well constitute the relevant product market. This will not be the case if it can be established that the same services may be offered to the same class of consumers by means of alternative, easily accessible competing networks. For example, in its Communication on unbundling the local loop (66), the Commission stated that although alternatives to the PSTN for providing high speed communications services to residential consumers exist (fibre optic networks, wireless local loops or upgradable TV networks), none of these alternatives may be considered as a substitute to the fixed local loop infrastructure (67). Future innovative and technological changes may, however, justify different conclusions (68).

69. Access to mobile networks may also be defined by reference to two potentially separate markets, one for call origination and another for call termination. In this respect, the question whether the access market to mobile infrastructure relates to access to an individual mobile network or to all mobile networks, in general, should be decided on the basis of an analysis of the structure and functioning of the market (69).

## 3. **ASSESSING SIGNIFICANT MARKET POWER (DOMINANCE)**

70. According to Article 14 of the framework Directive 'an undertaking shall be deemed to have significant market power if, either individually or jointly with others, it enjoys a position equivalent to dominance, that is to say a position of economic strength affording it the

power to behave to an appreciable extent independently of competitors customers and ultimately consumers'. This is the definition that the Court of Justice case-law ascribes to the concept of dominant position in Article 82 of the Treaty ([70]). The new framework has aligned the definition of SMP with the Court's definition of dominance within the meaning of Article 82 of the Treaty ([71]). Consequently, in applying the new definition of SMP, NRAs will have to ensure that their decisions are in accordance with the Commission's practice and the relevant jurisprudence of the Court of Justice and the Court of First Instance on dominance ([72]). However, the application of the new definition of SMP, *ex-ante*, calls for certain methodological adjustments to be made regarding the way market power is assessed. In particular, when assessing *ex-ante* whether one or more undertakings are in a dominant position in the relevant market, NRAs are, in principle, relying on different sets of assumptions and expectations than those relied upon by a competition authority applying Article 82, *ex post*, within a context of an alleged committed abuse ([73]). Often, the lack of evidence or of records of past behaviour or conduct will mean that the market analysis will have to be based mainly on a prospective assessment. The accuracy of the market analysis carried out by NRAs will thus be conditioned by information and data existing at the time of the adoption of the relevant decision.

71. The fact that an NRA's initial market predictions do not finally materialise in a given case does not necessarily mean that its decision at the time of its adoption was inconsistent with the Directive. In applying *ex ante* the concept of dominance, NRAs must be accorded discretionary powers correlative to the complex character of the economic, factual and legal situations that will need to be assessed. In accordance with the framework Directive, market assessments by NRAs will have to be undertaken on a regular basis. In this context, therefore, NRAs will have the possibility to react at regular intervals to any market developments and to take any measure deemed necessary.

### 3.1. Criteria for assessing SMP

72. As the Court has stressed, a finding of a dominant position does not preclude some competition in the market. It only enables the undertaking that enjoys such a position, if not to determine, at least to have an appreciable effect on the conditions under which that competition will develop, and in any case to act in disregard of any such competitive constraint so long as such conduct does not operate to its detriment ([74]).

73. In an *ex-post* analysis, a competition authority may be faced with a number of different examples of market behaviour each indicative of market power within the meaning of Article 82. However, in an *ex-ante* environment, market power is essentially measured by reference of the power of the undertaking concerned to raise prices by restricting output without incurring a significant loss of sales or revenues.

74. The market power of an undertaking can be constrained by the existence of potential competitors ([75]). An NRA should thus take into account the likelihood that undertakings not currently active on the relevant product market may in the medium term decide to enter the market following a small but significant non-transitory price increase. Undertakings which, in case of such a price increase, are in a position to switch or extend their line of production/services and enter the market should be treated by NRAs as potential market participants even if they do not currently produce the relevant product or offer the relevant service.

75. As explained in the paragraphs below, a dominant position is found by reference to a number of criteria and its assessment is based, as stated above, on a forward-looking market analysis based on existing market conditions. Market shares are often used as a proxy for market power. Although a high market share alone is not sufficient to establish the possession of significant market power (dominance), it is unlikely that a firm without a significant share of the relevant market would be in a dominant position. Thus, undertakings with market shares of no more than 25 % are not likely to enjoy a (single) dominant position on the market concerned ([76]). In the Commission's decision-making practice, single dominance concerns normally arise in the case of undertakings with market shares of over 40 %, although the Commission may in some cases have concerns about dominance even with lower market shares ([77]), as dominance may occur without the existence of a large market share. According to established case-law, very large market shares — in excess of 50 % — are in themselves, save in exceptional circumstances, evidence of the existence of a dominant position ([78]). An undertaking with a large market share may be presumed to have SMP, that is, to be in a dominant position, if its market share has remained stable over time ([79]). The fact that an undertaking with a significant position on the market is gradually losing market share may well indicate that the market is becoming more competitive, but it does not preclude a finding of significant market power. On the other hand, fluctuating market shares over time may be indicative of a lack of market power in the relevant market.

76. As regards the methods used for measuring market size and market shares, both volume sales and value sales provide useful information for market measurement ([80]). In the case of bulk products preference is given to volume whereas in the case of differentiated products (i.e. branded products) sales in value and their associated market share will often be considered to reflect better the relative position and strength of each provider. In bidding markets the number of bids won and lost may also be used as approximation of market shares ([81]).

77. The criteria to be used to measure the market share of the undertaking(s) concerned will depend on the characteristics of the relevant market. It is for NRAs to decide which are the criteria most appropriate for measuring market presence. For instance, leased lines revenues, leased capacity or numbers of leased line termination points are possible criteria for measuring an undertaking's relative strength on leased lines markets. As the Commission has indicated, the mere number of leased line termination points does not take into account the different types of leased lines that are available on the market — ranging from analogue voice quality to high-speed digital leased lines, short distance to long distance international leased lines. Of the two criteria, leased lines revenues may be more transparent and less complicated to measure. Likewise, retail revenues, call minutes or numbers of fixed telephone lines or subscribers of public telephone network operators are possible criteria for measuring the market shares of undertakings operating in these markets (82). Where the market defined is that of interconnection, a more realistic measurement parameter would be the revenues accrued for terminating calls to customers on fixed or mobile networks. This is so because the use of revenues, rather than for example call minutes, takes account of the fact that call minutes can have different values (i.e. local, long distance and international) and provides a measure of market presence that reflects both the number of customers and network coverage (83). For the same reasons, the use of revenues for terminating calls to customers of mobile networks may be the most appropriate means to measure the market presence of mobile network operators (84).

78. It is important to stress that the existence of a dominant position cannot be established on the sole basis of large market shares. As mentioned above, the existence of high market shares simply means that the operator concerned might be in a dominant position. Therefore, NRAs should undertake a thorough and overall analysis of the economic characteristics of the relevant market before coming to a conclusion as to the existence of significant market power. In that regard, the following criteria can also be used to measure the power of an undertaking to behave to an appreciable extent independently of its competitors, customers and consumers. These criteria include amongst others:

— overall size of the undertaking,

— control of infrastructure not easily duplicated,

— technological advantages or superiority,

— absence of or low countervailing buying power,

— easy or privileged access to capital markets/financial resources,

— product/services diversification (e.g. bundled products or services),

— economies of scale,

— economies of scope,

— vertical integration,

— a highly developed distribution and sales network,

— absence of potential competition,

— barriers to expansion.

79. A dominant position can derive from a combination of the above criteria, which taken separately may not necessarily be determinative.

80. A finding of dominance depends on an assessment of ease of market entry. In fact, the absence of barriers to entry deters, in principle, independent anti-competitive behaviour by an undertaking with a significant market share. In the electronic communications sector, barriers to entry are often high because of existing legislative and other regulatory requirements which may limit the number of available licences or the provision of certain services (i.e. GSM/DCS or 3G mobile services). Furthermore, barriers to entry exist where entry into the relevant market requires large investments and the programming of capacities over a long time in order to be profitable (85). However, high barriers to entry may become less relevant with regard to markets characterised by on-going technological progress. In electronic communications markets, competitive constraints may come from innovative threats from potential competitors that are not currently in the market. In such markets, the competitive assessment should be based on a prospective, forward-looking approach.

81. As regards the relevance of the notion of 'essential facilities' for the purposes of applying the new definition of SMP, there is for the moment no jurisprudence in relation to the electronic communications sector. However, this notion, which is mainly relevant with regard to the existence of an abuse of a dominant position under Article 82 of the EC Treaty, is less relevant with regard to the *ex-ante* assessment of SMP within the meaning of Article 14 of the framework Directive. In particular, the doctrine of 'essential facilities' is complementary to existing general obligations imposed on dominant undertaking, such as the obligation not to discriminate among customers and has been applied in cases under Article 82 in exceptional circumstances, such as where the refusal to supply or to grant access to third parties would limit or prevent the emergence of new markets, or new products, contrary to Article 82(b) of the Treaty. It has thus primarily been associated with

access issues or cases involving a refusal to supply or to deal under Article 82 of the Treaty, without the presence of any discriminatory treatment. Under existing case-law, a product or service cannot be considered 'necessary' or 'essential' unless there is no real or potential substitute. Whilst it is true that an undertaking which is in possession of an 'essential facility' is by definition in a dominant position on any market for that facility, the contrary is not always true. The fact that a given facility is not 'essential' or 'indispensable' for an economic activity on some distinct market, within the meaning of the existing case-law ([86]) does not mean that the owner of this facility might not be in a dominant position. For instance, a network operator can be in a dominant position despite the existence of alternative competing networks if the size or importance of its network affords him the possibility to behave independently from other network operators ([87]). In other words, what matters is to establish whether a given facility affords its owner significant market power in the market without thus being necessary to further establish that the said facility can also be considered 'essential' or 'indispensable' within the meaning of existing case-law.

82. It follows from the foregoing that the doctrine of the 'essential facilities' is less relevant for the purposes of applying *ex ante* Article 14 of the framework Directive than applying ex-post Article 82 of the EC Treaty.

3.1.1. *Leverage of market power*

83. According to Article 14(3) of the framework Directive, 'where an undertaking has significant market power on a specific market, it may also be deemed to have significant market power on a closely related market, where the links between the two markets are such as to allow the market power held in one market to be leveraged into the other market, thereby strengthening the market power of the undertaking'.

84. This provision is intended to address a market situation comparable to the one that gave rise to the Court's judgment in *Tetra Pak II* ([88]). In that case, the Court decided that an undertaking that had a dominant position in one market, and enjoyed a leading position on a distinct but closely associated market, was placed as a result in a situation comparable to that of holding a dominant position on the markets in question taken as a whole. Thanks to its dominant position on the first market, and its market presence on the associated, secondary market, an undertaking may thus leverage the market power which it enjoys in the first market and behave independently of its customers on the latter market ([89]). Although in Tetra Pak the markets taken as a whole in which Tetra Pak was found to be dominant were horizontal, close associative links, within the meaning of the Court's case-law, will most often be found in vertically integrated markets. This is often the case in the telecommunications sector, where an operator often has a dominant position on the infrastructure market and a significant presence on the downstream, services market ([90]). Under such circumstances, an NRA may consider it appropriate to find that such operator has SMP on both markets taken together. However, in practice, if an undertaking has been designated as having SMP on an upstream wholesale or access market, NRAs will normally be in a position to prevent any likely spill-over or leverage effects downstream into the retail or services markets by imposing on that undertaking any of the obligations provided for in the access Directive which may be appropriate to avoid such effects. Therefore, it is only where the imposition of *ex-ante* obligations on an undertaking which is dominant in the (access) upstream market would not result in effective competition on the (retail) downstream market that NRAs should examine whether Article 14(3) may apply.

85. The foregoing considerations are also relevant in relation to horizontal markets ([91]). Moreover, irrespective of whether the markets under consideration are vertical or horizontal, both markets should be electronic communications markets within the meaning of Article 2 of the framework Directive and both should display such characteristics as to justify the imposition of *ex-ante* regulatory obligations ([92]).

3.1.2. *Collective dominance*

86. Under Article 82 of the EC Treaty, a dominant position can be held by one or more undertakings ('collective dominance'). Article 14(2) of the framework Directive also provides that an undertaking may enjoy significant market power, that is, it may be in a dominant position, either individually or jointly with others.

87. In the Access notice, the Commission had stated that, although at the time both its own practice and the case-law of the Court were still developing, it would consider two or more undertakings to be in a collective dominant position when they had substantially the same position vis-à-vis their customers and competitors as a single company has if it is in a dominant position, provided that no effective competition existed between them. The lack of competition could be due, in practice, to the existence of certain links between those companies. The Commission had also stated, however, that the existence of such links was not a prerequisite for a finding of joint dominance ([93]).

88. Since the publication of the Access notice, the concept of collective dominance has been tested in a number of decisions taken by the Commission under Regulation No 17 and under the merger control Regulation. In addition, both the Court of First Instance (CFI) and the Court of Justice of the European Communities (ECJ) have given judgments which have contributed to further clarifying the exact scope of this concept.

3.1.2.1. The jurisprudence of the CFI/ECJ

89. The expression 'one or more undertakings' in Article 82 of the EC Treaty implies that a dominant position may be held by two or more economic entities which are legally and economically independent of each other (94).

90. Until the ruling of the ECJ in *Compagnie maritime belge* (95) and the ruling of the CFI in *Gencor* (96) (see below), it might have been argued that a finding of collective dominance was based on the existence of economic links, in the sense of structural links, or other factors which could give rise to a connection between the undertakings concerned (97). The question of whether collective dominance could also apply to an oligopolistic market, that is a market comprised of few sellers, in the absence of any kind of links among the undertakings present in such a market, was first raised in Gencor. The case concerned the legality of a decision adopted by the Commission under the merger control Regulation prohibiting the notified transaction on the grounds that it would lead to the creation of a duopoly market conducive to a situation of oligopolistic dominance (98). Before the CFI, the parties argued that the Commission had failed to prove the existence of 'links' between the members of the duopoly within the meaning of the existing case-law.

91. The CFI dismissed the application by stating, *inter alia*, that there was no legal precedent suggesting that the notion of 'economic links' was restricted to the notion of structural links between the undertakings concerned: According to the CFI, 'there is no reason whatsoever in legal or economic terms to exclude from the notion of economic links the relationship of interdependence existing between the parties to a tight oligopoly within which, in a market with the appropriate characteristics, in particular in terms of market concentration, transparency and product homogeneity, those parties are in a position to anticipate one another's behaviour and are therefore strongly encouraged to align their conduct in the market, in particular in such a way as to maximise their joint profits by restricting production with a view to increasing prices. In such a context, each trader is aware that highly competitive action on its part designed to increase its market share (for example a price cut) would provoke identical action by the others, so that it would derive no benefit from its initiative. All the traders would thus be affected by the reduction in price *levels*' (99). As the Court pointed out, market conditions may be such that 'each undertaking may become aware of common interests and, in particular, cause prices to increase without having to enter into an agreement or resort to concerted practice' (100).

92. The CFI's ruling in Gencor was later endorsed by the ECJ in Compagnie maritime belge, where the Court gave further guidance as to how the term of collective dominance should be understood and as to which conditions must be fulfilled before such finding can be made. According to the Court, in order to show that two or more undertakings hold a joint dominant position, it is necessary to consider whether the undertakings concerned together constitute a collective entity vis-à-vis their competitors, their trading partners and their consumers on a particular market (101). This will be the case when (i) there is no effective competition among the undertakings in question; and (ii) the said undertakings adopt a uniform conduct or common policy in the relevant market (102). Only when that question is answered in the affirmative, is it appropriate to consider whether the collective entity actually holds a dominant position (103). In particular, it is necessary to ascertain whether economic links exist between the undertakings concerned which enable them to act independently of their competitors, customers and consumers. The Court recognised that an implemented agreement, decision or concerted practice (whether or not covered by an exemption under Article 81(3) of the Treaty) may undoubtedly result in the undertakings concerned being linked in a such way that their conduct on a particular market on which they are active results in them being perceived as a collective entity vis-à-vis their competitors, their trading partners and consumers (104).

93. The mere fact, however, that two or more undertakings are linked by an agreement, a decision of associations of undertakings or a concerted practice within the meaning of Article 81(1) of the Treaty does not, of itself, constitute a necessary basis for such a finding. As the Court stated, 'a finding of a collective dominant position may also be based on other connecting factors and would depend on an economic assessment and, in particular, on an assessment of the structure of the market in question' (105).

94. It follows from the Gencor and Compagnie maritime belge judgments that, although the existence of structural links can be relied upon to support a finding of a collective dominant position, such a finding can also be made in relation to an oligopolistic or highly concentrated market whose structure alone in particular, is conducive to coordinated effects on the relevant market (106).

3.1.2.2. The Commission's decision-making practice and Annex II of the framework Directive

95. In a number of decisions adopted under the merger control Regulation, the Commission considered the concept of collective dominance. It sought in those cases to ascertain whether the structure of the oligopolistic markets in question was conducive to coordinated effects on those markets ([107]).

96. When assessing *ex-ante* the likely existence or emergence of a market which is or could become conducive to collective dominance in the form of tacit coordination, NRAs, should analyse:

(a) whether the characteristics of the market makes it conducive to tacit coordination; and

(b) whether such form of coordination is sustainable that is, (i) whether any of the oligopolists have the ability and incentive to deviate from the coordinated outcome, considering the ability and incentives of the non-deviators to retaliate; and (ii) whether buyers/fringe competitors/potential entrants have the ability and incentive to challenge any anti-competitive coordinated outcome ([108]).

97. This analysis is facilitated by looking at a certain number of criteria which are summarised in Annex II of the framework Directive, which have also been used by the Commission in applying the notion of collective dominance under the merger control Regulation. According to this Annex, 'two or more undertakings can be found to be in a joint dominant position within the meaning of Article 14 if, even in the absence of structural or other links between them, they operate in a market, the structure of which is considered to be conducive to coordinated effects ([109]). Without prejudice to the case-law of the Court of Justice on joint dominance, this is likely to be the case where the market satisfies a number of appropriate characteristics, in particular in terms of market concentration, transparency and other characteristics mentioned below:

— mature market,

— stagnant or moderate growth on the demand side,

— low elasticity of demand,

— homogeneous product,

— similar cost structures,

— similar market shares,

— lack of technical innovation, mature technology,

— absence of excess capacity,

— high barriers to entry,

— lack of countervailing buying power,

— lack of potential competition,

— various kind of informal or other links between the undertakings concerned,

— retaliatory mechanisms,

— lack or reduced scope for price competition'.

98. Annex II of the framework Directive expressly states that the above is not an exhaustive list, nor are the criteria cumulative. Rather, the list is intended to illustrate the sorts of evidence that could be used to support assertions concerning the existence of a collective (oligopolistic) dominance in the form of tacit coordination ([110]). As stated above, the list also shows that the existence of structural links among the undertakings concerned is not a prerequisite for finding a collective dominant position. It is however clear that where such links exist, they can be relied upon to explain, together with any of the other abovementioned criteria, why in a given oligopolistic market coordinated effects are likely to arise. In the absence of such links, in order to establish whether a market is conducive to collective dominance in the form of tacit coordination, it is necessary to consider a number of characteristics of the market. While these characteristics are often presented in the form of the abovementioned list, it is necessary to examine all of them and to make an overall assessment rather than mechanistically applying a 'check list'. Depending on the circumstances of the case, the fact that one or another of the structural elements usually associated with collective dominance may not be clearly established is not in itself decisive to exclude the likelihood of a coordinated outcome ([111]).

99. In an oligopolistic market where most, if not all, of the abovementioned criteria are met, it should be examined whether, in particular, the market operators have a strong incentive to converge to a coordinated market outcome and refrain from reliance on competitive conduct. This will be the case where the long-term benefits of an anti-competitive conduct outweigh any short-term gains resulting from a resort to a competitive behaviour.

100. It must be stressed that a mere finding that a market is concentrated does not necessarily warrant a finding that its structure is conducive to collective dominance in the form of tacit coordination ([112]).

C 165/20 EN Official Journal of the European Communities 11.7.2002

101. Ultimately, in applying the notion of collective dominance in the form of tacit coordination, the criteria which will carry the most sway will be those which are critical to a coordinated outcome in the specific market under consideration. For instance, in Case COMP/M.2499 — Norske Skog/Parenco/Walsum, the Commission came to the conclusion that even if the markets for newsprint and wood-containing magazine paper were concentrated, the products were homogeneous, demand was highly inelastic, buyer power was limited and barriers to entry were high, nonetheless the limited stability of market shares, the lack of symmetry in costs structures and namely, the lack of transparency of investments decisions and the absence of a credible retaliation mechanism rendered unlikely and unsustainable any possibility of tacit coordination among the oligopolists ([113]).

3.1.2.3. Collective dominance and the telecommunications sector

102. In applying the notion of collective dominance, NRAs may also take into consideration decisions adopted under the merger control Regulation in the electronic communications sector, in which the Commission has examined whether any of the notified transactions could give rise to a finding of collective dominance.

103. In *MCI WorldCom/Sprint*, the Commission examined whether the merged entity together with Concert Alliance could be found to enjoy a collective dominant position on the market for global telecommunications services (GTS). Given that operators on that market competed on a bid basis where providers were selected essentially in the first instances of the bidding process on the basis of their ability to offer high quality, tailor-made sophisticated services, and not on the basis of prices, the Commission's investigation was focused on the incentives for market participants to engage in parallel behaviour as to who wins what bid (and who had won what bids) ([114]). After having examined in depth the structure of the market (homogenous product, high barriers of entry, customers countervailing power, etc.) the Commission concluded that it was not able to show absence of competitive constraints from actual competitors, a key factor in examining whether parallel behaviour can be sustained, and thus decided not to pursue further its objections in relation to that market ([115]).

104. In *BT/Esat* ([116]), one of the issues examined by the Commission was whether market conditions in the Irish market for dial-up Internet access lent themselves to the emergence of a duopoly consisting of the incumbent operator, Eircom, and the merged entity. The Commission concluded that this was not the case for the following reasons. First, market shares were not stable; second, demand was doubling every six months; third, internet access products were not considered homogeneous; and finally, technological developments were one of the main characteristics of the market ([117]).

105. In *Vodafone/Airtouch* ([118]), the Commission found that the merged entity would have joint control of two of the four mobile operators present on the German mobile market (namely D2 and E-Plus, the other two being T-Mobil and VIAG Interkom). Given that entry into the market was highly regulated, in the sense that licences were limited by reference to the amount of available radio frequencies, and that market conditions were transparent, it could not be ruled out that such factors could lead to the emergence of a duopoly conducive to coordinated effects ([119]).

106. In *France Telecom/Orange* the Commission found that, prior to the entry of Orange into the Belgian mobile market, the two existing players, Proximus and Mobistar, were in a position to exercise joint dominance. As the Commission noted, for the four years preceding Orange's entry, both operators had almost similar and transparent pricing, their prices following exactly the same trends ([120]). In the same decision the Commission further dismissed claims by third parties as to the risk of a collective dominant position of Vodafone and France Telecom in the market for the provision of pan-European mobile services to internationally mobile customers. Other than significant asymmetries between the market shares of the two operators, the market was considered to be emerging, characterised by an increasing demand and many types of different services on offer and on price ([121]).

## 4. IMPOSITION, MAINTENANCE, AMENDMENT OR WITHDRAWAL OF OBLIGATIONS UNDER THE REGULATORY FRAMEWORK

107. Section 3 of these guidelines dealt with the analysis of relevant markets that NRAs must carry out under Article 16 of the framework Directive to determine whether a market is effectively competitive, i.e. whether there are undertakings in that market who are in a dominant position. This section aims to provide guidance for NRAs on the action they should take following that analysis, i.e. the imposition, maintenance, amendment or withdrawal, as appropriate, of specific regulatory obligations on undertakings designated as having SMP. This section also describes the circumstances in which similar obligations than those that can be imposed on SMP operators may, exceptionally, be imposed on undertakings who have not been designated as having SMP.

108. The specific regulatory obligations which may be imposed on SMP undertakings can apply both to wholesale and retail markets. In principle, the obligations related to wholesale markets are set out in Articles 9 to 13 of the access Directive. The obligations related to retail markets are set out in Articles 17 to 19 of the universal service Directive.

109. The obligations set out in the access Directive are: transparency (Article 9); non-discrimination (Article 10); accounting separation (Article 11), obligations for access to and use of specific network facilities (Article 12), and price control and cost accounting obligations (Article 13). In addition, Article 8 of the access Directive provides that NRAs may impose obligations outside this list. In order to do so, they must submit a request to the Commission, which will take a decision, after seeking the advice of the Communications Committee, as to whether the NRA concerned is permitted to impose such obligations.

110. The obligations set out in the universal service Directive are: regulatory controls on retail services (Article 17), availability of the minimum set of leased lines (Article 18 and Annex VII) and carrier selection and preselection (Article 19).

111. Under the regulatory framework, these obligations should only be imposed on undertakings which have been designated as having SMP in a relevant market, except in certain defined cases, listed in Section 4.3.

### 4.1. Imposition, maintenance, amendment or withdrawal of obligations on SMP operators

112. As explained in Section 1, the notion of effective competition means that there is no undertaking with dominance on the relevant market. In other words, a finding that a relevant market is effectively competitive is, in effect, a determination that there is neither single nor joint dominance on that market. Conversely, a finding that a relevant market is not effectively competitive is a determination that there is single or joint dominance on that market.

113. If an NRA finds that a relevant market is subject to effective competition, it is not allowed to impose obligations on any operator on that relevant market under Article 16. If the NRA has previously imposed regulatory obligations on undertaking(s) in that market, the NRA must withdraw such obligations and may not impose any new obligation on that undertaking(s). As stipulated in Article 16(3) of the framework Directive, where the NRA proposes to remove existing regulatory obligations, it must give parties affected a reasonable period of notice.

114. If an NRA finds that competition in the relevant market is not effective because of the existence of an undertaking or undertakings in a dominant position, it must designate in accordance with Article 16(4) of the framework Directive the undertaking or undertakings concerned as having SMP and impose appropriate regulatory obligations on the undertaking(s) concerned. However, merely designating an undertaking as having SMP on a given market, without imposing any appropriate regulatory obligations, is inconsistent with the provisions of the new regulatory framework, notably Article 16(4) of the framework Directive. In other words, NRAs must impose at least one regulatory obligation on an undertaking that has been designated as having SMP. Where an NRA determines the existence of more than one undertaking with dominance, i.e. that a joint dominant position exists, it should also determine the most appropriate regulatory obligations to be imposed, based on the principle of proportionality.

115. If an undertaking was previously subject to obligations under the 1998 regulatory framework, the NRA must consider whether similar obligations continue to be appropriate under the new regulatory framework, based on a new market analysis carried out in accordance with these guidelines. If the undertaking is found to have SMP in a relevant market under the new framework, regulatory obligations similar to those imposed under the 1998 regulatory framework may therefore be maintained. Alternatively, such obligations could be amended, or new obligations provided in the new framework might also be imposed, as the NRA considers appropriate.

116. Except where the Community's international commitments under international treaties prescribe the choice of regulatory obligation (see Section 4.4) or when the Directives prescribe particular remedies as under Article 18 and 19 of the universal service Directive, NRAs will have to choose between the range of regulatory obligations set out in the Directives in order to remedy a particular problem in a market found not to be effectively competitive. Where NRAs intend to impose other obligations for access and interconnection than those listed in the access Directive, they must submit a request for Commission approval of their proposed course of action. The Commission must seek the advice of the Communications Committee before taking its decision.

117. Community law, and in particular Article 8 of the framework Directive, requires NRAs to ensure that the measures they impose on SMP operators under Article 16 of the framework Directive are justified in relation to the objectives set out in Article 8 and are proportionate to the achievement of those objectives. Thus any obligation imposed by NRAs must be proportionate to the problem to be remedied. Article 7 of the framework Directive requires NRAs to set out the reasoning on which any proposed measure is based when they communicate that measure to other NRAs and to the Commission. Thus, in addition to the market analysis supporting the finding of SMP, NRAs need to include in their decisions a justification of the proposed measure in relation to the objectives of Article 8, as well as an explanation of why their decision should be considered proportionate.

118. Respect for the principle of proportionality will be a key criterion used by the Commission to assess measures proposed by NRAs under the procedure of Article 7 of framework Directive. The principle of proportionality is well-established in Community law. In essence, the principle of proportionality requires that the means used to attain a given end should be no more than what is appropriate and necessary to attain that end. In order to establish that a proposed measure is compatible with the principle of proportionality, the action to be taken must pursue a legitimate aim, and the means employed to achieve the aim must be both necessary and the least burdensome, i.e. it must be the minimum necessary to achieve the aim.

119. However, particularly in the early stages of implementation of the new framework, the Commission would not expect NRAs to withdraw existing regulatory obligations on SMP operators which have been designed to address legitimate regulatory needs which remain relevant, without presenting clear evidence that those obligations have achieved their purpose and are therefore no longer required since competition is deemed to be effective on the relevant market. Different remedies are available in the new regulatory framework to address different identified problems and remedies should be tailored to these specified problems.

120. The Commission, when consulted as provided for in Article 7(3) of the framework Directive, will also check that any proposed measure taken by the NRAs is in conformity with the regulatory framework as a whole, and will assess the impact of the proposed measure on the single market.

121. The Commission will assist NRAs to ensure that as far as possible they adopt consistent approaches in their choice of remedies where similar situations exist in different Member States. Moreover, as noted in Article 7(2) of the framework Directive, NRAs shall seek to agree on the types of remedies best suited to address particular situations in the marketplace.

### 4.2. Transnational markets: joint analysis by NRAs

122. Article 15(4) of the framework Directive gives the Commission the power to issue a Decision identifying product and service markets that are transnational, covering the whole of the Community or a substantial part thereof. Under the terms of Article 16(5) of the framework Directive, the NRAs concerned must jointly conduct the market analysis and decide whether obligations need to be imposed. In practice, the European Regulators Group is expected to provide a suitable forum for such a joint analysis.

123. In general, joint analysis by NRAs would follow similar procedures (e.g. for public consultation) to those required when a single national regulatory authority is conducting a market analysis. Precise arrangements for collective analysis and decision-making will need to be drawn up.

### 4.3. Imposition of certain specific regulatory obligations on non-SMP operators

124. The preceding parts of this section set out the procedures whereby certain specific obligations may be imposed on SMP undertakings, under Articles 7 and 8 of the access Directive and Article 16-19 of the universal service Directive. Exceptionally, similar obligations may be imposed on operators other than those that have been designated as having SMP, in the following cases, listed in Article 8(3) of the access Directive:

— obligations covering *inter alia* access to conditional access systems, obligations to interconnect to ensure end-to-end interoperability, and access to application program interfaces and electronic programme guides to ensure accessibility to specified digital TV and radio broadcasting services (Article 5(1), 5(2) and 6 of the access Directive),

— obligations that NRAs may impose for co-location where rules relating to environmental protection, health, security or town and country planning deprive other undertakings of viable alternatives to co-location (Article 12 of the framework Directive),

— obligations for accounting separation on undertakings providing electronic communications services who enjoy special or exclusive rights in other sectors (Article 13 of the framework Directive),

— obligations relating to commitments made by an undertaking in the course of a competitive or comparative selection procedure for a right of use of radio frequency (Condition B7 of the Annex to the authorisation Directive, applied via Article 6(1) of that Directive),

— obligations to handle calls to subscribers using specific numbering resources and obligations necessary for the implementation of number portability (Articles 27, 28 and 30 of the universal service Directive),

— obligations based on the relevant provisions of the data protection Directive, and

— obligations to be imposed on non-SMP operators in order to comply with the Community's international commitments.

4.4. **Relationship to WTO commitments**

125. The EC and its Member States have given commitments in the WTO in relation to undertakings that are 'major suppliers' of basic telecommunications services ([122]). Such undertakings are subject to all of the obligations set out in the EC's and its Member States' commitments in the WTO for basic telecommunications services. The provisions of the new regulatory framework, in particular relating to access and interconnection, ensure that NRAs continue to apply the relevant obligations to undertakings that are major suppliers in accordance with the WTO commitments of the EC and its Member States.

5. **POWERS OF INVESTIGATION AND COOPERATION PROCEDURES FOR THE PURPOSE OF MARKET ANALYSIS**

5.1. **Overview**

126. This section of the guidelines covers procedures in respect of an NRA's powers to obtain the information necessary to conduct a market analysis.

127. The regulatory framework contains provisions to enable NRAs to require undertakings that provide electronic communications networks and services to supply all the information, including confidential information, necessary for NRAs to assess the state of competition in the relevant markets and impose appropriate *ex-ante* obligations and thus to ensure compliance with the regulatory framework.

128. This section of the guidelines also includes guidance as to measures to ensure effective cooperation between NRAs and NCAs at national level, and among NRAs and between NRAs and the Commission at Community level. In particular this section deals with the exchange of information between those authorities.

129. Many electronic communication markets are fast-moving and their structures are changing rapidly. NRAs should ensure that the assessment of effective competition, the public consultation, and the designation of operators having SMP are all carried out within a reasonable period. Any unnecessary delay in the decision could have harmful effects on incentives for investment by undertakings in the relevant market and therefore on the interests of consumers.

5.2. **Market analysis and powers of investigation**

130. Under Article 16(1) of the framework Directive, NRAs must carry out an analysis of the relevant markets identified in the Recommendation and any Decision as soon as possible after their adoption or subsequent revision. The conclusions of the analysis of each of the relevant markets, together with the proposed regulatory action, must be published and a public consultation must be conducted, as described in Section 6.

131. In order to carry out their market analysis, NRAs will first need to collect all the information they consider necessary to assess market power in a given market. To the extent that such information needs to be obtained directly from undertakings, Article 11 of the authorisation Directive provides that undertakings are required by the terms of their general authorisation to supply the information necessary for NRAs to conduct a market analysis within the meaning of Article 16(2) of the framework Directive. This is reinforced by the more general obligation in Article 5(1) of the framework Directive which provides that Member States shall ensure that undertakings providing electronic communications networks and services provide all the information necessary for NRAs to ensure conformity with Community law.

132. When NRAs request information from an undertaking, they should state the reasons justifying the request and the time limit within which the information is to be provided. As provided for in Article 10(4) of the authorisation Directive, NRAs may be empowered to impose financial penalties on undertakings for failure to provide information.

133. In accordance with Article 5(4) of the framework Directive, NRAs must publish all information that would contribute to an open and competitive market, acting in accordance with national rules on public access to information and subject to Community and national rules on commercial confidentiality.

134. However, as regards information that is confidential in nature, the provisions of Article 5(3) of the framework Directive, require NRAs to ensure the confidentiality of such information in accordance with Community and national rules on business confidentiality. This confidentiality obligation applies equally to information that has been received in confidence from another public authority.

5.3. **Cooperation procedures**

*Between NRAs and NCAs*

135. Article 16(1) of the framework Directive requires NRAs to associate NCAs with the market analyses as appropriate. Member States should put in place the necessary procedures to guarantee that the analysis under Article 16 of the framework Directive is carried out effectively. As the NRAs conduct their market analyses in accordance with the methodologies of competition law, the views of NCAs in respect of the assessment of competition are highly relevant. Cooperation between NRAs and NCAs will be essential, but NRAs remain legally responsible for conducting the relevant analysis. Where under national law the tasks assigned under Article 16 of the framework Directive are carried out by two or more separate regulatory bodies, Member States should ensure clear division of tasks and set up procedures for consultation and cooperation between regulators in order to assure coherent analysis of the relevant markets.

136. Article 3(5) of the framework Directive requires NRAs and NCAs to provide each other with the information necessary for the application of the regulatory framework, and the receiving authority must ensure the same level of confidentiality as the originating authority. NCAs should therefore provide NRAs with all relevant information obtained using the former's investigatory and enforcement powers, including confidential information.

137. Information that is considered confidential by an NCA, in accordance with Community and national rules on business confidentiality, should only be exchanged with NRAs where such exchange is necessary for the application of the provisions of the regulatory framework. The information exchanged should be limited to that which is relevant and proportionate to the purpose of such exchange.

*Between the Commission and NRAs*

138. For the regulatory framework to operate efficiently and effectively, it is vital that there is a high level of cooperation between the Commission and the NRAs. It is particularly important that effective informal cooperation takes place. The European Regulators Group will be of great importance in providing a framework for such cooperation, as part of its task of assisting and advising the Commission. Cooperation is likely to be of mutual benefit, by minimising the likelihood of divergences in approach between different NRAs, in particular divergent remedies to deal with the same problem ([123]).

139. In accordance with Article 5(2) of the framework Directive, NRAs must supply the Commission with information necessary for it to carry out its tasks under the Treaty. This covers information relating to the regulatory framework (to be used in verifying compatibility of NRA action with the legislation), but also information that the Commission might require, for example, in considering compliance with WTO commitments.

140. NRAs must ensure that, where they submit information to the Commission which they have requested undertakings to provide, they inform those undertakings that they have submitted it to the Commission.

141. The Commission can also make such information available to another NRA, unless the original NRA has made an explicit and reasoned request to the contrary. Although there is no legal requirement to do so, the Commission will normally inform the undertaking which originally provided the information that it has been passed on to another NRA.

*Between NRAs*

142. It is of the utmost importance that NRAs develop a common regulatory approach across Member States that will contribute to the development of a true single market for electronic communications. To this end, NRAs are required under Article 7(2) of the framework Directive to cooperate with each other and with the Commission in a transparent manner to ensure the consistent application, in all Member States, of the new regulatory framework. The European Regulators' Group is expected to serve as an important forum for cooperation.

143. Article 5(2) of the framework Directive also foresees that NRAs will exchange information directly between each other, as long as there is a substantiated request. This will be particularly necessary where a transnational market needs to be analysed, but it will also be required within the framework of cooperation in the European Regulators' Group. In all exchanges of information, the NRAs are required to maintain the confidentiality of information received.

## 6. PROCEDURES FOR CONSULTATION AND PUBLICATION OF PROPOSED NRA DECISIONS

### 6.1. Public consultation mechanism

144. Except in the urgent cases as explained below, an NRA that intends to take a measure which would have a significant impact on the relevant market should give the interested parties the opportunity to comment on the draft measure. To this effect, the NRA must hold a public consultation on its proposed measure. Where the draft measure concerns a decision relating to an SMP designation or non-designation it should include the following:

— the market definition used and reasons therefor, with the exception of information that is confidential in accordance with European and national law on business confidentiality,

— evidence relating to the finding of dominance, with the exception of information that is confidential in accordance with European and national law on business confidentiality together with the identification of any undertakings proposed to be designated as having SMP,

— full details of the sector-specific obligations that the NRA proposes to impose, maintain, modify or withdraw on the abovementioned undertakings together with an assessment of the proportionality of that proposed measure.

145. The period of the consultation should be reasonable. However, NRAs' decisions should not be delayed excessively as this can impede the development of the market. For decisions related to the existence and designation of undertakings with SMP, the Commission considers that a period of two months would be reasonable for the public consultation. Different periods could be used in some cases if justified. Conversely, where a draft SMP decision is proposed on the basis of the results of an earlier consultation, the length of consultation period for these decisions may well be shorter than two months.

### 6.2. **Mechanisms to consolidate the internal market for electronic communications**

146. Where an NRA intends to take a measure which falls within the scope of the market definition or market analysis procedures of Articles 15 and 16 of the framework Directive, as well as when NRAs apply certain other specific Articles in the regulatory framework ([124]) and where the measures have an effect on trade between Member States, the NRAs must communicate the measures, together with their reasoning, to NRAs in other Member States and to the Commission in accordance with Article 7(3) of the framework Directive. It should do this at the same time as it begins its public consultation. The NRA must then give other NRAs and the Commission the chance to comment on the NRA's proposed measures, before adopting any final decision. The time available for other NRAs and the Commission to comment should be the same time period as that set by the NRA for its national public consultation, unless the latter is shorter than the minimum period of one month provided for in Article 7(3). The Commission may decide in justified circumstances to publish its comments.

147. With regard to measures that could affect trade between Member States, this should be understood as meaning measures that may have an influence, direct or indirect, actual or potential, on the pattern of trade between Member States in a manner which might create a barrier to the single European market ([125]). Therefore, the notion of an effect on trade between Member States is likely to cover a broad range of measures.

148. NRAs must make public the results of the public consultation, except in the case of information that is confidential in accordance with Community and national law on business confidentiality.

149. With the exception of two specific cases, explained in the following paragraph, the NRA concerned may adopt the final measure after having taken account of views expressed during its mandatory consultation. The final measure must then be communicated to the Commission without delay.

### 6.3. **Commission power to require the withdrawal of NRAs' draft measures**

150. Under the terms of Article 7(4) of the framework Directive, there are two specific situations where the Commission has the possibility to require an NRA to withdraw a draft measure which falls within the scope of Article 7(3):

— the draft measure concerns the definition of a relevant market which differs from that identified in the Recommendation, or

— the draft measure concerns a decision as to whether to designate, or not to designate, an undertaking as having SMP, either individually or jointly with others.

151. In respect of the above two situations, where the Commission has indicated to the NRA in the course of the consultation process that it considers that the draft measure would create a barrier to the single European market or where the Commission has serious doubts as to the compatibility of the draft measure with Community law, the adoption of the measure must be delayed by a maximum of an additional two months.

152. During this two-month period, the Commission may, after consulting the Communications Committee following the advisory procedure ([126]), take a decision requiring the NRA to withdraw the draft measure. The Commission's decision will be accompanied by a detailed and objective analysis of why it considers that the draft measure should not be adopted together with specific proposals for amending the draft measure. If the Commission does not take a decision within that period, the draft measure may be adopted by the NRA.

### 6.4. **Urgent cases**

153. In exceptional circumstances, NRAs may act urgently in order to safeguard competition and protect the interest of users. An NRA may therefore, exceptionally, adopt proportionate and provisional measures without consulting either interested parties, the NRAs in other Member States, or the Commission. Where an NRA has taken such urgent action, it must, without delay, communicate these measures, with full reasons, to the Commission, and to the other NRAs. The Commission will verify the compatibility of those measures with Community law and in particular will assess their proportionality in relation to the policy objectives of Article 8 of the framework Directive.

154. If the NRA wishes to make the provisional measures permanent, or extends the time for which it is applicable, the NRA must go through the normal consultation procedure set out above. It is difficult to foresee any circumstances that would justify urgent action to define a market or designate an SMP operator, as such measure are not those that can be carried out immediately. The Commission therefore does not expect NRAs to use the exceptional procedures in such cases.

C 165/26 EN Official Journal of the European Communities 11.7.2002

### 6.5. **Adoption of the final decision**

155. Once an NRA's decision has become final, NRAs should notify the Commission of the names of the undertakings that have been designated as having SMP and the obligations imposed on them, in accordance with the requirements of Article 36(2) of the universal service Directive and Articles 15(2) and 16(2) of the access Directive. The Commission will thereafter make this information available in a readily accessible form, and will transmit the information to the Communications Committee as appropriate.

156. Likewise, NRAs should publish the names of undertakings that they have designated as having SMP and the obligations imposed on them. They should ensure that up-to-date information is made publicly available in a manner that guarantees all interested parties easy access to that information.

(1) OJ L 108, 24.4.2002, p. 33.

(2) OJ L 108, 24.4.2002, p. 21.

(3) OJ L 108, 24.4.2002, p. 7.

(4) OJ L 108, 24.4.2002, p. 51.

(5) To be adopted.

(6) OJ L 24, 30.1.1998, p. 1.

(7) Except where the new regulatory framework expressly permits obligations to be imposed independently of the competitive state of the market.

(8) Article 14 of the framework Directive.

(9) In addition, transnational markets whose characteristics may be such as to justify sector-specific regulation may be identified by the Commission in a Decision on transnational markets.

(10) Recital 27 of the framework Directive.

(11) Regulation (EEC) No 4064/89 on the control of concentrations between undertakings (OJ L 395, 30.12.1989, p. 1), as last amended by Regulation (EC) No 1310/97 of 30 June 1997 (OJ L 180, 9.7.1997, p. 1) (hereafter the merger control Regulation).

(12) Guidelines on the application of EEC competition rules in the telecommunications sector (OJ C 233, 6.9.1991, p. 2).

(13) Commission notice on the definition of relevant market for the purposes of Community competition law (OJ C 372, 9.12.1997, p. 5).

(14) Notice on the application of the competition rules to access agreements in the telecommunications sector (OJ C 265, 22.8.1998, p. 2).

(15) It is expected that effective cooperation between NRAs and NCAs would prevent the duplication of procedures concerning identical market issues.

(16) Guidelines on the application of EEC competition rules in the telecommunications sector (OJ C 233, 6.9.1991, p. 2).

(17) Article 14(2) of the framework Directive.

(18) Case C-209/98, *Entreprenørforeningens Affalds* [2000] ECR I-3743, paragraph 57, and Case C-242/95 *GT-Link* [1997] ECR I-4449, paragraph 36. It should be recognised that the objective of market definition is not an end in itself, but part of a process, namely assessing the degree of a firm's market power.

(19) See Directive 97/33/EC of the European Parliament and of the Council of 30 June 1997 on interconnection in telecommunications with regard to ensuring universal service and interoperability through application of the principles of open network provision (ONP) (OJ L 199, 26.7.1997, p. 32) (the interconnection Directive); Council Directive 90/387/EEC of 28 June 1990 on the establishment of the internal market for telecommunications services through the implementation of open network provision (OJ L 192, 24.7.1990, p. 1) (the ONP framework Directive); Council Directive 92/44/EEC of 5 June 1992 on the application of open network provision to leased lines (OJ L 165, 19.6.1992, p. 27) (the leased lines Directive); Directive 95/62/EC of the European Parliament and of the Council of 13 December 1995 on the application of open network provision (ONP) to voice telephony (OJ L 321, 30.12.1995, p. 6), replaced by Directive 98/10/EC of the European Parliament and of the Council of 26 February 1998 on the application of open network provision (ONP) to voice telephony and on universal service for telecommunications in a competitive environment (OJ L 101, 1.4.1998, p. 24) (the ONP voice telephony Directive).

(20) Joined Cases C-68/94 and C-30/95, *France and Others* v *Commission* [1998] ECR I-1375. See, also, Notice on market definition, at paragraph 12.

(21) To the extent that the electronic communications sector is technology and innovation-driven, any previous market definition may not necessarily be relevant at a later point in time.

(22) Notice on market definition, paragraph 12.

(23) See, also, Notice on market definition, paragraphs 20-23, Case IV/M.1225 — Enso/Stora, (OJ L 254, 29.9.1999), paragraph 40.

(24) See Notice on market definition, paragraph 24. Distinguishing between supply-side substitution and potential competition in electronic communications markets may be more complicated than in other markets given the dynamic character of the former. What matters, however, is that potential entry from other suppliers is taken into consideration at some stage of the relevant market analysis, that is, either at the initial market definition stage or at the subsequent stage of the assessment of market power (SMP).

(25) It is not necessary that all consumers switch to a competing product; it suffices that enough or sufficient switching takes place so that a relative price increase is not profitable. This requirement corresponds to the principle of 'sufficient interchangeability' laid down in the case-law of the Court of Justice; see below, footnote 32.

(26) See, also, Access notice, paragraph 46, and Case T-83/91, *Tetra Pak* v *Commission*, [1994] ECR II-755, paragraph 68. This test is also known as 'SSNIP' (small but significant non transitory increase in price). Although the SSNIP test is but one example of methods used for defining the relevant market and notwithstanding its formal econometric nature, or its margins for errors (the so-called 'cellophane fallacy', see below), its importance lies primarily in its use as a conceptual tool for assessing evidence of competition between different products or services.

(27) See Notice on market definition, paragraphs 17-18.

(28) In other words, where the cross-price elasticity of demand between two products is high, one may conclude that consumers view these products as close substitutes. Where consumer choice is influenced by considerations other than price increases, the SSNIP test may not be an adequate measurement of product substitutability; see Case T-25/99, *Colin Arthur Roberts and Valerie Ann Roberts* v *Commission*, [2001] ECR II-1881.

(29) Within the context of market definition under Article 82 of the EC Treaty, a competition authority or a court would estimate the 'starting price' for applying the SSNIP on the basis of the price charged by the alleged monopolist. Likewise, under the prospective assessment of the effects which a merger may have on competition, the starting price would be based on the prevailing prices of the merging parties. However, where an NRA carries out a market analysis for the purposes of applying Article 14 of the framework Directive the service or product in question may be offered by several firms. In such a case, the starting price should be the industry 'average price'.

(30) It is worth noting that prices which result from price regulation which does not aim at ensuring that prices are cost-based, but rather at ensuring an affordable offer within the context of the provision of universal services, may not be presumed to be set at a competitive level, nor should they serve as a starting point for applying the SSNIP test.

(31) Indeed, one of the drawbacks of the application of the SSNIP test is that in some cases, a high-demand cross-price elasticity may mean that a firm has already exercised market power, a situation known in competition law and practice as the 'cellophane fallacy'. In such cases, the prevailing price does not correspond to a competitive price. Determining whether the prevailing price is set above the competitive level is admittedly one of the most difficult aspects of the SSNIP test. NRAs faced with such difficulties could rely on other criteria for assessing demand and supply substitution such as functionality of services, technical characteristics, etc. Clearly, if evidence exist to show that in the past a firm has engaged in anti-competitive behaviour (price-fixing) or has enjoyed market power, then this may serve as an indication that its prices are not under competitive constraint and accordingly are set above the competitive level.

(32) Case C-333/94 P, *Tetra Pak* v *Commission* [1996] ECR I-5951, paragraph 13, Case 31/80 *L'Oréal* [1980] ECR 3775, paragraph 25, Case 322/81, *Michelin* v *Commission* [1983] ECR 3461, paragraph 37, Case C-62/86, *AkzoChemie* v *Commission* [1991] ECR I-3359, Case T-504/93, *Tiercé Ladbroke* v *Commission* [1997] ECR II-923, paragraph 81, T-65/96, *Kish Glass* v *Commission* [2000] ECR II-1885, paragraph 62, Case C-475/99, *Ambulanz Glöckner and Landkreis Südwestpfalz*, [2001] ECR I-0000, paragraph 33. The test of sufficient substitutability or interchangeability was first laid down by the Court of Justice in Case 6/72, *Europemballage and Continental Can* v *Commission*, [1973] ECR 215, paragraph 32 and Case 85/76, *Hoffmann La-Roche* v *Commission* [1979] ECR 461, paragraph 23.

(33) Case C-333/94 P, *Tetra Pak* v *Commission* [1996] ECR I-5951, paragraph 13, Case 66/86, *Ahmed Saeed* [1989] ECR 803, paragraphs 39 and 40, Case *United Brands* v *Commission* [1978] ECR 207, paragraphs 22 and 29, and 12; Case T-229/94, *Deutsche Bahn* v *Commission* [1997] ECR II-1689, paragraph 54. In *Tetra Pak*, the Court confirmed that the fact that demand for aseptic and non-aseptic cartons used for packaging fruit juice was marginal and stable over time relative to the demand for cartons used for packaging milk was evidence of a very little interchangeability between the milk and the non-milk packaging sector, idem, paragraphs 13 and 15.

(34) For example, in the case of a relative price increase, consumers of a lower quality/price service may switch to a higher quality/price service if the cost of doing so (the premium paid) is offset by the price increase. Conversely, consumers of a higher quality product may no longer accept a higher premium and switch to a lower quality service. In such cases, low and high quality products would appear to be effective substitutes.

(35) Communication from the Commission — Status of voice on the Internet under Community law, and in particular, under Directive 90/388/EEC — Supplement to the Communication by the Commission to the European Parliament and the Council on the status and implementation of Directive 90/388/EEC on competition in the markets for telecommunications services (OJ C 369, 22.12.2000, p. 3). Likewise, it cannot be excluded that in the future. xDSL technology and multipoint video distribution services based on wireless local loops may be used for the transmission of TV materials in direct competition with other existing TV delivery systems based on cable systems, direct-to-home satellite transmission and terrestrial analogue or digital transmission platforms.

(36) Switching costs which stem from strategic choices by undertakings rather than from exogenous factors should be considered, together with some other form of entry barriers, at the subsequent stage of SMP assessment. Where a market is still growing, total switching costs for already 'captured' consumers may not be significant and may not thus deter demand or supply-side substitution.

(37) The time frame to be used to assess the likely responses of other suppliers in case of a relative price increase will inevitably depend on the characteristics of each market and should be decided on a case-by-case basis.

(38) See, also, Case C-333/94, *Tetra Pak* v *Commission*, op. cit., paragraph 19. As mentioned above, the required investments should also be undertaken within a reasonable time frame.

(39) See, also, Case COMP/M.2574 — Pirelli/Edizione/Olivetti/Telecom Italia, paragraph 58.

(40) *United Brands*, op. cit., paragraph 44, *Michelin*, op. cit., paragraph 26, Case 247/86 *Alsatel* v *Novasam* [1988] ECR 5987, paragraph 15; *Tiercé Ladbroke* v *Commission*, op. cit., paragraph 102.

(41) *Deutsche Bahn* v *Commission*, op. cit., paragraph 92. Case T-139/98 *AAMS* v *Commission*, [2001] ECR 0000-II, paragraph 39.

(42) See, for instance, Case IV/M.1025 — Mannesmann/Olivetti/Infostrada, paragraph 17, and Case COMP/JV.23 — Telefónica Portugal Telecom/Médi Telecom.

(43) In practice, this area will correspond to the limits of the area in which an operator is authorised to operate. In Case COMP/M.1650 — ACEA/Telefónica, the Commission pointed out that since the notified joint venture would have a licence limited to the area of Rome, the geographical market could be defined as local; at paragraph 16.

(44) The fact that mobile operators can provide services only in the areas where they have been authorised to and the fact that a network architecture reflects the geographical dimension of the mobile licences explains why mobile markets are considered to be national in scope. The extra connection and communications costs that consumers face when roaming abroad, coupled with the loss of certain additional service functionalities (i.e. lack of voice mail abroad) further supports this definition; see Case IV/M.1439 — Telia/Telenor, paragraph 124, Case IV/M.1430 — Vodafone/Airtouch, paragraphs 13-17, Case COMP/JV.17 — Mannesmann/Bell Atlantic/Omnitel, paragraph 15.

(45) Physical interconnection agreements may also be taken into consideration for defining the geographical scope of the market, Case IV/M.570 — TBT/BT/TeleDanmark/Telenor, paragraph 35.

(46) Case IV/M.856 — British Telecom/MCI (II), paragraph 19s., Case IV/JV.15 — BT/AT & T, paragraph 84 and 92, Case COMP/M.2257 — France Telecom/Equant, paragraph 32, It is highly unlikely that the provision of electronic communications services could be segmented on the basis of national (or local) bilateral routes.

(47) Reference may be made, for instance, to the market for backhaul capacity in international routes (i.e. cable station serving country A to country E) where a potential for substitution between cable stations serving different countries (i.e., cable stations connecting Country A to B, A to C and A to D) may exist where a supplier of backhaul capacity in relation to the route A to E is or would be constrained by the ability of consumers to switch to any of the other 'routes', also able to deal with traffic from or to country E.

(48) Where a market is defined on the basis of a bilateral route, its geographical scope could be wider than national if suppliers are present in both ends of the market and can satisfy demand coming from both ends of the relevant route.

(49) See Notice on market definition, paragraphs 57 and 58. For instance, chain substitutability could occur where an undertaking providing services at national level constraints the prices charged by undertakings providing services in separate geographical markets. This may be the case where the prices charged by undertakings providing cable networks in particular areas are constrained by a dominant undertaking operating nationally; see also, Case COMP/M.1628 — TotalFina/Elf (OJ L 143, 29.5.2001, p. 1), paragraph 188.

(50) Evidence should show clear price interdependence at the extremes of the chain and the degree of substitutability between the relevant products or geographical areas should be sufficiently strong.

(51) The Commission has, *inter alia*, made references in its decisions to the existence of the following markets: international voice-telephony services (Case IV/M.856 — British Telecommunications/MCI (II), OJ L 336, 8.12.1997), advanced telecommunications services to corporate users (Case IV/35.337, Atlas, OJ L 239, 19.9.1996, paragraphs 5-7, Case IV/35617, Phoenix/Global/One, OJ L 239, 19.9.1996, paragraph 6, Case IV/34.857, BT-MCI (I), OJ L 223, 27.8.1994), standardised low-level packet-switched data-communications services, resale of international transmission capacity (Case IV/M.975 — Albacom/BT/ENI, paragraph 24) audioconferencing (Albacom/BT/ENI, paragraph 17), satellite services (Case IV/350518 — Iridium, OJ L 16, 18.1.1997), (enhanced) global telecommunications services (Case IV/JV.15 — BT/AT & T, Case COMP/M.1741 — MCI WorldCom/Sprint, paragraph 84, Case COMP/M.2257 — France Telecom/Equant, paragraph 18), directory-assistance services (Case IV/M.2468 — SEAT Pagine Gialle/ENIRO, paragraph 19, Case COMP/M.1957 — VIAG Interkom/Telenor Media, paragraph 8), Internet-access services to end users (Case IV/M.1439 — Telia/Telenor, Case COMP/JV.46 — Blackstone/CDPQ/Kabel Nordrhein/Westfalen, paragraph 26, Case COMP/M.1838 — BT/Esat, paragraph 7), top-level or universal Internet connectivity (Case COMP/M.1741 — MCI WorldCom/Sprint, paragraph 52), seamless pan-European mobile telecommunications services to internationally mobile customers (Case COMP/M.1975 — Vodafone Airtouch/Mannesmann, Case COMP/M.2016 — France Telecom/Orange, paragraph 15), wholesale roaming services (Case COMP/M.1863 — Vodafone/Airtel, paragraph 17), and market for connectivity to the international signalling network (Case COMP/2598 — TDC/CMG/Migway JV, paragraphs 17-18).

(52) See, also, Joined Cases T-125/97 and T-127/97, *The Coca-Cola Company and Others* v *Commission* [2000] ECR II-1733, at paragraphs 81 and 82.

(53) See, also, Article 15 of the framework Directive.

(54) Access notice, paragraph 45.

(55) See Case COMP/M.1439 — Telia/Telenor.

(56) See Telia/Telenor, BT/AT & T, France Télécom/Equant, op. cit. See also Commission Decision of 20 May 1999, Cégétel + 4 (OJ L 218, 18.8.1999), paragraph 22. With regard to the emerging market for 'Global broadband data communications services — GBDS', the Commission has found that such services can be supported by three main network architectures: (i) terrestrial wireline systems; (ii) terrestrial wireless systems; and (iii) satellite-based systems, and that from a demand side, satellite-based GBDS can be considered as a separate market, Case COMP/M.1564 — Astrolink, paragraphs 20-23.

(57) Directive 96/19/EC, recital 20 (OJ L 74, 22.3.1996, p. 13). See, also, communication from the Commission, 'Unbundled access to the local loop: enabling the competitive provision of a full range of electronic communication services, including broadband multimedia and high speed Internet' (OJ C 272, 23.9.2000, p. 55). Pursuant to point 3.2, 'While categories of services have to be monitored closely, particularly given the speed of technological change, and regularly reassessed on a case-by-case basis, these services are presently normally not substitutable for one another, and would therefore be considered as forming different relevant markets'.

(58) The Commission has identified separate markets for services to large multinational corporations (MNCs) given the significant differences in the demand (and supply) of services to this group of customers compared to other retail (business) customers, see Case IV/JV.15 — BT/AT & T, Case COMP/M.1741 — MCI WorldCom/Sprint, Case COMP/M.2257 — France Télécom/Equant.

(59) See communication on 'Unbundled access to the local loop', op.cit, point 3.2. The market for 'high-speed' communications services could possibly be further divided into distinct segments depending on the nature of the services offered (i.e. Internet services, video-on-demand, etc.).

(60) Case COMP/M.2574 — Pirelli/Edizione/Olivetti/Telecom Italia, paragraph 33. It could also be argued that dial-up access to the Internet via existing 2G mobile telephones is a separate market from dial-up access via the public switched telecommunications network. According to the Commission, accessing the Internet via a mobile phone is unlikely to be a substitute for existing methods of accessing the Internet via a PC due to difference in sizes of the screen and the format of the material that can be obtained through the different platforms; see Case COMP/M.1982 — Telia/Oracle/Drutt, paragraph 15, and Case COMP/JV.48 Vodafone/Vivendi/Canal+.

([61]) Case COMP/M.2469 — Vodafone/Airtel, paragraph 7, Case IV/M.1430 — Vodafone/Airtouch, Case IV/M.1669, Deutsche Telecom/One2One, paragraph 7. Whether this market can be further segmented into a carrier (network operator) market and a downstream service market should be decided on a case-by-case basis; see Case IV/M.1760 — Mannesmann/Orange, paragraphs 8-10, and Case COMP/M.2053 — Telenor/BellSouth/Sonofon, paragraphs 9-10.

([62]) For instance, in British Interactive Broadcasting/Open, the Commission noted that for the provision of basic voice services to consumers, the relevant infrastructure market included not only the traditional copper network of BT but also the cable networks of the cable operators, which were capable of providing basic telephony services, and possibly wireless fixed networks, Case IV/36.359, (OJ L 312, 6.12.1999, paragraphs 33-38). In Case IV/M.1113 — Nortel/Norweb, the Commission recognised that electricity networks using 'digital power line' technology could provide an alternative to existing traditional local telecommunications access loop, paragraphs 28-29.

([63]) In assessing the conditions of network competition in the Irish market that would ensue following full liberalisation, the Commission also relied on the existence of what, at that period of time, were perceived as potential alternative infrastructure providers, namely, cable TV and electricity networks, Telecom Eireann, cit., paragraph 30. The Commission left open the question whether the provision of transmission capacity by an undersea network infrastructure constitutes a distinct market from terrestrial or satellite transmissions networks, Case COMP/M.1926 — Telefonica/Tyco/JV, at paragraph 8.

([64]) Case COMP/M.1439, Telia/Telenor, paragraph 79. For instance, an emerging pan-European market for wholesale access (SMS) to mobile infrastructure has been identified by the Commission in Case COMP/2598 — TDC/CMG/Migway JV, at paragraphs 28-29.

([65]) In applying these criteria, the Commission has found that, as far as the fixed infrastructure is concerned, demand for the lease of transmission capacity and the provision of related services to other operators occurs at wholesale level (the market for carrier's carrier services; see Case IV/M.683 — GTS-Hermes Inc./HIT Rail BV, paragraph 14, Case IV/M.1069 — WorldCom/MCI (OJ L 116, 4.5.1999, p. 1), Unisource (OJ L 318, 20.11.1997, p. 1), Phoenix/Global One (OJ L 239, 19.9.1996, p. 57), Case IV/JV.2 — Enel/FT/DT. In Case COMP/M.1439 — Telia/Telenor, the Commission identified distinct patterns of demand for wholesale and retail (subscriber) access to network infrastructure (provision or access to the local loop, and provision or access to long distance and international network infrastructure), paragraphs 75-83.

([66]) See footnote 58.

([67]) Fibre optics are currently competitive only on upstream transmission markets whereas wireless local loops which are still to be deployed will target mainly professionals and individuals with particular communications needs. With the exception of certain national markets, existing cable TV networks need costly upgrades to support two ways broadband communications, and, compared with xDLS technologies, they do not offer a guaranteed bandwidth since customers share the same cable channel.

([68]) See also Case IV/JV.11 — @Home Benelux BV.

([69]) For example, if a fixed operator wants to terminate calls to the subscribers of a particular network, in principle, it will have no other choice but to call or interconnect with the network to which the called party has subscribed. For instance, in light of the 'calling party pays' principle, mobile operators have no incentives to compete on prices for terminating traffic to their own network. See also, OECD, 'Competition issues in telecommunications-background note for the secretariat', DAFFE/CLP/WP2(2001)3, and Commission's press release IP/02/483.

([70]) Case 27/76 *United Brands* v *Commission* [1978] ECR 207.

([71]) See, also, recital 25 of the framework Directive.

([72]) See Article 14, paragraph 2, and recital 28 of the framework Directive.

([73]) It should be noted that NRAs do not have to find an abuse of a dominant position in order to designate an undertaking as having SMP.

([74]) Case 85/76, *Hoffmann-La Roche* v *Commission* [1979] ECR 461, paragraph 39. It should be stressed here that for the purposes of *ex-ante* regulation, if an undertaking has already been imposed regulatory obligations, the fact that competition may have been restored in the relevant market as a result precisely of the obligations thus imposed, this does not mean that that undertaking is no longer in a dominant position and that it should no longer continue being designated as having SMP.

([75]) The absence of any substitutable service or product may justify a finding of a situation of economic dependence which is characteristic of the existence of a dominant position. See Commission decisions, Decca Navigator System (OJ L 43, 15.2.1987, p. 27) and Magill TV Guide: ITP, BBC, RTE (OJ L 78, 21.3.1989, p. 43). See also, Case 22/78 *Hugin* v *Commission* 1979 [ECR] 1869, Case 226/84, *British Leyland* v *Commission* 1986 [ECR] p. 3263.

([76]) See, also, recital 15 of Council Regulation (EEC) No 4064/89.

([77]) *United Brands* v *Commission*, op. cit. The greater the difference between the market share of the undertaking in question and that of its competitors, the more likely will it be that the said undertaking is in a dominant position. For instance, in Case COMP/M.1741 — MCI WorldCom/Sprint it was found that the merged entity would have in the market for the provision of top-level Internet connectivity an absolute combined market share of more than [35-45] %, several times larger than its closest competitor, enabling it to behave independently of its competitors and customers (see paragraphs 114, 123, 126, 146, 155 and 196).

([78]) Case C-62/86, *AKZO* v *Commission*, [1991] ECR I-3359, paragraph 60; Case T-228/97, *Irish Sugar* v *Commission*, [1999] ECR II-2969, paragraph 70, Case *Hoffmann-La Roche* v *Commission*, op. cit, paragraph 41, Case T-139/98, *AAMS and Others* v *Commission* [2001 ECR II-0000, paragraph 51. However, large market shares can become accurate measurements only on the assumption that competitors are unable to expand their output by sufficient volume to meet the shifting demand resulting from a rival's price increase.

(79) Case *Hoffmann-La Roche* v *Commission*, op. cit., paragraph 41, Case C-62/86, *Akzo* v *Commission* [1991] ECR I-3359, paragraphs 56, 59. 'An undertaking which has a very large market share and holds it for some time, by means of the volume of production and the sale of the supply which it stands for — without holders of much smaller market shares being able to meet rapidly the demand from those who would like to break away from the undertaking which has largest market share — is by virtue of that share in a position of strength which makes it an unavoidable trading partner and which, because of this alone, secures for it, at the very least during relatively long periods, that freedom of action which is the special feature of a dominant position', Case *AAMS and Others* v *Commission*, op. cit., paragraph 51.

(80) Notice on market definition, op. cit., at p. 5.

(81) See Case COMP/M.1741 — MCI WorldCom/Sprint, paragraph 239-240. In bidding markets, however, it is important not to rely only on market shares as they in themselves may not be representative of the undertakings actual position, for further discussion, see, also, Case COMP/M.2201 — MAN/Aüwarter.

(82) See, Determination of organisations with significant power (SMP) for the implementation of the ONP Directive, DG XIII, 1 March 1999, at http://europa.eu.int/ISPO/infosoc/telecompolicy/en/SMPdeter.pdf, at paragraph 3.2.

(83) Idem, at paragraph 5.2.

(84) With regard to the interconnection market of fixed and mobile networks, the termination traffic to be measured should include own network traffic and interconnection traffic received from all other fixed and mobile networks, national or international.

(85) *Hoffmann-La Roche* v *Commission*, op. cit., at paragraph 48. One of the most important types of entry barriers is sunk costs. Sunk costs are particularly relevant to the electronic communications sector in view of the fact that large investments are necessary to create, for instance, an efficient electronic communications network for the provision of access services and it is likely that little could be recovered if a new entrant decides to exit the market. Entry barriers are exacerbated by further economies of scope and density which generally characterise such networks. Thus, a large network is always likely to have lower costs than a smaller one, with the result that an entrant in order to take a large share of the market and be able to compete would have to price below the incumbent, making it thus difficult to recover sunk costs.

(86) Joined Cases C-241/91 P and C-242/91 P, *RTE and ITP* v *Commission*, [1995] ECR I-743, Case C-7/97, *Oscar Bronner* [1998] ECR I-7791, and Joined Cases T-374/94, T-375/94, T-384/94 and T-388/94, *European Night Services and others* v *Commission* [1998] ECR II-3141.

(87) Case COMP/M.1741 — MCI WorldCom/Sprint, paragraph 196.

(88) Case C-333/94 P, *Tetra Pak* v *Commission* [1996] ECR I-5951.

(89) See, also, Case COMP/M.2146 — Tetra Laval/Sidel, paragraphs 325-389, *sub judice*, T-5/02.

(90) See Access notice, paragraph 65.

(91) In the case of horizontal markets, the market analysis should focus on establishing the existence of close associative links which will enable an undertaking dominant in one market to behave independently of its competitors in a neighbouring market. Such links may be found to exist by reference to the type of conduct of suppliers and users in the markets under consideration (same customers and/or suppliers in both markets, i.e. customers buying both retail voice calls and retail Internet access) or the fact that the input product or service is essentially the same (i.e. provision by a fixed operator of network infrastructure to ISPs for wholesale call origination and wholesale call termination); see, also, Case T-83/91, *Tetra Pak* v *Commission*, op. cit., paragraph 120 and Case COMP/M.2416 — *Tetra Laval/Sidel*.

(92) Article 14(3) of the framework Directive is not intended to apply in relation to market power leveraged from a 'regulated' market into an emerging, 'non-regulated' market. In such cases, any abusive conduct in the 'emerging' market would normally be dealt with under Article 82 of the EC Treaty.

(93) See Access notice, paragraph 79.

(94) Joined cases C-395/96 P and C-396/96 P, *Compagnie maritime belge and others* v *Commission* [2000] ECR I-1365.

(95) Idem, at paragraph 39.

(96) Case T102/96, *Gencor* v *Commission* [1999] ECR II-753.

(97) See Joined Cases T-68/89, T-77/89 and T-78/89, *SIV and Others* v *Commission* [1992] ECR II-1403, paragraph 358, Case C-393/92 *Almelo* [1994] ECR I-1477, paragraph 43, Case C-96/94, *Centro Servizi Spediporto* [1995] ECR I-2883, paragraph 33, Joined Cases C-140/94, 141/94, and C-142/94, *DIP*, [1995] ECR I-3257, paragraph 62, Case C-70/95, *Sodemare* [1997] ECR I-3395, paragraph 46, and Joined Cases C-68/94 and C-30/95 *France and Others* v *Commission* [1998] ECR I-1375, paragraph 221.

(98) Case IV/M.619 — Gencor Lonhro (OJ L 11, 14.1.1997, p. 30).

(99) *Gencor* v *Commission*, op. cit., at paragraph 276.

(100) Idem, at paragraph 277.

(101) *Compagnie maritime belge transports and Others*, op. cit., at paragraph 39, see, also, Case T-342/99 *Airtours/Commission* [2002] ECR II-0000, paragraph 76.

(102) See, in particular, *France and Others* v *Commission*, op. cit., paragraph 221.

(103) Compagnie maritime belge, at paragraph 39.

(104) Idem at paragraph 44.

(105) Idem at paragraph 45.

(106) The use here of the term 'coordinated effects' is no different from the term 'parallel anticompetitive behaviour' also used in Commission's decisions applying the concept of collective (oligopolistic) dominance.

(107) See in particular, Cases COMP/M.2498 — UPM-Kymmene/Haindl, and COMP/M.2499 — Norske Skog/Parenco/Walsum, Case COMP/M.2201 — MAN/Auwärter, Case COMP/M.2097 — SCA/Matsä Tissue, Case COMP/M.1882 — Pirelli/BICC, Case COMP/M.1741 — MCI WorldCom/Sprint, *sub judice*, T-310/00 Case IV/M.1524 — Airtours/First Choice (OJ L 93, 13.4.2000, p. 1), *sub judice* T-342/99, Case IV/M.1383 — Exxon/Mobil, Case IV/M.1313 — Danish Crown/Vestjyske Slagterier (OJ L 20, 25.1.2000, p. 1), Case IV/M.1225 — Enso/Stora (OJ L 254, 29.9.1999, p. 9), Case IV/M.1016 — Price Waterhouse/Coopers & Lybrand (OJ L 50, 26.2.1999, p. 27), Case IV/M.619 — Gencor/Lonrho, cit., Case IV/M.308, Kali + Salz/MdK/Treuhand (OJ L 186, 21.7.1994, p. 38) and Case IV/M.190 — Nestlé/Perrier (OJ L 356, 5.12.1992, p. 1).

(108) This is in essence the type of analysis carried out by the Commission in past decisions related to collective dominance, see, for instance, Case IV/M.190 — Nestlé/Perrier, (OJ L 356, 5.12.1992, p. 1), Gencor/Lonrho, cit., Case IV/M.1383 — Exxon/Mobil, paragraph 259, Case IV/M.1524 — Airtours/First Choice (OJ L 93, 13.4.2000, p. 1), and Case COMP/M.2499 — Norske Skog/Parenco/Walsum, paragraph 76; see, also, *Airtours v Commission*, op. cit., paragraph 62.

(109) See, also, recital 26 of the framework Directive: 'two or more undertakings can be found to enjoy a joint dominant position not only where there exist structural or other links between them but also where the structure of the relevant market is conducive to coordinated effects, that is, it encourages parallel or aligned anticompetitive behaviour on the market'.

(110) See Case COMP/M.2498 — UPM-Kymmene/Haindl, and Case COMP/M.2499 — Norske Skog/Parenco/Walsum, at paragraph 77.

(111) See, for instance, Case COMP/M.2097 — SCA/Metsä Tissue.

(112) For instance, in Case COMP/M.2201 — MAN/Auwärter, despite the fact that two of the parties present in the German city-bus market in Germany, MAN/Auwärter and EvoBus, would each supply just under half of that market, the Commission concluded that there was no risk of joint dominance. In particular, the Commission found that any tacit division of the market between EvoBus and MAN/Auwärter was not likely as there would be no viable coordination mechanism. Secondly, significant disparities between EvoBus and MAN/Auwärter, such as different cost structures, would make it likely that the companies would compete rather than collude. Likewise, in the Alcoa/British Aluminium case, the Commission found that despite the fact that two of the parties present in the relevant market accounted for almost 80 % of the sales, the market could not be said to be conducive to oligopolistic dominance since (i) market shares were volatile and unstable; and (ii) demand was quite irregular making it difficult for the parties to be able to respond to each other's action in order to tacitly coordinate their behaviour. Furthermore, the market was not transparent in relation to prices and purchasers had significant countervailing power. The Commission's conclusions were further reinforced by the absence of any credible retaliation mechanism likely to sustain any tacit coordination and the fact that competition in the market was not only based on prices but depended to a large extent on technological innovation and after-sales follow-up, Case COMP/M.2111 — Alcoa/British Aluminium.

(113) Likewise, in Case COMP/M.2348 — Outokumpu/Norzink, the Commission found that even if the zinc market was composed of few players, entry barriers were high and demand growth perspectives low, the likelihood of the emergence of a market structure conducive to coordinated outcome was unlikely if it could be shown that (i) parties could not manipulate the formation of prices; (ii) producers had asymmetric cost structures and there was no credible retaliation mechanism in place.

(114) See Case COMP/M.1741 — MCI WorldCom/Sprint, paragraph 263.

(115) Idem, paragraphs 257-302.

(116) Case COMP/M.1838 — BT/Esat.

(117) Idem, paragraphs 10 to 14.

(118) Case IV/M.1430 — Vodafone/Airtouch.

(119) Idem, at paragraph 28. The likely emergence of a duopolistic market concerned only the three largest mobile operators, that is D2 and E-Plus, on the one hand, and T-Mobil on the other hand, given that VIAG Interkom's market share was below 5 %. The Commission's concerns were finally removed after the parties proposed to divest Vodafone's entire stake in E-Plus.

(120) Case COMP/M.2016 — France Telecom/Orange, at paragraph 26.

(121) Idem, at paragraphs 39-40. In its working document 'On the initial findings of the sector inquiry into mobile roaming charges', the Commission made reference to (i) the likely existence of a number of economic links between mobile operators, namely through their interconnection agreements, their membership of the GSM Association, the WAP and the UMTS forum, the fact that terms and conditions of roaming agreements were almost standardised; and (ii) the likely existence of high barriers to entry. In its preliminary assessment the Commission also stressed that the fact that the mobile market is, in general, technology driven, did not seem to have affected the conditions of competition prevailing on the wholesale international roaming market, see: http://europa.eu.int/comm/competition/antitrust/others/sector_inquiries/roaming/, at pages. 24 and 25.

(122) GATS commitments taken by EC on telecommunications: http://gats-info.eu.int/gats-info/swtosvc.pl?&SECCODE=02.C.

(123) The Communications Committee in Article 22 of the framework Directive also aims at ensuring effective cooperation between the Commission and the Member States.

(124) The specific Articles covered are as follows: Articles 15 and 16 of the framework Directive (the latter of which refers to Articles 16-19 of the universal service Directive and Articles 7 and 8 of the access Directive), Articles 5 and 8 of the access Directive (the latter of which refers to the obligations provided for in Articles 9-13 of the access Directive) and Article 16 of the universal service Directive (which refers to Articles 17-19 of universal service Directive). In addition, Article 6 of the access Directive, although not explicitly referenced in Article 7 of the framework Directive, itself contains cross-reference to Article 7 of the framework Directive and is therefore covered by the procedures therein.

(125) Recital 38 of the framework Directive.

(126) As provided for in Article 3 of Council Decision 1999/468/EC laying the procedure for the exercising of implementing powers conferred on the Commission, the Commission shall take the utmost account of the opinion delivered by the Committee, but shall not be bound by the opinion.

# Commission notice on the definition of relevant market for the purposes of Community competition law*

## I. INTRODUCTION

1. The purpose of this notice is to provide guidance as to how the Commission applies the concept of relevant product and geographic market in its ongoing enforcement of Community competition law, in particular the application of Council Regulation No. 17 and (EEC) No. 4064/89, their equivalents in other sectoral applications such as transport, coal and steel, and agriculture, and the relevant provisions of the EEA Agreement.[1] Throughout this notice, references to Articles 85 and 86 of the Treaty and to merger control are to be understood as referring to the equivalent provisions in the EEA Agreement and the ECSC Treaty.
2. Market definition is a tool to identify and define the boundaries of competition between firms. It serves to establish the framework within which competition policy is applied by the Commission. The main purpose of market definition is to identify in a systematic way the competitive constraints that the undertakings involved[2] face. The objective of defining a market in both its product and geographic dimension is to identify those actual competitors of the undertakings involved that are capable of constraining those undertakings' behaviour and of preventing them from behaving independently of effective competitive pressure. It is from this perspective that the market definition makes it possible inter alia to calculate market shares that would convey meaningful information regarding market power for the purposes of assessing dominance or for the purposes of applying Article 85.
3. It follows from point 2 that the concept of 'relevant market' is different from other definitions of market often used in other contexts. For instance, companies often use the term 'market' to refer to the area where it sells its products or to refer broadly to the industry or sector where it belongs.
4. The definition of the relevant market in both its product and its geographic dimensions often has a decisive influence on the assessment of a competition case. By rendering public the procedures which the Commission follows when considering market definition and by indicating the criteria and evidence on which it relies to reach a decision, the Commission expects to increase the transparency of its policy- and decision-making in the area of competition policy.
5. Increased transparency will also result in companies and their advisers being able to better anticipate the possibility that the Commission may raise competition concerns

* (97/C 372/03); (text with EEA relevance).

in an individual case. Companies could, therefore, take such a possibility into account in their own internal decision-making when contemplating, for instance, acquisitions, the creation of joint ventures, or the establishment of certain agreements. It is also intended that companies should be in a better position to understand what sort of information the Commission considers relevant for the purposes of market definition.

6. The Commission's interpretation of 'relevant market' is without prejudice to the interpretation which may be given by the Court of Justice or the Court of First Instance of the European Communities.

## II. DEFINITION OF RELEVANT MARKET

### Definition of Relevant Product Market and Relevant Geographic Market

7. The Regulations based on Article 85 and 86 of the Treaty, in particular in section 6 of Form A/B with respect to Regulation No. 17, as well as in section 6 of Form CO with respect to Regulation (EEC) No. 4064/89 on the control of concentrations having a Community dimension have laid down the following definitions, 'Relevant product markets' are defined as follows:
   'A relevant product market comprises all those products and/or services which are regarded as interchangeable or substitutable by the consumer, by reason of the products' characteristics, their prices and their intended use.'
8. 'Relevant geographic markets' are defined as follows:
   'The relevant geographic market comprises the area in which the undertakings concerned are involved in the supply and demand of products or services, in which the conditions of competition are sufficiently homogeneous and which can be distinguished from neighbouring areas because the conditions of competition are appreciably different in those areas'.
9. The relevant market within which to assess a given competition issue is therefore established by the combination of the product and geographic markets. The Commission interprets the definitions in paragraphs 7 and 8 (which reflect the case-law of the Court of Justice and the Court of First Instance as well as its own decision-making practice) according to the orientations defined in this notice.

### Concept of Relevant Market and Objectives of Community Competition Policy

10. The concept of relevant market is closely related to the objectives pursued under Community competition policy. For example, under the Community's merger control, the objective in controlling structural changes in the supply of a product/service is to prevent the creation or reinforcement of a dominant position as a result of which effective competition would be significantly impeded in a substantial part of the common market. Under the Community's competition rules, a dominant position is such that a firm or group of firms would be in a position to behave to an appreciable extent independently of its competitors, customers and ultimately of its consumers.[3] Such a position would usually arise when a firm or group of firms accounted for a large share of the supply in any given market, provided that other factors analysed in

the assessment (such as entry barriers, customers' capacity to react, etc.) point in the same direction.

11. The same approach is followed by the Commission in its application of Article 86 of the Treaty to firms that enjoy a single or collective dominant position. Within the meaning of Regulation No. 17, the Commission has the power to investigate and bring to an end abuses of such a dominant position, which must also be defined by reference to the relevant market. Markets may also need to be defined in the application of Article 85 of the Treaty, in particular, in determining whether an appreciable restriction of competition exists or in establishing if the condition pursuant to Article 85 (3) (b) for an exemption from the application of Article 85 (1) is met.
12. The criteria for defining the relevant market are applied generally for the analysis of certain types of behaviour in the market and for the analysis of structural changes in the supply of products. This methodology, though, might lead to different results depending on the nature of the competition issue being examined. For instance, the scope of the geographic market might be different when analysing a concentration, where the analysis is essentially prospective, from an analysis of past behaviour. The different time horizon considered in each case might lead to the result that different geographic markets are defined for the same products depending on whether the Commission is examining a change in the structure of supply, such as a concentration or a cooperative joint venture, or examining issues relating to certain past behaviour.

## Basic Principles for Market Definition

### Competitive constraints

13. Firms are subject to three main sources or competitive constraints: demand substitutability, supply substitutability and potential competition. From an economic point of view, for the definition of the relevant market, demand substitution constitutes the most immediate and effective disciplinary force on the suppliers of a given product, in particular in relation to their pricing decisions. A firm or a group of firms cannot have a significant impact on the prevailing conditions of sale, such as prices, if its customers are in a position to switch easily to available substitute products or to suppliers located elsewhere. Basically, the exercise of market definition consists in identifying the effective alternative sources of supply for the customers of the undertakings involved, in terms both of products/services and of geographic location of suppliers.
14. The competitive constraints arising from supply side substitutability other than those described in paragraphs 20 to 23 and from potential competition are in general less immediate and in any case require an analysis of additional factors. As a result such constraints are taken into account at the assessment stage of competition analysis.

### Demand substitution

15. The assessment of demand substitution entails a determination of the range of products which are viewed as substitutes by the consumer. One way of making this determination can be viewed as a speculative experiment, postulating a hypothetical small, lasting change in relative prices and evaluating the likely reactions of customers to that increase. The exercise of market definition focuses on prices for operational and practical purposes, and more precisely on demand substitution arising from small,

permanent changes in relative prices. This concept can provide clear indications as to the evidence that is relevant in defining markets.

16. Conceptually, this approach means that, starting from the type of products that the undertakings involved sell and the area in which they sell them, additional products and areas will be included in, or excluded from, the market definition depending on whether competition from these other products and areas affect or restrain sufficiently the pricing of the parties' products in the short term.
17. The question to be answered is whether the parties' customers would switch to readily available substitutes or to suppliers located elsewhere in response to a hypothetical small (in the range 5 % to 10 %) but permanent relative price increase in the products and areas being considered. If substitution were enough to make the price increase unprofitable because of the resulting loss of sales, additional substitutes and areas are included in the relevant market. This would be done until the set of products and geographical areas is such that small, permanent increases in relative prices would be profitable. The equivalent analysis is applicable in cases concerning the concentration of buying power, where the starting point would then be the supplier and the price test serves to identify the alternative distribution channels or outlets for the supplier's products. In the application of these principles, careful account should be taken of certain particular situations as described within paragraphs 56 and 58.
18. A practical example of this test can be provided by its application to a merger of, for instance, soft-drink bottlers. An issue to examine in such a case would be to decide whether different flavours of soft drinks belong to the same market. In practice, the question to address would be whether consumers of flavour A would switch to other flavours when confronted with a permanent price increase of 5 % to 10 % for flavour A. If a sufficient number of consumers would switch to, say, flavour B, to such an extent that the price increase for flavour A would not be profitable owing to the resulting loss of sales, then the market would comprise at least flavours A and B. The process would have to be extended in addition to other available flavours until a set of products is identified for which a price rise would not induce a sufficient substitution in demand.
19. Generally, and in particular for the analysis of merger cases, the price to take into account will be the prevailing market price. This may not be the case where the prevailing price has been determined in the absence of sufficient competition. In particular for the investigation of abuses of dominant positions, the fact that the prevailing price might already have been substantially increased will be taken into account.

**Supply substitution**

20. Supply-side substitutability may also be taken into account when defining markets in those situations in which its effects are equivalent to those of demand substitution in terms of effectiveness and immediacy. This means that suppliers are able to switch production to the relevant products and market them in the short term[4] without incurring significant additional costs or risks in response to small and permanent changes in relative prices. When these conditions are met, the additional production that is put on the market will have a disciplinary effect on the competitive behaviour of the companies involved. Such an impact in terms of effectiveness and immediacy is equivalent to the demand substitution effect.

21. These situations typically arise when companies market a wide range of qualities or grades of one product; even if, for a given final customer or group of consumers, the different qualities are not substitutable, the different qualities will be grouped into one product market, provided that most of the suppliers are able to offer and sell the various qualities immediately and without the significant increases in costs described above. In such cases, the relevant product market will encompass all products that are substitutable in demand and supply, and the current sales of those products will be aggregated so as to give the total value or volume of the market. The same reasoning may lead to group different geographic areas.
22. A practical example of the approach to supply-side substitutability when defining product markets is to be found in the case of paper. Paper is usually supplied in a range of different qualities, from standard writing paper to high quality papers to be used, for instance, to publish art books. From a demand point of view, different qualities of paper cannot be used for any given use, i.e. an art book or a high quality publication cannot be based on lower quality papers. However, paper plants are prepared to manufacture the different qualities, and production can be adjusted with negligible costs and in a short time-frame. In the absence of particular difficulties in distribution, paper manufacturers are able, therefore, to compete for orders of the various qualities, in particular if orders are placed with sufficient lead time to allow for modification of production plans. Under such circumstances, the Commission would not define a separate market for each quality of paper and its respective use. The various qualities of paper are included in the relevant market, and their sales added up to estimate total market value and volume.
23. When supply-side substitutability would entail the need to adjust significantly existing tangible and intangible assets, additional investments, strategic decisions or time delays, it will not be considered at the stage of market definition. Examples where supply-side substitution did not induce the Commission to enlarge the market are offered in the area of consumer products, in particular for branded beverages. Although bottling plants may in principle bottle different beverages, there are costs and lead times involved (in terms of advertising, product testing and distribution) before the products can actually be sold. In these cases, the effects of supply-side substitutability and other forms of potential competition would then be examined at a later stage.

**Potential competition**

24. The third source of competitive constraint, potential competition, is not taken into account when defining markets, since the conditions under which potential competition will actually represent an effective competitive constraint depend on the analysis of specific factors and circumstances related to the conditions of entry. If required, this analysis is only carried out at a subsequent stage, in general once the position of the companies involved in the relevant market has already been ascertained, and when such position gives rise to concerns from a competition point of view.

# III. EVIDENCE RELIED ON TO DEFINE RELEVANT MARKETS

## The Process of Defining the Relevant Market in Practice

### Product dimension

25. There is a range of evidence permitting an assessment of the extent to which substitution would take place. In individual cases, certain types of evidence will be determinant, depending very much on the characteristics and specificity of the industry and products or services that are being examined. The same type of evidence may be of no importance in other cases. In most cases, a decision will have to be based on the consideration of a number of criteria and different items of evidence. The Commission follows an open approach to empirical evidence, aimed at making an effective use of all available information which may be relevant in individual cases. The Commission does not follow a rigid hierarchy of different sources of information or types of evidence.
26. The process of defining relevant markets may be summarized as follows: on the basis of the preliminary information available or information submitted by the undertakings involved, the Commission will usually be in a position to broadly establish the possible relevant markets within which, for instance, a concentration or a restriction of competition has to be assessed. In general, and for all practical purposes when handling individual cases, the question will usually be to decide on a few alternative possible relevant markets. For instance, with respect to the product market, the issue will often be to establish whether product A and product B belong or do not belong to the same product market. It is often the case that the inclusion of product B would be enough to remove any competition concerns.
27. In such situations it is not necessary to consider whether the market includes additional products, or to reach a definitive conclusion on the precise product market. If under the conceivable alternative market definitions the operation in question does not raise competition concerns, the question of market definition will be left open, reducing thereby the burden on companies to supply information.

### Geographic dimension

28. The Commission's approach to geographic market definition might be summarized as follows: it will take a preliminary view of the scope of the geographic market on the basis of broad indications as to the distribution of market shares between the parties and their competitors, as well as a preliminary analysis of pricing and price differences at national and Community or EEA level. This initial view is used basically as a working hypothesis to focus the Commission's inquiries for the purposes of arriving at a precise geographic market definition.
29. The reasons behind any particular configuration of prices and market shares need to be explored. Companies might enjoy high market shares in their domestic markets just because of the weight of the past, and conversely, a homogeneous presence of companies throughout the EEA might be consistent with national or regional geographic markets. The initial working hypothesis will therefore be checked against an analysis of demand characteristics (importance of national or local preferences, current patterns of purchases of customers, product differentiation/brands, other) in order to

establish whether companies in different areas do indeed constitute a real alternative source of supply for consumers. The theoretical experiment is again based on substitution arising from changes in relative prices, and the question to answer is again whether the customers of the parties would switch their orders to companies located elsewhere in the short term and at a negligible cost.

30. If necessary, a further check on supply factors will be carried out to ensure that those companies located in differing areas do not face impediments in developing their sales on competitive terms throughout the whole geographic market. This analysis will include an examination of requirements for a local presence in order to sell in that area the conditions of access to distribution channels, costs associated with setting up a distribution network, and the presence or absence of regulatory barriers arising from public procurement, price regulations, quotas and tariffs limiting trade or production, technical standards, monopolies, freedom of establishment, requirements for administrative authorizations, packaging regulations, etc. In short, the Commission will identify possible obstacles and barriers isolating companies located in a given area from the competitive pressure of companies located outside that area, so as to determine the precise degree of market interpenetration at national, European or global level.

31. The actual pattern and evolution of trade flows offers useful supplementary indications as to the economic importance of each demand or supply factor mentioned above, and the extent to which they may or may not constitute actual barriers creating different geographic markets. The analysis of trade flows will generally address the question of transport costs and the extent to which these may hinder trade between different areas, having regard to plant location, costs of production and relative price levels.

**Market integration in the Community**

32. Finally, the Commission also takes into account the continuing process of market integration, in particular in the Community, when defining geographic markets, especially in the area of concentrations and structural joint ventures. The measures adopted and implemented in the internal market programme to remove barriers to trade and further integrate the Community markets cannot be ignored when assessing the effects on competition of a concentration or a structural joint venture. A situation where national markets have been artifically isolated from each other because of the existence of legislative barriers that have now been removed will generally lead to a cautious assessment of past evidence regarding prices, market shares or trade patterns. A process of market integration that would, in the short term, lead to wider geographic markets may therefore be taken into consideration when defining the geographic market for the purposes of assessing concentrations and joint ventures.

**The Process of Gathering Evidence**

33. When a precise market definition is deemed necessary, the Commission will often contact the main customers and the main companies in the industry to enquire into their views about the boundaries of product and geographic markets and to obtain the necessary factual evidence to reach a conclusion. The Commission might also contact the relevant professional associations, and companies active in upstream markets, so as to be able to define, in so far as necessary, separate product and geographic markets,

for different levels of production or distribution of the products/services in question. It might also request additional information to the undertakings involved.

34. Where appropriate, the Commission will address written requests for information to the market players mentioned above. These requests will usually include questions relating to the perceptions of companies about reactions to hypothetical price increases and their views of the boundaries of the relevant market. They will also ask for provision of the factual information the Commission deems necessary to reach a conclusion on the extent of the relevant market. The Commission might also discuss with marketing directors or other officers of those companies to gain a better understanding on how negotiations between suppliers and customers take place and better understand issues relating to the definition of the relevant market. Where appropriate, they might also carry out visits or inspections to the premises of the parties, their customers and/or their competitors, in order to better understand how products are manufactured and sold.

35. The type of evidence relevant to reach a conclusion as to the product market can be categorized as follows:

**Evidence for defining markets – product dimension**

36. An analysis of the product characteristics and its intended use allows the Commission, as a first step, to limit the field of investigation of possible substitutes. However, product characteristics and intended use are insufficient to show whether two products are demand substitutes. Functional interchangeability or similarity in characteristics may not, in themselves, provide sufficient criteria, because the responsiveness of customers to relative price changes may be determined by other considerations as well. For example, there may be different competitive constraints in the original equipment market for car components and in spare parts, thereby leading to a separate delineation of two relevant markets. Conversely, differences in product characteristics are not in themselves sufficient to exclude demand substitutability, since this will depend to a large extent on how customers value different characteristics.

37. The type of evidence the Commission considers relevant to assess whether two products are demand substitutes can be categorized as follows:

38. Evidence of substitution in the recent past. In certain cases, it is possible to analyse evidence relating to recent past events or shocks in the market that offer actual examples of substitution between two products. When available, this sort of information will normally be fundamental for market definition. If there have been changes in relative prices in the past (all else being equal), the reactions in terms of quantities demanded will be determinant in establishing substitutability. Launches of new products in the past can also offer useful information, when it is possible to precisely analyse which products have lost sales to the new product.

39. There are a number of quantitative tests that have specifically been designed for the purpose of delineating markets. These tests consist of various econometric and statistical approaches, estimates of elasticities and cross-price elasticities[5] for the demand of a product, tests based on similarity of price movements over time, the analysis of causality between price series and similarity of price levels and/or their convergence. The Commission takes into account the available quantitative evidence capable of withstanding rigorous scrutiny for the purposes of establishing patterns of substitution in the past.

40. Views of customers and competitors. The Commission often contacts the main customers and competitors of the companies involved in its enquiries, to gather their views on the boundaries of the product market as well as most of the factual information it requires to reach a conclusion on the scope of the market. Reasoned answers of customers and competitors as to what would happen if relative prices for the candidate products were to increase in the candidate geographic area by a small amount (for instance of 5 % to 10 %) are taken into account when they are sufficiently backed by factual evidence.
41. Consumer preferences. In the case of consumer goods, it may be difficult for the Commission to gather the direct views of end consumers about substitute products. Marketing studies that companies have commissioned in the past and that are used by companies in their own decision-making as to pricing of their products and/or marketing actions may provide useful information for the Commission's delineation of the relevant market. Consumer surveys on usage patterns and attitudes, data from consumers' purchasing patterns, the views expressed by retailers and more generally, market research studies submitted by the parties and their competitors are taken into account to establish whether an economically significant proportion of consumers consider two products as substitutable, also taking into account the importance of brands for the products in question. The methodology followed in consumer surveys carried out ad hoc by the undertakings involved or their competitors for the purposes of a merger procedure or a procedure pursuant to Regulation No. 17 will usually be scrutinized with utmost care. Unlike pre-existing studies, they have not been prepared in the normal course of business for the adoption of business decisions.
42. Barriers and costs associated with switching demand to potential substitutes. There are a number of barriers and costs that might prevent the Commission from considering two prima facie demand substitutes as belonging to one single product market. It is not possible to provide an exhaustive list of all the possible barriers to substitution and of switching costs. These barriers or obstacles might have a wide range of origins, and in its decisions, the Commission has been confronted with regulatory barriers or other forms of State intervention, constraints arising in downstream markets, need to incur specific capital investment or loss in current output in order to switch to alternative inputs, the location of customers, specific investment in production process, learning and human capital investment, retooling costs or other investments, uncertainty about quality and reputation of unknown suppliers, and others.
43. Different categories of customers and price discrimination. The extent of the product market might be narrowed in the presence of distinct groups of customers. A distinct group of customers for the relevant product may constitute a narrower, distinct market when such a group could be subject to price discrimination. This will usually be the case when two conditions are met: (a) it is possible to identify clearly which group an individual customer belongs to at the moment of selling the relevant products to him, and (b) trade among customers or arbitrage by third parties should not be feasible.

**Evidence for defining markets – geographic dimension**

44. The type of evidence the Commission considers relevant to reach a conclusion as to the geographic market can be categorized as follows:
45. Past evidence of diversion of orders to other areas. In certain cases, evidence on changes in prices between different areas and consequent reactions by customers

might be available. Generally, the same quantitative tests used for product market definition might as well be used in geographic market definition, bearing in mind that international comparisons of prices might be more complex due to a number of factors such as exchange rate movements, taxation and product differentiation.

46. Basic demand characteristics. The nature of demand for the relevant product may in itself determine the scope of the geographical market. Factors such as national preferences or preferences for national brands, language, culture and life style, and the need for a local presence have a strong potential to limit the geographic scope of competition.

47. Views of customers and competitors. Where appropriate, the Commission will contact the main customers and competitors of the parties in its enquiries, to gather their views on the boundaries of the geographic market as well as most of the factual information it requires to reach a conclusion on the scope of the market when they are sufficiently backed by factual evidence.

48. Current geographic pattern of purchases. An examination of the customers' current geographic pattern of purchases provides useful evidence as to the possible scope of the geographic market. When customers purchase from companies located anywhere in the Community or the EEA on similar terms, or they procure their supplies through effective tendering procedures in which companies from anywhere in the Community or the EEA submit bids, usually the geographic market will be considered to be Community-wide.

49. Trade flows/pattern of shipments. When the number of customers is so large that it is not possible to obtain through them a clear picture of geographic purchasing patterns, information on trade flows might be used alternatively, provided that the trade statistics are available with a sufficient degree of detail for the relevant products. Trade flows, and above all, the rationale behind trade flows provide useful insights and information for the purpose of establishing the scope of the geographic market but are not in themselves conclusive.

50. Barriers and switching costs associated to divert orders to companies located in other areas. The absence of trans-border purchases or trade flows, for instance, does not necessarily mean that the market is at most national in scope. Still, barriers isolating the national market have to be identified before it is concluded that the relevant geographic market in such a case is national. Perhaps the clearest obstacle for a customer to divert its orders to other areas is the impact of transport costs and transport restrictions arising from legislation or from the nature of the relevant products. The impact of transport costs will usually limit the scope of the geographic market for bulky, low-value products, bearing in mind that a transport disadvantage might also be compensated by a comparative advantage in other costs (labour costs or raw materials). Access to distribution in a given area, regulatory barriers still existing in certain sectors, quotas and custom tariffs might also constitute barriers isolating a geographic area from the competitive pressure of companies located outside that area. Significant switching costs in procuring supplies from companies located in other countries constitute additional sources of such barriers.

51. On the basis of the evidence gathered, the Commission will then define a geographic market that could range from a local dimension to a global one, and there are examples of both local and global markets in past decisions of the Commission.

52. The paragraphs above describe the different factors which might be relevant to define markets. This does not imply that in each individual case it will be necessary to obtain evidence and assess each of these factors. Often in practice the evidence provided by a subset of these factors will be sufficient to reach a conclusion, as shown in the past decisional practice of the Commission.

## IV. CALCULATION OF MARKET SHARE

53. The definition of the relevant market in both its product and geographic dimensions allows the identification of the suppliers and the customers/consumers active on that market. A total market size and market shares for each supplier can be calculated on the basis of their sales of the relevant products in the relevant area. In practice, the total market size and market shares are often available from market sources, i.e. companies' estimates, studies commissioned from industry consultants and/or trade associations. When this is not the case, or when available estimates are not reliable, the Commission will usually ask each supplier in the relevant market to provide its own sales in order to calculate total market size and market shares.
54. If sales are usually the reference to calculate market shares, there are nevertheless other indications that, depending on the specific products or industry in question, can offer useful information such as, in particular, capacity, the number of players in bidding markets, units of fleet as in aerospace, or the reserves held in the case of sectors such as mining.
55. As a rule of thumb, both volume sales and value sales provide useful information. In cases of differentiated products, sales in value and their associated market share will usually be considered to better reflect the relative position and strength of each supplier.

## V. ADDITIONAL CONSIDERATIONS

56. There are certain areas where the application of the principles above has to be undertaken with care. This is the case when considering primary and secondary markets, in particular, when the behaviour of undertakings at a point in time has to be analysed pursuant to Article 86. The method of defining markets in these cases is the same, i.e. assessing the responses of customers based on their purchasing decisions to relative price changes, but taking into account as well, constraints on substitution imposed by conditions in the connected markets. A narrow definition of the market for secondary products, for instance, spare parts, may result when compatibility with the primary product is important. Problems of finding compatible secondary products together with the existence of high prices and a long lifetime of the primary products may render relative price increases of secondary products profitable. A different market definition may result if significant substitution between secondary products is possible or if the characteristics of the primary products make quick and direct consumer responses to relative price increases of the secondary products feasible.
57. In certain cases, the existence of chains of substitution might lead to the definition of a relevant market where products or areas at the extreme of the market are not directly

substitutable. An example might be provided by the geographic dimension of a product with significant transport costs. In such cases, deliveries from a given plant are limited to a certain area around each plant by the impact of transport costs. In principle, such an area could constitute the relevant geographic market. However, if the distribution of plants is such that there are considerable overlaps between the areas around different plants, it is possible that the pricing of those products will be constrained by a chain substitution effect, and lead to the definition of a broader geographic market. The same reasoning may apply if product B is a demand substitute for products A and C. Even if products A and C are not direct demand substitutes, they might be found to be in the same relevant product market since their respective pricing might be constrained by substitution to B.

58. From a practical perspective, the concept of chains of substitution has to be corroborated by actual evidence, for instance related to price interdependence at the extremes of the chains of substitution, in order to lead to an extension of the relevant market in an individual case. Price levels at the extremes of the chains would have to be of the same magnitude as well.

## NOTES

1. The focus of assessment in State aid cases is the aid recipient and the industry/sector concerned rather than identification of competitive constraints faced by the aid recipient. When consideration of market power and therefore of the relevant market are raised in any particular case, elements of the approach outlined here might serve as a basis for the assessment of State aid cases.
2. For the purposes of this notice, the undertakings involved will be, in the case of a concentration, the parties to the concentration; in investigations within the meaning of Article 86 of the Treaty, the undertaking being investigated or the complainants; for investigations within the meaning of Article 85, the parties to the Agreement.
3. Definition given by the Court of Justice in its judgment of 13 February 1979 in Case 85/76, *Hoffmann-La Roche* [1979] ECR 461, and confirmed in subsequent judgments.
4. That is such a period that does not entail a significant adjustment of existing tangible and intangible assets (see paragraph 23).
5. Own-price elasticity of demand for product X is a measure of the responsiveness of demand for X to percentage change in its own price. Cross-price elasticity between products X and Y is the responsiveness of demand for product X to percentage change in the price of product Y.

# Commission Recommendation of 11/02/2003*

THE COMMISSION OF THE EUROPEAN COMMUNITIES,

Having regard to the Treaty establishing the European Community,

Having regard to Directive 2002/21/EC of the European Parliament and of the Council on a common regulatory framework for electronic communication networks and services,[1] and in particular Article 15 thereof,

Whereas:

(1) Directive 2002/21/EC (hereinafter the Framework Directive) establishes a new legislative framework for the electronic communications sector that seeks to respond to convergence trends by covering all electronic communications networks and services within its scope. The aim is to reduce *ex-ante* sector-specific rules progressively as competition in the market develops.

(2) The purpose of this Recommendation is to identify those product and service markets in which *ex-ante* regulation may be warranted. However, this first Recommendation has to be consistent with the transition from the 1998 regulatory framework to the new regulatory framework. Directive 2002/19/EC of the European Parliament and of the Council on access to, and interconnection of, electronic communications networks and associated facilities,[2] hereinafter the Access Directive, and Directive 2002/22/EC of the European Parliament and of the Council on universal service and users' rights relating to electronic communications networks and services,[3] hereinafter the Universal Service Directive, already identify specific market areas which need to be analysed by national regulatory authorities in addition to the markets listed in this Recommendation. In accordance with the Framework Directive, it is for national regulatory authorities to define relevant geographic markets within their territory.

(3) Under the 1998 regulatory framework, several areas in the telecommunications sector are subject to *ex-ante* regulation. These areas have been delineated in the applicable directives, but are not always 'markets' within the meaning of competition law and practice. Annex I of the Framework Directive provides a list of such market areas to be included in the initial version of the Recommendation.

* on Relevant Product and Service Markets within the electronic communications sector susceptible to ex ante regulation in accordance with Directive 2002/21/EC of the European Parliament and of the Council on a common regulatory framework for electronic communication networks and services (Text with EEA relevance). Brussels, C(2003)497

(4) As the title of Annex I of the Framework Directive makes clear, all the market areas listed therein need to be included in the initial version of the Recommendation in order that NRAs can carry out a review of existing obligations imposed under the 1998 regulatory framework.

(5) Article 15(1) of the Framework Directive requires the Commission to define markets in accordance with the principles of competition law. The Commission has therefore defined markets (corresponding to the market areas listed in Annex I of the Framework Directive) in accordance with competition law principles.

(6) There are in the electronic communications sector at least two main types of relevant markets to consider: markets for services or products provided to end users (retail markets), and markets for the inputs which are necessary for operators to provide services and products to end users (wholesale markets). Within these two types of markets, further market distinctions may be made depending on demand and supply-side characteristics.

(7) The starting point for the definition and identification of markets is a characterisation of retail markets over a given time horizon, taking into account demand-side and supply-side substitutability. Having characterised and defined retail markets which are markets involving the supply and demand of end users, it is then appropriate to identify relevant wholesale markets which are markets involving the demand of products of, and supply of products to, a third party wishing to supply end users.

(8) Defining markets in accordance with the principles of competition law means that some of the market areas in Annex I of the Framework Directive comprise a number of separate individual markets on the basis of demand-side characteristics. This is the case of products for retail access to the public telephone network at a fixed location and for telephone services provided at a fixed location. The market area in Annex I referring to wholesale leased lines is defined as separate markets for wholesale terminating segments and wholesale trunk segments on the basis of both demand-side and supply-side characteristics.

(9) In identifying markets in accordance with competition law principles, recourse should be had to the following three criteria. The first criterion is the presence of high and non-transitory entry barriers whether of structural, legal or regulatory nature. However, given the dynamic character and functioning of electronic communications markets, possibilities to overcome barriers within a relevant time horizon have also to be taken into consideration when carrying out a prospective analysis to identify the relevant markets for possible *ex-ante* regulation. Therefore the second criterion admits only those markets the structure of which does not tend towards effective competition within the relevant time horizon. The application of this criterion involves examining the state of competition behind the barriers of entry. The third criterion is that application of competition law alone would not adequately address the market failure(s) concerned.

(10) In particular, as far as entry barriers are concerned, two types of entry barriers are relevant for the purpose of this Recommendation: structural barriers and legal or regulatory barriers.

(11) Structural barriers to entry result from original cost or demand conditions that create asymmetric conditions between incumbents and new entrants impeding or preventing market entry of the latter. For instance, high structural barriers may be found to exist when the market is characterised by substantial economies of scale and/or economies of scope and high sunk cost. To date, such barriers can still be identified with respect to the widespread deployment and/or provision of local access networks to fixed locations. A related structural barrier can also exist where the provision of service requires a network component that cannot be technically duplicated or only duplicated at a cost that makes it uneconomic for competitors.

(12) Legal or regulatory barriers are not based on economic conditions, but result from legislative, administrative or other state measures that have a direct effect on the conditions of entry and/or the positioning of operators on the relevant market. Examples are legal or regulatory barriers preventing entry into a market where there is a limit on the number of undertakings that have access to spectrum for the provision of underlying services. Other examples of legal or regulatory barriers are price controls or other price-related measures imposed on undertakings, which affect not only entry but also the positioning of undertakings on the market.

(13) Entry barriers may also become less relevant with regard to innovation-driven markets characterised by ongoing technological progress. In such markets, competitive constraints often come from innovative threats from potential competitors that are not currently in the market. In such innovation-driven markets, dynamic or longer-term competition can take place among firms that are not necessarily competitors in an existing 'static' market. This Recommendation does not identify markets where entry barriers are not expected to persist over a foreseeable period.

(14) Even when a market is characterised by high barriers to entry, other structural factors in that market may mean that the market tends towards an effectively competitive outcome within the relevant time horizon. This may for instance be the case in markets with a limited – but sufficient – number of undertakings having diverging cost structures and facing price-elastic market demand. There may also be excess capacity in a market that would allow rival firms to expand output very rapidly in response to any price increase. In such markets, market shares may change over time and/or falling prices may be observed.

(15) The decision to identify a market as justifying possible *ex-ante* regulation should also depend on an assessment of the sufficiency of competition law in reducing or removing such barriers or in restoring effective competition. Furthermore, new and emerging markets, in which market power may be found to exist because of 'first-mover' advantages, should not in principle be subject to *ex-ante* regulation.

(16) In undertaking periodic reviews of the markets identified in this Recommendation, the three criteria should be used. These criteria should be applied cumulatively, so that failing any one of them means that the market should not be identified in subsequent recommendations. Thus, whether an electronic communications market continues to be identified by subsequent versions of the Recommendation as justifying possible *ex-ante* regulation would depend on the persistence of high entry barriers, on the second criterion measuring the dynamic state of competitiveness and thirdly on the sufficiency of competition law (absent *ex-ante* regulation) to address persistent market failures. A market could also be removed from a recommendation once there is evidence of sustainable and effective competition on that market within the Community, provided that the removal of existing regulation obligations would not reduce competition on that market.

(17) The Annex to this Recommendation indicates how each market in the Recommendation is linked to the market areas in Annex I to the Framework Directive. When reviewing existing obligations imposed under the previous regulatory framework, in order to determine whether to maintain, amend or withdraw them, NRAs should undertake the analysis on the basis of the markets identified in this Recommendation, in order to give effect to the requirement that market definition for the purposes of *ex-ante* regulation should be based on competition law principles. Pending the first market analysis by NRAs under the new regulatory framework, existing obligations remain in force.

(18) The identification of markets in this Recommendation is without prejudice to markets that may be defined in specific cases under competition law.

(19) The range of different network topologies and technologies deployed across the Community means that in some cases national regulatory authorities must decide the precise boundaries between, or elements within, particular markets identified in the Recommendation, while adhering to competition law principles. National regulatory authorities may identify markets that differ from those of the Recommendation, provided they act in accordance with Article 7 of the Framework Directive. Since the imposition of *ex-ante* regulation on a market could affect trade between Member States as described in recital 38 of the Framework Directive, the Commission considers that the identification of any market that differs from those of the Recommendation are likely to be subject to the appropriate procedure in Article 7 of the Framework Directive. Failure to notify a market which affects trade between Member States may result in infringement proceedings being taken. Any market identified by national regulatory authorities should be based on the competition principles developed in the Commission Notice on the definition of relevant market for the purposes of Community competition law,[4] and be consistent with the Commission Guidelines on market analysis and the assessment of significant market power and satisfy the three criteria set out above. Should an NRA consider that demand and supply patterns may justify an alternative market definition of a market listed in this Recommendation, it should then follow the appropriate procedures set out in Articles 6 and 7 of the Framework Directive.

(20) The fact that this Recommendation identifies those product and service markets in which *ex-ante* regulation may be warranted does not mean that regulation is always warranted or that these markets will be subject to the imposition of regulatory obligations set out in the specific Directives. Regulation will not be warranted if there is effective competition on these markets. In particular, regulatory obligations must be appropriate and be based on the nature of the problem identified, proportionate and justified in the light of the objectives laid down in the Framework Directive, in particular maximising benefits for users, ensuring no distortion or restriction of competition, encouraging efficient investment in infrastructure and promoting innovation, and encouraging efficient use and management of radio frequencies and numbering resources.

(21) The Commission will review the need for any update of this Recommendation no later than 30 June 2004 on the basis of market developments.

(22) This Recommendation has been subject to a public consultation and to consultation with national regulatory authorities and national competition authorities.

## HAS ADOPTED THIS RECOMMENDATION:

1. In defining relevant markets in accordance with Article 15(3) of Directive 2002/21/EC, national regulatory authorities are recommended to analyse the product and service markets identified in the Annex.

2. This Recommendation is addressed to the Member States.

*For the Commission*
*Erkki Liikanen*
*Member of the Commission*

## NOTES

1. OJ L 108, 24.4.2002, p. 33.
2. OJ L 108, 24.4.2002, p. 7.
3. OJ L 108, 24.4.2002, p. 51.
4. OJ C 372, 9.12.1997, p. 5.

# ANNEX

## Retail Level

1. Access to the public telephone network at a fixed location for residential customers.

2. Access to the public telephone network at a fixed location for non-residential customers.

3. Publicly available local and/or national telephone services provided at a fixed location for residential customers.

4. Publicly available international telephone services provided at a fixed location for residential customers.

5. Publicly available local and/or national telephone services provided at a fixed location for non-residential customers.

6. Publicly available international telephone services provided at a fixed location for non-residential customers.

   These six markets are identified for the purpose of analysis in respect of Article 17 of the Universal Service Directive.

   Together, markets 1 through 6 correspond to 'the provision of connection to and use of the public telephone network at fixed locations', referred to in Annex I (1) of the Framework Directive. This combined market is also referred to in Article 19 of the Universal Service Directive (for possible imposition of carrier call-by-call selection or carrier selection).

7. The minimum set of leased lines (which comprises the specified types of leased lines up to and including 2Mb/sec as referenced in Article 18 and Annex VII of the Universal Service Directive).

   This market is referred to in Annex I (1) of the Framework Directive in respect of Article 16 of the Universal Service Directive ('the provision of leased lines to end users').

   A market analysis must be undertaken for the purposes of Article 18 of the Universal Service Directive which covers regulatory controls on the provision of the minimum set of leased lines.

## Wholesale Level

8. Call origination on the public telephone network provided at a fixed location. For the purposes of this Recommendation, call origination is taken to include local call conveyance and delineated in such a way as to be consistent with the delineated boundaries

for the markets for call transit and for call termination on the public telephone network provided at a fixed location.

This market corresponds to that referred to in Annex I (2) of the Framework Directive in respect of Directive 97/33/EC ('call origination in the fixed public telephone network').

9. Call termination on individual public telephone networks provided at a fixed location.

For the purposes of this Recommendation, call termination is taken to include local call conveyance and delineated in such a way as to be consistent with the delineated boundaries for the markets for call origination and for call transit on the public telephone network provided at a fixed location.

This market corresponds to the one referred to in Annex I (2) of the Framework Directive in respect of Directive 97/33/EC ('call termination in the fixed public telephone network').

10. Transit services in the fixed public telephone network.

For the purposes of this Recommendation, transit services are taken as being delineated in such a way as to be consistent with the delineated boundaries for the markets for call origination and for call termination on the public telephone network provided at a fixed location.

This market corresponds to the one referred to in Annex I (2) of the Framework Directive in respect of Directive 97/33/EC ('transit services in the fixed public telephone network').

11. Wholesale unbundled access (including shared access) to metallic loops and sub-loops for the purpose of providing broadband and voice services.

This market corresponds to that referred to in Annex I (2) of the Framework Directive in respect of Directive 97/33/EC and Directive 98/10/EC ('access to the fixed public telephone network, including unbundled access to the local loop') and to that referred to in Annex I (3) of the Framework Directive in respect of Regulation No. 2887/2000.

12. Wholesale broadband access.

This market covers 'bit-stream' access that permits the transmission of broadband data in both directions and other wholesale access provided over other infrastructures, if and when they offer facilities equivalent to bit-stream access. It includes 'Network access and special network access' referred to in Annex I (2) of the Framework Directive, but does not cover the market in point 11 above, nor the market in point 18.

13. Wholesale terminating segments of leased lines.

14. Wholesale trunk segments of leased lines.

    Together, the wholesale markets 13 and 14 correspond to those referred to in Annex I (2) of the Framework Directive in respect of Directive 97/33/EC and Directive 98/10/EC ('leased line interconnection') and to those referred to in Annex I (2) of the Framework Directive in respect of Directive 92/44/EEC ('wholesale provision of leased line capacity to other suppliers of electronic communications networks or services').

15. Access and call origination on public mobile telephone networks.

    Referred to (separately) in Annex I (2) of the Framework Directive in respect of Directives 97/33/EC and 98/10/EC.

16. Voice call termination on individual mobile networks.

    This market corresponds to the one referred to in Annex I (2) of the Framework Directive in respect of Directive 97/33/EC ('call termination on public mobile telephone networks').

17. The wholesale national market for international roaming on public mobile networks.

    This market corresponds to the one referred to in Annex I (4) of the Framework Directive.

18. Broadcasting transmission services, to deliver broadcast content to end users.

**Note**

National regulatory authorities have discretion with respect to the analysis of the market for 'Conditional access systems to digital television and radio services broadcast' in accordance with Article 6(3) of the Access Directive. Article 6(3) of the Access Directive provides that Member States may permit their NRAs to review the market for conditional access system to digital television and radio services broadcast, irrespective of the means of transmission.

# EXPLANATORY MEMORANDUM

## 1. Introduction

The Lisbon European Council of March 2000 highlighted the potential for growth, competitiveness and job creation of the shift to a digital, knowledge-based economy. In particular it emphasised the importance of access to inexpensive, world-class communications infrastructure and services. The legislative package for the electronic communications sector aims to establish a harmonised regulatory framework for networks and services across the EU and seeks to respond to convergence trends by covering all electronic communications networks and services within its scope. The national measures implementing the new legislation will apply from 25 July 2003, when it will supersede the national measures implementing the existing 1998 legislative framework. The Barcelona European Council of March 2002 encouraged Member States to ensure timely implementation of the package.

The new regulatory framework for electronic communications networks and services comprises five Directives:

1. Directive of the European Parliament and of the Council on a common regulatory framework for electronic communications networks and services,[1] hereinafter the Framework Directive
2. Directive of the European Parliament and of the Council on the authorisation of electronic communications networks and services,[2] hereinafter the Authorisation Directive
3. Directive of the European Parliament and of the Council on access to, and interconnection of, electronic communications networks and associated facilities,[3] hereinafter the Access Directive
4. Directive of the European Parliament and of the Council on universal service and users' rights relating to electronic communications networks and services,[4] hereinafter the Universal Service Directive
5. Directive of the European Parliament and of the Council concerning the processing of personal data and the protection of privacy in the electronic communications sector.[5]

Article 15(1) of the Framework Directive requires the adoption of a Recommendation on Relevant Product and Service Markets, hereinafter 'this Recommendation'. This Recommendation will identify those product and service markets within the electronic communications sector, the characteristics of which may be such as to justify the imposition of regulatory obligations set out in the specific directives. This is without prejudice to markets that may be defined in specific cases under competition law. The markets identified in this Recommendation will be markets defined in accordance with the principles of competition law.

This Recommendation should be considered in conjunction with the 'Guidelines for market analysis and the assessment of market power' referred to in Article 15(2) of the Directive[6] (hereinafter, 'the Guidelines'). National regulatory authorities ('NRAs') are required, taking utmost account of this Recommendation and the Guidelines, to define relevant markets appropriate to national circumstances, in particular relevant geographic markets within their territory, in accordance with the principles of competition law and to analyse those product and service markets, taking the utmost account of the Guidelines.

On the basis of this market analysis, NRAs will determine whether these markets are effectively competitive or not and impose, amend, or withdraw regulatory obligations accordingly.

The new regulatory framework aims at ensuring harmonisation across the single market and guaranteeing legal certainty. This Recommendation plays an important role in achieving both of these objectives, as it seeks to ensure that the same product and services markets will be subject to a market analysis in all Member States and that market players will be aware in advance of the markets to be analysed. It will only be possible for NRAs to regulate markets which differ from those identified in this Recommendation where this is justified by national circumstances and where the Commission does not raise any objections, in accordance with the procedures referred to in Articles 7(4) and 15(3) of the Framework Directive. As competition and convergence develop, it is expected that the range of markets identified in this Recommendation will in the future be reduced.

Looking ahead, the eEurope 2005 Action Plan promotes a multi-platform approach to broadband deployment, driven by strong competition between services provided over competing platforms. Competing network infrastructures are essential for achieving sustainable competition in networks and services in the long term. When there is effective competition, the new framework requires ex ante regulatory obligations to be lifted. In the meantime, however, those undertakings that were privileged to install infrastructure facilities under special or exclusive rights continue to benefit from those earlier investments – in particular those relating to long-lived facilities in the local access network. Granting others access to these facilities in a way that levels the playing field but does not remove incentives for new infrastructure investment ensures that users enjoy choice and competition during the transition to a fully competitive market. Investment in new and competing infrastructure will bring forward the day when such transitional access obligations can be relaxed.

In addition to the markets identified by the Commission in this Recommendation, the Directives also require that NRAs analyse certain specific markets. For instance, the Universal Service Directive identifies the markets for the minimum set of leased lines,[7] and for the provision of access to and use of the public telephone network at a fixed location.[8] The Access Directive also foresees the possibility for NRAs to limit obligations concerning conditional access systems to undertakings with significant market power, following a market analysis, but this limitation is at the discretion of individual Member States.[9]

NRAs define relevant markets appropriate to national circumstances, taking utmost account of the product markets listed in this Recommendation, in particular relevant geographic markets within their territory. The definition of relevant markets can and does change over time as the characteristics of products and services evolve and the possibilities for demand and supply substitution change. This is particularly important where the characteristics of products and services are continually evolving, where new products and services appear and where the way in which such products and services are produced and delivered evolves as a result of technological evolution. The convergence phenomenon where similar services can be delivered over different types of network is one example. This means that it will be necessary periodically to re-examine the markets identified in this Recommendation. At the same time the underlying purpose of the new regulatory framework (and its ex ante market analysis and possible regulation) is to deal with predictable problems of lack of effective competition. The fact that the new framework deals with situations where any lack of effective competition is durable means that a degree of continuity (as opposed to frequent revisions of this Recommendation) is warranted.

In accordance with Article 5 of the Framework Directive 2002/21/EC, Member States should ensure that NRAs have at their disposal whatever powers are necessary to gather data to allow a market analysis of any market which is covered by the new framework, regardless of whether it is identified in this Recommendation or not, or whether it is a wholesale or retail market.

## 2. Transition from the Current to the New Framework

Provisions in the new regulatory framework allow a smooth transition from the 1998 regulatory framework to the new regulatory framework. In accordance with Article 16 of the Framework Directive, NRAs are required to carry out an analysis of relevant markets as soon as possible after the Recommendation is adopted.

In doing so, NRAs will be reviewing obligations imposed under the 1998 regulatory framework, as required by Article 7 of the Access Directive and Article 16 of the Universal Service Directive. Article 16(1) of the Framework Directive requires NRAs to begin their market analysis 'as soon as possible after the adoption of the recommendation or any updating thereof'. The Commission would expect these analyses to be completed in most cases and where possible by 25 July 2003, the date of application of the new regulatory framework. If the analyses are not completed, or the resulting measures not yet in place, by that date, Article 27 of the Framework Directive ensures continuity with the existing regulatory framework by requiring NRAs to maintain the obligations imposed under national law in implementation of the 1998 regulatory framework until such time as they have been reviewed under the new regulatory framework.

Under the 1998 regulatory framework, several areas in the telecommunications sector are subject to ex ante regulation. These areas have been delineated in the applicable directives, but are not always 'markets' within the meaning of competition law and practice. Annex I of the Framework Directive provides a list of such market areas to be included in the initial version of the Recommendation.

Table 1 sets out the market areas in which undertakings are subject to existing obligations under the 1998 regulatory framework, and lists the corresponding market areas included in Annex I of the Framework Directive.

Where an NRA concludes that a market identified under the new framework is effectively competitive, that is, no operator is found to enjoy significant market power in the relevant market, the NRA removes all existing specific obligations on operators previously designated as having significant market power.[10] Conversely, where an NRA determines that a relevant market is not effectively competitive, it must impose appropriate regulatory obligations or alternatively maintain or amend existing obligations on undertakings with significant market power. In some cases, therefore it may be that regulatory obligations imposed under the 1998 regulatory framework are re-imposed under the new regulatory framework, albeit in respect of designations of significant market power on relevant markets defined and analysed in accordance with the new regulatory framework.

*Table 1 Market areas subject to obligations under the 1998 framework – link to Annex I Framework Directive under the new regulatory framework*

| Market area subject to obligations under the 1998 regulatory framework | Market area listed in Annex I of the Framework Directive |
|---|---|
| **Retail** | |
| Fixed public telephone networks and voice telephony services<br>Directive 98/10/EC[11] | Provision of connection to and use of the public telephone network at fixed locations |
| Leased lines, including the minimum set of leased lines<br>Directive 92/44/EEC[12] | Provision of leased lines to end users |
| **Wholesale** | |
| Fixed public switched telecommunications networks – interconnection<br>Directive 97/33/EC[13] Art 7. 1, Annex I part 1 | Call origination in the fixed public telephone network<br>Call termination in the fixed public telephone network<br>Transit services in the fixed public telephone network |
| Fixed public switched telecommunications networks – unbundling of the local loop<br>Regulation 2887/2000/EC[14] | Access to the fixed public telephone network, including unbundled access to the local loop |
| Fixed public switched telecommunications networks – carrier (pre)selection<br>Directive 97/33/EC Art. 12(7), introduced by 98/61/EC | Call origination in the fixed public telephone network |
| Leased lines services – interconnection<br>Directive 97/33/EC Art 7(1), Annex I part 2 | Wholesale provision of leased line capacity to other suppliers of electronic communications networks or services<br>Leased line interconnection (interconnection of part circuits) |
| Mobile public switched telecommunications networks – interconnection (linked to national market for interconnection)<br>Directive 97/33/EC Art. 7(2), Annex I part 3 | Call origination on public mobile telephone networks<br>Call termination on public mobile telephone networks |

## 3. Market Definition, Identifying Markets and Definition of Other Markets

### 3.1 Methodologies used to define markets

In the new regulatory framework, markets are defined in accordance with the principles of competition law, as explained in the Commission Notice on Market Definition[15] and the Guidelines.

The main purpose of market definition is to identify in a systematic way the competitive constraints that the undertakings face. The objective is to identify those actual and potential competitors of the undertakings that are capable of constraining their behaviour and of preventing them from behaving independently of an effective competitive pressure. The market definition arrived at can depend on the relative weight given to demand-side and supply-side substitutability, and can also depend on the prospective time horizon considered. It is important to recall that market definition for the purposes of the Recommendation is not an end in itself but is a means to assessing effective competition for the purposes of ex ante regulation.

As stated in the Commission's Guidelines and Access Notice,[16] there are in the electronic communications sector at least two main types of relevant markets to consider, that of services or facilities provided to end users (retail markets) and that of access to facilities necessary to provide such services provided to operators (wholesale markets). Within these two types of markets, further market distinctions may be made depending on demand- and supply-side characteristics.

The starting point for the definition and identification of markets is a characterisation of retail markets over a given time horizon,[17] taking into account demand-side and supply-side substitutability.[18] Having characterised and defined retail markets, which are markets involving the supply and demand of end users, it is then appropriate to identify relevant wholesale markets, which are markets involving the demand and supply of products to a third party wishing to supply end users.

As the market analyses carried out by NRAs have to be forward-looking, markets are defined prospectively.[19] Their definitions take account of expected or foreseeable technological or economic developments over a reasonable horizon linked to the timing of the next market review. Moreover, given the possibility to review a market at regular intervals, an NRA would be justified in taking into account past performance and existing market position as well as expectations concerning forthcoming developments.[20]

Markets defined in the Recommendation are without prejudice to the markets defined in specific cases under competition law. Markets identified in the Recommendation, while based on competition law methodologies, will not necessarily be identical to markets defined in individual competition law cases. As explained in paragraph 27 of the Guidelines, the starting point for carrying out a market analysis for the purpose of Article 15 of the Framework Directive is not the existence of an agreement or concerted practice within the scope of Article 81 EC Treaty, nor a concentration within the scope of the Merger Regulation, nor an alleged abuse of dominance within the scope of Article 82 EC Treaty, but is based on an overall forward-looking assessment of the structure and the functioning of the market under examination. NRAs and competition authorities, when examining the same issues in the same circumstances and with the same objectives, should in principle reach the same conclusions. However, it cannot be excluded that given the differences outlined above, markets defined for the purposes of competition law and markets defined for the purpose of sector-specific regulation may not always be identical.

### 3.2 The basis for identifying markets in recommendations

Article 15(1) of the Framework Directive requires that the Recommendation identify, in accordance with Annex I of that directive, those product and service markets within the electronic communications sector, the characteristics of which may be such as to justify the imposition of regulatory obligations set out in the specific directives. The starting point for identifying markets in the initial Recommendation is therefore the list of markets in Annex I.[21]

One question that arises for the initial Recommendation is whether any markets should be added to the Annex I list. Article 15 of the Framework Directive also requires the Commission to regularly review the Recommendation.

It should be recalled that the Framework Directive is based on the premise that there is a need for ex ante obligations in certain circumstances in order to ensure the development of a competitive market (see e.g. recital 25). Section 2 of this document indicates the ex ante obligations that have been put in place as part of the 1998 liberalisation in order to create the conditions for effective competition under the old definition of significant market power.

In that regard, recital 27 of the Framework Directive indicates that ex ante regulatory obligations (with respect to electronic communications networks and services) should only be imposed where there is not effective competition, i.e. in markets where there are one or more undertakings with significant market power, within the meaning of Article 14 of that Directive, and where national and Community competition law remedies are not sufficient to address the problem.[22]

Consequently, ex ante regulation and competition law serve as complementary instruments in achieving policy objectives[23] in the electronic communications sector and in dealing with lack of effective competition. At the same time, a principle underlying the new regulatory framework is that ex ante regulation should be rolled back when it is no longer needed.

It is appropriate, therefore, to identify as clearly as possible the circumstances in which ex ante regulation is justified (or conversely the circumstances in which competition law is adequate to deal on its own with problems arising) and to translate this into a basis or set of criteria for identifying relevant markets.

The experience of the initial phase of liberalisation in the European Union is that ex ante regulation has served as an important complement to competition law in dealing with two particular types of problems that market entrants have faced following liberalisation. The first is the need for a market entrant or potential competitor to obtain access to existing networks belonging to other operators in order to serve customers. The second is the fact that the competitive provision of many electronic communications services also depends on negotiating interconnection with other networks to facilitate the necessary connectivity.

This experience does not necessarily imply that these two problems will persist indefinitely or in all circumstances. The evidence arising from the establishment of mobile cellular networks (with respect to the problem of establishing networks) and from the establishment of widespread voluntary agreements to exchange Internet traffic on a best efforts basis (with respect to the interconnection problem) indicates that there have not been persistent obstacles to market entry in these two areas. In addition, technology and markets in this sector can change rapidly, so that an obstacle to the development of competition at one point in time can subsequently be removed or disappear over a given horizon.

It is considered that the criteria for identifying markets for the purposes of the new regulatory framework should include an overall assessment of the effectiveness of competition law alone in addressing the market failures concerned. Such an assessment will draw on the experience gained from the application of competition law and the imposition of ex ante regulatory obligations in the electronic communications sector as a complementary instrument. Only markets where national and Community competition law is not considered sufficient by itself to redress market failures and to ensure effective and sustainable competition over a foreseeable time horizon should be identified for potential ex ante regulation.

It is therefore considered that the following specific cumulative criteria are appropriate to identify relevant markets.

The first criterion is whether a market is subject to high and non-transitory entry barriers. The presence of high and non-transitory entry barriers, although a necessary condition, is not of itself a sufficient condition to warrant inclusion of a given defined market. Given the dynamic character of electronic communications markets, possibilities for the market to tend towards a competitive outcome, in spite of high and non-transitory barriers to entry, need also to be taken into consideration.

The second criterion, therefore, is whether a market has characteristics such that it will tend over time towards effective competition. This criterion is a dynamic one and takes into account a number of structural and behavioural aspects which on balance indicate whether or not, over the time period considered, the market has characteristics which may be such as to justify the imposition of regulatory obligations as set out in the specific directives of the new regulatory framework.

The third criterion considers the sufficiency of competition law by itself (absent ex ante regulation), taking account of the particular characteristics of the electronic communications sector.

1. Barriers to entry and to the development of competition

   With respect to the first criterion, two types of barriers to entry and to the development of competition in the electronic communications sector appear to be relevant: structural barriers and legal or regulatory barriers.

   A structural barrier to entry exists when the state of the technology, and its associated cost structure, and/or the level of the demand, are such that they create asymmetric conditions between incumbents and new entrants impeding or preventing market entry of the latter. For instance, high structural barriers may be found to exist when the market is characterised by substantial economies of scale, scope and density and high sunk costs. Such barriers can still be identified with respect to the widespread deployment and/or provision of local access networks to fixed locations.

   A specific and different type of barrier to the development of effective competition can also occur in the electronic communications sector where interconnection is required to enable a calling party to make a call to a specific subscriber number. In cases where a charge is levied for terminating the call (which is passed on as a retail charge to the calling party), the terminating network operator can affect competition adversely by raising a rival's costs or by passing on inefficiencies to competitors. This barrier by itself need not lead to an absence of competition. For example, where the receiver rather than the calling party is responsible for paying any charge associated with incoming calls or traffic, the incentive to raise termination charges above costs is absent. Technological solutions might also provide a way round the technical barrier.

Legal or regulatory barriers are not based on economic conditions, but result from legislative, administrative or other state measures that have a direct effect on the conditions of entry and/or the positioning of operators on the relevant market.

One example is the case of a legal limit on the number of undertakings that have access to spectrum. Such a limitation is typically linked to a related technical or technological barrier, e.g., a constraint on the amount of spectrum that can be assigned and consequently a limit on the number of licences given to undertakings seeking to enter a market. Additional entry is blocked unless additional spectrum becomes available or secondary trading of spectrum is permitted.

A significant legal or regulatory barrier to entry may also exist when entry into a particular market is rendered non-viable as a result of regulatory requirements, and in addition this situation is expected to persist for a foreseeable period. Regulatory requirements may lead to some services being provided at below cost or at rates of return that deter entry. One example is the retail pricing of access to the public telephone network (and local calls) at a fixed location or address. In cases where services fail to cover their forward-looking incremental costs, entry into local access is deterred. Tariff rebalancing will address such a barrier. However, broader policy concerns and objectives may mean that the situation persists for a significant period.

2. Dynamic aspects
The second criterion is whether or not the market has characteristics such that it will tend towards effective competition without the need for ex ante regulatory intervention. The application of this criterion involves examining the state of competition behind the barrier to entry, taking account of the fact that even when a market is characterised by high barriers to entry, other structural factors or market characteristics may mean that the market tends towards effective competition. This is for instance the case in markets with a limited, but sufficient, number of undertakings behind the entry barrier having diverging cost structures and facing price-elastic market demand. In such markets, market shares may change over time and/or falling prices may be observed. It is this structural dynamic element which may push the market to an effectively competitive outcome which is important and can be principally observed in a historical and current context. Future developments are also important but must be developments which are considered in a meaningful timeframe rather than something which may be theoretically possible. This timeframe should be linked to the period between NRA reviews, so as to capture only developments taking effect during that period.

   Entry barriers may also become less relevant with regard to innovation-driven markets characterised by ongoing technological progress. In such markets, competitive constraints often come from innovative threats from potential competitors that are not currently in the market. In such innovation-driven markets, dynamic or longer term competition can take place among firms that are not necessarily competitors in an existing 'static' market.

3. Relative efficiency of competition law and complementary ex ante regulation
The final decision to identify a market that fulfils the first two criteria (high and persistent entry barriers and absence of characteristics such that the market would tend towards effective competition) as justifying possible ex ante regulation, should depend on an assessment of the sufficiency of competition law by itself (absent ex ante regulation) in reducing or removing such barriers or in restoring effective competition.

> Ex ante regulation would be considered to constitute an appropriate complement to competition law in circumstances where the application of competition law would not adequately address the market failures concerned. Such circumstances would for example include situations where the compliance requirements of an intervention to redress a market failure are extensive (e.g. the need for detailed accounting for regulatory purposes, assessment of costs, monitoring of terms and conditions including technical parameters etc.) or where frequent and/or timely intervention is indispensable, or where creating legal certainty is of paramount concern.
>
> In practical application NRAs should consult with their NCA and take into account that body's opinion when deciding whether use of both complementary regulatory tools is appropriate to deal with a specific issue, or whether competition law instruments are sufficient.

In summary, whether an electronic communications market continues to be identified by subsequent versions of the Recommendation as justifying possible ex ante regulation would depend on the persistence of high entry barriers, on the second criterion measuring the dynamic state of competitiveness and thirdly on the sufficiency of competition law by itself (absent ex ante regulation) to address persistent market failures. A market could also be removed from a recommendation once there is evidence of sustainable and effective competition on that market within the Community, provided that removal of ex ante regulatory obligations will not reduce effective competition on the market. For instance it may be found that, through the use of carrier pre-selection and carrier selection, the national market for residential calls has become competitive. However, since it is the regulatory requirement to grant pre-selection that has allowed effective competition to emerge, it may not be appropriate to remove that regulatory requirement without first analysing the broader implications of such a move.

> The Commission expects NRAs to follow the same basic criteria and principles when they identify markets other than those appearing in this Recommendation. These criteria need not necessarily remain unchanged over time; they could for instance be adapted in the light of the experience gained with the implementation of the regulatory framework in respect of identified markets. The Commission will use these criteria when making future revisions to this Recommendation.

### 3.3 The definition by NRAs of other relevant markets

In this Recommendation, care has been taken to identify (in accordance with Annex I of the Framework Directive) markets whose characteristics may be such as to justify the imposition of regulatory obligations as set out in the specific directives. This list of relevant markets may not be exhaustive in the context of national circumstances, which may vary from Member State to Member State. Pursuant to Article 15(3) of the Framework Directive, NRAs may therefore identify markets that differ from those identified in the Recommendation provided they follow the procedures referred to in Articles 6 and 7 of the Framework Directive.

There may be a number of ways in which market definition at a national level, for the purposes of market analysis, might differ from the markets identified in the Recommendation.[24]

The issue of the appropriateness of defining narrower markets should be kept separate from the issue of the identification and application of remedies. In dealing with lack of effective competition in an identified market, it may be necessary to impose several obligations to achieve an overall solution. For instance, it may often be the case that adjacent or related remedies are applied to technical areas as part of the overall obligation that addresses SMP on the analysed market. If specific remedies are thought to be necessary in a specific narrow technical area, it is not necessary or appropriate to identify each technical area as a relevant market in order to place obligations in that area. An example would be where an obligation to provide unbundled access to the local loop is complemented by related obligations concerning access to co-location facilities. Proportionate and justified obligations in related areas may simply be part of the overall remedy.

Conversely, an obligation in a related area that is not necessary and proportionate to solve a problem on an identified market cannot be considered to be part of the proposed remedy. If the remedy on the related area went beyond what is necessary to remedy the problem on the identified market, the related area would need to be identified as a relevant market and analysed separately in order to impose such an obligation.

When NRAs consider redefining markets more narrowly or more broadly for reasons related to national market circumstances they need to follow the principles and the methodology set out in the Guidelines and in this Recommendation. There may for example be evidence that demand or supply substitution is possible in part of the market but not (over a given horizon) in the remainder, so that a narrower market definition is justified. In the same vein, if market conditions are such that an undertaking may remain dominant in one area of a broad market definition and not in another, there is reason to believe that the use of narrow market definitions is warranted. One example could be where retail competition for certain well defined and separate categories of residential user is much better developed than for other categories.

Where a market identified by an NRA differs from those identified in the Recommendation, NRAs should follow the procedure set out in Article 6 and 7 of the Framework Directive. At the same time, it may be that the co-operation procedures between NRAs and between NRAs and the Commission (pursuant to Article 7(2) of the Framework Directive) and/or the harmonisation procedures (pursuant to Article 19) result in the adoption of more detailed approaches concerning narrower or broader market definition practices across the Community.

In accordance with Article 15(3) of the Framework Directive, NRAs notify pursuant to Article 7(4) of the Framework Directive the draft measure which proposes to define a market different from those identified in the Recommendation. This requirement in Article 15(3) of the Framework Directive for NRAs to notify markets differing from those identified in the Recommendation refers solely to those markets affecting trade between Member States, in accordance with Article 7(3)(b). If a market was not notified on the grounds that an NRA did not foresee an effect on trade between Member States, but an effect on trade was subsequently demonstrated, the failure to notify that market could constitute an infringement of the Directive. Moreover, failure to notify could also have adverse legal consequences in national proceedings.

### 3.4 Remedies

Remedies are the final part of a process which starts with market definition and identification, which is followed by market analysis and, in the event of an SMP designation, moves to corrective action. Markets susceptible to ex ante regulation are selected on the basis of the criteria set out in section 3.2. The identification of a market for analysis does not of itself mean that that market requires regulatory intervention. It is only where NRAs find that there is the absence of effective competition on that market that they impose remedies. Even then there needs to be careful consideration of which remedy should be applied. The new regulatory framework is very flexible. NRAs have a suite of regulatory tools at their disposal, as set out in Directive 2002/19/EC and Directive 2002/22/EC. When imposing a specific obligation on an undertaking with significant market power, the NRA will need to demonstrate the obligation in question is based on the nature of the problem identified, proportionate and justified in the light of the NRA's basic objectives as set out in the Framework Directive.

These basic objectives require NRAs to:

- promote competition in the provision of electronic communications networks, electronic communications services and associated facilities and services;
- contribute to the development of the internal market;
- promote the interests of the citizens of the European Union.

The Framework Directive also requires NRAs to seek to agree between themselves on the types of instruments and remedies best suited to address particular types of situations in the marketplace. In particular, as noted in the Guidelines on market analysis, in order to establish that a proposed remedy is compatible with the principle of proportionality, the action to be taken must pursue a legitimate aim and the means employed to achieve the aim must be both necessary and the least burdensome, i.e. it must be the minimum necessary to achieve the aim.

A number of considerations are set out in the Directives qualifying the use of specific remedies. In particular, before imposing the more onerous remedies, NRAs need to be mindful of the initial investment by the facility owner, bearing in mind the risks involved in making the investment. The NRAs have a duty to safeguard competition in the long term which will inter alia be a function of the need to assess the technical and economic viability of using or installing competing facilities and the effect of such an intervention on possible investment in such competing facilities. This is especially important where new technologies or networks are being deployed in unproven markets.

The availability of a competition law remedy corresponding to that which might be imposed by the NRA does not preclude the NRA imposing a similar remedy. Where, for instance, a firm is in a dominant position under competition law, Article 82 of the Treaty obliges that firm to behave in a non-discriminatory manner; it may for example be appropriate for an NRA to also impose specific ex ante transparency and reporting requirements designed to facilitate the monitoring of the non-discrimination requirement by that firm (for example as regards delivery periods). In addition, imposing an ex ante remedy would mean that the provisions of the new regulatory framework for dispute resolution and NRA intervention can be invoked in case of problems in that market, in addition to an appeal to the relevant competition authorities.

## 4. Examination of Markets in Order to Identify Relevant Markets for the Purposes of the Recommendation

This section examines the broad market areas within the electronic communications sector, analyses briefly the relevant retail and wholesale markets within those broad areas, and identifies the specific relevant markets for the purpose of ex ante analysis of effective competition.

The starting point is a characterisation of retail markets, followed by a description and definition of related wholesale markets, as indicated in section 3.1.

A key aim of the new regulatory framework is to enhance user and consumer benefits in terms of choice, price and quality by promoting and ensuring effective competition. NRAs have powers as a last resort and after due consideration to impose retail regulation on an undertaking with significant market power. However, regulatory controls on retail services should only be imposed where NRAs consider that relevant wholesale or related measures would fail to achieve the objective of ensuring effective competition.[25] In principle, lack of effective competition may occur at the retail level or the wholesale level or both. That means that NRAs may need to examine the overall degree of market power of undertakings and the impact on effective competition. The identification of a retail market for the purposes of ex ante market analysis does not imply, where there is a finding of a lack of effective competition by an NRA, that regulatory remedies would be applied to a retail market. Regulatory controls on retail services can only be imposed where relevant wholesale or related measures would fail to achieve the objective of ensuring effective competition.

Retail markets should in principle be examined, for market definition purposes, in a way that is independent of the network or infrastructure being used to provide services, as well as in accordance with the principles of competition law. At the same time, for the purposes of the initial Recommendation, the starting point for market definition and identification is Annex I of the Framework Directive. Included in Annex I are market areas that are described in relation to specific existing networks and their technical configurations.

Over time, investments will be made to develop new networks and infrastructure as well as to upgrade and change existing networks and infrastructure (e.g. by using fibre or wireless technologies in the access network). The fact that an obligation is applied to an SMP undertaking with regard to a specific type of infrastructure (e.g. metallic local loops) does not imply that a parallel obligation will be applied when networks and infrastructure are upgraded, for example to deliver (or be capable of delivering) new services. Where new types of infrastructure are used to provide existing services, any regulatory intervention (in respect of that new type of infrastructure) would depend on a revised market analysis. Indeed, encouraging efficient investment in infrastructure and promoting innovation are explicit objectives for regulators.[26]

### 4.1 Product and service markets in the electronic communications sector

Electronic communications networks and services are defined in the Framework Directive. Electronic communications services include telecommunications services and transmission services in networks used for broadcasting, but exclude services providing or exercising editorial control over content transmitted using electronic communications networks and services. They do not include information society services, as defined in Directive 98/34/EC, which do not consist wholly or mainly in the conveyance of signals on electronic communications networks.[27]

In examining retail and wholesale markets for the purposes of the Recommendation, a general division is made between services provided at fixed locations and those provided to non-fixed locations. A general distinction is also made between voice services and non-voice (data) services. These distinctions for the purposes of analysing markets do not imply an advance judgement that these services constitute separate markets. However, at the current time voice and data services are considered overall to be sufficiently distinct in terms of demand substitution that they are analysed separately. At a wholesale level, this distinction between voice and non-voice services may be less easy. For example a transmission channel may carry (or be capable of carrying) both voice and non-voice services.[28] These issues are dealt with in the relevant analysis sections.

Across the EU different Member States have communication network topologies which differ significantly from each other. In that context the distinction between what constitutes a service such as call origination or what constitutes a transit service in terms of what parts of the network are involved will differ across Member States. It is left to the NRAs to decide on such technical aspects, in the light of national circumstances within the constraints set out in this Recommendation.

## 4.2 Services provided at fixed locations

### *4.2.1 Public telephone services provided at fixed locations*

The aim of this section is to (i) describe and define relevant retail markets for voice services provided at fixed locations,[29] (ii) define the linked wholesale markets and (iii) identify markets in accordance with Annex I of the Framework Directive.

#### RETAIL MARKETS

The retail market at a general level can be described as the provision of a connection or access (at a fixed location or address) to the public telephone network for the purpose of making and/or receiving telephone calls and related services. Such access and services may be supplied by several possible means in respect of the undertaking providing the service and the technology that is used. The most common technology currently employed is via traditional telephone networks using metallic twisted pairs. Current alternatives include cable TV networks offering telephone service, mobile cellular networks that have been adapted to provide service to fixed locations, and other wireless-based networks. Of these alternatives, only cable networks have any extensive roll-out in any Member States. The extent to which they are currently capable of supplying voice services varies greatly across Member States and overall is still limited. For locations where there is demand for a large number of user connections, some form of dedicated access may be used. In the future a wider range of technologies is likely to be used commercially to provide local access and the provision of services including the public telephone service.

It is common to distinguish between business and residential markets because the contractual terms of access and service may vary. In addition, it will not normally be possible for a supplier to the business market to respond to price increases by a hypothetical monopolist in the residential market because the economics of serving customers in the two locations may be significantly different. Therefore it is appropriate to distinguish end user categories in the markets defined.

Telephone service is usually supplied as an overall package of access and usage. Various options and packages may be available to end users depending on their typical usage or calling patterns.[30] Although many end users appear to prefer to purchase both access and outgoing calls from the same undertaking, others choose alternative undertakings from the one providing access (and the receipt of calls) in order to make some or all of their outgoing calls. An undertaking that attempted to raise the price of outgoing calls (in particular for national and international calls) above the competitive level would face the prospect of end users substituting alternative service providers. End users can relatively easily choose alternative undertakings by means of short access codes (via contractual or pre-paid means) or by means of carrier pre-selection. Whilst undertakings that provide access compete on the market for outgoing calls, it does not appear to be the case that undertakings supplying outgoing calls via carrier selection or pre-selection would systematically enter the access market in response to a small but significant non-transitory increase in the price of access. Therefore, it is possible to identify separate retail markets for access and outgoing calls.

In respect of outgoing telephone calls, end users do not perceive local, national and international calls to be substitutes for each other. Therefore, it can be argued that these constitute separate retail markets. However, an increase in the price of calls in one of these categories by a hypothetical monopolist could induce service providers in one of the other categories to purchase the wholesale elements needed, provided that they are available, to supply the relevant services. There are, however, particular restrictions which limit this ability to purchase all the wholesale elements in an international calls market.

Calls from fixed locations to mobile telephones or indeed to other non-geographic locations are likewise not perceived by end users to be sufficiently substitutable to be considered in the same market in terms of demand. A user wishing to call an individual on a mobile phone is unlikely to consider a call to a fixed line number an equivalent. However, just as with geographic equivalents, service providers providing one category of services can be expected to be in a position to purchase the wholesale elements that are needed including mobile termination if a hypothetical monopolist attempts to increase prices.

Therefore, on the basis of supply substitution, it will normally be appropriate to define a broader market for outgoing calls at the retail level. However where supply-side substitution is not possible or is very limited it may be appropriate to define narrower markets.

## RELATED WHOLESALE MARKETS

### *Wholesale access and call origination*

At the retail level, a distinction has been made between access and outgoing calls. An undertaking may make a decision to enter the combined market for access and calls or simply enter part or all of the calls market. In assessing the relevant linked wholesale markets, it is necessary, therefore, to bear in mind that there are a number of means of addressing the retail markets.

With respect to access, the main alternatives are between building and buying, as indicated below. The latter option potentially includes any transmission path that is capable of supporting voice services, e.g. a leased circuit, an unbundled local loop or the wholesale provision of digital subscriber line or bit-stream services. Such alternatives are also capable of supporting the provision of data services or multiple voice channels and are considered in more detail in sections 4.1.2 and 4.1.3.

The main elements required to produce or supply retail telephone service are call origination, call conveyance (including routing and switching) of varying kinds and call termination. Related elements include signalling and the ancillary services needed, for example, for billing purposes. An undertaking that supplies retail telephone service could purchase these inputs separately or together, or produce all of them by constructing an extensive network, or purchase some and produce others. One direct alternative to the purchase of call origination is to establish network access to the end user location. Another alternative is to purchase or lease an established network connection to the end user location. Neither of these alternatives, which entail considerable investments, is likely to be widely possible in response to a price rise that is non-transitory but small by a hypothetical monopolist in the market for call origination. The alternatives may however be viable in the context of a broader retail market (including the provision of non-voice services) or in circumstances where end user expenditure is sufficiently high.

An indirect alternative to call origination is for an undertaking providing service to obtain retail connections to the public telephone network in relevant areas in order that end users can dial-in to a point of presence of the undertaking in order to use onward service. Such an alternative is not considered to be a close substitute to call origination.

Wholesale call origination services (originating access or interconnection) can be provided in the form of minutes or in the form of capacity. They may also be supplied together with switching and/or call conveyance services (see below). The market identified for the purpose of this Recommendation is wholesale call origination services. The relevant market is considered to comprise call origination for telephone calls and for the purposes of accessing dial-up Internet service provision. Therefore the market is defined as call origination on the public telephone network provided at a fixed location.

***Wholesale transit services***

In addition to wholesale call origination and call termination, call conveyance or transit will be needed in order to complete a call. Call conveyance or transit interconnection involves transmission and/or switching or routing. For an undertaking providing services to a limited number of end users, an alternative to using wholesale call conveyance services could be to use interconnected leased lines or dedicated trunk capacity. Transit services refer to the (long-distance) conveyance of switched calls on the public telephone network provided at a fixed location. This is a different product from say the provision of dedicated capacity of itself, even if some transit services are provided over leased circuits or links. The difference is that transit services are equivalent as a product to (long-distance) voice paths per minute whereas leased circuits provide capacity per month (that can for instance be filled up with individual voice paths). Transit services therefore compromise both conveyance between tandem switches on a given network, between tandem switches on different networks and including pure conveyance across a third network. Some parts of this transit service market are likely to become more competitive more quickly than others, but there cannot be a presumption that some switched call conveyance (from an incumbent to an entrant's network) is automatically different from other switched call conveyance (between two entrants' networks).

The provision of transit interconnection can either be bought directly or the elements necessary for the provision of such services can be bought separately and the services can be self provided. The range of operators providing services or indeed the necessary network elements is almost entirely dependent on the traffic volumes on particular transit routes. While it is clear that certain busy routes have experienced large scale entry, other less busy or thinner routes have not, meaning that geographically ubiquitous transit services still normally depend on incumbent provided transit services.

Thin routes often have little or no available capacity so that an entrant wishing to self provide would not be in a position to buy the wholesale elements necessary to self provide transit services in a speedy fashion, having instead to build out the necessary infrastructures. Often, the existing capacity on these thin routes is sufficient to supply all operators' requirements for transit services on such routes. It is clear that in a large number of Member States new entrants are dependent on the incumbent for the provision of transit services on many routes.

Along with wholesale call origination services (originating access or interconnection), switched transit and/or call conveyance services is also identified for the purposes of this Recommendation. The market identified for the purpose of this Recommendation is Wholesale Transit Services in the Public Telephone Network. Depending on network topologies, the delineation between call origination and transit services can vary and it is left to NRAs to define those elements constituting each part. It should be noted by the NRAs that while there is a degree of discretion in deciding the appropriate elements constituting call origination, call termination and transit services, these elements are additive, the sum of the three making the whole. This means for instance that if call origination and call termination are already defined, then transit is also defined by default.

Despite the presence of certain thin routes, transit is one of the areas of fixed line telecommunications where competition can be expected to develop. Depending on experience gained over the implementation period of this first Recommendation, the need for the transit market's continued inclusion will be assessed.

***Wholesale call termination***

Wholesale call termination is required in order to terminate calls to called locations or subscribers. Undertakings owning or operating networks to provide telephone services may interconnect at relatively high levels in the network, i.e. at a few interconnect points. Consequently call termination arrangements may in practice comprise call conveyance as well as local call termination. However, undertakings faced by a price increase in say national call termination could purchase local call termination separately from the call conveyance part. Therefore, it makes sense to focus on local call termination as the relevant call termination market.

In using demand and supply substitution possibilities to define a relevant market for competition analysis, it is normal to start with a narrow definition and expand it as appropriate. In this case, the starting point is call termination to a specified location, subscriber or number. However, it is difficult for an undertaking that supplies wholesale call termination to other undertakings wishing to terminate calls to price discriminate between termination charges to different subscribers or locations on its network. Therefore the relevant market is at least as wide as each network operator.

In considering whether a wider definition is appropriate, it is necessary to examine the possibilities for demand and supply substitution that might constrain the setting of termination charges on a given network.[31] If all (or at least a substantial number of) fixed locations or subscribers in a given geographical area were connected by two or more networks, then alternative possibilities would exist for terminating calls to given locations. Another possible source of supply substitution would occur if it was possible technically for calls to a given location or end user to be terminated by an undertaking other than the one operating the network that serves the given location. Currently no such supply substitution is possible.

Call termination charges at a wholesale level on a given network might be constrained via demand substitution but there is currently no potential for demand substitution at the wholesale level. However, there are possibilities for demand substitution at the retail level. Examples could comprise any means of communication that constituted a reasonable alternative to making a call to the location or subscriber number concerned. Such alternatives might include terminating the call to the person concerned via a mobile network, a call using a call-back arrangement, a call that does not involve a specific call termination arrangement (e.g. where parties themselves arrange to speak via IP based links) or communication via messages of varying kinds (e.g. email, voicemail, paging). It is also necessary that the alternative possibility leads to an effective constraint on the setting of call termination charges by making it unprofitable for a network to raise call termination charges.

Such alternatives for demand or supply substitution do not appear currently to provide sufficient discipline on call termination at fixed locations or an argument in favour of a wider market definition, so that the relevant market appears to be call termination on individual networks. The key aspect that gives rise to the incentive to raise call termination charges is that there is no technical alternative by which a call can be terminated (other than by the operator of the network to which the called party subscribes) and that in general the calling party pays entirely for the call. Such an incentive would not arise or would be limited if no charge was levied for incoming traffic or if the receiver rather than the caller paid any charge or if there was a very close financial relationship between the calling and the called party.

Such a market definition – call termination on individual networks – does not automatically mean that every network operator has significant market power; this depends on the degree of any countervailing buyer power[32] and other factors potentially limiting that market power. Small networks will normally face some degree of buyer power that will limit greatly the associated market power. Absent any regulatory rules on interconnection, a small network may have very little market power relative to a larger one in respect of call termination. The existence of a regulatory requirement to negotiate interconnection in order to ensure end-to-end connectivity (as required by the regulatory framework) redresses this imbalance of market power. However, such a regulatory requirement would not endorse any attempt by a small network to set excessive termination charges. Consequently, there is still likely to be an imbalance of market power between large and small networks because it would be easier for a large network than a small network to initiate the step of raising call termination charges and would be more difficult for a small network to resist a move by a large network to lower termination charges.

### CONCLUSION

Therefore it is considered that the following specific markets related to the provision of public telephone services at fixed locations should be included in the Recommendation:

***Retail level***

1. Access to the public telephone network at a fixed location for residential customers.
2. Access to the public telephone network at a fixed location for non-residential customers.
3. Publicly available local and/or national telephone services provided at a fixed location for residential customers.
   Publicly available local and/or national telephone services provided at a fixed location for non-residential customers.
4. Publicly available international telephone services provided at a fixed location for residential customers.
   Publicly available international telephone services provided at a fixed location for non-residential customers.

***Wholesale level***

Call origination or capacity (on all networks serving a fixed location).
Call termination on individual networks.
Transit services in the public telephone network.

### *4.2.2 Access to data and related services at fixed locations*

The aim of this section is to (i) describe and define relevant markets for access to generic data services (in particular the provision of Internet service) at fixed locations at a retail level, (ii) define the linked wholesale markets and (iii) identify markets in accordance with Annex I of the Framework Directive.

#### RETAIL MARKETS

The increased use of Internet for a mix of communications services has created potentially wide-ranging retail markets for access to data and related services at fixed locations. In general, the provision of retail Internet access consists of two parts: (i) the network or transmission service to and from the end user's location and (ii) the provision of Internet services, in particular end-to-end connectivity with other end users or hosts. These two services may be bundled together.

At the current time, it is possible to identify three commonly available forms of Internet access: (i) dial-up service, (ii) higher bandwidth services using digital subscriber line technologies (or equivalents) or cable modems and (iii) dedicated access.[33]

For residential subscribers and small business users accessing Internet from fixed locations, the most common form of access continues to occur using narrow-band connections, including dial-up access via analogue telephone lines and modems, and dial-up via ISDN. However, a range of other access possibilities (including those listed above) at a fixed location exist and many become increasingly attractive as the demand for data traffic, in terms of either peak or average bandwidth or connection time, increases and as the retail prices of these services decline. Satellite and terrestrial TV networks (provided they have adequate capacity and are linked to a return path) are also capable of providing data services and access to Internet.[34] In the future, other wireless technologies and power-line technologies may be widely exploited for access at fixed locations.

One of the key questions is whether there are distinct relevant markets at a retail level and what are the consequent implications for the definition of relevant wholesale markets. In particular, it is necessary to establish whether dial-up or narrow-band access is a market that is distinct from higher-speed or broadband access. In addition, it is useful to examine the question of whether metered and unmetered retail services constitute a single or separate markets within the dial-up market.

An examination of available services and packages at a retail level reveals a wide range of service characteristics in terms of bandwidth, on-line time, price and pricing characteristics, permitted quantity of data received and sent, quality and also accompanying services. Basic Internet connectivity brings the ability to use email and messenger services, establish links with other end users, exchange files, consult and use web-sites and make transactions. But other specialised services may also be available, including the ability to use various voice and video services.

An end user with a preference for long usage time but less sensitivity to bandwidth might choose a flat-rate narrow-band offer or a broader-band (flat-rate) product, if the two are available, whilst an end user with high bandwidth requirements but limited time needs might choose a higher bandwidth offer via DSL or cable modem access (or equivalent). In response to a non-transitory small increase in price or equivalent deterioration in quality, an end user (or significant increments of end users) may well migrate to another retail service. Basic cable modem or DSL service could potentially constrain the pricing of unmetered dial-up service, although the reverse appears less likely to be true.

However, a number of features appear to indicate that separate retail markets do exist. One is that there are a number of technical characteristics of higher-speed or broadband access that imply that certain applications are just not viable on dial-up access. From the standpoint of broadband, therefore, narrow-band is a separate market, because the services and/or the quality features of those services (including their uplink and downlink speed) which can be offered over a narrow-band connection will not be seen as viable substitutes from the point of view of an end user which is making use of a broadband connection.[35] In addition, a flat-rate or unmetered narrow-band dial-up service cannot be considered to be an 'always-on' service in the way that a broadband service can be, as the service is likely to be interrupted if unused for a given period.

Price differentials can be observed between narrow-band and broadband access but these can vary and they may be a function of the specific data-rate or qualitative features of services offered, whether flat-rate narrow-band offers are available or not, whether there is competition between different forms of broadband access or for other reasons. It is not therefore easy, at the current time, simply on the basis of price differentials, to discern whether separate retail markets exist.

At the same time, for the purposes of deriving wholesale markets, there are sufficient distinguishing characteristics from a demand perspective for the market for dial-up or narrow-band service to be considered distinct from the market for broadband services. At a retail level customers in the broadband market have a range of options to purchase connectivity at these speeds. Consumers can buy service from cable operators with upgraded networks using cable modems, they can buy service from new entrants using either unbundled local loops that the entrant has modified themselves or which have been modified for them or the customer can buy these services directly from the incumbent. Other technologies such as wireless local loops are not widely available but are capable of providing equivalent services. Between these options, provided prices are comparable, a consumer will be indifferent.

In the narrow-band market, dial-up services may be paid for on the basis of a subscription, usage or a combination of the two. Unmetered or flat-rate retail (subscription only) service is becoming more widely available in the Community and take-up by end users where such service is available is substantial. In general a user would switch from metered to unmetered service when the price for an unmetered product becomes low enough relative to the metered price.[36] Hence more intensive users will switch first.

Metered and unmetered (flat-rate) access can be considered to be part of the same market for a number of reasons. Firstly, the only difference between the products is the way in which tariffs are structured. Secondly, the two products appear to be substitutable for end users. Thirdly, if obligations exist to allow operators to buy wholesale call origination on an unmetered basis supply substitution will be possible in that a hypothetical monopolist that raised the price of unmetered access would induce other providers (of metered products) to offer an unmetered product at a lower, competitive price level. At the same time, there appears to be little evidence of end users substituting metered service in response to price increases in unmetered service. Therefore metered and unmetered call origination do not constitute distinct markets.

It would be open to NRAs to impose requirements on the broader call origination market which included a requirement to offer unmetered call origination, provided this obligation is proportionate as stated in Article 8(4) of the Access Directive.

### RELATED WHOLESALE MARKETS

Given the preceding description of retail markets, it is possible to define and identify wholesale markets for ex ante regulatory analysis. In order for access to data and related services to be supplied to an end user at a fixed location, an appropriate transmission channel is required that is capable of passing data in both directions and at rates that are appropriate for the service demanded. Therefore, an undertaking providing services to end users needs to build, establish or obtain access to a transmission channel to an end user location.

Current DSL and cable modem technologies or equivalents are not for the moment typically capable of providing higher speed service in excess of 2 Mbit/sec in both directions. Therefore the main options for delivering such services involve building fibre access or exploiting leased lines capable of supporting such higher speeds. The relevant markets for wholesale dedicated capacity (including higher bandwidth access in excess of 2 Mbit/sec) are treated in section 4.2.3.

To address a market intermediate between the higher-speed service described above and dial-up Internet access requires the use of either a leased line, DSL (or equivalent) service or a cable modem on a coaxial cable network. In future, other technologies are likely to become widely available, although not in the current one to three year timeframe. Such technologies could include wireless local loops, digital broadcast systems and power-line systems. At the current time, upgraded cable systems are not sufficiently widely developed or deployed to address such a market in a widespread fashion, although the situation is likely to change in many parts of the Community over the timeframe of the current Recommendation. Leased lines do not provide an adequate wholesale demand substitute to address the market given their high cost. Consequently the only reasonable widespread means of supplying this market is through the use of unbundled local loops which can be enhanced or a related wholesale product where the local infrastructure has already been enhanced or upgraded to provide wholesale broadband services. Wholesale broadband services includes what is traditionally referred to as bit-stream services but is not technology specific. It should be noted that NRAs can define different geographic markets within their jurisdiction provided the conditions of demand and supply are conducive to such a definition.[37]

An undertaking can therefore address the retail broadband market by using a range of technologies applied to unbundled loops or to alternative infrastructures such as a cable infrastructure. Fully unbundled loops give an entrant more flexibility to deploy its desired technology (including the bandwidth of services supplied), and potentially to develop more sophisticated retail services. This implies that there is a two-tier wholesale market serving the relevant downstream retail market. The overall result is a linked market structure between local loops which is technology specific and a wholesale market consisting of a range of technologies applied to loop paths or completely alternative infrastructures which are capable of supplying the market for broadband services.

An operator using unbundled local loops will not normally consider another form of wholesale broadband service to be a substitute even if the service provided by the broadband service provider allowed the supply of all the same services that were provided over the unbundled loops. In this example, it would require that the DSL technologies or equivalents used over the loops were compatible at every level of the network. Even if such a synchronous deployment of technology existed it would require continued synchronous technology deployment in the future making it difficult for service differentiation at a technical level to evolve. Equally, it is questionable as to whether an entrant using wholesale broadband service to deliver service to the final user market could easily switch to using unbundled local loops to provide an equivalent service. From a demand perspective a wholesale provider using wholesale broadband service will only consider unbundled local loops a substitute if the wholesale broadband service operator has all the other network elements needed to self provide an equivalent service. The supply substitution possibilities depend on the same condition. It is clear therefore that unbundled local loops and wholesale broadband service constitute distinct markets.

The aim of the new regulatory framework is ultimately to achieve a situation where there is full infrastructure competition between a number of different infrastructures. This can occur within or between platforms. Regulation mandating access to existing networks serves as a transitional measure to ensure services competition and consumer choice until such time as sufficient infrastructural competition exists. Investment in new network infrastructure will hasten the day when ex ante regulation can be withdrawn from this market.

NRAs will need to conduct a particularly detailed review of this market before deciding what the relevant geographic market or markets are, and what the appropriate intervention might be. This market is likely to be particularly susceptible to geographic variations in competition; in general alternative competing infrastructures are more prevalent in higher density urban areas. The levels of investment required to build new infrastructures and indeed to upgrade existing infrastructures need to be given careful consideration when examining the need for obligations of access to, and use of, specific network facilities, and if granted, the terms for such access.

This market concerns transmission systems capable of passing data in both directions and at rates that are appropriate for the service demanded, the services involved do not cover traditional broadcast services, which are dealt with in Section 4.4.

For wholesale broadband services there are clearly very high barriers to entry since a network will need to be deployed in order to provide the service. The dynamic of competition behind the barrier is more complicated. In countries where there are widely deployed cable networks which have been upgraded to provide broadband services and are capable of providing wholesale broadband services they are in direct competition with wholesale broadband services deployed over more traditional networks. For there to be the possibility of dynamic competition on the wholesale broadband services market depends on the existence of competing networks regardless of questions regarding the nature of competition between these networks and their possible susceptibility to tacit collusion. The question is whether the alternative infrastructures capable of competing are sufficiently widespread to justify the inclusion of this market in this Recommendation, that is whether cable, fibre optic and wireless networks capable of providing wholesale broadband services are sufficiently widely deployed in the Community. While these networks are well developed in some Member States, in most they are not. If alternative infrastructures continue to be developed and upgraded and competition increases, this market could become effectively competitive.

The relevant market for the purposes of this Recommendation is unbundled access (including shared access) to metallic loops and sub-loops (including shared access) for the purpose of providing broadband services and the market for wholesale broadband services. The point in the network at which the wholesale broadband services market will need to be supplied will depend on the market analysis and in particular on the network topology and the state of network competition.

#### WHOLESALE INPUTS TO DIAL-UP INTERNET ACCESS AND SERVICES – WHOLESALE CALL ORIGINATION

An Internet service provider (ISP) supplying dial-up Internet access requires wholesale call origination and wholesale call termination as inputs as well as wholesale Internet connectivity. A wholesale market corresponding to the retail market for access to the public telephone network at a fixed location would be necessary for the provision of dial-up Internet services or other narrow-band services depends on a number of factors. Users encountering a hypothetical monopolist on the call origination market would be able to easily switch service provider through the use of carrier pre-selection (CPS) or carrier selection (CS). Switching call origination service providers is in general both easy and cheap. This may result in there being more separate bills to be paid as the access provider and the service provider(s) cease to be the same entity or entities. While there is undoubtedly a cadre of customers who value the ease of single billing, it is not clear that this population would be significantly large to mitigate the disciplining role of those not concerned with single billing. Whether service is supplied on a metered basis or on an unmetered basis (or a combination of the two), call origination frequently takes place using appropriate number ranges which route calls to the network used by an ISP for onward connectivity with the public Internet. Depending on the specific call origination arrangements used, ISPs may compensate originating network operators on behalf of their end users or call origination may be paid for directly by end users.

In general, end users accessing the Internet via dial-up means at a fixed location use the undertaking that provides access to the public telephone network. The relevant market includes both call origination for the purpose of speech communications and for other forms of communication such as fax or data. Therefore, the relevant market for wholesale call origination for dial-up Internet service is call origination on the public telephone network provided at a fixed location (the same market defined in section 4.2.1).

CALL TERMINATION

In order to provide dial-up end users with Internet access and connectivity, ISPs need to ensure that dial-up calls are terminated, i.e. go through a terminating operator en route to the ISPs server.

Wholesale call termination as part of Internet service provision is different from call termination on fixed or mobile networks for the completion of calls between two end users. In the case of call termination for Internet service provision, end users have a contractual relationship (implicit or explicit) with an ISP but normally have no notion of the undertaking terminating dial-up calls. The ISP chooses the terminating operator (or operators) that receives the dial-in calls and may itself pay the terminating charge.[38] Since any terminating charge is incorporated into the overall amount that is charged by the ISP (and faced by the end user), and end users can switch between competing ISPs, ISPs have an incentive to minimise the termination charges that they pay.

In general, ISPs will have a wide choice with respect to terminating operators and there is evidence of ISPs switching terminating operators. However, in certain Member States it may be that there is less choice of terminating operators or that one or more operators that have market power on originating access are in a position to more fully exert that market power with respect to call termination. The more limited choice may occur because operators may need to build out networks in order to terminate dial-up calls under unmetered arrangements. Therefore if NRAs consider it necessary to define Internet termination market they can do so by following the Art. 7 procedure.

Whilst the relevant wholesale call origination market fulfils the criteria to warrant identification in the Recommendation, the relevant wholesale call termination market does not for the purpose of this Recommendation.

WHOLESALE INTERNET CONNECTIVITY

Irrespective of whether end users access Internet via dial-up or broadband means, ISPs still need to ensure connectivity with other ISPs and their end users.

To ensure that data packets sent by end users reach the intended destinations and also that incoming traffic is received, undertakings need to make the necessary arrangements to permit connectivity with all other Internet end users or at least with the networks that they use. This global connectivity can be arranged in a number of ways. It can be purchased from a network that is in a position by its own arrangements to guarantee such connectivity. It can be obtained by interconnecting and exchanging traffic with a sufficiently large number of networks that all possible destinations are covered. Or it can be arranged by a combination of interconnecting with certain networks and purchasing the remaining connectivity that is needed.

Two questions arise for the purposes of the Recommendation. Is it necessary to identify a market for Internet connectivity or packet delivery for the purposes of ex ante market analysis, and if so, what is the relevant market? There are a number of differences between the typical arrangements for terminating calls on the public telephone network and delivering packets to destination addresses on the public Internet. In the latter case, end users are implicitly paying to both send and receive packets. It is not automatically or typically the case that incoming traffic is charged for and that this charge is passed to the traffic sender via the sender's network. As indicated above, traffic connectivity can be arranged in a number of ways.

Entry barriers to this market are low and although there is evidence of economies of scale and that the ability to strike mutual traffic exchange (peering) agreements is helped by scale, this alone can not be construed as inhibiting competition. Therefore, unlike the case of call termination in section 4.2.1, there is no a priori presumption that ex ante market analysis is required. Therefore, no market for wholesale Internet connectivity (or delivery of incoming packets) is identified for the purposes of the Recommendation.

CONCLUSION

Therefore it is considered that the following specific markets related to access to data and related services at fixed locations should be included in the Recommendation:

***Wholesale level***
Wholesale unbundled access (including shared access) to metallic loops and sub-loops for the purpose of providing broadband and voice services.
Wholesale broadband services.
Call origination on the public telephone network at a fixed location.

### *4.2.3 Dedicated connections and capacity (leased lines)*

The markets related to dedicated connections and capacity have a link to some of the markets defined with respect to access at fixed locations and the provision of services at fixed locations. For example dedicated connections may be an alternative to unbundled local loops and vice versa in certain circumstances. Also dedicated trunk or long-distance connections may be an alternative to long-distance call conveyance.

Dedicated capacity or leased lines may be required by end users to construct networks or link locations or be required by undertakings that in turn provide services to end users. Therefore it is possible to define retail and wholesale markets that are broadly parallel.

The key elements in the demand and supply for dedicated connections are bandwidth, distance and the location or locations to be served. There may also be qualitative characteristics because in some cases distinctions are still made between voice grade and data grade circuits.

At the retail level, specific reference is made in the Universal Service Directive to the provision of the minimum set of leased lines.[39] The minimum set of leased lines refers to specified leased circuits with harmonised characteristics that must be made available under particular conditions throughout the national territory. Therefore this minimum set of leased lines is identified in the Recommendation as a specific set of retail markets. It is not felt necessary to identify specific markets for each category of leased line in the minimum set since it is likely that the market structure will be similar for each sub-set. It is not necessary to expand the retail leased line categories to capacities beyond the minimum set since there must always be a presumption that an intervention at a wholesale level will be sufficient to address any problems that arise. There is no a priori reason to believe that this would not be the case in a leased line context.

At the wholesale level, it is possible to distinguish separate markets, in particular between the terminating segments of a leased circuit (sometimes called local tails or local segments) and the trunk segments. What constitutes a terminating segment will depend on the network topology specific to particular Member States and will be decided upon by the relevant NRA. In addition while many trunk segment markets on major routes are likely to be effectively competitive in certain geographic areas in Member States, trunk segments may not support alternative suppliers. For this reason trunk segments are also identified. Additional market segmentation is possible between high and low capacity leased lines.

CONCLUSION

Therefore it is considered that the following specific markets related to the provision of dedicated connections and capacity (leased lines) should be included in the Recommendation:

***Retail level***
Retail leased lines in the minimum set.[40]

***Wholesale level***
Wholesale terminating segments of leased lines.
Wholesale trunk segments of leased lines.

## 4.3 Services provided at non-fixed locations

### *4.3.1 Mobile voice services*

The aim of this section is to (i) describe relevant retail markets for mobile voice services, (ii) define the linked wholesale markets and (iii) identify markets in accordance with Annex I of the Framework Directive.

RETAIL MARKETS

With respect to the overall retail mobile market, it is not clear whether residential and business customers can be considered to be part of the same market or not as there does not appear to be a clear way to separate them even if there may be significant differential pricing of services in order to attract certain types of customer or use.[41] With respect to demand substitution, end users may be indifferent between tariff packages designed for business or residential users provided the terms suit their usage profile. With respect to supply substitution, an undertaking serving the business market may easily switch to supplying residential users in response to a small but non-transitory price increase by a hypothetical monopolist.

Such ability contrasts with the situation for the supply of services to fixed locations where lack of network coverage (and the economics of extending networks to serve fixed locations in other geographical areas) constrains such supply substitution.

Pre- and post-pay mobile services can also be considered to be part of the same market. Supply substitutability is relatively easy, as is demand substitutability (in particular from pre-pay to contractual terms).

Several separate mobile markets can be defined at a retail level. Mobile telephone users have no apparent substitute for mobile access. And there is no supply substitute unless new spectrum becomes available. Therefore it seems that access can be considered as a market that is separate from the supply of services over the network at a retail level. However, every end user purchases access to a mobile network with the objective of making calls or receiving calls or both (nationally or whilst roaming internationally). Even if a user purchasing service chose not to originate calls, their decision to have service must be based on a need for call termination (to receive calls) otherwise access would be meaningless. This has implications for the definition of corresponding wholesale markets.

***Retail services for roaming on other national networks***

Similar considerations exist for international roaming at a retail level. Retail international roaming services include the ability to make and to receive calls whilst in a country other than the one where the end user has established his or her network subscription. From a demand perspective, the retail provision of international roaming services appears to be a separate market.

However, a domestic supplier could respond to a price increase by a hypothetical monopolist by making agreements with foreign operators so as to supply retail roaming services. Therefore it is possible to define a broader outgoing calls market at the retail level that includes national, international and roaming calls. This does not prevent a specific national market for international roaming to be defined at a wholesale level (see below).

***Retail short message services***

Given the widespread development and growing use of new retail services such as SMS (short messaging service), the question arises as to whether additional retail markets can be defined. In certain circumstances SMS can be considered as a reasonably close (demand) substitute to a mobile call.[42]

In addition, a supplier of mobile voice calls could switch resources into supplying SMS in response to a price rise by a hypothetical monopolist. However, a supplier of SMS may not be able to similarly switch to provide voice services, so this asymmetry may argue in favour of separating the retail markets, although most users seem to buy SMS services as part of bundles of mobile telephony services. No specific retail SMS market is defined for the purposes of this Recommendation.

RELATED WHOLESALE MARKETS

***Access and call origination***

The key elements required to produce a retail service are network access, call origination and call conveyance (of varying kinds) and where necessary call termination on other networks. Network access and call origination are typically supplied together by a network operator so that both services can be considered as part of the same market at a wholesale level.[43] This might not necessarily be the case in the future, if for instance call selection were introduced for mobile networks to mirror what has happened with respect to incumbent fixed network operators.

The relevant wholesale market appears in general to be access and call origination on mobile networks. This market is still subject to entry barriers, because undertakings without spectrum assignments can only enter the market on the basis of future spectrum allocations and assignment, secondary trading of spectrum or by purchasing a licensed operator. While in principle this is not an absolute entry barrier since there are various possibilities to share spectrum including the development of national roaming or indirect access relationships, such structures have not evolved to date in this market. However, the level of competition generally observed in this market at the retail level indicates that ex ante regulatory intervention at a wholesale level may not be warranted.

***Wholesale international roaming***

Wholesale international roaming services provide access and capacity (airtime minutes) to a foreign mobile network operator for the purposes of enabling its subscribers to make and receive calls while on another operator's network abroad. International wholesale roaming services are thus provided by a domestic mobile network operator (visited network) to a mobile network operator in another country (home network).

Wholesale international roaming satisfies a demand by foreign mobile network operators whose main objective is to offer their own subscribers a seamless service, not limited to the territory in which they have their own physical network. This operators' demand results from a demand from their subscribers to be able to make and receive calls on their mobile terminals abroad without having to acquire a new SIM card, or enter into another subscription with a foreign GSM operator, while enabling them to keep the number they use on their home network. In addition, international mobile roaming is characterised by requirements of coverage (having signal), accessibility (being reached by use of a given number on a particular SIM card) and mobility (ability to be reached when on the move), three features that are important in considering potential demand- and supply-side substitutes.

There are entry barriers to this market because only licensed mobile network operators are able to supply international roaming services in any given national market. In addition, there appear to be entry barriers because of restrictions governing the negotiation of roaming agreements. On the other hand, there are clear incentives for mobile network operators to negotiate lower-priced wholesale agreements for roaming if it means that they can generate more traffic on their networks or attract and retain end users. However, the ability of mobile network operators to direct roaming traffic is reported as being limited for the time being.

National roaming provided to mobile network operators within the same Member State (usually based on regulatory measures), although likewise provided on a wholesale basis, is a distinct service from wholesale international roaming for two main reasons. First, because national roaming arises in most cases from a temporary regulatory obligation[44] imposed on existing mobile network operators to provide roaming to subscribers of a new entrant network operator in the initial phase of its network roll-out, outside of the new entrant's coverage area.

Secondly, even if a product of commercial negotiations, national roaming does not involve foreign mobile licensed operators and does not entail a cross-border element, as international roaming always does.

Domestic wholesale mobile services involving access and provision of airtime ('wholesale airtime access'), provided by licensed mobile operators to service providers in their home country, cannot be viewed as alternatives to international roaming by foreign mobile operators. The main reason is that such services do not include the service that enables identification and recognition of a foreign operator's subscribers. If the foreign operator entered into a wholesale airtime agreement with an operator in the territory which its customers were visiting, that would require the foreign operator's subscriber having to acquire a new SIM card and a new mobile number issued by the visited operator's network.

Wholesale international roaming cannot be substituted by any other form of access to fixed telephone services abroad because such access cannot ensure accessibility (in being reached on the usual mobile number) and mobility.

Indirect access through carrier selection (i.e. wholesale call origination and access) would not be an effective supply substitute on the supply side, because in the current state of technology and commercial reality this service cannot provide the features of coverage, accessibility and mobility inherent to international roaming provided by mobile operators. This conclusion is likewise valid for carrier pre-selection on a call-by-call basis and wholesale airtime access for the same reasons.

Service providers and mobile virtual network operators (MVNOs) could not viably switch to providing wholesale international roaming on their own, as for the time being mobile network operators sell wholesale roaming services to other licensed mobile network operators only. They do not conclude international roaming agreements with service providers and MVNOs.

Therefore, it appears that there is a very limited scope for either demand- or supply-side substitution for wholesale international roaming. The Framework Directive (Annex I) also identifies an additional market for inclusion in this initial Recommendation, 'the national market for international roaming services on public mobile telephone networks'. Therefore, this market is identified for the purposes of this Recommendation.

***Wholesale call termination on mobile networks***

Two questions are examined. What is the appropriate relevant market in respect of call termination on mobile markets? Should an appropriately defined relevant market be identified for the purposes of this Recommendation?

Mobile call termination is an input both to the provision of mobile calls (that terminate on other mobile networks) but also to calls that are originated by callers on networks serving fixed locations, that terminate on mobile networks. Since the termination charge is set by the called network, which is chosen by the called subscriber, the calling party in general does not have the ability to affect or influence termination charges. This is the case under the calling party pays principle which is currently common in Europe.

This externality, whereby the called party may independently and adversely affect the calling party, can potentially be internalised, so that the ability for a network to set excessive call termination charges is constrained. Whether such a process can be expected to occur does affect both how a relevant call termination market is defined and whether a relevant call termination market should be identified for the purposes of the Recommendation. These issues are examined in more detail below.

At a retail level a call to a given user or user's terminal is not a substitute for a call to another user and this limitation on demand substitution follows through at the wholesale level. In addition there is a legal obligation at the wholesale level to supply any-to-any interconnection so operators are legally obliged to conclude interconnect agreements. In respect of supply substitution, if the supplier of call termination raises its price, it is not easy for alternative suppliers to switch to supply that market because they would need the SIM card details of that user to do so. However, the market is wider than call termination on a given user terminal because it is not possible for an operator to readily price discriminate between termination charges to different users across their network. Therefore the relevant market is at least as wide as termination for each operator.

However, with such a starting point in market definition, the supplier and the product are perfectly linked. It is important therefore to consider the possibilities for demand and supply substitution that might constrain termination charges and also the behaviour of network operators in setting termination charges. A constraint would exist if when a network operator tried to raise termination charges (or resisted lowering them), the overall impact were unprofitable. Such supply-side substitution is not currently possible but may become feasible at some point in the future.

This could become the case with software-enabled SIM cards, something that is likely to be exploited first in cases where operators establish preferred arrangements for their end users when they are roaming internationally.

Mobile termination charges might be constrained via demand substitution. There is no potential for demand substitution at a wholesale level. Demand at the wholesale level is inextricably linked to supply. The operator (of the caller) is unable to purchase call termination on a given network from an alternative source (as indicated above).

However, there are various possibilities for demand substitution at the retail level. It may be that other forms of calls or communications are reasonably close substitutes for the calls considered above, such as call-back, call forwarding and SMS messaging but in order for that potential substitution to broaden the market it would need to constrain the behaviour of the operator setting termination charges by lowering its overall profitability.

There may be substitutes for different classes of call, for instance a possible substitute for a fixed to mobile call is a mobile to mobile call.[45] The substitute call would need to be on net to lower profitability and constrain behaviour. In conjunction with the possibility for closed groups of users to exert buyer power (as described below), the potential substitution has a stronger impact because it could lead not only to the loss of termination charges but also to the loss of subscribers from one network to another.

A possible substitute for an off-net mobile call could be a mobile to fixed call. This would result in the loss of the termination charges but it is likely that the alternative call is only a close substitute in specific circumstances (e.g. knowing that the called party is close to a given fixed phone).

To summarise, some of these potential substitutions could constrain termination charges but whether they do so in practice also depends on empirical circumstances. There is no reason to believe that any of the demand substitutes above would operate a level that would constrain the operator's behaviour.

Another specific way in which end users and their operators can avoid excessive termination charges is by tromboning (traffic re-file) or re-routing. In particular, it is possible to re-originate traffic so that it appears that it is coming from the mobile network on which calls are due to terminate. The latter practice is only viable for end users that originate a significant amount of traffic for termination on a mobile network. In addition, it is possible for mobile operators to design differentiated tariff services in order to separate such user groups.

Another possible constraint on the ability of operators to set excessive termination charges may come from buyer power at the retail level. Two main types of buyer power may arise.

The first is where users of mobile phones are sufficiently concerned about receiving incoming calls that the price of incoming calls affects their choice of supplier. For this to exert a constraint on the pricing of call termination it is necessary that such a factor is as important to users as the pricing of other services such as outgoing calls, rental subscriptions etc. Under the calling party pays (CPP) principle, the calling party pays for the call, and the called party does not, therefore there is no direct relationship between the charges applied and demand for the service by the user of the mobile network who receives the call. Mobile users have shown little price sensitivity to termination charges on their own network, i.e. how much it costs others to call them.

A second type of buyer power can come from closed user groups where a particular group of users (whether or not they pay for part of the bill associated with incoming calls) make sufficient calls between them that intra-group calls constitute a significant proportion of their bill. If a given network raised termination charges and thereby increased the price of incoming calls, group members could switch networks to be on a given network and take advantage of lower on-net prices. However, mobile operators are able to price discriminate among the various categories of users and use of on-net tariffs and offering closed economic groups discounts for calls to a particular mobile number for instance are used to such an end.

In general, therefore, whilst it is apparent that end users who subscribe to mobile services have a choice about the network to which they subscribe and that it is relatively easy to switch between networks, there is limited evidence of widespread constraints on the pricing of wholesale call termination. The first option for the scope of the market definition is one for call termination on each mobile network. This would imply that currently each mobile network operator is a single supplier on each market. However, whether every operator then has market power still depends on whether there is any countervailing buyer power, which would render any non-transitory price increase unprofitable.

Another option for the scope of the market definition would be a national market for (mobile) call termination but the supply-side substitution necessary for such a definition does not currently exist.

A third option for the scope of the market definition would consist of linked national markets for mobile services. For this definition to be valid, mobile subscribers must be concerned about the price of calling mobiles and therefore termination charges (as an important determinant of such charges). It is also necessary that the services are strong complements so that subscribers do not consider the prices of the services separately when choosing a network but rather the price of the overall bundle or package. This would mean that a mobile operator could only raise termination charges and thereby the retail price of incoming calls (without losing subscribers) if at the same time it reduced prices for other services in the bundle. In this case the assessment of market power in call termination would be similar to that for other services in the bundle. If call termination were less important, there might be scope for a higher degree of market power in call termination.

The conclusion at the current time (under a calling party pays system) is that call termination on individual networks is the appropriate relevant market. However, such a definition would be undermined by (i) technical possibilities to terminate via other networks (this would broaden the market definition to call termination on all networks) (ii) evidence that users employ alternative means to circumvent high termination charges or (iii) evidence that users subscribe to networks on the basis of what it costs to be called (the last two would imply a linked market definition, comprising access, call origination and termination).

A market definition for call termination on each mobile network would imply that currently each mobile network operator is a single supplier on each market. However, whether every operator then has market power still depends on whether there is any countervailing buyer power, which would render any non-transitory price increase unprofitable.

WHOLESALE DATA SERVICES

### *SMS*

Wholesale SMS, including both SMS origination and SMS termination, or more generally narrow-band mobile data services, exhibit a number of features which justify separate treatment from voice call origination and termination dealt with above. At the retail level, it is not clear whether users consider SMS a sufficient substitute for a voice call in sufficient numbers for short voice calls and SMS to be considered part of the same market. While there are undoubtedly some substitution possibilities these are unlikely to operate to the extent necessary to place SMS in the same market as voice. In order for two products to be in the same market it is not necessary that these products are perfect alternatives for all users, only that there is a sufficient number of users who find them sufficiently close substitutes to constrain a hypothetical monopolist from raising prices profitably. It seems likely that from both a demand and supply perspective, instant messaging services over mobile networks, multi-media messaging services and general new data services that are currently available on 2.5G networks and which will be developed further on 3G networks will act as effective substitutes for SMS services. This would imply that SMS might be considered part of a broader data market, at least from the perspective of narrow-band services.

From an origination perspective there is, in addition to the normal competitive forces at work in voice call origination, a choice of different media to initiate SMS messages. These are developing and in addition to Internet other fixed line operators are entering this market.

However, it is on the termination side that differences appear between mobile voice origination and SMS. In the first instance SMS is not time dependent in the same way as voice calls, and this creates a number of different routing possibilities for operators. Furthermore, it also creates opportunities for different operators to enter the market and by aggregating demand to possibly create the kind of countervailing buyer power that is absent on the voice termination market. This may mitigate the degree of market power which MNOs enjoy in terminating SMS over their own networks. It is apparent that a large number of SMS origination generates an SMS response through for example, value added services which can typically entail the receiver of the message having to pay a price for the content of the message. In such circumstances senders of SMS are concerned with the cost of receiving an SMS since it is very often an explicit portion of the price paid for the service bought. It is clear that in this particular circumstance a receiving party pays dynamic is created and may restrain operators' ability to exercise market power.[46]

No wholesale SMS market is identified for the purposes of this Recommendation.

***Other mobile data services***
In addition to voice and SMS services mobile or wireless cellular networks can be used to access data and related services, including Internet and such services are developing. Current generation networks are in the process of being modified and upgraded to enable data rates in excess of those possible using dial-up access, and next or third generation networks will be capable of providing yet higher rates.

Retail services are currently less developed than their equivalent provision to fixed locations and it remains to be seen how services will be supplied and priced in the context of third generation networks. It is difficult at this stage to foresee how data services and access to Internet will develop, and also how voice and non-voice services will develop in the context of third generation networks.

For example, although the basic technologies for deploying upgraded second generation networks and third generation networks are now established, there is still uncertainty about the way in which services will be deployed, the protocols that will be used and the way in which services will be tariffed and sold. It might be for instance that voice services will be priced in a way that resembles the approach used in packet data networks, where receivers as well as senders pay for part of a communication. Such a development would modify the incentives to set excessive termination charges.

Most of these issues can currently be dealt with only with a high degree of uncertainty, and it is not possible to evaluate substitution patterns until more services are offered and establish themselves on the market. On this basis, no retail or wholesale markets are identified for the purposes of the Recommendation.

CONCLUSION

Therefore it is considered that the following specific markets related to the provision of voice services provided at non-fixed locations should be included in the Recommendation:

***Retail level***
No markets are identified.

***Wholesale level***
Access and call origination on public mobile telephone networks.
Voice call termination on individual mobile networks.
National market for international roaming services on public mobile telephone networks.

### 4.4 Markets related to broadcasting transmission

Electronic communications services include transmission services in networks used for broadcasting but exclude services providing or exercising control over content transmitted using electronic communications networks and services. The provision of broadcasting services therefore lies outside the scope of this regulatory framework, but the networks and associated facilities used for the delivery of broadcast services are within its scope.

Consequently no retail market is identified for the purposes of the Recommendation. However, a brief characterisation of retail services is made in order to derive wholesale markets for transmission services and assess whether any wholesale markets should be identified for the purposes of the Recommendation.

The overall retail market or markets supplied to end users consists of the delivery of radio and television broadcasting and includes free-to-air broadcasting as well as subscription and pay broadcasting and also the delivery or transmission of interactive services.

Radio and television broadcasting, including subscription broadcasting and also pay broadcasting, can be delivered over analogue or digital platforms. To deliver subscription and pay broadcasting (and in some cases free-to-air broadcasting), some means of both accessing end users and controlling end user access is required. Conditional access systems (and other associated facilities) are used by broadcasters to deliver services to end users that are authorised to receive them. In order to provide fully interactive services (beyond end-user choice of broadcasting or related content), a return path is required. The return path may be provided as part of the network or separately, for example, using a telephone line. Delivery and control of interactive services is typically undertaken by means of software and related applications for navigating and interaction, incorporated in a set-top box (along with conditional access systems and where necessary appropriate decoders).

Currently, end users, depending on their particular circumstance, may receive radio and television broadcasting via terrestrial, cable or satellite transmission networks. In the future other transmission channels are likely to become more common, e.g. telecommunications networks whose bandwidth is increased using DSL or other technologies.

It is possible to characterise the market both in terms of the services that are delivered and the transmission network over which they are delivered. Whether services broadcast over these transmission systems constitute separate markets or not (and thereby affect the definition of wholesale markets) depends on a number of factors. An important factor is whether they can be considered to constitute effective demand substitutes in terms of price and product for a significant number of end users. A related factor is the coverage or availability of the different transmission networks. Also relevant is the ability of end users (or a significant proportion of them) to be able to switch between broadcasting or transmission platforms.

In respect of the linked wholesale markets, several general market links are possible in respect of the delivery of broadcast content over transmission networks to end users. The undertaking owning or operating a transmission network may seek broadcast content that it delivers or transmits to its retail customers or end users. The programme provider or broadcaster may provide content to a transmission or broadcast network or negotiate conditional access arrangements by which it can transmit broadcasting content to the end user via an access arrangement. The broadcaster that has content or rights to content may own or operate a broadcasting network or transmission network and also supply conditional access services by which other broadcasters or content providers may supply services to end users.[47]

As indicated above, three main platforms are currently available (terrestrial, cable and satellite) for the digital transmission or delivery of content and others are likely to become more common in the future. The existence and coverage of digitally capable broadcasting transmission platforms varies between Member States of the European Union. In some, cable networks have not been built or widely developed. In many, digital terrestrial networks have not developed. The upgrading of telecommunications networks, such that broadcasting services could potentially be transmitted, is in its early stages.

In the absence of some limitation on switching between platforms, the first wholesale market that is identified is broadcasting transmission services and distribution networks in so far as they provide the means to deliver broadcast content to end users. Demand and supply substitution conditions between the different delivery platforms may be such that the feasibility of switching between platforms is limited. In that situation a hypothetical monopolist on one platform may not necessarily be constrained by the activities of operators of other platforms.

In order to deliver subscription or pay broadcasting or interactive services to end users, the undertaking supplying such services (unless it owns or operates its own transmission network) typically needs access to ancillary technical broadcasting services such as conditional access systems, application programme interfaces and electronic programme guides on one or more transmission networks.

However, no wholesale market for ancillary technical broadcasting services (including conditional access systems) is identified for the purposes of the Recommendation. Articles 5 and 6, in conjunction with Annex I of the Access Directive, permit Member States to require all operators of conditional access systems to offer access on fair, reasonable and non-discriminatory terms or to apply such access terms only in respect of operators found to have SMP on the relevant market. Therefore, by virtue of the Access Directive, there is no requirement to identify relevant product and service markets in this area in the Recommendation.

When a Member State chooses to permit its NRA to undertake a market analysis for the purpose of determining whether to apply such access terms only in respect of operators found to have SMP on the relevant market, which it may do under Article 6.3 of the Access Directive, the following considerations apply.

The relevant issue for undertakings seeking access to ancillary technical broadcasting services (including conditional access systems) is considered to be the ability to deliver to, or negotiate access to, sufficient end users to sustain a viable business, rather than to negotiate access to all delivery platforms or to all possible end users. In addition, end users may switch between delivery platforms in response to the services offered and expectations about overall prices. Only where there was a non-transitory price increase by a hypothetical monopolist across all delivery platforms would an undertaking (seeking to deliver to end users and demanding wholesale access) have no substitution possibilities. However, if switching costs for a substantial proportion of users between alternative delivery platforms were sufficiently high,[48] it might be argued that the relevant market could be for ancillary technical broadcasting services on *a given delivery platform*.

The relevant market for the purposes of ex ante market analysis, where a Member State so decides, is therefore considered to be the market for wholesale ancillary technical broadcasting services across all relevant delivery platforms, unless specific national situations in respect of switching costs and available delivery platforms justify a narrower market definition.

CONCLUSION

Therefore it is considered that the following specific markets related to broadcasting transmission should be included in the Recommendation:

***Wholesale level***

Broadcasting transmission services and distribution networks in so far as they provide the means to deliver broadcast content to end users.

## 5. Transition to the New Framework – Defining Markets in Accordance with Annex I of the Framework Directive

Section 2 describes how the new regulatory framework provides for the transition from the 1998 framework, and in particular for the review of existing obligations on the basis of the provisions and principles in the new regulatory framework.

As the title of Annex I of the Framework Directive makes clear, all the market areas listed therein need to be included in the initial version of the Recommendation in order that NRAs can carry out a review of existing obligations imposed under the 1998 regulatory framework. If they were not so included, Article 15(3) of the Framework Directive would require NRAs to notify them individually to the Commission under the procedure set out in Article 7 of the Framework Directive.

However Article 15(1) of the Framework Directive requires the Commission to define markets in accordance with the principles of competition law. The market areas set out in Annex I of the Framework Directive cannot all be regarded as such on the basis of the examination undertaken in section 4. The Commission has therefore defined markets (corresponding to the market areas listed in Annex I of the Framework Directive) in accordance with competition law principles. Where NRAs have imposed obligations on SMP operators in market areas under the 1998 regulatory framework, they should therefore analyse whether there is effective competition on the corresponding competition law markets identified in the Annex to the Recommendation.

The provisions concerning the review of existing obligations which are contained in Article 7(2) Access Directive and Article 16(2) Universal Service Directive state that the Commission shall indicate relevant markets for the obligations under the old framework specified in those Articles in the initial recommendation on product and service markets. This requirement is satisfied by the list of markets set out in Annex I of this document. The correspondence between market areas regulated under the old framework and Annex I of the Framework Directive is made in Table 1. The correspondence between the Framework Directive Annex I markets and the markets identified in this Recommendation is made in Table 2 below.

Provided that NRAs analyse relevant markets using the precise product or service market definitions listed in the Annex, there is no need to notify the market definition to the Commission in accordance with Article 15(3) and Article 7(4a) of the Framework Directive. However, measures to designate or not designate undertakings as having significant market power on these markets do need to be notified in accordance with Article 7 of the Framework Directive.

As section 4 of the SMP Guidelines explains in more detail, Article 16 requires that where an NRA concludes that a market identified under the new regulatory framework is effectively competitive, the NRA removes any existing specific SMP obligations. Where an NRA determines that a relevant market is not effectively competitive, it must identify undertakings with SMP and either impose appropriate regulatory obligations or alternatively maintain or amend existing obligations. In some cases therefore it may be that regulatory obligations imposed under the 1998 regulatory framework are re-imposed under the new regulatory framework, albeit in respect of designations of significant market power on relevant markets defined and analysed in accordance with the new regulatory framework.

A summary of the market areas listed in Annex I, together with their corresponding competition law markets and an indication as to where they appear in the Annex to the Recommendation, is set out in the following table.

*Table 2 Treatment of Annex I market areas in this Recommendation*

| Market areas listed in Annex I of the Framework Directive | Corresponding competition law market set out in Annex I of the Recommendation | |
|---|---|---|
| **Retail** | | |
| The provision of connection to and use of the public telephone network at fixed locations | 1. | Access to the public telephone network at a fixed location – residential |
| | 2. | Access to the public telephone network at a fixed location – business |
| | 3. | Publicly available local/and or national telephone services provided at a fixed location – residential |
| | 4. | Publicly available local/and or national telephone services provided at a fixed location – business |
| | 5. | Publicly available international telephone services provided at a fixed location – business |
| | 6. | Publicly available international telephone services provided at a fixed location – residential |
| The provision of leased lines to end users. | 7. | The minimum set of leased lines |
| **Wholesale** | | |
| Call origination in the fixed public telephone network | 8. | Call origination on the public telephone network provided at a fixed location |
| Call termination in the fixed public telephone network | 9. | Call termination on individual public telephone networks provided at a fixed location |
| Transit services in the fixed public telephone network | 10. | Transit services in the public telephone network provided at fixed locations |
| Call origination on public mobile telephone networks | 15. | Access and call origination on public mobile telephone networks |
| Call termination on public mobile telephone networks | 16. | Voice call termination on individual mobile networks |
| Leased line interconnection (interconnection of part circuits) | 13. | Wholesale terminating segments of leased lines |
| | 14. | Wholesale trunk segments of leased lines |

*Table 2 continued*

| Market areas listed in Annex I of the Framework Directive | Corresponding competition law market set out in Annex I of the Recommendation |
|---|---|
| Access to the fixed public telephone network, including unbundled access to the local loop | 11. Wholesale unbundled access (including shared access) to metallic loops and sub-loops, for the purpose of providing broadband and voice services.<br>12. Wholesale broadband services |
| Access to public mobile telephone networks, including carrier selection | 15. Access and call origination on public mobile telephone networks |
| Wholesale provision of leased line capacity to other suppliers of electronic communications networks or services | 13. Wholesale or terminating segments of leased lines<br>14. Wholesale trunk segments of leased lines |
| Services provided over unbundled (twisted metallic pair) loops | 11. Wholesale unbundled access (including shared access) to metallic loops and sub-loops, for the purpose of providing broadband Internet services. |
| The national market for international roaming services on public mobile telephone networks | 17. The wholesale national market for international roaming on public mobile networks |

## 6. Publication of Recommendation and Subsequent Revision

This Recommendation was adopted following a public consultation on a draft text that took place in July 2002 and a further consultation with NRAs and NCAs that took place in October 2002. The Recommendation will be periodically reviewed by the Commission depending on the speed of market developments, the period needed by NRAs to undertake market analysis, the principle set out in section 1 that the imposition of ex ante regulation to address lack of effective competition implies a degree of continuity, and the need for predictability and legal security for market players.

National regulatory authorities will regularly review their market analysis on the basis of the market identified in any updating of the Recommendation, as stated in Article 16 of the Framework Directive.

In reviewing the Recommendation, the Commission will consult Member States, NRAs and NCAs, and all interested parties via a public consultation.

The Commission will consider no later than 30 June 2004 the need to review this Recommendation in the light of market developments.

# NOTES

1. OJ L 108, 24.4.2002, p. 33.
2. OJ L 108, 24.4.2002, p. 21.
3. OJ L 108, 24.4.2002, p. 7.
4. OJ L 108, 24.4.2002, p. 51.
5. OJ L 201, 31.7.2000, p. 37.
6. OJ C 165, 11.7.2002, p. 6.
7. Article 18 of the Universal Service Directive.
8. Article 19 of the Universal Service Directive.
9. Article 6 of the Access Directive.
10. All obligations imposed as SMP obligations must be removed, but certain obligations may be imposed on non-SMP operators in accordance with Article 8(3) of the Access Directive. These latter obligations would not be affected by the withdrawal of a SMP designation.
11. OJ L 101, 1.4.1998, p. 24.
12. OJ L 295, 29.10.1997, p. 23.
13. OJ L 199, 26.7.1997, p. 32.
14. OJ L 336, 30.12.2000, p. 4.
15. OJ C 372, 9.12.1997, p. 5.
16. OJ C 265, 22.8.1998, p. 2.
17. Ex ante regulation addresses lack of effective competition that is expected to persist over a given horizon. Therefore, the time horizon for market definition and identification for the purposes of this Recommendation should be commensurate with the period during which possible ex ante regulatory remedies are likely to be imposed. The period may depend on whether an existing obligation is being maintained or reviewed, or a new obligation is being imposed.
18. See section 2 of the SMP Guidelines.
19. Framework Directive recital 27.
20. See paragraph 20 of the SMP Guidelines.
21. The Framework Directive requires that markets are defined in accordance with competition law. This necessitates the market areas listed in Annex I to be re-defined accordingly.
22. Recital 27 also indicates that newly emerging markets, where de facto the market leader is likely to have a substantial market share, should not be subjected to inappropriate obligations.
23. Article 8 of the Framework Directive requires NRAs to pursue a number of objectives including: ensuring users derive maximum benefits in terms of choice, price and quality; ensuring there is no distortion or restriction of competition; encouraging efficient investment in infrastructure and promoting innovation; encouraging efficient use of and effective management of radio frequencies and numbering resources.
24. For example, there might be narrower market definition at a national level of an identified market. According to national circumstances, there might be a reason to broaden the market definition such that two or more markets that are identified separately in this Recommendation are identified as a single market. Similarly, there might be a reason to identify a newly developed market and combine it with a market identified in this Recommendation. According to national circumstances, there might be a reason to identify additional markets to those identified in this Recommendation.
25. Universal Service Directive recital 26.
26. Framework Directive, Article 8(2)c.
27. Framework Directive, Article 2.
28. This raises the question of technical neutrality with respect to the treatment of services and the means by which they are delivered. As well as recognising that some services may constitute substitutes, irrespective of technical provision, it is also necessary to recognise that different services may be characterised by different technical requirements within a given network, for example in terms of delay (real-time or not) and bandwidth (and the level and variance of these technical requirements).
29. Dial-up Internet services are treated in section 4.2.2 on access to data and related services.
30. The question of whether metered and unmetered (flat-rate) access are in the same or separate markets is considered in section 4.2.2.
31. It is also important to examine countervailing market power, in this case countervailing buyer power in negotiating call termination charges, but this is part of the effective competition analysis once the relevant market is defined.
32. Considerations of relative market power are not limited to networks (of differing size or coverage) serving end users at a fixed location or address but also networks such as mobile cellular networks serving non-fixed locations. In circumstances where a 'fixed' network with significant market power is subject to a regulatory remedy (beyond the basic one to negotiate interconnect) such as regulated prices for call termination, market power relative to mobile networks would be affected.

33. Higher bandwidth or broadband Internet services may be characterised as allowing downstream capacity to end users in excess of 128 kbits/sec. The bandwidth of the service supplied may be asymmetric or symmetric. Dedicated access would typically involve the provision of symmetric bandwidth.
34. Internet access via the TV is becoming more common, although there are often limitations with respect to the content that can be accessed and the applications that can be used. In most cases a standard modem on a telephone line is used. However, the broadcast path could also be used, in which case access would more closely resemble other higher-speed access methods.
35. The above analysis may well lead to different results were the starting point to be services offered on narrow-band connections. In other words, asymmetric substitutability may exist for this type of market: for example, under certain conditions a broadband connection may be a viable substitute for a narrow-band connection, since it offers additional features, whereas a narrow-band connection may not be a viable substitute for a broadband connection.
36. The quality of an unmetered product is essentially the same as a metered one. At the same time, the unmetered subscription can be seen as the purchase of an option to use additional access time at a zero price.
37. See in particular section 2.2.2. of 'Guidelines for market analysis and the assessment of market power'.
38. A number of actual business models may exist. In the metered approach, a portion of the retail usage charge may be passed from the originating to the terminating operator and on to the ISP. In a subscription model, the terminating operator may compensate the originating operator and charge this to the ISP.
39. OJ C 339, 7.11.1998 P. 0006–0024.
40. The leased lines in the minimum set are identified in the List of Standards published in the Official Journal of the European Communities in accordance with Article 16 of Directive [Framework].
41. One area where a specific business market might be identified is in the retail provision of national and international services (including international roaming) for large corporate customers. Such a market is not identified for the purpose of ex ante regulatory analysis.
42. In certain circumstances (e.g. off-peak and in lieu of a short mobile call), the price of an SMS is close to the price of a mobile call. However, it is difficult to ascertain whether these prices represent competitive prices, which is an important consideration in assessing whether the two products are in the same market. At the same time, it is possible for products for which there is a price differential to be part of the same market if such a differential reflects a qualitative difference and end users might still switch between the products in response to a small price change in one or the other.
43. In fact it could be argued that access, call origination and call termination constitute one wholesale market and on the other hand that call termination is a separate stand-alone wholesale product. The former is sold to the retail arm of a network operator; the latter is sold to other networks.
44. Finland is an exception because of some remote areas that justify national roaming not limited in time.
45. It is possible for these alternatives to be substitutes (as well as complements) even if broadly speaking the fixed market is defined separately from the mobile market.
46. The receiving party pays principle in this particular case would apply only to the relation between a content provider or value added service provider and the final client. However, it would appear that the ability of terminating operators to exploit their market power over content delivered to their own subscribers would still persist. In particular, it may happen that content providers would not find it profitable to offer services to end users given the wholesale SMS termination price set by the mobile operator, and that the mobile operator itself could diversify its offering by commercialising content provision and possibly by applying discriminatory practices to prevent service providers from competing effectively.
47. In a fully converged environment, the market is likely to be one that is more generally described as the delivery of broadband content over a broadband capable network.
48. Switching costs may be taken into account by end users when subscribing to a given delivery platform. In addition, depending on the size of switching costs relative to expected profitability, suppliers may compensate end users for switching costs in order to build a subscriber base.

# Index